Springer Proceedings in Mathematics & Statistics

Volume 267

Springer Proceedings in Mathematics & Statistics

This book series features volumes composed of selected contributions from workshops and conferences in all areas of current research in mathematics and statistics, including operation research and optimization. In addition to an overall evaluation of the interest, scientific quality, and timeliness of each proposal at the hands of the publisher, individual contributions are all refereed to the high quality standards of leading journals in the field. Thus, this series provides the research community with well-edited, authoritative reports on developments in the most exciting areas of mathematical and statistical research today.

More information about this series at http://www.springer.com/series/10533

Kurusch Ebrahimi-Fard · María Barbero Liñán
Editors

Discrete Mechanics, Geometric Integration and Lie–Butcher Series

DMGILBS, Madrid, May 2015

Springer

Editors
Kurusch Ebrahimi-Fard
Department of Mathematical Sciences
Norwegian University of Science and
Technology—NTNU
Trondheim, Norway

María Barbero Liñán
Department of Applied Mathematics
Universidad Politécnica de Madrid (UPM)
Madrid, Spain

and

Institute for the Mathematical Sciences
(ICMAT)
Madrid, Spain

ISSN 2194-1009 ISSN 2194-1017 (electronic)
Springer Proceedings in Mathematics & Statistics
ISBN 978-3-030-01396-7 ISBN 978-3-030-01397-4 (eBook)
https://doi.org/10.1007/978-3-030-01397-4

Library of Congress Control Number: 2018956279

Mathematics Subject Classification (2010): 65D30, 34A26, 15A16, 34C40, 16T05, 70G75, 37C10, 70G65, 93B25, 17B99, 22E65, 65P10

This Springer imprint is published by the registered company Springer Nature Switzerland AG
The registered company address is: Gewerbestrasse 11, 6330 Cham, Switzerland

Preface

Geometric numerical integration is a rather important research topic in numerical analysis of differential equations. In the introductory chapter of this volume, two distinguished mathematicians, Arieh Iserles and Reinout Quispel, explore recent and ongoing developments, as well as new research directions in geometric integration methods for differential equations. The collection of manuscripts following Iserles' and Quispel's contribution display a combination of research and overview chapters including detailed presentations of many of the mathematical tools necessary in the areas of geometric integration theory, nonlinear systems theory, and discrete mechanics. The scope and high quality of this volume is maybe best exemplified by briefly mentioning the topics it contains. Many mechanical systems evolve on Lie groups, that is why Lie group integrators are essential for numerically solving differential equations. A comprehensive overview on Lie group integrators is provided by Brynjulf Owren. The algebraic, geometric, and computational aspects relevant to numerical integration methods, such as Lie–Butcher series and word series algorithms, are described extensively in the following chapters by Munthe-Kaas and Føllesdal, Murua and Sanz-Serna, Ebrahimi-Fard and Mencattini, and Casas. The contribution by Duffaut Espinosa, Ebrahimi-Fard, and Gray explores interconnections of nonlinear systems with a view towards discretisation. The following chapters by Bogfjellmo, Dahmen, and Schmeding, Barbero Liñán and Martín de Diego, Vermeeren, and Verdier are shorter and address more specific research questions, with the exception of the paper by Bogfjellmo et al., which also includes a timely overview of Lie theoretic and Hopf algebraic aspects relevant to geometric numerical integration. Indeed, a common thread underlying those works is the fruitful use of modern algebraic and combinatorial structures common to those topics.

The contributions are written in a self-contained style to make the volume accessible to a broader audience, including in particular researchers and graduate students interested in theoretical and applied aspects in geometric integration theory, nonlinear control theory, and discrete mechanics.

These chapters are based on extended lectures and research talks presented at the international "Brainstorming Workshop on New Developments in Discrete Mechanics, Geometric Integration and Lie–Butcher Series". The event took place at the Instituto de Ciencias Matemáticas (ICMAT) in Madrid, Spain, and was one of the main activities organised by the Norwegian–Spanish NILS–ABEL 2014–2015 research project "Discrete Mechanics, Geometric Integration and Lie–Butcher Series". The two partners of the NILS–ABEL project (Bergen–Madrid) were very eager to consult with experts (Elena Celledoni and Brynjulf Owren) from the Norwegian University of Science and Technology (NTNU) in Trondheim. In fact, we were convinced that combining the expertise from researchers from these three institutions and including other invited participants would certainly lead to a substantial boost of the perspectives of this mathematics research project. This was one of the main motivations to organise this brainstorming workshop back in 2015. The meeting brought together senior experts as well as young researchers, from Germany, Norway, Spain, and the USA. Its central aim was to provide a platform for discussing theoretical and applied aspects of computational solutions of differential equations describing dynamical systems in natural sciences and technology as well as nonlinear control systems. We particularly appreciate Profs. Iserles and Quispel for their valuable contribution to this volume. Although they could not join us in the workshop, they were very enthusiastic in preparing the introductory chapter.

Last but not least, this volume would not have been possible without the commitment of all the speakers in the workshop. They prepared excellent expositions which made this event rather successful. The event received funding from the EEA grant provided by Norway, Iceland, Liechtenstein, and Spain (NILS), as well as from the Fundación BBVA, the ICMAT Severo Ochoa Excellence Programme, and Universidad Carlos III de Madrid. We also thank the referees for helping us in preparing this volume and the ICMAT for providing the facilities and the human resources to make the event a success.

Trondheim, Norway — Kurusch Ebrahimi-Fard
Madrid, Spain — María Barbero Liñán

Contents

Contributors

María Barbero Liñán Departamento de Matemática Aplicada, Universidad Politécnica de Madrid (UPM), Madrid, Spain; Instituto de Ciencias Matemáticas (CSIC-UAM-UC3M-UCM), Madrid, Spain

Geir Bogfjellmo Chalmers Technical University & Gothenburg University, Gothenburg, Sweden

Fernando Casas Institut de Matemàtiques i Aplicacions de Castelló and Departament de Matemàtiques, Universitat Jaume I, Castellón, Spain

Rafael Dahmen TU Darmstadt, Darmstadt, Germany

Luis A. Duffaut Espinosa Department of Electrical and Biomedical Engineering, University of Vermont, Burlington, VT, USA

Kurusch Ebrahimi-Fard Department of Mathematical Sciences, Norwegian University of Science and Technology—NTNU, Trondheim, Norway

Kristoffer K. Føllesdal Department of Mathematics, University of Bergen, Bergen, Norway

W. Steven Gray Department of Electrical and Computer Engineering, Old Dominion University, Norfolk, VA, USA

Arieh Iserles Department of Applied Mathematics and Theoretical Physics Centre for Mathematical Sciences, University of Cambridge, Cambridge, UK

David Martín de Diego Instituto de Ciencias Matemáticas (CSIC-UAM-UC3M-UCM), Madrid, Spain

Igor Mencattini Instituto de Ciências Matemáticas e de Computação, Universidade de São Paulo (USP), São Carlos, SP, Brazil

Hans Z. Munthe-Kaas Department of Mathematics, University of Bergen, Bergen, Norway

Ander Murua Konputazio Zientziak eta A. A. Saila, Informatika Fakultatea, UPV/EHU, Donostia–San Sebastián, Spain

Brynjulf Owren Norwegian University of Science and Technology, Trondheim, Norway

G. R. W. Quispel La Trobe University, Melbourne, Australia

Jesús M. Sanz-Serna Departamento de Matemáticas, Universidad Carlos III de Madrid, Leganés, Madrid, Spain

Alexander Schmeding Department of Mathematical Sciences, Norwegian University of Science and Technology—NTNU, Trondheim, Norway

Olivier Verdier Department of Computing, Mathematics and Physics, Western Norway University of Applied Sciences, Bergen, Norway; Department of Mathematics, KTH - Royal Institute of Technology, Stockholm, Sweden

Mats Vermeeren Institut für Mathematik, Technische Universität Berlin, Berlin, Germany

Why Geometric Numerical Integration?

Arieh Iserles and G. R. W. Quispel

Abstract Geometric numerical integration (GNI) is a relatively recent discipline, concerned with the computation of differential equations while retaining their geometric and structural features exactly. In this paper we review the rationale for GNI and review a broad range of its themes: from symplectic integration to Lie-group methods, conservation of volume and preservation of energy and first integrals. We expand further on four recent activities in GNI: highly oscillatory Hamiltonian systems, W. Kahan's 'unconventional' method, applications of GNI to celestial mechanics and the solution of dispersive equations of quantum mechanics by symmetric Zassenhaus splittings. This brief survey concludes with three themes in which GNI joined forces with other disciplines to shed light on the mathematical universe: abstract algebraic approaches to numerical methods for differential equations, highly oscillatory quadrature and preservation of structure in linear algebra computations.

Keywords Symplectic methods · Lie-group methods · Splittings · Exponential integrators · Kahan's method · Variational integrators · Preservation of integrals Preservation of volume

MSC numbers Primary 65L05 · Secondary 65P10 · 65M99 · 17B45 · 57S50

A. Iserles (✉)
Department of Applied Mathematics and Theoretical Physics Centre for Mathematical Sciences, University of Cambridge, Cambridge, UK
e-mail: ai10@cam.ac.uk

G. R. W. Quispel
La Trobe University, Melbourne, Australia

K. Ebrahimi-Fard and M. Barbero Liñán (eds.), *Discrete Mechanics, Geometric Integration and Lie–Butcher Series*, Springer Proceedings in Mathematics & Statistics 267, https://doi.org/10.1007/978-3-030-01397-4_1

1 The Purpose of GNI

Geometric numerical integration (GNI) emerged as a major thread in numerical mathematics some 25 years ago. Although it has had antecedents, in particular the concerted effort of the late Feng Kang and his group in Beijing to design structure-preserving methods, the importance of GNI has been recognised and its scope delineated only in the 1990s.

But we are racing ahead of ourselves. At the beginning, like always in mathematics, there is the definition and the rationale of GNI. The rationale is that all-too-often mathematicians concerned with differential equations split into three groups that have little in common. Firstly, there are the applied mathematicians, the model builders, who formulate differential equations to describe physical reality. Secondly, there are those pure mathematicians investigating differential equations and unravelling their qualitative features. Finally, the numerical analysts who flesh out the numbers and the graphics on the bones of mathematical formulation. Such groups tended to operate in mostly separate spheres and, in particular, this has been true with regards to computation. Discretisation methods were designed (with huge creativity and insight) to produce rapidly and robustly numerical solutions that can be relied to carry overall small error. Yet, such methods have often carried no guarantee whatsoever to respect qualitative features of the underlying system, the very same features that had been obtained with such effort by pure and applied mathematicians.

Qualitative features come basically in two flavours, the *dynamical* and the *geometric*. Dynamical features—sensitivity with respect to initial conditions and other parameters, as well as the asymptotic behaviour—have been recognised as important by numerical analysts for a long time, not least because they tend to impinge directly on accuracy. Thus, sensitivity with respect to initial conditions and perturbations comes under 'conditioning' and the recovery of correct asymptotics under 'stability', both subject to many decades of successful enquiry. Geometric attributes are invariants, constants of the flow. They are often formulated in the language of differential geometry (hence the name!) and mostly come in three varieties: *conservation laws*, e.g. Hamiltonian energy or angular momentum, which geometrically mean that the solution, rather than evolving in some large space $\mathbb{R}^d$, is restricted to a lower-dimensional manifold $\mathcal{M}$, *Lie point symmetries*, e.g. scaling invariance, which restrict the solution to the tangent bundle of some manifold, and quantities like *symplecticity* and *volume*, conservation laws for the derivative of the flow. *The design and implementation of numerical methods that respect geometric invariants is the business of GNI.*

Since its emergence, GNI has become the new paradigm in numerical solution of ODEs, while making significant inroads into numerical PDEs. As often, yesterday's revolutionaries became the new establishment. This is an excellent moment to pause and take stock. Have all the major challenges been achieved, all peaks scaled, leaving just a tidying-up operation? Is there still any point to GNI as a separate activity or should it be considered as a victim of its own success and its practitioners depart to fields anew—including new areas of activity that have been fostered or enabled by GNI?

These are difficult questions and we claim no special authority to answer them in an emphatic fashion. Yet, these are questions which, we believe, must be addressed. This short article is an attempt to foster a discussion. We commence with a brief survey of the main themes of GNI *circa* 2015. This is followed by a review of recent and ongoing developments, as well as of some new research directions that have emerged from GNI but have acquired a life of their own.

2 The Story So Far

2.1 Symplectic Integration

The early story of GNI is mostly the story of symplectic methods. A Hamiltonian system

$$\dot{\boldsymbol{p}} = -\frac{\partial H(\boldsymbol{p},\boldsymbol{q})}{\partial \boldsymbol{q}}, \qquad \dot{\boldsymbol{q}} = \frac{\partial H(\boldsymbol{p},\boldsymbol{q})}{\partial \boldsymbol{p}}, \tag{2.1}$$

where $H : \mathbb{R}^{2d} \to \mathbb{R}$ is a *Hamiltonian energy,* plays a fundamental role in mechanics and is known to possess a long list of structural invariants, e.g. the conservation of the Hamiltonian energy. Yet, arguably its most important feature is the conservation of the *symplectic form* $\sum_{k=1}^{d} \mathrm{d}\boldsymbol{p}_k \wedge \mathrm{d}\boldsymbol{q}_k$ because symplecticity is equivalent to Hamiltonicity—in other words, every solution of a Hamiltonian system is a symplectic flow and every symplectic flow is locally Hamiltonian with respect to an appropriate Hamiltonian energy [33].

The solution of Hamiltonian problems using symplectic methods has a long history, beautifully reviewed in [32], but modern efforts can be traced to the work of Feng and his collaborators at the Chinese Academy of Sciences, who have used generating-function methods to solve Hamiltonian systems [21]. And then, virtually simultaneously, [46, 77, 83] proved that certain Runge–Kutta methods, including the well-known Gauss–Legendre methods, preserve symplecticity and they presented an easy criterion for the symplecticity of Runge–Kutta methods. GNI came of age!

Symplectic methods readily found numerous uses, from particle accelerator physics [23] and celestial mechanics [47] to molecular dynamics [49] and beyond.

Subsequent research into symplectic Runge–Kutta methods had branched out into a number of directions, each with its own important ramifications outside the Hamiltonian world:

- *Backward error analysis.* The idea of backward error analysis (BEA) can be traced to Wilkinson's research into linear algebra algorithms in the 1950ties. Instead of asking "what is the numerical error for our problem", Wilkinson asked "which nearby problem is solved *exactly* by our method?". The difference between the original and the nearby problem can tell us a great deal about the nature of the error in a numerical algorithm.

A generalisation of BEA to the field of differential equations is fraught with difficulties. Perhaps the first successful attempt to analyse Hamiltonian ODEs in this setting was by [70] and it was followed by many, too numerous to list: an excellent exposition (like for many things GNI) is the monograph of [33]. A major technical tool is the B-series, an expansion of composite functions in terms of forests of rooted trees, originally pioneered by [7]. (We mention in passing that the Hopf algebra structure of this *Butcher group* has been recently exploited by mathematical physicists to understand the renormalisation group [15]—as the authors write, "We regard Butcher's work on the classification of numerical integration methods as an impressive example that concrete problem-oriented work can lead to far-reaching conceptual results".) It is possible to prove that, subject to very generous conditions, the solution of a Hamiltonian problem by a symplectic method, implemented with constant step size, is exponentially near to the *exact* solution of a nearby Hamiltonian problem for an exponentially long time. This leads to considerably greater numerical precision, as well as to the conservation on average (in a strict ergodic sense) of Hamiltonian energy.

B-series fall short in a highly oscillatory and multiscale setting, encountered frequently in practical Hamiltonian systems. The alternative in the BEA context is an expansion into *modulated Fourier series* [29], as well as expansions into *word series* [68], to which we return in Sect. 4.1.

- *Composition and splitting.*

  Many Hamiltonians of interest can be partitioned into a sum of kinetic and potential energy, $H(\boldsymbol{p},\boldsymbol{q}) = \boldsymbol{p}^\top M\boldsymbol{p} + V(\boldsymbol{q})$. It is often useful to take advantage of this in the design of symplectic methods. While conventional symplectic Runge–Kutta methods are implicit, hence expensive, *partitioned Runge–Kutta methods,* advancing separately in $\boldsymbol{p}$ and $\boldsymbol{q}$, can be explicit and are in general much cheaper. While perhaps the most important method, the Störmer–Verlet scheme, has been known for many years, modern theory has led to an entire menagerie of composite and partitioned methods [79].

  Splitting methods[1] have been used in the numerical solution of PDEs since 1950s. Thus, given the equation $u_t = \mathcal{L}_1(u) + \mathcal{L}_2(u)$, where the $\mathcal{L}_k$s are (perhaps nonlinear) operators, the idea is to approximate the solution in the form

$$u(t+h) \approx \mathrm{e}^{\alpha_1 h\mathcal{L}_1}\mathrm{e}^{\beta_1 h\mathcal{L}_2}\mathrm{e}^{\alpha_2 h\mathcal{L}_1}\cdots\mathrm{e}^{\alpha_s h\mathcal{L}_1}\mathrm{e}^{\beta_s \mathcal{L}_2}u(t), \tag{2.2}$$

  where $v(t_0+h) =: \mathrm{e}^{h\mathcal{L}_1}v(t_0)$ and $w(t_0+h) =: \mathrm{e}^{h\mathcal{L}_2}w(t_0)$ are, formally, the solutions of $\dot{v} = \mathcal{L}_1(v)$ and $\dot{w} = \mathcal{L}_2(w)$ respectively, with suitable boundary conditions. The underlying assumption is that the solutions of the latter two equations are either available explicitly or are easy to approximate, while the original equation is more difficult.

  A pride of place belongs to *palindromic compositions* of the form

$$\mathrm{e}^{\alpha_1 h\mathcal{L}_1}\mathrm{e}^{\beta_1 h\mathcal{L}_2}\mathrm{e}^{\alpha_2 h\mathcal{L}_1}\cdots\mathrm{e}^{\alpha_q h\mathcal{L}_1}\mathrm{e}^{\beta_q h\mathcal{L}_2}\mathrm{e}^{\alpha_q h\mathcal{L}_1}\cdots\mathrm{e}^{\alpha_2 h\mathcal{L}_1}\mathrm{e}^{\beta_1 h\mathcal{L}_2}\mathrm{e}^{\alpha_1 h\mathcal{L}_1}, \tag{2.3}$$

[1] Occasionally known in the PDE literature as *alternate direction methods.*

invariant with respect to a reversal of the terms. They constitute a *time-symmetric map,* and this has a number of auspicious consequences. Firstly, they are always of an even order. Secondly—and this is crucial in the GNI context—they respect both structural invariants whose integrators are closed under composition, i.e. form a group (for example integrators preserving volume, symmetries, or first integrals), as well as invariants whose integrators are closed under symmetric composition, i.e. form a symmetric space (for example integrators that are self-adjoint, or preserve reversing symmetries). A basic example of (2.3) is the second-order *Strang composition*

$$e^{\frac{1}{2}h\mathcal{L}_1}e^{h\mathcal{L}_2}e^{\frac{1}{2}h\mathcal{L}_1} = e^{h(\mathcal{L}_1+\mathcal{L}_2)} + \mathcal{O}\left(h^3\right).$$

Its order—and, for that matter, the order of any time-symmetric method—can be boosted by the *Yoshida device* [88]. (Cf. also [85].) Let Φ be a time-symmetric approximation to $e^{t\mathcal{L}}$ of order $2P$, say. Then

$$\Phi((1+\alpha)h)\Phi(-(1+2\alpha)h)\Phi((1+\alpha)h), \qquad \text{where} \qquad \alpha = \frac{2^{1/(2P+1)}-1}{2-2^{1/(2P+1)}}$$

is also time symmetric and of order $2P+2$. Successive applications of the Yoshida device allow to increase arbitrarily the order of the Strang composition, while retaining its structure-preserving features. This is but a single example of the huge world of splitting and composition methods, reviewed in [4, 57].

- *Exponential integrators.*
  Many 'difficult' ODEs can be written in the form $\dot{\boldsymbol{y}} = A\boldsymbol{y} + \boldsymbol{b}(\boldsymbol{y})$ where the matrix A is 'larger' (in some sense) than $\boldsymbol{b}(\boldsymbol{y})$—for example, A may be the Jacobian of an ODE (which may vary from step to step). Thus, it is to be expected that the 'nastiness' of the ODE under scrutiny—be it stiffness, Hamiltonicity or high oscillation—is somehow 'hardwired' into the matrix A. The exact solution of the ODE can be written in terms of the variation-of-constants formula,

$$\boldsymbol{y}(t+h) = e^{hA}\boldsymbol{y}(t) + \int_0^h e^{(h-\xi)A}\boldsymbol{b}(\boldsymbol{y}(t+\xi))\,d\xi, \tag{2.4}$$

except that, of course, the right-hand side includes the unknown function $\boldsymbol{y}$. Given the availability of very effective methods to compute the matrix exponential, we can exploit this to construct *exponential integrators,* explicit methods that often exhibit favourable stability and structure-preservation features. The simplest example, the *exponential Euler* method, freezes $\boldsymbol{y}$ within the integral in (2.4) at its known value at t, the outcome being the first-order method

$$\boldsymbol{y}_{n+1} = e^{hA}\boldsymbol{y}_n + A^{-1}(e^{hA} - I)\boldsymbol{b}(\boldsymbol{y}_n).$$

The order can be boosted by observing that (in a loose sense which can be made much more precise) the integral above is discretised by the Euler method, which is a one-stage explicit Runge–Kutta scheme, discretising it instead by multistage schemes of this kind leads to higher-order methods [35].

Many Hamiltonian systems of interest can be formulated as second-order systems of the form $\ddot{\boldsymbol{y}} + \Omega^2 \boldsymbol{y} = \boldsymbol{g}(\boldsymbol{y})$. Such systems feature prominently in the case of highly oscillatory mechanical systems, where Ω is positive definite and has some large eigenvalues. The variation of constants formula (2.4) now reads

$$\begin{bmatrix} \boldsymbol{y}(t+h) \\ \dot{\boldsymbol{y}}(t+h) \end{bmatrix} = \begin{bmatrix} \cos(h\Omega) & \Omega^{-1}\sin(h\Omega) \\ -\Omega\sin(h\Omega) & \cos(h\Omega) \end{bmatrix} \begin{bmatrix} \boldsymbol{y}(t) \\ \dot{\boldsymbol{y}}(t) \end{bmatrix} + \int_t^{t+h} \begin{bmatrix} \cos((h-\xi)\Omega) & \Omega^{-1}\sin((h-\xi)\Omega) \\ -\Omega\sin((h-\xi)\Omega) & \cos((h-\xi)\Omega) \end{bmatrix} \begin{bmatrix} \boldsymbol{0} \\ \boldsymbol{g}(\boldsymbol{y}(t+\xi)) \end{bmatrix} \mathrm{d}\xi$$

and we can use either standard exponential integrators or exponential integrators designed directly for second-order systems and using Runge–Kutta–Nyström methods on the nonlinear part [87].

An important family of exponential integrators for second-order systems are *Gautschi-type methods*

$$\boldsymbol{y}_{n+1} - 2\boldsymbol{y}_n + \boldsymbol{y}_{n-1} = h^2\Psi(h\Omega)(\boldsymbol{g}_n - \Omega^2\boldsymbol{y}_n), \tag{2.5}$$

which are of second order. Here $\Psi(x) = 2(1-\cos x)/x$ while, in Gautschi's original method, $\boldsymbol{g}_n = \boldsymbol{g}(\boldsymbol{y}_n)$ [35]. Unfortunately, this choice results in resonances and a better one is $\boldsymbol{g}_n = \boldsymbol{g}(\Phi(h\Omega)\boldsymbol{y}_n)$, where the *filter* Φ eliminates resonances: $\Phi(0) = I$ and $\Phi(k\pi) = 0$ for $k \in \mathbb{N}$. We refer to [35] for further discussion of such methods in the context of symplectic integration.

- *Variational integrators. Lagrangian formulation* recasts a large number of differential equations as extrema of nonlinear functionals. Thus, for example, instead of the Hamiltonian problem $M\ddot{\boldsymbol{q}} + \nabla V(\boldsymbol{q}) = \boldsymbol{0}$, where the matrix M is positive definite, we may consider the equivalent variational formulation of extremising the positive-definite nonlinear functional $L(\boldsymbol{q}, \dot{\boldsymbol{q}}) = \frac{1}{2}\dot{\boldsymbol{q}}^\top M\dot{\boldsymbol{q}} - V(\boldsymbol{q})$. With greater generality, Hamiltonian and Lagrangian formulations are connected via the familiar Euler–Lagrange equations and, given the functional L, the corresponding second-order system is

$$\frac{\partial L(\boldsymbol{q}, \dot{\boldsymbol{q}})}{\partial \boldsymbol{q}} - \frac{\mathrm{d}}{\mathrm{d}t}\left[\frac{\partial L(\boldsymbol{q}, \dot{\boldsymbol{q}})}{\partial \dot{\boldsymbol{q}}}\right] = \boldsymbol{0}.$$

The rationale of variational integrators parallels that of the *Ritz method* in the theory of finite elements. We first reformulate the Hamiltonian problem as a Lagrangian one, project it to a finite-dimensional space, solve it there and transform back. The original symplectic structure is replaced by a finite-dimensional symplectic structure, hence the approach is by design symplectic [64]. Marsden and West [54] review the implementation of variational integrators.

2.2 Lie-Group Methods

Let $\mathcal{G}$ be a Lie group and $\mathcal{M}$ a differentiable manifold. We say that $\Lambda : \mathcal{G} \times \mathcal{M} \to \mathcal{M}$ is a *group action* if

a. $\Lambda(\iota, y) = y$ for all $y \in \mathcal{M}$ (where ι is the identity of $\mathcal{G}$) and
b. $\Lambda(p, \Lambda(q, y)) = \Lambda(p \cdot q, y)$ for all $p, q \in \mathcal{G}$ and $y \in \mathcal{M}$.

If, in addition, for every $x, y \in \mathcal{M}$ there exists $p \in \mathcal{G}$ such that $y = \Lambda(p, x)$, the action is said to be transitive and $\mathcal{M}$ is a *homogeneous space,* acted upon by $\mathcal{G}$.

Every Lie group acts upon itself, while the orthogonal group $\mathrm{O}(n)$ acts on the $(n-1)$-sphere by multiplication, $\Lambda(p, y) = py$. The orthogonal group also acts on the *isospectral manifold* of all symmetric matrices similar to a specific symmetric matrix by similarity, $\Lambda(p, y) = pyp^\top$. Given $1 \le m \le n$, the *Grassmann manifold* $\mathbb{G}(n, m)$ of all m-dimensional subspaces of $\mathbb{R}^n$ is a homogeneous space acted upon by $\mathrm{SO}(m) \times \mathrm{SO}(n-m)$, where $\mathrm{SO}(m)$ is the special orthogonal group—more precisely, $\mathbb{G}(n, m) = \mathrm{SO}(n)/(\mathrm{SO}(m) \times \mathrm{SO}(n-m))$.

Faced with a differential equation evolving in a homogeneous space, we can identify its flow with a group action: Given an initial condition $y_0 \in \mathcal{M}$, instead of asking "what is the value of y at time $t > 0$" we might pose the equivalent question "what is the group action that takes the solution from y_0 to $y(t)$?". This is often a considerably more helpful formulation because a group action can be further related to an *algebra action.* Let $\mathfrak{g}$ be the Lie algebra corresponding to the matrix group $\mathcal{G}$, i.e. the tangent space at $\iota \in \mathcal{G}$, and denote by $\mathfrak{X}(\mathcal{M})$ the set of all Lipschitz vector fields over $\mathcal{M}$. Let $\lambda : \mathfrak{g} \to \mathfrak{X}(\mathcal{M})$ and $a : \mathbb{R}_+ \times \mathcal{M} \to \mathfrak{g}$ be both Lipschitz. In particular, we might consider

$$\lambda(a, y) = \frac{\mathrm{d}}{\mathrm{d}s}\Lambda(\rho(s, y), y)\bigg|_{s=0},$$

where Λ is a group action and $\rho : \mathbb{R}_+ \to \mathcal{G}$, $\rho(s, y(s)) = \iota + a(s, y(s))s + \mathcal{O}\left(s^2\right)$ for small $|s|$. The equation $\dot{y} = \lambda(a(t, y), y)$, $y(0) = y_0 \in \mathcal{M}$ represents *algebra action* and its solution evolves in $\mathcal{M}$. Moreover,

$$y(t) = \Lambda(v(t), y_0) \qquad \text{where} \qquad \dot{v} = a(t, \Lambda(v, y_0))v, \quad v(0) = \iota \in \mathcal{G} \tag{2.6}$$

is a *Lie-group equation.* Instead of solving the original ODE on $\mathcal{M}$, it is possible to solve (2.6) and use the group action Λ to advance the solution to the next step: this is the organising principle of most *Lie-group methods* [42]. It works because a Lie-group equation can be solved in the underlying Lie algebra, which is a *linear space.* Consider an ODE[2] $\dot{y} = f(y)$, $y(0) \in \mathcal{M}$, such that $f : \mathcal{M} \to \mathfrak{X}$—the solution $y(t)$ evolves on the manifold. While conventional numerical methods are highly unlikely to stay in $\mathcal{M}$, this is not the case for Lie-group methods. We can travel safely between $\mathcal{M}$ and $\mathcal{G}$ using a group action. The traffic between $\mathcal{G}$ and $\mathfrak{g}$ is slightly more complicated and we need to define a *trivialisation,* i.e. an invertible map taking smoothly a

[2]Or, for that matter, a PDE, except that formalities are somewhat more complicated.

neighbourhood of $0 \in \mathfrak{g}$ to a neighbourhood of $\iota \in \mathcal{G}$ and taking zero to identity. The most ubiquitous example of trivialisation is the exponential map, which represents the solution of (2.6) as $v(t) = \mathrm{e}^{\omega(t)}$, where ω is the solution of the *dexpinv equation*

$$\dot{\omega} = \sum_{m=0}^{\infty} \frac{\mathrm{B}_m}{m!} \mathrm{ad}^m_{a(t,\mathrm{e}^{\omega})} \omega, \qquad \omega(0) = 0 \in \mathfrak{g} \tag{2.7}$$

[42]. Here the B_ms are Bernoulli numbers, while ad_b^m is the *adjoint operator* in $\mathfrak{g}$,

$$\mathrm{ad}_b^0 c = c, \qquad \mathrm{ad}_b^m c = [b, \mathrm{ad}_b^{m-1} c], \quad m \in \mathbb{N}, \qquad b, c \in \mathfrak{g}.$$

Because $\mathfrak{g}$ is closed under linear operations and commutation, solving (2.7) while respecting Lie-algebraic structure is straightforward. Mapping back, first to $\mathcal{G}$ and finally to $\mathcal{M}$, we keep the numerical solution of $\dot{y} = f(t)$ on the manifold.

Particularly effective is the use of explicit Runge–Kutta methods for (2.7), the so-called Runge–Kutta–Munthe-Kaas (RKMK) methods [65]. To help us distinguish between conventional Runge–Kutta methods and RKMK, consider the three-stage, third-order method with the Butcher tableau[3]

$$\begin{array}{c|ccc} 0 & & & \\ \frac{1}{2} & \frac{1}{2} & & \\ 1 & -1 & 2 & \\ \hline & \frac{1}{6} & \frac{2}{3} & \frac{1}{6} \end{array} \; . \tag{2.8}$$

Applied to the ODE $\dot{y} = f(t, y)$, $y(t_n) = y_n \in \mathcal{M}$, evolving on the manifold $\mathcal{M} \subset \mathbb{R}^d$, it becomes

$$\begin{aligned} k_1 &= f(t_n, y_n), \\ k_2 &= f(t_{n+\frac{1}{2}}, y_n + \tfrac{1}{2} h k_1), \\ k_3 &= f(t_{n+1}, y_n - h k_1 + 2 h k_2), \\ \Delta &= h(\tfrac{1}{6} k_1 + \tfrac{2}{3} k_2 + \tfrac{1}{6} k_3), \\ y_{n+1} &= y_n + \Delta. \end{aligned}$$

Since we operate in $\mathbb{R}^d$, there is absolutely no reason for y_{n+1} to live in $\mathcal{M}$. However, once we implement (2.8) at an algebra level (truncating first the dexpinv equation (2.7)),

[3]For traditional concepts such as Butcher tableaux, Runge-Kutta methods and B-series, the reader is referred to [34].

$$\begin{aligned}
k_1 &= a(t_n, \iota),\\
k_2 &= a(t_{n+\frac{1}{2}}, \mathrm{e}^{hk_1/2}),\\
k_3 &= a(t_{n+1}, \mathrm{e}^{-hk_1+2hk_2}),\\
\Delta &= h(\tfrac{1}{6}k_1 + \tfrac{2}{3}k_2 + \tfrac{1}{6}k_3),\\
\omega_{n+1} &= \Delta + \tfrac{1}{6}h[\Delta, k_1]\\
y_{n+1} &= \Lambda(\mathrm{e}^{\omega_{n+1}}, y_n),
\end{aligned}$$

the solution is guaranteed to stay in $\mathcal{M}$.

An important special case of a Lie-group equation is the linear ODE $\dot{v} = a(t)v$, where $a : \mathbb{R}_+ \to \mathfrak{g}$. Although RKMK works perfectly well in a linear case, special methods do even better. Perhaps the most important is the *Magnus expansion* [53], $v(t) = \mathrm{e}^{\omega(t)}v(0)$, where

$$\begin{aligned}
\omega(t) = &\int_0^t a(\xi)\,\mathrm{d}\xi - \frac{1}{2}\int_0^t\int_0^{\xi_1}[a(\xi_2), a(\xi_1)]\,\mathrm{d}\xi_2\,\mathrm{d}\xi_1\\
&+ \frac{1}{4}\int_0^t\int_0^{\xi_1}\int_0^{\xi_2}[[a(\xi_3), a(\xi_2)], a(\xi_1)]\,\mathrm{d}\xi_3\,\mathrm{d}\xi_2\,\mathrm{d}\xi_1\\
&+ \frac{1}{12}\int_0^t\int_0^{\xi_1}\int_0^{\xi_2}[a(\xi_3), [a(\xi_2), a(\xi_1)]]\,\mathrm{d}\xi_3\,\mathrm{d}\xi_2\,\mathrm{d}\xi_1 + \cdots .
\end{aligned} \tag{2.9}$$

We refer to [6, 40, 42] for explicit means to derive expansion terms, efficient computation of multivariate integrals that arise in this context and many other implementation details. Magnus expansions are important in a number of settings when preservation of structure is not an issue, not least in the solution of linear stochastic ODEs [51].

There are alternative means to expand the solution of (2.7) in a linear case, not least the *Fer expansion* [22, 38], that has found recently an important application in the computation of Sturm–Liouville spectra [76].

Another approach to Lie-group equations uses *canonical coordinates of the second kind* [72].

2.3 Conservation of Volume

An ODE $\dot{\boldsymbol{x}} = \boldsymbol{f}(\boldsymbol{x})$ is divergence-free if $\nabla \cdot \boldsymbol{f}(\boldsymbol{x}) = 0$. The flows of divergence-free ODEs are volume-preserving (VP). Volume is important to preserve, as it leads to KAM-tori, incompressibility, and, most importantly, is a crucial ingredient for ergodicity. Unlike symplecticity, however, phase space volume can generically *not* be preserved by Runge–Kutta methods, or even by their generalisations, B-series methods. This was proved independently in [13] and in [43]. Since B-series methods cannot preserve volume, we need to look to other methods.

There are essentially two known numerical integration methods that preserve phase space volume. The first volume-preserving method is based on splitting [20]. As an example, consider a 3D volume preserving vector field:

$$\begin{aligned} \dot{x} &= u(x, y, z) \\ \dot{y} &= v(x, y, z) \\ \dot{z} &= w(x, y, z) \end{aligned} \tag{2.10}$$

with

$$u_x + v_y + w_z = 0.$$

We split this 3D VP vector field into two 2D VP vector fields as follows

$$\begin{aligned} \dot{x} &= u(x, y, z), & \dot{x} &= 0, \\ \dot{y} &= -\int u_x(x, y, z)\,\mathrm{d}y, & \dot{y} &= v(x, y, z) + \int u_x(x, y, z)\,\mathrm{d}y, \\ \dot{z} &= 0; & \dot{z} &= w(x, y, z). \end{aligned} \tag{2.11}$$

The vector field on the left is divergence-free by construction, and since both vector fields add up to (2.1), it follows that the vector field on the right is also volume-preserving.

Having split the original vector field into 2D VP vector fields, we need to find VP integrators for each of these 2D VP vector fields. But that is easy, because in 2D volume-preserving and symplectic vector fields are the same—this, of course, holds also for symplectic Runge–Kutta methods.

The above splitting method is easily generalised to n dimensions, where one splits into $n - 1$ 2D VP vector fields, and integrates each using a symplectic Runge–Kutta method.

An alternative VP integration method was discovered independently by Shang and by Quispel [74, 80]. We again illustrate this method in 3D.

We will look for an integrator of the form

$$\begin{aligned} x_1 &= g_1(x_1', x_2, x_3) \\ x_2' &= g_2(x_1', x_2, x_3) \\ x_3' &= g_1(x_1', x_2', x_3) \end{aligned} \tag{2.12}$$

where (here and below) $x_i = x_i(nh)$, and $x_i' = x_i((n+1)h)$. The reason the form (2.12) is convenient, is because any such map is VP iff

$$\frac{\partial x_1}{\partial x_1'} = \frac{\partial x_2'}{\partial x_2}\frac{\partial x_3'}{\partial x_3}. \tag{2.13}$$

To see how to construct a VP integrator of the form (2.12), consider as an example the ODE

$$\begin{aligned}\dot{x}_1 &= x_2 + x_1^2 + x_3^3 \\ \dot{x}_2 &= x_3 + x_1 x_2 + x_1^4 \\ \dot{x}_3 &= x_1 - 3x_1 x_3 + x_2^5\end{aligned} \tag{2.14}$$

It is easy to check that it is divergence-free.

Now consistency requires that any integrator for (2.14) should satisfy

$$\begin{aligned}x_1' &= x_1 + h(x_2 + x_1^2 + x_3^3) + \mathcal{O}\left(h^2\right) \\ x_2' &= x_2 + h(x_3 + x_1 x_2 + x_1^4) + \mathcal{O}\left(h^2\right) \\ x_3' &= x_3 + h(x_1 - 3x_1 x_3 + x_2^5) + \mathcal{O}\left(h^2\right)\end{aligned} \tag{2.15}$$

and therefore

$$x_1 = x_1' - h(x_2 + x_1'^2 + x_3^3) + \mathcal{O}\left(h^2\right) \tag{2.16}$$
$$x_2' = x_2 + h(x_3 + x_1' x_2 + x_1'^4) + \mathcal{O}\left(h^2\right) \tag{2.17}$$
$$x_3' = x_3 + h(x_1' - 3x_1' x_3 + x_2'^5) + \mathcal{O}\left(h^2\right) \tag{2.18}$$

Since we are free to choose any consistent g_2 and g_3 in (2.12), provided g_1 satisfies (2.13), we choose the terms designated by $\mathcal{O}\left(h^2\right)$ in (2.15) and (2.16) to be identically zero. Equation (2.13) then yields

$$\frac{\partial x_1}{\partial x_1'} = (1 + h x_1')(1 - 3h x_1'). \tag{2.19}$$

This can easily be integrated to give

$$x_1 = x_1' - h x_1'^2 - h^2 x_1'^3 + k(x_2, x_3; h). \tag{2.20}$$

where the function k denotes an integration constant that we can choose appropriately. The simplest VP integrator satisfying both (2.14) and (2.20) is therefore:

$$\begin{aligned}x_1 &= x_1' - h(x_2 + x_1'^2 + x_3^3) - h^2 x_1'^3 \\ x_2' &= x_2 + h(x_3 + x_1' x_2 + x_1'^4) \\ x_3' &= x_3 + h(x_1' - 3x_1' x_3 + x_2'^5)\end{aligned} \tag{2.21}$$

A nice aspect of the integrator (2.21) (and (2.12)) is that it is essentially only implicit in one variable. Once x_1' is computed from the first (implicit) equation, the other two equations are essentially explicit.

Of course the method just described also generalises to any divergence-free ODE in any dimension.

2.4 Preserving Energy and Other First Integrals

As mentioned, Hamiltonian systems exhibit two important geometric properties simultaneously, they conserve both the symplectic form and the energy. A famous no-go theorem by Ge and Marsden [25] has shown that it is generically impossible to construct a geometric integrator that preserves both properties at once. One therefore must choose which one of these two to preserve in any given application. Particularly in low dimensions and if the energy surface is compact, there are often advantages in preserving the energy.

An energy-preserving B-series method was discovered in [75] cf. also [59].

For any ODE $\dot{\boldsymbol{x}} = \boldsymbol{f}(\boldsymbol{x})$, this so-called average vector field method is given by

$$\frac{\boldsymbol{x}' - \boldsymbol{x}}{h} = \int_0^1 \boldsymbol{f}(\xi \boldsymbol{x}' + (1 - \xi)\boldsymbol{x})\,\mathrm{d}\xi. \tag{2.22}$$

If the vector field $\boldsymbol{f}$ is Hamiltonian, i.e. if there exists a Hamiltonian function $H(\boldsymbol{x})$ and a constant skew-symmetric matrix S such that $\boldsymbol{f}(\boldsymbol{x}) = S\nabla H(\boldsymbol{x})$, then it follows from (2.22) that energy is preserved, i.e. $H(\boldsymbol{x}') = H(\boldsymbol{x})$.

While the B-series method (2.22) can generically preserve all types of Hamiltonians H, it can be shown that no Runge–Kutta method is energy-preserving for all H. (In other words, this can only be done using B-series methods that are *not* RK methods.) For a given *polynomial* H however, Runge–Kutta methods preserving that H do exist [37]. This can be seen as follows.

Note that the integral in (2.22) is one-dimensional. This means that e.g. for cubic vector fields (and hence for quartic Hamiltonians) an equivalent method is obtained by replacing the integral in (2.22) using Simpson's rule:

$$\int_0^1 g(\xi)\,\mathrm{d}\xi \approx \frac{1}{6}\left[g(0) + 4g(\tfrac{1}{2}) + g(1)\right]. \tag{2.23}$$

yielding the Runge–Kutta method

$$\frac{\boldsymbol{x}' - \boldsymbol{x}}{h} = \frac{1}{6}\left[\boldsymbol{f}(\boldsymbol{x}) + 4\boldsymbol{f}\left(\frac{\boldsymbol{x} + \boldsymbol{x}'}{2}\right) + \boldsymbol{f}(\boldsymbol{x}')\right], \tag{2.24}$$

preserving all quartic Hamiltonians.

We note that (2.22) has second order accuracy. Higher order generalisations have been given in [27]. We note that the average vector field method has also been applied to a slew of semi-discretised PDEs in [9].

While energy is one of the most important constants of the motion in applications, many other types of first integrals do occur. We note here that all B-series methods preserve all linear first integrals, and that all symplectic B-series methods preserve all quadratic first integrals. So, for example, the implicit midpoint rule

$$\frac{\boldsymbol{x}' - \boldsymbol{x}}{h} = \boldsymbol{f}\left(\frac{\boldsymbol{x} + \boldsymbol{x}'}{2}\right)$$

(which is symplectic) preserves all linear and quadratic first integrals. There are however many cases not covered by any of the above.

How does one preserve a cubic first integral that is not energy? And what about Hamiltonian systems whose symplectic structure is not constant? It turns out that generically, any ODE $\dot{\boldsymbol{x}} = \boldsymbol{f}(\boldsymbol{x})$ that preserves an integral $I(\boldsymbol{x})$, can be written in the form

$$\dot{\boldsymbol{x}} = S(\boldsymbol{x})\nabla I(\boldsymbol{x}), \tag{2.25}$$

where $S(\boldsymbol{x})$ is a skew-symmetric matrix.[4]

An integral-preserving discretisation of (2.25) is given by

$$\frac{\boldsymbol{x}' - \boldsymbol{x}}{h} = \bar{S}(\boldsymbol{x}, \boldsymbol{x}')\bar{\nabla} I(\boldsymbol{x}, \boldsymbol{x}'), \tag{2.26}$$

where $\bar{S}(\boldsymbol{x}, \boldsymbol{x}')$ is any consistent approximation to $S(\boldsymbol{x})$ (e.g. $\bar{S}(\boldsymbol{x}, \boldsymbol{x}') = S(\boldsymbol{x})$), and the *discrete gradient* $\bar{\nabla} I$ is defined by

$$(\boldsymbol{x}' - \boldsymbol{x}) \cdot \bar{\nabla} I(\boldsymbol{x}, \boldsymbol{x}') = I(\boldsymbol{x}') - I(\boldsymbol{x}) \tag{2.27}$$

and

$$\lim_{\boldsymbol{x}' \to \boldsymbol{x}} \bar{\nabla} I(\boldsymbol{x}, \boldsymbol{x}') = \nabla I(\boldsymbol{x}). \tag{2.28}$$

There are many different discrete gradients that satisfy (2.27) and (2.28). A particularly simple one is given by the Itoh–Abe discrete gradient, which for example in 3D reads

$$\bar{\nabla} I(\boldsymbol{x}, \boldsymbol{x}') = \begin{bmatrix} \dfrac{I(x_1', x_2, x_3) - I(x_1, x_2, x_3)}{x_1' - x_1} \\ \dfrac{I(x_1', x_2', x_3) - I(x_1', x_2, x_3)}{x_2' - x_2} \\ \dfrac{I(x_1', x_2', x_3') - I(x_1', x_2', x_3)}{x_3' - x_3} \end{bmatrix}. \tag{2.29}$$

Other examples of discrete gradients, as well as constructions of the skew-symmetric matrix $S(\boldsymbol{x})$ for a given vector field $\boldsymbol{f}$ and integral I may be found in [59].

We note that the discrete gradient method can also be used for systems with any number of integrals. For example an ODE $\dot{\boldsymbol{x}} = \boldsymbol{f}(\boldsymbol{x})$ possessing two integrals $I(\boldsymbol{x})$ and $J(\boldsymbol{x})$ can be written

$$\dot{x}_i = S_{ijk}(\boldsymbol{x})\frac{\partial I(\boldsymbol{x})}{\partial x_j}\frac{\partial J(\boldsymbol{x})}{\partial x_k}, \tag{2.30}$$

[4]Note that in general $S(\boldsymbol{x})$ need not satisfy the so-called Jacobi identity.

where the summation convention is assumed over repeated indices and $S(\boldsymbol{x})$ is a completely antisymmetric tensor. A discretisation of (2.30) which preserves both I and J is given by

$$\frac{x_i' - x_i}{h} = \bar{S}_{ijk}(\boldsymbol{x}, \boldsymbol{x}')\bar{\nabla} I(\boldsymbol{x}, \boldsymbol{x}')\Big|_j \bar{\nabla} J(\boldsymbol{x}, \boldsymbol{x}')\Big|_k \tag{2.31}$$

with $\bar{S}$ any completely skew approximation of S and $\bar{\nabla} I$ and $\bar{\nabla} J$ discrete gradients as defined above. Discrete gradient methods have recently found an intriguing application in variational image regularisation [26].

3 Four Recent Stories of GNI

The purpose of this section is not to present a totality of recent research into GNI, a subject that would have called for a substantially longer paper. Instead, we wish to highlight a small number of developments with which the authors are familiar and which provide a flavour of the very wide range of issues on the current GNI agenda.

3.1 Highly Oscillatory Hamiltonian Systems

High oscillation occurs in many Hamiltonian systems. Sometimes, e.g. in the integration of equations of celestial mechanics, the source of the problem is that we wish to compute the solution across a very large number of periods and the oscillation is an artefact of the time scale in which the solution has physical relevance. In other cases oscillation is implicit in the multiscale structure of the underlying problem. A case in point are the (modified) *Fermi–Pasta–Ulam (FPU) equations*, describing a mechanical system consisting of alternating stiff harmonic and soft nonlinear springs. The soft springs impart fast oscillation, while the hard springs generate slow transfer of energy across the system: good numerical integration must capture both!

A good point to start (which includes modified FPU as a special case) is the second-order ODE

$$\ddot{\boldsymbol{q}} + \Omega^2 \boldsymbol{q} = \boldsymbol{g}(\boldsymbol{q}), \qquad t \geq 0, \qquad \boldsymbol{q}(0) = \boldsymbol{u}_0, \quad \dot{\boldsymbol{q}}(0) = \boldsymbol{v}_0, \tag{3.1}$$

where $\boldsymbol{g}(\boldsymbol{q}) = -\boldsymbol{\nabla} U(\boldsymbol{q})$ and

$$\Omega = \begin{bmatrix} O & O \\ O & \omega I \end{bmatrix}, \qquad \omega \gg 1, \qquad \boldsymbol{q} = \begin{bmatrix} \boldsymbol{q}_0 \\ \boldsymbol{q}_1 \end{bmatrix}, \qquad \boldsymbol{q}_0 \in \mathbb{R}^{n_0}, \quad \boldsymbol{q}_1 \in \mathbb{R}^{n_1}.$$

An important aspect of systems of the form (3.1) is that the exact solution, in addition to preserving the total Hamiltonian energy

$$H(\boldsymbol{p},\boldsymbol{q}) = \frac{1}{2}(\|\boldsymbol{p}_1\|^2 + \omega^2\|\boldsymbol{q}_1\|^2) + \frac{1}{2}\|\boldsymbol{p}_0\|^2 + U(\boldsymbol{q}_0,\boldsymbol{q}_1), \tag{3.2}$$

where $\dot{\boldsymbol{q}} = \boldsymbol{p}$, also preserves the *oscillatory energy*

$$I(\boldsymbol{p},\boldsymbol{q}) = \frac{1}{2}\|\boldsymbol{p}_1\|^2 + \frac{\omega^2}{2}\|\boldsymbol{q}_1\|^2 \tag{3.3}$$

for intervals of length $\mathcal{O}\left(\omega^N\right)$ for any $N \geq 1$. This has been proved using the *modulated Fourier expansions*

$$\boldsymbol{q}(t) = \sum_{m=-\infty}^{\infty} e^{im\omega t} z_m(t).$$

The solution of (3.1) exhibits oscillations at frequency $\mathcal{O}(\omega)$ and this inhibits the efficiency of many symplectic methods, requiring step size of $\mathcal{O}\left(\omega^{-1}\right)$, a situation akin to stiffness in more conventional ODEs. However, by their very structure, exponential integrators (and in particular Gautschi-type methods (2.5)) are particularly effective in integrating the linear part, which gives rise to high oscillation. The problem with Gautschi-type methods, though, might be the occurrence of resonances and we need to be careful to avoid them, both in the choice of the right filter (cf. the discussion in Sect. 2.1) and step size h.

Of course, one would like geometric numerical integrators applied to (3.1) to exhibit favourable preservation properties with respect to both total energy (3.2) and oscillatory energy (3.3). Applying modulated Fourier expansions to trigonometric and modified trigonometric integrators, this is indeed the case provided that the step size obeys the *non-resonance condition* with respect to the frequency ω,

$$|\sin(\tfrac{1}{2}mh\omega)| \geq ch^{1/2}, \qquad m = 1,\ldots,N, \quad N \geq 2,$$

cf. Hairer and Lubich [30].

All this has been generalised to systems with multiple frequencies, with the Hamiltonian function

$$H(\boldsymbol{p},\boldsymbol{q}) = \overbrace{\frac{1}{2}\sum_{j=1}^{s}\left(\|\boldsymbol{p}_j\|^2 + \omega_j^2\|\boldsymbol{q}_j\|^2\right)}^{\text{oscillatory}} + \overbrace{\frac{1}{2}\|\boldsymbol{p}_0\|^2 + U(\boldsymbol{q})}^{\text{slow}},$$

where

$$\boldsymbol{p} = \begin{bmatrix} \boldsymbol{p}_0 \\ \boldsymbol{p}_1 \\ \vdots \\ \boldsymbol{p}_s \end{bmatrix}, \quad \boldsymbol{q} = \begin{bmatrix} \boldsymbol{q}_0 \\ \boldsymbol{q}_1 \\ \vdots \\ \boldsymbol{q}_s \end{bmatrix}, \qquad 0 < \min_{j=1,\ldots,s} \omega_j, \quad 1 \ll \max_{j=1,\ldots,s} \omega_j$$

for both the exact solution [24] and for discretisations obtained using trigonometric and modified trigonometric integrators [14].

Further achievements and open problem in the challenging area of marrying symplectic integration and high oscillation are beautifully described in [28] and [31].

3.2 Kahan's 'Unconventional' Method

A novel discretisation method for quadratic ODEs was introduced and studied in [44, 45] and analysed first from the GNI standpoint in [78]. This new method discretised the vector field

$$\dot{x}_i = \sum_{j,k} a_{ijk} x_j x_k + \sum_j b_{ij} x_j + c_i \tag{3.4}$$

as follows,

$$\frac{x_i' - x_i}{h} = \sum_{j,k} a_{ijk} \left(\frac{x_j x_k' + x_j' x_k}{2} \right) + \sum_j b_{ij} \left(\frac{x_j + x_j'}{2} \right) + c_i. \tag{3.5}$$

Kahan called the method (3.5) 'unconventional', because it treats the quadratic terms different from the linear terms. He also noted some nice features of (3.5), e.g. that it often seemed to be able to integrate through singularities.

Properties of Kahan's method:

1. *Kahan's method is (the reduction of) a Runge–Kutta method.*
   Celledoni et al. [12] showed that (3.5) is the reduction to quadratic vector fields of the Runge–Kutta method
   $$\frac{\boldsymbol{x}' - \boldsymbol{x}}{h} = 2\boldsymbol{f}\left(\frac{\boldsymbol{x} + \boldsymbol{x}'}{2} \right) - \frac{1}{2}\boldsymbol{f}(\boldsymbol{x}) - \frac{1}{2}\boldsymbol{f}(\boldsymbol{x}') \tag{3.6}$$
   The RK method (3.6) (which is defined for all vector fields f), once applied to quadratic vector fields, coincides with Kahan's method (which is only defined in the quadratic case).
   This explains *inter alia* why Kahan's method preserves all linear first integrals.
2. *Kahan's method preserves a modified energy and measure.*
   For any Hamiltonian vector field of the form
   $$\dot{\boldsymbol{x}} = \boldsymbol{f}(x) = S\nabla H(\boldsymbol{x}), \tag{3.7}$$
   with cubic Hamiltonian $H(\boldsymbol{x})$ and constant symplectic (or Poisson) structure S, Kahan's method preserves a modified energy as well as a modified measure exactly [12].

The modified volume is

$$\frac{\mathrm{d}x_1 \wedge \cdots \wedge \mathrm{d}x_n}{\det\left(I - \frac{1}{2}hf'(\boldsymbol{x})\right)}, \tag{3.8}$$

while the modified energy is

$$\tilde{H}(\boldsymbol{x}) := H(\boldsymbol{x}) + \frac{1}{3}h\nabla H(\boldsymbol{x})^\top \left(I - \frac{1}{2}hf'(\boldsymbol{x})\right)^{-1} \boldsymbol{f}(\boldsymbol{x}). \tag{3.9}$$

3. *Kahan's method preserves the integrability of many integrable systems of quadratic ODEs.*
   Beginning with the work of Hirota and Kimura, subsequently extended by Suris and collaborators [73], and by Quispel and collaborators [10, 12, 86], it was shown that Kahan's method preserves the complete integrability of a surprisingly large number of quadratic ODEs. In most cases this means that, in n dimensions, Kahan's method preserves a (modified) volume form, as well as $n-1$ (modified) first integrals.

Here we list some 2D vector fields whose integrability is preserved by Kahan's method:

- Quadratic Hamiltonian systems in 2D:
  The 9-parameter family

$$\begin{bmatrix} \dot{x} \\ \dot{y} \end{bmatrix} = \begin{bmatrix} bx^2 + 2cxy + dy^2 + fx + gy + i \\ -ax^2 - 2bxy - cy^2 - ex - fy - h \end{bmatrix}; \tag{3.10}$$

- Suslov systems in 2D:
  The 9-parameter family

$$\begin{bmatrix} \dot{x} \\ \dot{y} \end{bmatrix} = l(x, y) \begin{bmatrix} 0 & 1 \\ -1 & 0 \end{bmatrix} \nabla H(x, y), \tag{3.11}$$

  where $l(x, y) = ax + by + c$; $H(x, y) = dx^2 + exy + fy^2 + gx + hy + i$;
- Reduced Nahm equations in 2D:
  Octahedral symmetry:

$$\begin{bmatrix} \dot{x} \\ \dot{y} \end{bmatrix} = \begin{bmatrix} 2x^2 - 12y^2 \\ -6x^2 - 4y^2 \end{bmatrix}; \tag{3.12}$$

  Icosahedral symmetry:

$$\begin{bmatrix} \dot{x} \\ \dot{y} \end{bmatrix} = \begin{bmatrix} 2x^2 - y^2 \\ -10xy + y^2 \end{bmatrix}. \tag{3.13}$$

The modified energy and measure for the Kahan discretisations of these 2D systems, as well as of many other (higher-dimensional) integrable quadratic vector fields are given in [10, 12, 73].

Generalisations to higher degree polynomial equations using polarisation are presented in [11].

3.3 Applications to Celestial Mechanics

GNI methods particularly come into their own when the integration time is large compared to typical periods of the system. Thus long-term integrations of e.g. solar-type systems and of particle accelerators typically need symplectic methods. In this subsection we focus on the former.[5]

In those studies where dissipation can be neglected, a common approach to solar system type dynamics is to split the N-body Hamiltonian H in the form

$$H = H_1 + \varepsilon H_2, \tag{3.14}$$

where H_1, representing the Keplerian motion of the $N-1$ planets, is integrable, H_2 represents the interaction between the planets and $\varepsilon > 0$ is a small parameter. In this manner, special methods for near-integrable Hamiltonian dynamics can and have been used, cf. e.g. [56].

One of the first symplectic integrations of the solar system was done in [84] where it was confirmed that the solar system has a positive Lyapunov exponent, and hence exhibits chaotic behaviour cf [47].

More recently these methods have been improved and extended [5, 17, 48]. Several symplectic integrators of high order were tested in [19], in order to determine the best splitting scheme for long-term studies of the solar system.

These various methods have resulted in the fact that numerical algorithms for solar system dynamics are now so accurate that they can be used to define the geologic time scales in terms of the initial conditions and parameters of solar system models (or vice versa). For related work cf [82].

3.4 Symmetric Zassenhaus Splitting and the Equations of Quantum Mechanics

Equations of quantum mechanics in the semiclassical regime represent a double challenge of structure conservation and high oscillation. A good starting point is the linear Schrödinger equation

[5] A very readable early review of integrators for solar system dynamics is [63], cf also [62].

$$\frac{\partial u}{\partial t} = \mathrm{i}\varepsilon \frac{\partial^2 u}{\partial x^2} - \mathrm{i}\varepsilon^{-1} V(x)u \tag{3.15}$$

(for simplicity we restrict our discourse to a single space dimension), given in $[-1, 1]$ with periodic boundary conditions. Here V is the potential energy of a quantum system, $|u(x,t)|^2$ is a position density of a particle and $0 < \varepsilon \ll 1$ represents the difference in mass between an electron and nuclei. It is imperative to preserve the unitarity of the solution operator (otherwise $|u(\cdot\,,t)|^2$ is no longer a probability function), but also deal with oscillation at a frequency of $\mathcal{O}\left(\varepsilon^{-1}\right)$. A conventional approach advances the solution using a palindromic splitting (2.3), but this is suboptimal for a number of reasons. Firstly, the number of splittings increases exponentially with order. Secondly, error constants are exceedingly large. Thirdly, quantifying the quality of approximation in terms of the step-size h is misleading, because there are three small quantities at play: the step size h, N^{-1} where N is the number of degrees of freedom in space discretisation (typically either a spectral method or spectral collocation) and, finally, $\varepsilon > 0$ which, originating in physics rather than being a numerical artefact, is the most important. We henceforth let $N = \mathcal{O}\left(\varepsilon^{-1}\right)$ (to resolve the high-frequency oscillations) and $h = \mathcal{O}\left(\varepsilon^{\sigma}\right)$ for some $\sigma > 0$—obviously, the smaller σ, the larger the time step.

Bader et al. [1] have recently proposed an alternative approach to the splitting of (3.15), of the form

$$\mathrm{e}^{\mathrm{i}h(\varepsilon\partial_x^2 - \varepsilon^{-1}V)} \approx \mathrm{e}^{\mathcal{R}_0}\mathrm{e}^{\mathcal{R}_1}\cdots\mathrm{e}^{\mathcal{R}_s}\mathrm{e}^{\mathcal{T}_{s+1}}\mathrm{e}^{\mathcal{R}_s}\cdots\mathrm{e}^{\mathcal{R}_1}\mathrm{e}^{\mathcal{R}_0} \tag{3.16}$$

such that $\mathcal{R}_k = \mathcal{O}\left(\varepsilon^{\alpha_k}\right)$, $\mathcal{T}_{s+1} = \mathcal{O}\left(\varepsilon^{\alpha_{s+1}}\right)$, where $\alpha_0 \le \alpha_1 < \alpha_2 < \alpha_3 < \cdots$—the *symmetric Zassenhaus splitting*. Here $\partial_x = \partial/\partial x$.

The splitting (3.16) is derived at the level of differential operators (i.e., prior to space discretisation), applying the symmetric Baker–Campbell–Hausdorff formula to elements in the free Lie algebra spanned by ∂_x^2 and V. For $\sigma = 1$, for example, this yields

$$\begin{aligned}
\mathcal{R}_0 &= -\tfrac{1}{2}\tau\varepsilon^{-1}V = \mathcal{O}(1),\\
\mathcal{R}_1 &= \tfrac{1}{2}\tau\varepsilon\partial_x^2 = \mathcal{O}(1),\\
\mathcal{R}_2 &= \tfrac{1}{24}\tau^3\varepsilon^{-1}(\partial_x V)^2 + \tfrac{1}{12}\tau^3\varepsilon\{(\partial_x^2 V)\partial_x^2 + \partial_x^2[(\partial_x^2 V)\cdot]\} = \mathcal{O}\left(\varepsilon^2\right),\\
\mathcal{R}_3 &= -\tfrac{1}{120}\tau^5\varepsilon^{-1}(\partial_x^2 V)(\partial_x V)^2 - \tfrac{1}{24}\tau^3\varepsilon(\partial_x^4 V) + \tfrac{1}{240}\tau^5\varepsilon\left(7\{(\partial_x^2 V)^2\partial_x^2\right.\\
&\quad \left. + \partial_x^2[(\partial_x^2 V)^2\cdot] + \{(\partial_x^3 V)(\partial_x V)\partial_x^2 + \partial_x^2[(\partial_x^3 V)(\partial_x V)\cdot]\}\right)\\
&\quad + \tfrac{1}{120}\tau^5\varepsilon^{-3}\{(\partial_x^4 V)\partial_x^4 + \partial_x^4[(\partial_x^4 V)\cdot]\} = \mathcal{O}\left(\varepsilon^4\right),
\end{aligned}$$

where $\tau = \mathrm{i}h$. Note that all the commutators, ubiquitous in the BCH formula, have disappeared: in general, the commutators in this free Lie algebra can be replaced by linear combinations of derivatives, with the remarkable property of *height reduction:* each commutator 'kills' one derivative, e.g.

$$[V, \partial_x^2] = -(\partial_x^2 V) - 2(\partial_x V)\partial_x, \qquad [[V, \partial_x^2], \partial_x^2] = (\partial_x^4 V) + 4(\partial_x^3 V)\partial_x + 4(\partial_x^2 V)\partial_x^2.$$

Once we discretise with spectral collocation, $\mathcal{R}_0$ becomes a diagonal matrix and its exponential is trivial, while $e^{\mathcal{R}_1}\boldsymbol{v}$ can be computed in two FFTs for any vector $\boldsymbol{v}$ because $\mathcal{R}_1$ is a Toeplitz circulant matrix. Neither $\mathcal{R}_2$ nor $\mathcal{R}_3$ possess useful structure, except that they are *small!* Therefore we can approximate $e^{\mathcal{R}_k}\boldsymbol{v}$ using the Krylov–Arnoldi process in just 3 and 2 iterations for $k = 2$ and $k = 3$, respectively, to attain an error of $\mathcal{O}\left(\varepsilon^6\right)$ [1].

All this has been generalised to time-dependent potentials and is applicable to a wider range of quantum mechanics equations in the semiclassical regime [2].

4 Beyond GNI

Ideas in one area of mathematical endeavour often inspire work in another area. This is true not just because new mathematical research equips us with a range of innovative tools but because it provides insight that casts new light not just on the subject in question but elsewhere in the mathematical universe. GNI has thus contributed not just toward its own goal, better understanding of structure-preserving discretisation methods for differential equations, but in other, often unexpected, directions.

4.1 GNI Meets Abstract Algebra

The traditional treatment of discretisation methods for differential equations was wholly analytic, using tools of functional analysis and approximation theory. (Lately, also tools from algebraic topology.) GNI has added an emphasis on geometry and this leads in a natural manner into concepts and tools from abstract algebra. As often in such mathematical dialogues, while GNI borrowed much of its conceptual background from abstract algebra, it has also contributed to the latter, not just with new applications but also new ideas.

- *B-series and beyond.* Consider numerical integration methods that associate to each vector field $\boldsymbol{f}$ a map $\boldsymbol{\psi}_h(\boldsymbol{f})$. A method $\boldsymbol{\psi}_h$ is called g-covariant[6] if the following diagram commutes,

[6] Also called equivariant.

$$\begin{array}{ccc}
\dot{x} = f(x) & \xrightarrow{\quad x = g(y) \quad} & \dot{y} = \tilde{f}(y) \\
\Big\downarrow & & \Big\downarrow \\
\tilde{x} = \psi_h(f)(x) & \xrightarrow{\quad x = g(y) \quad} & \tilde{y} = \psi_h(\tilde{f})(y)
\end{array}$$

It follows that if g is a symmetry of the vector field f and ψ is g-covariant, then ψ preserves the symmetry g. It seems that this concept of covariance for integration methods was first introduced in [55, 60].

It is not hard to check that all B-series methods are covariant with respect to the group of affine transformations. A natural question to ask then, was "are B-series methods the only numerical integration methods that preserve the affine group?". This question was open for many years, until it was answered in the negative by [66], who introduced a more general class of integration methods dubbed "aromatic Butcher series", and showed that (under mild assumptions) this is the most general class of methods preserving affine covariance. Expansions of methods in this new class contain both rooted trees (as in B-series), as well as products of rooted trees and so-called k-loops [43].

Whereas it may be said that to date the importance of aromatic B-series has been at the formal rather than at the constructive level, these methods may hold the promise of the construction of affine-covariant volume-preserving integrators, cf also [58].

- *Word expansions.* Classical B-series can be significantly generalised by expanding in *word series* [69]. This introduced an overarching framework for Taylor expansions, Fourier expansions, modulated Fourier expansions and splitting methods. We consider an ODE of the form

$$\dot{\boldsymbol{x}} = \sum_{a \in \mathcal{A}} \lambda_a(t) \boldsymbol{f}_a(\boldsymbol{x}), \qquad \boldsymbol{x}(0) = \boldsymbol{x}_0, \tag{4.1}$$

where $\mathcal{A}$ is a given *alphabet.* The solution of (4.1) can be formally expanded in the form

$$\boldsymbol{x}(t) = \sum_{n=0}^{\infty} \sum_{w \in \mathcal{W}_n} \alpha_w(t) f_w(\boldsymbol{x}_0),$$

where $\mathcal{W}_n$ is the set of all words with n letters from $\mathcal{A}$. The coefficients α_w and functions $\boldsymbol{f}_w$ can be obtained recursively from the λ_as and $\boldsymbol{f}_a$s in a manner similar

to B-series. Needless to say, exactly like with B-series, word series can be interpreted using an algebra over rooted trees.
The concept of word series is fairly new in numerical mathematics but it exhibits an early promise to provide a powerful algebraic tool for the analysis of dynamical systems and their discretisation.

- *Extension of Magnus expansions.* Let $\mathcal{W}$ be a *Rota–Baxter algebra*, a commutative unital algebra equipped with a linear map R such that

$$R(x)R(y) = R(R(x)y + xR(y) + \theta xy), \qquad x, y \in \mathcal{W},$$

where θ is a parameter. The inverse ∂ of R obeys

$$\partial(xy) = \partial(x)y + x\partial(y) + \theta\partial(x)\partial(y)$$

and is hence a generalisation of a derivation operator: a neat example, with clear numerical implications, is the backward difference $\partial(x) = [x(t) - x(t - \theta)]/\theta$. [18] generalised Magnus expansions to this and similar settings, e.g. dendriform algebras. Their work uses the approach in [40], representing individual 'Magnus terms' as rooted trees, but generalises it a great deal.

- *The algebra of the Zassenhaus splitting.* The success of the Zassenhaus splitting (3.16) rests upon two features. Firstly, the replacement of commutators by simpler, more tractable expressions and, secondly, height reduction of derivatives under commutation. Singh [81] has derived an algebraic structure $\mathfrak{J}$ which, encoding these two features, allows for a far-reaching generalisation of the Zassenhaus framework. The elements of $\mathfrak{J}$ are operators of the form $\langle f\rangle_k = f \circ \partial_x^k + \partial_x^k \circ f$, where $k \in \mathbb{Z}_+$ and f resides in a suitable function space. $\mathfrak{J}$ can be endowed with a Lie-algebraic structure and, while bearing similarities with the Weyl algebra and the Heisenberg group, is a new and intriguing algebraic concept.

4.2 Highly Oscillatory Quadrature

Magnus expansions (2.9) are particularly effective when the matrix $A(t)$ oscillates rapidly. This might seem paradoxical—we are all conditioned to expect high oscillation to be 'difficult'—but actually makes a great deal of sense. Standard numerical methods are based on Taylor expansions, hence on *differentiation*, and their error typically scales as a high derivative of the solution. Once a function oscillates rapidly, differentiation roughly corresponds to multiplying amplitude by frequency, high derivatives become large and so does the error. However, the Magnus expansion does not differentiate, it *integrates!* This has an opposite effect: the more we integrate, the smaller the amplitude and the series (2.9) converges more rapidly. Indeed, often it pays to render a linear system highly oscillatory by a change of variables, in a manner described in [39], and then solve it considerably faster and cheaper. Yet, once we contemplate the discretisation of (2.9) for a highly oscillatory matrix function $A(t)$, we soon come up another problem, usually considered difficult, if not insurmountable: computing multivariate integrals of highly oscillatory functions.

In a long list of methods for highly oscillatory quadrature (HOQ) *circa* 2002, ranging from the useless to the dubious, one method stood out: [50] proposed to calculate univariate integrals by converting the problem to an ODE and using collocation. This was the only effective method around, yet incompletely understood.

The demands of GNI gave the initial spur to the emergence in the last ten years to a broad swath of new methods for HOQ: Filon-type methods, which replace the *non-oscillatory* portion of the integrand by an interpolating polynomial [41], improved Levin-type methods [71] and the method of numerical stationary phase cf [36]. The common characteristic of all these methods is that they are based on asymptotic expansions. This means that high oscillation is no longer the enemy—indeed, the faster the oscillation, the smaller the error!

Highly oscillatory integrals occur in numerous applications, from electromagnetic and acoustic scattering to fluid dynamics, quantum mechanics and beyond. Their role in GNI is minor. However, their modern numerical theory was originally motivated by a problem in GNI [16]. This is typical to how scholarship progresses and it is only natural that HOQ has severed its GNI moorings and has become an independent area on its own.

4.3 *Structured Linear Algebra*

GNI computations often lead to specialised problems in numerical linear algebra. However, structure preservation has wider impact in linear algebraic computations. Often a matrix in an algebraic problem belongs to an algebraic structure, e.g. a specific Lie algebra or a symmetric space, and it is important to retain this in computation—the sobriquet "Geometric Numerical Algebra" might be appropriate! Moreover, as in GNI so in GNA, respecting structure often leads to better, more accurate and cheaper numerical methods. Finally, structured algebraic computation is often critical to GNI computations.

- Matrix factorization is the lifeblood of numerical algebra, the basis of the most effective algorithms for the solution of linear systems, computation of eigenvalues and solution of least-squares problems. A major question in GNA is "Suppose that $A \in \mathcal{A}$, where $\mathcal{A}$ is a set of matrices of given structure. Given a factorization $A = BC$ according to some set of rules, what can we say about the structure of B or C?". [52] addressed three such 'factorization rules': the *matrix square root,* $B = C$, the *matrix sign,* where the elements of B are ± 1, and the *polar decomposition,* with unitary B and positive semidefinite C. They focussed on sets $\mathcal{A}$ generated by a sesquilinear form $\langle \cdot\, , \cdot \rangle$. Such sets conveniently fit into three classes:

(a) Automorphisms G, such that $\langle Gx, Gy\rangle = \langle x, y\rangle$, generate a *Lie group;*
(b) Self-adjoint matrices S, such that $\langle Sx, y\rangle = \langle x, Sy\rangle$, generate a *Jordan algebra;* and
(c) Skew-adjoint matrices H such that $\langle Hx, y\rangle = -\langle x, Hy\rangle$, generate a *Lie algebra.*

It is natural to expect that conservation of structure under factorization would depend on the nature of the underlying inner product. The surprising outcome of [52] is that, for current purposes, it is sufficient to split sesquilinear forms into just two classes, unitary and orthosymmetric, each exhibiting similar behaviour.

- Many algebraic eigenvalue problems are structured, the simplest example being that the eigenvalues of a symmetric matrix are real and of a skew-symmetric are pure imaginary: all standard methods for the computation of eigenvalues respect this. However, many other problems might have more elaborate structure, and this is the case in particular for nonlinear eigenvalue problems. An important example, with significant applications in mechanics, is

$$(\lambda^2 M + \lambda G + K)x = \mathbf{0}, \tag{4.2}$$

where both M and K are symmetric, while G is skew symmetric. The eigenvalues λ of (4.2) exhibit *Hamiltonian* pattern: if λ is in the spectrum then so are $-\lambda$, $\bar{\lambda}$ and $-\bar{\lambda}$.[7] As often in numerical algebra, (4.2) is particularly relevant when the underlying matrices are large and sparse.
Numerical experiments demonstrate that standard methods for the computation of a quadratic eigenvalue problem may fail to retain the Hamiltonian structure of the spectrum but this can be obtained by bespoke algorithms, using a symplectic version of the familiar Lanczos algorithm, cf. [3].
This is just one example of the growing field of structured eigenvalue and inverse eigenvalue problems.
- The exponential from an algebra to a group: Recall Lie-group methods from Sect. 2.2: a critical step, e.g. in the RKMK methods, is the exponential map from a Lie algebra to a Lie group. Numerical analysis knows numerous effective ways to approximate the matrix exponential [61], yet most of them fail to map a matrix from a Lie algebra to a Lie group! There is little point to expand intellectual and computational effort to preserve structure, only to abandon the latter in the ultimate step, and this explains the interest in the computation of the matrix exponential which is assured to map A in a Lie algebra to an element in the corresponding Lie group.
While early methods have used structure constants and, for maximal sparsity, Lie-algebraic bases given by space-root decomposition [8], the latest generation of algorithms is based upon *generalised polar decomposition* [67].

---

[7] To connect this to the GNI narrative, such a pattern is displayed by matrices in the *symplectic Lie algebra* $\mathfrak{sp}(2n)$.

Acknowledgements This work has been supported by the Australian Research Council. The authors are grateful to David McLaren for assistance during the preparation of this paper, as well as to Philipp Bader, Robert McLachlan and Marcus Webb, whose comments helped to improve this paper.

References

1. Bader, P., Iserles, A., Kropielnicka, K., Singh, P.: Effective approximation for the semiclassical Schrödinger equation. Found. Comput. Math. **14**(4), 689–720 (2014)
2. Bader, P., Iserles, A., Kropielnicka, K., Singh, P.: Efficient methods for linear Schrödinger equation in the semiclassical regime with time-dependent potential. Proc. R. Soc. A **472**(2193), 20150733 (2016)
3. Benner, P., Fassbender, H., Stoll, M.: Solving large-scale quadratic eigenvalue problems with Hamiltonian eigenstructure using a structure-preserving Krylov subspace method. Electron. Trans. Numer. Anal. **29**, 212–229 (2007/08)
4. Blanes, S., Casas, F., Murua, A.: Splitting and composition methods in the numerical integration of differential equations. Bol. Soc. Esp. Mat. Apl. **45**, 89–145 (2008)
5. Blanes, S., Casas, F., Farrés, A., Laskar, J., Makazaga, J., Murua, A.: New families of symplectic splitting methods for numerical integration in dynamical astronomy. Appl. Numer. Math. **68**, 58–72 (2013)
6. Blanes, S., Casas, F., Oteo, J.A., Ros, J.: The Magnus expansion and some of its applications. Phys. Rep. **470**(5–6), 151–238 (2009)
7. Butcher, J.C.: Coefficients for the study of Runge-Kutta integration processes. J. Austral. Math. Soc. **3**, 185–201 (1963)
8. Celledoni, E., Iserles, A.: Methods for the approximation of the matrix exponential in a Lie-algebraic setting. IMA J. Numer. Anal. **21**(2), 463–488 (2001)
9. Celledoni, E., Grimm, V., McLachlan, R.I., McLaren, D.I., O'Neale, D., Owren, B., Quispel, G.R.W.: Preserving energy resp. dissipation in numerical PDEs using the "average vector field" method. J. Comput. Phys. **231**(20), 6770–6789 (2012)
10. Celledoni, E., McLachlan, R.I., McLaren, D.I., Owren, B., Quispel, G.R.W.: Integrability properties of Kahan's method. J. Phys. A **47**(36), 365202, 20 (2014)
11. Celledoni, E., McLachlan, R.I., McLaren, D.I., Owren, B., Quispel, G.R.W.: Discretization of polynomial vector fields by polarization. Proc. R. Soc. A **471**, 20150390 (2015)
12. Celledoni, E., McLachlan, R.I., Owren, B., Quispel, G.R.W.: Geometric properties of Kahan's method. J. Phys. A **46**(2), 025201, 12 (2013)
13. Chartier, P., Murua, A.: Preserving first integrals and volume forms of additively split systems. IMA J. Numer. Anal. **27**(2), 381–405 (2007)
14. Cohen, D., Gauckler, L., Hairer, E., Lubich, C.: Long-term analysis of numerical integrators for oscillatory Hamiltonian systems under minimal non-resonance conditions. BIT **55**(3), 705–732 (2015)
15. Connes, A., Kreimer, D.: Lessons from quantum field theory: Hopf algebras and spacetime geometries. Lett. Math. Phys. **48**(1), 85–96 (1999)
16. Deaño, A., Huybrechs, D., Iserles, A.: Computing Highly Oscillatory Integrals, SIAM (2018)
17. Duncan, M., Levison, H.F., Lee, M.H.: A multiple time step symplectic algorithm for integrating close encounters. Astron. J. **116**, 2067–2077 (1998)
18. Ebrahimi-Fard, K., Manchon, D.: A Magnus- and Fer-type formula in dendriform algebras. Found. Comput. Math. **9**(3), 295–316 (2009)
19. Farrés, A., Laskar, J., Blanes, S., Casas, F., Makazaga, J., Murua, A.: High precision symplectic integrators for the Solar System. Celest. Mech. Dyn. Astron. **116**(2), 141–174 (2013)
20. Feng, K., Shang, Z.J.: Volume-preserving algorithms for source-free dynamical systems. Numer. Math. **71**(4), 451–463 (1995)

21. Feng, K., Wu, H.M., Qin, M.Z., Wang, D.L.: Construction of canonical difference schemes for Hamiltonian formalism via generating functions. J. Comput. Math. **7**(1), 71–96 (1989)
22. Fer, F.: Résolution de l'équation matricielle $dU/dt = pU$ par produit infini d'exponentielles matricielles. Acad. Roy. Belg. Bull. Cl. Sci. **5**(44), 818–829 (1958)
23. Forest, É.: Geometric integration for particle accelerators. J. Phys. A **39**(19), 5321–5377 (2006)
24. Gauckler, L., Hairer, E., Lubich, C.: Energy separation in oscillatory Hamiltonian systems without any non-resonance condition. Commun. Math. Phys. **321**(3), 803–815 (2013)
25. Ge, Z., Marsden, J.E.: Lie-Poisson Hamilton-Jacobi theory and Lie-Poisson integrators. Phys. Lett. A **133**(3), 134–139 (1988)
26. Grimm, V., McLachlan, R.I., McLaren, D.I., Quispel, G.R.W., Schönlieb, C.-B.: Discrete gradient methods for solving variational image regularization models. J. Phys. A **50**, 295201 (2017)
27. Hairer, E.: Energy-preserving variant of collocation methods. JNAIAM. J. Numer. Anal. Ind. Appl. Math. **5**(1–2), 73–84 (2010)
28. Hairer, E.: Challenges in geometric numerical integration. In: Trends in Contemporary Mathematics, pp. 125–135. Springer (2014)
29. Hairer, E., Lubich, C.: Long-time energy conservation of numerical methods for oscillatory differential equations. SIAM J. Numer. Anal. **38**(2), 414–441 (electronic) (2000)
30. Hairer, E., Lubich, C.: Oscillations over long times in numerical Hamiltonian systems. In: Highly oscillatory problems, Vol. 366 of London Mathematical Society Lecture Note Series, pp. 1–24. Cambridge University Press, Cambridge (2009)
31. Hairer, E., Lubich, C.: Long-term analysis of the Störmer-Verlet method for Hamiltonian systems with a solution-dependent frequency. Numerische Mathematik **134**(1), 119–138 (2016)
32. Hairer, E., Lubich, C., Wanner, G.: Geometric numerical integration illustrated by the Störmer-Verlet method. Acta Numer. **12**, 399–450 (2003)
33. Hairer, E., Lubich, C., Wanner, G.: Geometric Numerical Integration, Vol. 31 of Springer Series in Computational Mathematics, 2nd edn. Springer, Berlin (2006)
34. Hairer, E., Nørsett, S.P., Wanner, G.: Solving Ordinary Differential Equations. I, Vol. 8 of Springer Series in Computational Mathematics, 2nd edn. Springer, Berlin (1993)
35. Hochbruck, M., Ostermann, A.: Exponential integrators. Acta Numer. **19**, 209–286 (2010)
36. Huybrechs, D., Vandewalle, S.: On the evaluation of highly oscillatory integrals by analytic continuation. SIAM J. Numer. Anal. **44**(3), 1026–1048 (2006)
37. Iavernaro, F., Trigiante, D.: High-order symmetric schemes for the energy conservation of polynomial Hamiltonian problems. JNAIAM J. Numer. Anal. Ind. Appl. Math. **4**(1–2), 87–101 (2009)
38. Iserles, A.: Solving linear ordinary differential equations by exponentials of iterated commutators. Numer. Math. **45**(2), 183–199 (1984)
39. Iserles, A.: On the global error of discretization methods for highly-oscillatory ordinary differential equations. BIT **42**(3), 561–599 (2002)
40. Iserles, A., Nørsett, S.P.: On the solution of linear differential equations in Lie groups. R. Soc. Lond. Philos. Trans. Ser. A Math. Phys. Eng. Sci. **357**(1754), 983–1019 (1999)
41. Iserles, A., Nørsett, S.P.: Efficient quadrature of highly oscillatory integrals using derivatives. Proc. R. Soc. Lond. Ser. A Math. Phys. Eng. Sci. **461**(2057), 1383–1399 (2005)
42. Iserles, A., Munthe-Kaas, H.Z., Nørsett, S.P., Zanna, A.: Lie-group methods. Acta Numer. **9**, 215–365 (2000)
43. Iserles, A., Quispel, G.R.W., Tse, P.S.P.: B-series methods cannot be volume-preserving. BIT **47**(2), 351–378 (2007)
44. Kahan, W.: Unconventional methods for trajectory calculations. Department of Mathematics, University of California at Berkeley (1993)
45. Kahan, W., Li, R.-C.: Unconventional schemes for a class of ordinary differential equations-with applications to the Korteweg-de Vries equation. J. Comput. Phys. **134**(2), 316–331 (1997)
46. Lasagni, F.M.: Canonical Runge-Kutta methods. Z. Angew. Math. Phys. **39**(6), 952–953 (1988)
47. Laskar, J.: Chaos in the solar system. Ann. Henri Poincaré **4**(suppl. 2), S693–S705 (2003)

48. Laskar, J., Fienga, A., Gastineau, M., Manche, H.: La2010: a new orbital solution for the long-term motion of the Earth. Astron. Astrophys. **532**, A89 (2011)
49. Leimkuhler, B., Reich, S.: Simulating Hamiltonian Dynamics, Vol. 14 of Cambridge Monographs on Applied and Computational Mathematics. Cambridge University Press, Cambridge (2004)
50. Levin, D.: Procedures for computing one- and two-dimensional integrals of functions with rapid irregular oscillations. Math. Comput. **38**(158), 531–538 (1982)
51. Lord, G., Malham, S.J.A., Wiese, A.: Efficient strong integrators for linear stochastic systems. SIAM J. Numer. Anal. **46**(6), 2892–2919 (2008)
52. Mackey, D.S., Mackey, N., Tisseur, F.: Structured factorizations in scalar product spaces. SIAM J. Matrix Anal. Appl. **27**(3), 821–850 (2005)
53. Magnus, W.: On the exponential solution of differential equations for a linear operator. Commun. Pure Appl. Math. **7**, 649–673 (1954)
54. Marsden, J.E., West, M.: Discrete mechanics and variational integrators. Acta Numer. **10**, 357–514 (2001)
55. McLachlan, R.: Comment on: Poisson schemes for Hamiltonian systems on Poisson manifolds [Comput. Math. Appl. **27** (1994), no. 12, 7–16; MR1284126 (95d:65069)] by W.J. Zhu and M.Z. Qin, Comput. Math. Appl. **29**(3), 1 (1995)
56. McLachlan, R.I.: Composition methods in the presence of small parameters. BIT **35**(2), 258–268 (1995)
57. McLachlan, R.I., Quispel, G.R.W.: Splitting methods. Acta Numer. **11**, 341–434 (2002)
58. McLachlan, R.I., Modin, K., Munthe-Kaas, H., Verdier, O.: B-series are exactly the affine equivariant methods. Numerische Mathematik **133**(3), 599–622 (2016)
59. McLachlan, R.I., Quispel, G.R.W., Robidoux, N.: Geometric integration using discrete gradients. R. Soc. Lond. Philos. Trans. Ser. A Math. Phys. Eng. Sci. **357**(1754), 1021–1045 (1999)
60. McLachlan, R.I., Quispel, G.R.W., Turner, G.S.: Numerical integrators that preserve symmetries and reversing symmetries. SIAM J. Numer. Anal. **35**(2), 586–599 (1998)
61. Moler, C., Van Loan, C.: Nineteen dubious ways to compute the exponential of a matrix, twenty-five years later. SIAM Rev. **45**(1), 3–49 (electronic) (2003)
62. Morbidelli, A.: Modern Celestial Mechanics. Gordon & Breach, London (2002a)
63. Morbidelli, A.: Modern integrations of solar system dynamics. Annu. Rev. Earth Planet. Sci. **30**, 89–112 (2002b)
64. Moser, J., Veselov, A.P.: Discrete versions of some classical integrable systems and factorization of matrix polynomials. Commun. Math. Phys. **139**(2), 217–243 (1991)
65. Munthe-Kaas, H.: Runge-Kutta methods on Lie groups. BIT **38**(1), 92–111 (1998)
66. Munthe-Kaas, H., Verdier, O.: Aromatic Butcher series. Found. Comput. Math. **16**, 183–215 (2016)
67. Munthe-Kaas, H.Z., Quispel, G.R.W., Zanna, A.: Generalized polar decompositions on Lie groups with involutive automorphisms. Found. Comput. Math. **1**(3), 297–324 (2001)
68. Murua, A., Sanz-Serna, J.M.: Word series for dynamical systems and their numerical integrators, Technical report, Universidad Carlos III de Madrid (2015). arXiv:1502.05528v2 [math.NA]
69. Murua, A., Sanz-Serna, J.M.: Word series for dynamical systems and their numerical integrators. Foundations of Computational Mathematics **17**(3), 675–712 (2017)
70. Neĭshtadt, A.I.: The separation of motions in systems with rapidly rotating phase. Prikl. Mat. Mekh. **48**(2), 197–204 (1984)
71. Olver, S.: On the quadrature of multivariate highly oscillatory integrals over non-polytope domains. Numer. Math. **103**(4), 643–665 (2006)
72. Owren, B., Marthinsen, A.: Integration methods based on canonical coordinates of the second kind. Numer. Math. **87**(4), 763–790 (2001)
73. Petrera, M., Pfadler, A., Suris, Y.B.: On integrability of Hirota-Kimura type discretizations. Regul. Chaotic Dyn. **16**(3–4), 245–289 (2011)
74. Quispel, G.R.W.: Volume-preserving integrators. Phys. Lett. A **206**(1–2), 26–30 (1995)

75. Quispel, G.R.W., McLaren, D.I.: A new class of energy-preserving numerical integration methods. J. Phys. A **41**(4), 045206, 7 (2008)
76. Ramos, A.G.C.P., Iserles, A.: Numerical solution of Sturm-Liouville problems via Fer streamers. Numer. Math. **131**(3), 541–565 (2015)
77. Sanz-Serna, J.M.: Runge-Kutta schemes for Hamiltonian systems. BIT **28**(4), 877–883 (1988)
78. Sanz-Serna, J.M.: An unconventional symplectic integrator of W. Kahan. Appl. Numer. Math. **16**(1–2), 245–250. A Festschrift to honor Professor Robert Vichnevetsky on his 65th birthday (1994)
79. Sanz-Serna, J.M., Calvo, M.P.: Numerical Hamiltonian Problems, Vol. 7 of Applied Mathematics and Mathematical Computation. Chapman & Hall, London (1994)
80. Shang, Z.J.: Generating functions for volume-preserving mappings and Hamilton-Jacobi equations for source-free dynamical systems. Sci. China Ser. A **37**(10), 1172–1188 (1994)
81. Singh, P.: Algebraic theory for higher-order methods in computational quantum mechanics, Technical report, DAMTP, University of Cambridge (2015). arXiv:1510.06896v1 [math.NA]
82. Skokos, S.K., Gottwald, G., Laskar, J.: Chaos, Detection and Predictability, Springer, p. 18 (2016)
83. Suris, Y.B.: Preservation of symplectic structure in the numerical solution of Hamiltonian systems. In: Numerical Solution of Ordinary Differential Equations (Russian), Akad. Nauk SSSR, Inst. Prikl. Mat., Moscow, pp. 148–160, 232, 238–239 (1988)
84. Sussman, G.J., Wisdom, J.: Chaotic evolution of the solar system. Science **257**(5066), 56–62 (1992)
85. Suzuki, M.: Fractal decomposition of exponential operators with applications to many-body theories and Monte Carlo simulations. Phys. Lett. A **146**(6), 319–323 (1990)
86. van der Kamp, P.H., Kouloukas, T.E., Quispel, G.R.W., Tran, D.T., Vanhaecke, P.: Integrable and superintegrable systems associated with multi-sums of products. Proc. R. Soc. Lond. Ser. A Math. Phys. Eng. Sci. **470**(2172), 20140481, 23 (2014)
87. Wu, X., You, X., Wang, B.: Structure-Preserving Algorithms for Oscillatory Differential Equations. Springer, Heidelberg; Science Press Beijing, Beijing (2013)
88. Yoshida, H.: Construction of higher order symplectic integrators. Phys. Lett. A **150**(5–7), 262–268 (1990)

Lie Group Integrators

Brynjulf Owren

Abstract In this survey we discuss a wide variety of aspects related to Lie group integrators. These numerical integration schemes for differential equations on manifolds have been studied in a general and systematic manner since the 1990s and the activity has since then branched out in several different subareas, focussing both on theoretical and practical issues. From two alternative setups, using either frames or Lie group actions on a manifold, we here introduce the most important classes of schemes used to integrate nonlinear ordinary differential equations on Lie groups and manifolds. We describe a number of different applications where there is a natural action by a Lie group on a manifold such that our integrators can be implemented. An issue which is not well understood is the role of isotropy and how it affects the behaviour of the numerical methods. The order theory of numerical Lie group integrators has become an advanced subtopic in its own right, and here we give a brief introduction on a somewhat elementary level. Finally, we shall discuss Lie group integrators having the property that they preserve a symplectic structure or a first integral.

Keywords Lie group integrators · Geometric integration · Symplectic integrators
Geometric mechanics · Structure preserving integrators · Energy preservation
Numerical methods on manifolds · Lie groups · Homogeneous spaces

MSC 65L05 · 34C40 · 34G20 · 37M15

1 Introduction

Leonhard Euler is undoubtedly one of the most accomplished mathematicians of all times, and the modern theme *Numerical methods for Lie groups* can be traced back to Euler in more than one sense. Indeed, the simplest and possibly mostly

B. Owren (✉)
Norwegian University of Science and Technology, Alfred Getz vei 1,
7491 Trondheim, Norway
e-mail: bryn@math.ntnu.no

K. Ebrahimi-Fard and M. Barbero Liñán (eds.), *Discrete Mechanics, Geometric Integration and Lie–Butcher Series*, Springer Proceedings in Mathematics & Statistics 267, https://doi.org/10.1007/978-3-030-01397-4_2

used numerical approximation method for ordinary differential equations was first described in Euler's work (*Institutionum Calculi Integralis (1768), Volumen Primum, Ch VII*) and bears the name Euler's method. And undoubtedly, the most used test case for Lie group integrators is the Euler's free rigid body system, which was derived in his amazing treatise on Mechanics in 1736.

In the literature on structural mechanics, Lie group integrators have been around for a long time, but general, systematic studies of numerical integrators for differential equations on Lie groups and homogeneous manifolds began as recent as the 1990s. Some notable early contributions were those of Crouch and Grossman [25] and Lewis and Simo [41]. A series of papers by Munthe-Kaas [53–55] caused an increased activity from the late nineties when a large number of papers appeared over a short period of time. Many of these early results were summarised in a survey paper by Iserles et al. [35]. The work on Lie group integrators has been inspired by many subfields of mathematics. Notably, the study of order conditions and backward error analysis uses results from algebraic combinatorics, Hopf algebras, and has more recently been connected to post Lie algebras by Munthe-Kaas and coauthors, see e.g. [56]. In connection with the search for inexpensive coordinate representations of Lie groups as well as their tangent maps, the classical theory of Lie algebras has been put to use in many different ways. The theory of free Lie algebras [68] has been used to find optimal truncations of commutator expansions for general Lie algebras, see e.g. [57]. For coordinate maps taking advantage of properties of a particular Lie algebra, tools such as root space decomposition [66] and generalised polar decompositions [38] have been applied. Also there is of course a strong connection between numerical methods for Lie groups and the area of Geometric Mechanics. This connection is often used in the setup or formulation of differential equations in Lie groups or homogeneous manifolds where it provides a natural way of choosing a group action, and in order to construct Lie group integrators which are symplectic or conserve a particular first integral.

In this paper we shall discuss several aspects of Lie group integrators, we shall however not attempt to be complete. Important subjects related to Lie group integrators not covered here include the case of linear differential equations in Lie groups and the methods based on Magnus expansions, Fer expansions, and Zassenhaus splitting schemes. These are methods that could fit well into a survey on Lie group integrators, but for information on these topics we refer the reader to excellent expositions such as [4, 34]. Another topic we leave out here is that of stochastic Lie group integrators, see e.g. [43]. We shall also focus on the methods and the theory behind them rather than particular applications, of which there are many. The interested reader may check out the Refs. [10, 11, 20, 36, 71, 73].

In the next section we shall define a compact notation for differential equations on differential manifolds with a Lie group action. Then in Sect. 3, we discuss some of the most important classes of Lie group integrators and give a few examples of methods. Section 4 briefly treats a selection of group actions which are interesting in applications. Then in Sect. 5 we shall address the issue of isotropy in Lie group integrators, in particular how the freedom offered by the isotropy group can either be used to reduce the computational cost of the integrator or be used to improve the quality of the solution. We take our own look at order theory and expansions in terms of a generalised form of B-series in Sect. 6. Finally, in Sects. 7 and 8 we consider Lie group integrators which preserve a symplectic form or a first integral.

2 The Setup

Let M be a differentiable manifold of dimension $d < \infty$ and let the set of smooth vector fields on M be denoted $\mathcal{X}(M)$. Nearly everything we do in this paper is concerned with the approximation of the h-flow of a vector field $F \in \mathcal{X}(M)$ for some small parameter h usually called the *stepsize*. In other words, we approximate the solution to the differential equation

$$\dot{y} = \frac{\mathrm{d}}{\mathrm{d}t} y = F|_y, \qquad F \in \mathcal{X}(M). \tag{1}$$

Crouch and Grossman [25] used a set of smooth frame vector fields $E_1, \ldots, E_\nu$, $\nu \geq d$ on M, assuming

$$\operatorname{span}(E_1|_x, E_2|_x, \ldots, E_\nu|_x) = T_x M, \quad \text{for each } x \in M.$$

It can be assumed that the frame vector fields are linearly independent as derivations of the ring $\mathcal{F}(M)$ of smooth functions on M, and we denote their $\mathbb{R}$-span by V. Any smooth vector field $F \in \mathcal{X}(M)$ can be represented by ν functions $f_i \in \mathcal{F}(M)$

$$F|_x = \sum_{i=1}^{\nu} f_i(x) E_i|_x \tag{2}$$

where the f_i are not necessarily unique. We then have a natural affine connection

$$\nabla_F G = \sum_i F(g_i) E_i$$

which is flat with constant torsion $\tau(\sum_j f_j E_j, \sum_i g_i E_i) = \sum_{i,j} f_j g_i [E_i, E_j]$. For later, we shall need the notion of a frozen vector field relative to the frame The freeze operator $\mathrm{Fr} : M \times \mathcal{X}(M) \to V$ is defined as

$$\mathrm{Fr}(x, F) := F_x = \sum_i f_i(x) E_i$$

We note that the torsion can be defined by freezing the vector fields and then take the Lie-Jacobi bracket, i.e.

$$\tau(F, G)|_x = [\mathrm{Fr}(x, F), \mathrm{Fr}(x, G)]$$

Another setup is obtained by using a Lie group G acting transitively from the left on M [55]. The Lie algebra of G is denoted $\mathfrak{g}$. Any vector field F can now be represented via a map $f : M \to \mathfrak{g}$ and the infinitesimal action $\rho : \mathfrak{g} \to \mathscr{X}(M)$

$$F|_x = \rho \circ f(x)|_x, \qquad \rho : \mathfrak{g} \to \mathscr{X}(M), \quad \rho(\xi)|_x = \left.\frac{\mathrm{d}}{\mathrm{d}t}\right|_{t=0} \exp(t\xi) \cdot x \tag{3}$$

We note that the map f is not necessarily unique.

3 Types of Schemes

There is now a large variety of numerical integration schemes available, typically formulated with either of the setups of the previous section. In what follows, we assume that a finite dimensional Lie group G acts transitively on a manifold M and the Lie algebra of G is denoted $\mathfrak{g}$.

3.1 Schemes of Munthe-Kaas Type

Using the second setup, a powerful way of deriving numerical integrators devised by Munthe-Kaas [55] follows these steps:

1. In a neighborhood $U \subset \mathfrak{g}$ of 0 introduce a local diffeomorphism $\psi : U \to G$, such that
$$\psi(0) = 1 \in G, \qquad T_0\psi = \mathrm{Id}_{\mathfrak{g}}$$
2. Observe that the map $\lambda_{y_0}(v) = \psi(v) \cdot y_0$ is surjective on a neighborhood of the initial value $y_0 \in M$.
3. Compute the pullback of the vector field $F = \rho \circ f$ along λ_{y_0}.
4. Apply one step of a standard numerical integrator to the resulting problem on U.
5. Map the obtained approximation back to M by λ_{y_0}.

Even though the idea is very simple, there are several difficulties that need to be resolved in order to obtain fast and accurate integration schemes from this procedure.

Observe that the derivative of ψ can be trivialised by right multiplication of the Lie group, such that

$$T_u\psi = TR_{\psi(u)} \circ \mathrm{d}\psi_u, \quad \mathrm{d}\psi_u : \mathfrak{g} \to \mathfrak{g}.$$

With this in mind we set out to characterise the vector field on $U \subset \mathfrak{g}$, this is a simple generalisation of a result in [55].

Lemma 1 *Let M be a smooth manifold with a left Lie group action $\Lambda : G \times M \to M$ and let $\mathfrak{g}$ be the Lie algebra of G. Let $\psi : \mathfrak{g} \to G$ be a smooth map, $\psi(0) = 1$. Fix a point $m \in M$, and set $\Lambda_m = \Lambda(\cdot, m)$ so that*

$$\rho(\xi)|_m = T_1 \Lambda_m(\xi)$$

Suppose $F \in \mathscr{X}(M)$ is of the form

$$F|_m = \rho(\xi(m))|_m, \quad \text{for some } \xi : M \to \mathfrak{g}.$$

Define $\lambda_m(u) = \Lambda(\psi(u), m)$. Then there is an open set $U \subseteq \mathfrak{g}$ containing 0 such that the vector field $\eta \in \mathscr{X}(U)$ defined as

$$\eta|_u = \mathrm{d}\psi_u^{-1}(\xi \circ \lambda_m(u))$$

is λ_m-related to F.

The original proof in [55] where $\psi = \exp$ was adapted to general coordinate maps in [66]. One step of a Lie group integrator is obtained just by applying a classical integrator, such as a Runge–Kutta method, to the corresponding locally defined vector field η on $\mathfrak{g}$. A Runge–Kutta method with coefficients (A, b) applied to the problem $\dot{y} = \eta(y)$ in a linear space is of the following form

$$\begin{aligned}
Y_r &= y_0 + h \sum_{j=1}^{s} a_{rj} k_j, \quad r = 1, \ldots, s, \\
k_r &= \eta|_{Y_r}, \quad r = 1, \ldots, s, \\
y_1 &= y_0 + h \sum_{r=1}^{s} b_r k_r
\end{aligned}$$

Here y_0 is the initial value, h is the step size, and $y_1 \approx y(t_0 + h)$ is the approximate solution at time $t_0 + h$. The parameter $s \geq 1$ is called the number of stages of the method. If the matrix $A = (a_{rj})$ is strictly lower triangular, then the method is called *explicit*. The application of such a method to the transformed vector field of Lemma 1 can be written out as follows

$$u_r = h \sum_{j=1}^{s} a_{ij} \tilde{k}_j, \qquad k_r = \xi \circ \lambda_{y_0}(u_r), \qquad \tilde{k}_r = \mathrm{d}\psi_{u_r}^{-1}(k_r), \quad r = 1, \ldots, s, \tag{4}$$

$$v = h \sum_{r=1}^{s} b_r \tilde{k}_r \qquad y_1 = \lambda_{y_0}(v). \tag{5}$$

A major advantage of this approach compared to other types of Lie group schemes is its interpretation as a smooth change of variables which causes the convergence order to be (at least) preserved from the underlying classical integrator. As we shall see later, one generally needs to take into account additional order conditions to account for the fact that the phase space is not a linear space.

3.1.1 Choosing the Exponential Map as Coordinates, $\psi = \exp$

The first papers by Munthe-Kaas on Lie group integrators [53–55] all used $\psi = \exp$ as coordinates on the Lie group. In this case, there are several difficulties that need to be addressed in order to obtain efficient implementations of the methods. One is the computation of the exponential map itself. For matrix groups, there are a large number of algorithms that can be applied, see e.g. [50, 51]. In [17, 18] the authors developed approximations to the exponential which exactly map matrix Lie subalgebras to their corresponding Lie subgroups. Another issue to be dealt with is the differential of the exponential map

$$\mathrm{dexp}_u := T_{\exp(-u)} R_{\exp(-u)} \circ T_u \exp$$

Lemma 2 (Tangent map of $\exp : \mathfrak{g} \to G$) *Let* $u \in \mathfrak{g}$, $v \in T_u\mathfrak{g} \simeq \mathfrak{g}$. *Then*

$$T_u \exp(v) = \left.\frac{\mathrm{d}}{\mathrm{d}s}\right|_{s=0} \exp(u + sv) = T R_{\exp(u)} \circ \mathrm{dexp}_u\, v = \mathrm{dexp}_u(v) \cdot \exp(u)$$

where

$$\mathrm{dexp}_u(v) = \int_0^1 \exp(r\, \mathrm{ad}_u)(v)\, \mathrm{d}r = \left.\frac{e^z - 1}{z}\right|_{z=\mathrm{ad}_u} (v)$$

Proof Let $y_s(t) = \exp(t(u + sv))$ such that

$$T_u \exp(v) = \left.\frac{\mathrm{d}}{\mathrm{d}s}\right|_{s=0} y_s(1)$$

But for now we differentiate with respect to t to obtain

$$\dot{y}_s := \frac{\mathrm{d}}{\mathrm{d}t} y_s(t) = (u + sv) y_s(t) \tag{6}$$

We also note that $y_s(t) = \exp(tu) + \mathscr{O}(s)$ as $s \to 0$. From (6) we then get

$$\dot{y} - u y_s = sve^{tu} + \mathscr{O}(s^2)$$

and the integrating factor e^{-tu} yields

$$\frac{\mathrm{d}}{\mathrm{d}t}\left(e^{-tu} y_s\right) = se^{-tu} v e^{tu} + \mathscr{O}(s^2)$$

Integrating, and using that $y_s(0) = \mathrm{Id}$, we get

$$y_s(t) = e^{tu} + s\int_0^t e^{ru} v e^{-ru} e^{tu}\,\mathrm{d}r + \mathcal{O}(s^2)$$

and so

$$\left.\frac{\mathrm{d}}{\mathrm{d}s}\right|_{s=0} y_s(1) = \int_0^1 e^{ru} v e^{-ru}\,\mathrm{d}r\, e^u = \int_0^1 e^{r\,\mathrm{ad}_u}(v)\,\mathrm{d}r\, e^u$$

where we have used the well-known identity $\mathrm{Ad}_{\exp(u)} = \exp(\mathrm{ad}_u)$ in the last equality. Formally, we can write

$$\int_0^1 e^{r\,\mathrm{ad}_u}(v)\,\mathrm{d}r = \int_0^1 e^{rz}|_{z=\mathrm{ad}_u}(v)\,\mathrm{d}r = \left.\frac{e^z-1}{z}\right|_{z=\mathrm{ad}_u}(v)$$

It is often useful to consider the dexp-map as an infinite series of nested commutators

$$\mathrm{dexp}_u(v) = \left(I + \frac{1}{2!}\,\mathrm{ad}_u + \frac{1}{3!}\,\mathrm{ad}_u^2 + \cdots\right)(v) = v + \frac{1}{2}[u,v] + \frac{1}{6}[u,[u,v]] + \cdots$$

In (4) it is the inverse of dexp which is needed. Note that the function

$$\phi_1(z) = \frac{e^z-1}{z}$$

is entire, this means that its reciprocal

$$\frac{z}{e^z-1}$$

is analytic whenever $\phi_1(z) \neq 0$. In particular this means that $\frac{1}{\phi_1(z)}$ has a converging Taylor series about $z = 0$ in the open disk $|z| < 2\pi$. This series expansion is

$$\frac{z}{e^z-1} = 1 - \frac{z}{2} + \sum_{k=1}^{\infty}\frac{B_{2k}}{(2k)!}z^{2k}$$

where B_{2k} are the *Bernoulli numbers*, the first few of them are: $B_2 = \frac{1}{6}$, $B_4 = -\frac{1}{30}$, $B_6 = \frac{1}{42}$, $B_8 = -\frac{1}{30}$, $B_{10} = \frac{5}{66}$. The map

$$v = \mathrm{dexp}_u^{-1}(w) \qquad (\text{whenever } w = \mathrm{dexp}_u(v))$$

is given precisely as

$$\mathrm{dexp}_u^{-1}(w) = \left.\frac{z}{e^z-1}\right|_{z=\mathrm{ad}_u}(w) = w - \frac{1}{2}[u,w] + \frac{B_2}{2!}[u,[u,w]] + \cdots \tag{7}$$

We observe from (4) that one needs to compute $\mathrm{dexp}_{u_r}^{-1}(k_r)$ and that each $u_r = \mathcal{O}(h)$. This means that one may approximate the series in (7) by a finite sum,

$$\texttt{dexpinv}(u,w,m) = w - \frac{1}{2}[u,w] + \sum_{k=1}^{m} \frac{B_{2k}}{(2k)!}\,\mathrm{ad}_u^{2k}(w).$$

One has $\texttt{dexpinv}(u,w,m) \in \mathfrak{g}$ for every $m \geq 0$ and furthermore

$$\mathrm{dexp}_{u_r}^{-1}(k_r) - \texttt{dexpinv}(u_r,k_r,m) = \mathcal{O}(h^{2m+1})$$

As long as the classical integrator has order $p \leq 2m+1$, the resulting Munthe-Kaas scheme will also have order p. There exists however a clever way to substantially reduce the number of commutators that need to be computed in each step. Munthe-Kaas and Owren [57] realised that one could form linear combinations of the stage derivatives $\tilde{k}_r$ in (4) such that

$$Q_r = \sum_{j=1}^{r} \sigma_{r,j}\tilde{k}_j = \mathcal{O}(h^{q_r})$$

for q_r as large as possible for each r. Then these new quantities Q_r were each given the grade q_r and one considered the graded free Lie algebra based on this set. The result was a significant reduction in the number of commutators needed. Also Casas and Owren [15] provided a way to organise the commutator calculations to reduce even further the computational cost. Here is a Runge–Kutta Munthe–Kaas method of order four with four stages and minimal set of commutators

$$\begin{aligned}
k_1 &= hf(y_0),\\
k_2 &= hf(\exp(\tfrac{1}{2}k_1)\cdot y_0),\\
k_3 &= hf(\exp(\tfrac{1}{2}k_2 - \tfrac{1}{8}[k_1,k_2])\cdot y_0),\\
k_4 &= hf(\exp(k_3)\cdot y_0),\\
y_1 &= \exp(\tfrac{1}{6}(k_1+2k_2+2k_3+k_4-\tfrac{1}{2}[k_1,k_4]))\cdot y_0.
\end{aligned}$$

For later reference, we also give the Lie-Euler method, a first order Lie group integrator generalising the classical Euler method

$$y_1 = \exp(hf(y_0))\cdot y_0 \tag{8}$$

3.1.2 Canonical Coordinates of the Second Kind

The exponential map is generally expensive to compute exactly. For matrix Lie algebras $\mathfrak{g} \subseteq \mathfrak{gl}(n, \mathbb{F})$ where $\mathbb{F}$ is either $\mathbb{R}$ or $\mathbb{C}$, standard software for computing exp numerically has a computational cost of n^3 to the leading order, and the constant in front of n^3 may be as large as 20–30. Another, yet completely general alternative to the exponential function is constructed as follows: Fix a basis for $\mathfrak{g}$, say $e_1, \ldots, e_d$ and consider the map

$$\psi : v_1 e_1 + \cdots + v_d e_d \mapsto \exp(v_1 e_1) \cdot \exp(v_2 e_2) \cdots \exp(v_d e_d) \tag{9}$$

Although it might seem unnatural to replace one exponential by many, one needs to keep in mind that if the basis can be chosen such that its exponential can be computed explicitly, it may still be an efficient method. For instance, in the general linear matrix Lie algebra $\mathfrak{gl}(m, \mathbb{F})$ one may take the basis to be $e_{ij} = \mathbf{e}_i \mathbf{e}_j^T$ where $\mathbf{e}_i$ is the ith canonical unit vector in $\mathbb{R}^n$. Then

$$\exp(\alpha e_{ij}) = 1 + \alpha e_{ij}, \ i \neq j, \qquad \exp(\alpha e_{ii}) = 1 + (\mathrm{e}^{\alpha} - 1) e_{ii}$$

So computing (9) takes approximately nd operations which is much cheaper than computing the exponential of a general matrix.

The difficulty lies however in computing the map $\mathrm{d}\psi_u^{-1}$ in an efficient manner. A method for this was developed in [66]. The methodology is slightly different for solvable and semisimple Lie algebras. We here outline the main idea, for details we refer to the original paper [66]. Differentiate (9) to obtain

$$\mathrm{d}\psi_u(v) = v_1 e_1 + \sum_{i=2}^{d} v_i \ \mathrm{Ad}_{\mathrm{e}^{u_1 e_1}} \circ \cdots \circ \mathrm{Ad}_{\mathrm{e}^{u_{i-1} e_{i-1}}}(e_i)$$

The main idea is to find an equivalent expression which is a composition of cheaply invertible operators. For this, we introduce a projector onto the span of the last $d - k$ basis vectors as follows

$$P_k : \sum_{i=1}^{d} v_i e_i \mapsto \sum_{i=k+1}^{d} v_i e_i$$

where we let P_0 and P_d equal the identity operator and zero operator on $\mathfrak{g}$ respectively. We may now define a modified version of the Ad-operator, for any $u = \sum u_i e_i \in \mathfrak{g}$, let

$$\widehat{\mathrm{Ad}}_{\mathrm{e}^{u_k e_k}} = (\mathrm{Id} - P_k) + \mathrm{Ad}_{\mathrm{e}^{u_k e_k}} P_k$$

This is a linear operator which acts as the identity operator on basis vectors e_i, $i \leq k$, and on basis vectors e_i, $i \geq k$ it coincides with $\mathrm{Ad}_{\mathrm{e}^{u_k e_k}}$.

Definition 1 An ordered basis $(e_1, \ldots, e_d)$ is called an admissible ordered basis (AOB) if, for each $u = \sum u_j e_j \in \mathfrak{g}$ and for each $i = 1, \ldots, d-1$, we have

$$\mathrm{Ad}_{\mathrm{e}^{u_1 e_1}} \circ \cdots \circ \mathrm{Ad}_{\mathrm{e}^{u_i e_i}} P_i = \widehat{\mathrm{Ad}}_{\mathrm{e}^{u_1 e_1}} \circ \cdots \circ \widehat{\mathrm{Ad}}_{\mathrm{e}^{u_i e_i}} P_i \tag{10}$$

This definition is exactly what is needed to write $\mathrm{d}\psi_u$ as a composition of operators.

Proposition 1 *If the basis $(e_1, \ldots, e_d)$ is an AOB, then*

$$d\psi_u = \widehat{\mathrm{Ad}}_{\mathrm{e}^{u_1 e_1}} \circ \cdots \circ \widehat{\mathrm{Ad}}_{\mathrm{e}^{u_d e_d}}$$

Another important simplification can be obtained if an abelian subalgebra $\mathfrak{h}$ of dimension $d - d^*$ can be identified. In this case the ordered basis can be chosen such that $\mathfrak{h} = \mathrm{span}(e_{d^*+1}, \ldots, e_d)$. Then $\mathrm{Ad}_{\mathrm{e}^{u_i e_i}}|_{\mathfrak{h}}$ for $i > d^*$ is the identity operator and therefore $\widehat{\mathrm{Ad}}_{\mathrm{e}^{u_i e_i}}$ is the identity operator on all of $\mathfrak{g}$. Summarizing, we have the following expression

$$\mathrm{d}\psi_u^{-1} = \widehat{\mathrm{Ad}}^{-1}_{\mathrm{e}^{u_{d^*} e_{d^*}}} \circ \cdots \circ \widehat{\mathrm{Ad}}^{-1}_{\mathrm{e}^{u_1 e_1}}$$

Choosing typically a basis consisting of ad-nilpotent elements, the inversion of each $\widehat{\mathrm{Ad}}_{\mathrm{e}^{u_i e_i}}$ can be done cheaply by making use of the formula

$$\mathrm{Ad}_{\mathrm{e}^{u_i e_i}} = 1 + \sum_{k=1}^{K} \frac{u_i^k}{k!} \mathrm{ad}_{e_k}, \quad \mathrm{ad}_{e_i}^{K+1} = 0.$$

For choosing the basis one may, for semisimple Lie algebras, use a basis known as the Chevalley basis. This arises from the root space decomposition of the Lie algebra

$$\mathfrak{g} = \mathfrak{h} \oplus \coprod_{\alpha \in \Phi} \mathfrak{g}_\alpha. \tag{11}$$

Here Φ is the set of roots, and $\mathfrak{g}_\alpha$ is the one-dimensional subspace of $\mathfrak{g}$ corresponding to the root $\alpha \in \mathfrak{h}^*$, see e.g. Humphreys [33]. $\mathfrak{h}$ is the maximal toral subalgebra of $\mathfrak{g}$ and it is abelian. In the previous notation, the number of roots is d^* and the dimension of $\mathfrak{h}$ is $d - d^*$. The following result whose proof can be found in [66] provides a tool for determining whether an ordered Chevalley basis is an AOB.

Theorem 1 *Let $\{\beta_1, \ldots, \beta_{d_*}\}$, $d_* = d - \ell$, be the set of roots Φ for a semisimple Lie algebra $\mathfrak{g}$. Suppose that a Chevalley basis is ordered as*

$$(e_{\beta_1}, \ldots, e_{\beta_{d*}}, h_1, \ldots, h_\ell)$$

where $e_{\beta_i} \in \mathfrak{g}_{\beta_i}$, and $(h_1, \ldots, h_\ell)$ is a basis for $\mathfrak{h}$. Such an ordered basis is an AOB if

$$k\beta_i + \beta_s = \beta_m,\ m < i < s \le d_*, k \in \mathbb{N} \quad \Rightarrow \quad \beta_m + \beta_n \notin \overline{\Phi},\ m < n \le i - 1. \tag{12}$$

Here $\overline{\Phi} = \Phi \cup \{0\}$.

Example 1 As an example, we consider $A_\ell = \mathfrak{sl}(\ell + 1, \mathbb{C})$, commonly realized as the set of $(\ell + 1) \times (\ell + 1)$-matrices with vanishing trace. The maximal toral subalgebra is then the set of diagonal matrices in $\mathfrak{sl}(\ell + 1, \mathbb{C})$. The positive roots are denoted

$$\{\beta_{i,j}, 1 \le i \le j \le \ell\}.$$

Letting $\mathbf{e}_i$ be the ith canonical unit vector in $\mathbb{C}^{\ell+1}$, the root space corresponding to $\beta_{i,j}$ has a basis vector

$$\mathbf{e}_i\mathbf{e}_{j+1}^T \in \mathfrak{g}_{\beta_{i,j}}, \quad 1 \le i \le j \le \ell$$

whereas the negative roots are associated to the basis vectors

$$\mathbf{e}_{j+1}\mathbf{e_i}^T \in \mathfrak{g}_{-\beta_{i,j}}.$$

As a basis for $\mathfrak{h}$, one may choose the matrices $\mathbf{e}_i\mathbf{e}_i^T - \mathbf{e}_{i+1}\mathbf{e}_{i+1}^T,\ 1 \le i \le \ell$. The remaining difficulty now is to choose an ordering of the basis so that an AOB results. As indicated earlier, the basis for $\mathfrak{h}$ may be ordered as the last ones, i.e. with indices ranging from $d^* + 1 = \ell^2 + \ell + 1$ to $d = \ell^2 + 2\ell$. With the convention $e_\beta \in \mathfrak{g}_\beta,\ \beta \in \Phi, \mathfrak{h} = \mathrm{span}(e_{\mathfrak{h}_1}, \ldots, e_{\mathfrak{h}_\ell})$, let

$$B = (e_{\beta_{i_1,j_1}}, \ldots, e_{\beta_{i_m,j_m}}, e_{-\beta_{i_1,j_1}}, \ldots, e_{-\beta_{i_m,j_m}}, e_{\mathfrak{h}_1}, \ldots, e_{\mathfrak{h}_\ell}),$$

where $i_1 \le i_2 \le \cdots \le i_m$ and $m = \ell(\ell + 1)/2$. One can then prove by using Theorem 1 that B is an AOB.

Similar details as for A_ℓ were also given for the other classical Lie algebras, B_ℓ, C_ℓ, D_ℓ and the exceptional case G_2 in [66]. Also the case of solvable Lie algebras was considered.

3.1.3 Other Coordinate Maps and Retractions

We have discussed two ways to choose coordinates on a Lie group as an ingredient in the Lie group integrators, these are canonical coordinates of the first and second kind. These choices are generic in the sense that they can be used for any finite dimensional Lie group with a corresponding Lie algebra. But if one allows for maps $\Psi : \mathfrak{g} \to G$ that may only work for particular Lie groups there might be more options. Considering subgroups of the general linear group, a common type are those that can be embedded in $GL(n, \mathbb{R})$ via quadratic constraints, i.e. matrix groups of the form

$$G = \{A \in GL(n, \mathbb{R}) : A^T J A = J\},$$

for some $n \times n$-matrix J. If $J = \mathrm{Id}$, the identity matrix, then $G = SO(n, \mathbb{R})$ whereas if J equals the constant Poisson structure matrix, then we recover the symplectic group $SP(2d, \mathbb{R})$. The Lie algebra of such a group consists of matrices

$$\mathfrak{g} = \{a \in \mathfrak{gl}(n, \mathbb{R}) : a^T J + Ja = 0\}.$$

As an alternative to the exponential map, while still keeping a map of the form $A = \chi(a)$ where $\chi(z)$ is analytic in a neighborhood of $z = 0$ is the Cayley transformation

$$\chi(z) = \frac{1 + z/2}{1 - z/2}.$$

In fact, for any function $\chi(z)$ such that $\chi(-z)\chi(z) = 1$ one has

$$a^T J + Ja = 0 \quad \Rightarrow \quad \chi(a)^T J \chi(a) = J.$$

General software for computing $\chi(a)$ for an $n \times n$-matrix has a computational cost of $\mathcal{O}(n^3)$, but the constant in front of n^3 is much smaller than what is required for the exponential map. The computation of the (inverse) differential of the Cayley transformation is also relatively inexpensive to compute, the right trivialised version is

$$\mathrm{d}\chi_y(u) = (1 - \frac{y}{2})^{-1} u (I + \frac{y}{2})^{-1}, \qquad \mathrm{d}\chi_y^{-1}(v) = (I - \frac{y}{2}) v (I + \frac{y}{2}).$$

Retractions

In cases of Lie group integrators where the Lie group has much higher dimension than the manifold it acts upon the computational cost may become too high for doing arbitrary calculations in the Lie algebra which is the way the Munthe-Kaas methods work. An option is then to replace the Lie algebra by a *retraction* which is a map retracting the tangent bundle TM of the manifold into M

$$\phi : TM \to M.$$

We let ϕ_x be the restriction of ϕ to $T_x M$ and denote by 0_x the zero-vector in $T_x M$. Following [1] we impose the following conditions on ϕ

1. ϕ_x is smooth and defined in an open ball $B_{r_x}(0_x) \subset T_x M$ of radius r_x around 0_x.
2. $\phi_x(v) = x$ if and only if $v = 0_x$.
3. $T_{0_x}\phi_x = 1_{T_x M}$.

Thus, ϕ_x is a diffeomorphism from some neighborhood $\mathcal{U}$ of 0_x to its image $\mathcal{W} = \phi_x(\mathcal{U}) \subset M$.

In a similar way as for the Munthe-Kaas type methods, the idea is now to make a local change of coordinates, setting for a starting point y_0,

$$y(t) = \phi_{y_0}(\sigma(t)), \quad \sigma(0) = 0.$$

This implies

$$\dot{y} = T_\sigma \phi_{y_0} \dot{\sigma} = F \circ \phi_{y_0}(\sigma).$$

In some neighborhood of $0_{y_0} \in T_{y_0}M$ we have

$$\dot{\sigma} = (T_\sigma \phi_{y_0})^{-1} F \circ \phi_{y_0}(\sigma). \tag{13}$$

The ODE on the vector space $T_{y_0}M$ can be solved by a standard integrator, and the resulting approximation over one step σ_1 to (13) is mapped back to $y_1 = \varphi_{y_0}(\sigma_1)$ and the succeeding step is taken in coordinates from the tangent space $T_{y_1}M$ etc. This way of introducing local coordinates for computation is in principle very simple, though it does not take into account the representation of the vector field as is the case with the Munthe-Kaas and Crouch–Grossman frameworks. Several examples of computationally efficient retractions can be found in [21], for instance in the orthogonal group by means of (reduced) matrix factorisations.

Retractions on Riemannian Manifolds

Geodesics can be used to construct geodesics on a Riemannian manifold. We define

$$\phi_x(v) = \exp_x(c) = \gamma_v(1),$$

where $\gamma_v(t)$ is the geodesic emanating from x with $\dot{\gamma}(0) = v$. The map $\exp_x$ is defined and of maximal rank in a neighborhood of $0_x \in T_xM$. The derivative of ϕ_x is related to the Jacobi field satisfying the Jacobi equation, see e.g. [24, p. 70–82]. Let ∇ be the Levi–Civita connection with respect to the metric on M and let $\mathbf{R}$ be the curvature tensor. Consider the vector field Y defined along the geodesic γ, $\gamma(0) = x$, $\dot{\gamma}(0) = v$ satisfying the boundary value problem

$$\nabla_t^2 Y + \mathbf{R}(\dot{\gamma}, Y)\dot{\gamma} = 0, \quad Y(0) = 0, \quad Y(1) = w.$$

Then

$$(T_v \phi_x)^{-1}(w) = (\nabla_t Y)(0).$$

Of particular interest in many applications is the case when there is a natural embedding of the Riemannian manifold into Euclidean space, say $V = \mathbb{R}^n$. In this case, one has for every $x \in M$ a decomposition of $V = T_xM \oplus N_xM$, where N_xM is the orthogonal complement of T_xM in V. We may define a retraction

$$\phi_x(v) = x + v + n_x(v),$$

where $n_x(v)$ is defined in such away that $\phi_x(v) \in M$ for every v belonging to a sufficiently small neighborhood of $0_x \in T_xM$. The derivative can be computed as

$$W := T_v\phi_x(w) = w + T_v n_x(w), \quad T_v n_x(w) \perp T_xM$$

so the image of the derivative in $T_{\phi_x(v)}M$ is naturally split into components in T_xM and N_xM. Now we can just apply the orthogonal projector $\mathbf{P}_{T_xM}$ onto T_xM on each side to obtain

$$w = (T_v\phi_x)^{-1}W = \mathbf{P}_{T_xM}W.$$

3.2 *Integrators Based on Compositions of Flows*

3.2.1 Crouch–Grossman Methods

What characterises the Munthe-Kaas type schemes is that they can be represented via a change of variables. But there are also Lie group integrators which do not have this property. Crouch and Grossman [25] suggested method formats generalising both Runge–Kutta methods and linear multistep methods which they expressed in terms of frame vector fields, $E_1, \ldots, E_\nu$ as follows. Using the notation of [65] we present here the Runge–Kutta version of Crouch–Grossman methods.

$$Y_r = \exp(ha_{r,s}F_s) \circ \cdots \circ \exp(ha_{r,1}F_1)\, y_0, \tag{14}$$

$$F_r = \mathrm{Fr}(Y_r, F) = \sum_{i=1}^{\nu} f_i(Y_r)E_i, \tag{15}$$

$$y_1 = \exp(hb_sF_s) \circ \exp(hb_{s-1}F_{s-1}) \circ \cdots \circ \exp(hb_1F_1)y_0, \tag{16}$$

where we have assumed that the vector field on M has been written in the form (2). Here, the method coefficients $a_{r,j}$ and b_j correspond to the usual coefficients of Runge–Kutta methods. In fact, whenever the frame is chosen to be the standard basis of $\mathbb{R}^n$, the method reduces to the familiar Runge–Kutta methods. Note that regardless of the ordering of exponentials in (14, 15) the method will reduce to the same method whenever the flows are commuting. But in general a reordering of flows will alter the behaviour from order h^3 on, this indicates that it is not sufficient to enforce only the standard order conditions for Runge–Kutta methods on the $a_{r,j}$ and b_r coefficients. For classical explicit Runge–Kutta methods it is possible to obtain methods of order p with $s = p$ stages for $p = 1, 2, 3, 4$, but for order $p \geq 5$ it is necessary that $s > p$. For Crouch–Grossman methods one can obtain $p = s$ for $p = 1, 2, 3$, but for $p = 4$ it is necessary to have at least five stages. Crouch and Grossman [25] devised Runge–Kutta generalisations of methods of orders up to three, and Owren and Marthinsen [65] gave also and example of an explicit method of order four. We here give a third order method with three stages

$$\begin{aligned}
Y_1 &= y_0, & F_1 &= \mathrm{Fr}(Y_1, F),\\
Y_2 &= \exp(\tfrac{3}{4}hF_1)\, y_0, & F_2 &= \mathrm{Fr}(Y_2, F),\\
Y_3 &= \exp(\tfrac{17}{108}hF_2)\exp(\tfrac{119}{216}hF_1)\, y_0, & F_3 &= \mathrm{Fr}(Y_3, F),\\
y_1 &= \exp(\tfrac{24}{17}hF_3)\exp(-\tfrac{2}{3}hF_2)\exp(\tfrac{13}{51}hF_1)\, y_0. & &
\end{aligned}$$

3.2.2 Commutator-Free Lie Group Integrators

A disadvantage of the Crouch–Grossman methods is that they use a high number of exponentials or flow calculations which is usually among the most costly operations of the method. In fact, for an explicit method with s stages one needs to compute $(s+1)s/2$ exponentials. To improve this situation, Celledoni et al. [19] proposed a generalisation of the Crouch–Grossman Runge–Kutta style method which they called commutator-free methods

$$Y_r = \exp\left(\sum_k \alpha^k_{r,J_r} h F_k\right) \cdots \exp\left(\sum_k \alpha^k_{r,1} F_k\right), \qquad F_r = \mathrm{Fr}(Y_r, F), \tag{17}$$

$$y_1 = \exp\left(\sum_k \beta^k_J h F_k\right) \cdots \exp\left(\sum_k \beta^k_1 F_k\right). \tag{18}$$

The intention here was to choose the number of flow calculations as small as possible by minimising the J_r, J. One may also here conveniently define

$$a^k_r = \sum_{j=1}^{J_r} \alpha^k_{r,j}, \quad b^k = \sum_{j=1}^{J} \beta^k_j, \tag{19}$$

which will be the corresponding classical Runge–Kutta coefficients used when M is Euclidean space. There are actually commutator-free methods of order four with four stages which need the calculation of only five exponentials in total per step. In fact, the fourth order method presented in [19] needs effectively only four exponentials, because it reuses one exponential from a previous stage. Writing as before $F_r = \mathrm{Fr}(Y_r, F)$ the method reads

$$\begin{aligned}
Y_1 &= y_0, \\
Y_2 &= \exp(\tfrac{1}{2} h F_1) \cdot y_0, \\
Y_3 &= \exp(\tfrac{1}{2} h F_2) \cdot y_0 \\
Y_4 &= \exp(h F_3 - \tfrac{1}{2} h F_1) \cdot Y_2, \\
y_{\frac{1}{2}} &= \exp(\tfrac{1}{12} h (3F_1 + 2F_2 + 2F_3 - F_4)) \cdot y_0, \\
y_1 &= \exp(\tfrac{1}{12} h (-F_1 + 2F_2 + 2F_3 + 3F_4)) \cdot y_{\frac{1}{2}}.
\end{aligned} \tag{20}$$

Note in particular in this example how the expression for Y_4 involves Y_2 and thereby one exponential calculation has been saved. Details on order conditions for commutator-free schemes can be found in [64].

4 Choice of Lie Group Actions

The choice of frames or Lie group action is not unique and may have significant impact on the properties of the resulting Lie group integrator. It is not obvious which action is the best or suits the purpose in the problem at hand. Most examples we know from the literature are using matrix Lie groups $G \subseteq \mathrm{GL}(n)$, but the choice of group action depends on the problem and the objectives of the simulation. We give here some examples of situations where Lie group integrators can be used. Some more details are given in [20].

4.1 Lie Group Acting on Itself by Multiplication

In the case $M = G$, it is natural to use either left or right multiplication

$$L_g(m) = g \cdot m \quad \text{or} \quad R_g(m) = m \cdot g^{-1}, \qquad g, m \in G.$$

are both left Lie group actions by G on G. The corresponding maps ρ_L and ρ_R defined in (3) are

$$\rho_L(\xi)|_m = T_1 R_m(\xi) = \xi \cdot m, \quad \rho_R(\xi)|_m = -T_1 L_m(\xi) = -m \cdot \xi$$

For a vector field $F \in \mathcal{X}(G)$ we use functions f or $\tilde{f}$, $M \to \mathfrak{g}$. Using matrix notation we have $F|_g = f(g) \cdot g$ or $F|_g = -g \cdot \tilde{f}(g)$. Note that $\tilde{f}(g)$ is related to $f(g)$ through the adjoint representation of G, $\mathrm{Ad}\colon G \to \mathrm{Aut}(\mathfrak{g})$,

$$f(g) = -\mathrm{Ad}_g \tilde{f}(g), \qquad \mathrm{Ad}_g = T_1 L_g \circ T_1 R_g^{-1}.$$

4.2 The Affine Group and Its Use in Semilinear PDE Methods

Consider the semilinear partial differential equation

$$u_t = Lu + N(u), \tag{21}$$

where L is a linear differential operator and $N(u)$ is some nonlinear map, typically containing derivatives of lower order than L. Discretising (21) we obtain a system of n_d ODEs, for some large n_d, L becomes an $n_d \times n_d$-matrix, and $N\colon \mathbb{R}^{n_d} \to \mathbb{R}^{n_d}$ a nonlinear function. We follow [55] and introduce a Lie group action on $\mathbb{R}^{n_d}$ by some subgroup of the affine group represented as the semidirect product $G = \mathrm{GL}(n_d) \ltimes \mathbb{R}^{n_d}$. The group product, identity 1_G, and inverse are given as

$$(A_1, b_1) \cdot (A_2, b_2) = (A_1 A_2, A_1 b_2 + b_1), \quad 1_G = (1_{GL}, 0), \quad (A, b)^{-1} = (A^{-1}, -A^{-1} b).$$

where 1_{GL} is the $n_d \times n_d$ identity matrix, and $0 \in \mathbb{R}^{n_d}$. The action on $\mathbb{R}^{n_d}$ is

$$(A, b) \cdot x = Ax + b, \qquad (A, b) \in G, \ x \in \mathbb{R}^{n_d},$$

and the Lie algebra and commutator are given as

$$\mathfrak{g} = (\xi, c), \ \xi \in \mathfrak{gl}(n_d), \ c \in \mathbb{R}^{n_d}, \quad [(\xi_1, c_1), (\xi_2, c_2)] = ([\xi_1, \xi_2], \xi_1 c_2 - \xi_2 c_1 + c_1).$$

In many interesting PDEs, the operator L is constant, so it makes sense to consider the $n_d + 1$-dimensional subalgebra $\mathfrak{g}_L$ of $\mathfrak{g}$ consisting of elements $(\alpha L, c)$ where $\alpha \in \mathbb{R}$, $c \in \mathbb{R}^{n_d}$, so that the map $f : \mathbb{R}^{n_d} \to \mathfrak{g}_L$ is given as

$$f(u) = (L, N(u)).$$

One parameter subgroups are obtained through the exponential map as follows

$$\exp(t(L, b)) = (\exp(tL), \phi(tL)tb).$$

Here the entire function $\phi(z) = (\exp(z) - 1)/z$ familiar from the theory of exponential integrators appears. As an example, one could now consider the Lie–Euler method (8) in this setting, which coincides with the exponential Euler method

$$u_1 = \exp(h(L, N(u_0)) \cdot u_0 = \exp(hL)u_0 + h\phi(hL)N(u_0).$$

There is a large body of literature on exponential integrators, going almost half a century back in time, see [32] and the references therein for an extensive account. A rather general framework for exponential integrators were defined and studied in terms of order conditions in [3].

4.3 The Coadjoint Action and Lie–Poisson Systems

Lie group integrators for this interesting case were studied by Engø and Faltinsen [29]. Suppose G is a Lie group and the manifold under consideration is the dual space $\mathfrak{g}^*$ of its Lie algebra $\mathfrak{g}$. The coadjoint action by G on $\mathfrak{g}^*$ is denoted Ad_g^* defined for any $g \in G$ as

$$\langle \mathrm{Ad}_g^* \mu, \xi \rangle = \langle \mu, \mathrm{Ad}_g \xi \rangle, \quad \forall \xi \in \mathfrak{g}, \mu \in \mathfrak{g}^*, \tag{22}$$

for a duality pairing $\langle \cdot, \cdot \rangle$ between $\mathfrak{g}^*$ and $\mathfrak{g}$. It is well known (see e.g. Sect. 13.4 in [46]) that mechanical systems formulated on the cotangent bundle T^*G with a left or right invariant Hamiltonian can be reduced to a system on $\mathfrak{g}^*$ given as

$$\dot{\mu} = \pm \operatorname{ad}^*_{\frac{\partial H}{\partial \mu}} \mu,$$

where the negative sign is used in case of right invariance. The solution to this system preserves coadjoint orbits, which makes it natural to suggest the group action

$$g \cdot \mu = \operatorname{Ad}^*_{g^{-1}} \mu,$$

so that the resulting Lie group integrator also respects this invariant. For Euler's equations for the free rigid body, the Hamiltonian is left invariant and the coadjoint orbits are spheres in $\mathfrak{g}^* \cong \mathbb{R}^3$.

4.4 Homogeneous Spaces and the Stiefel and Grassmann Manifolds

The situation when G acts on itself by left of right multiplication is a special case of a homogeneous space [60], where the assumption is only that G acts freely, transitively and continuously on some manifold M. Homogeneous spaces are isomorphic to the quotient G/G_x where G_x is the *isotropy group* for the action at an arbitrarily chosen point $x \in M$

$$G_x = \{h \in G \mid h \cdot x = x\}.$$

Note that if x and z are two points on M, then by transitivity of the action, $z = g \cdot x$ for some $g \in G$. Therefore, whenever $h \in G_z$ it follows that $g^{-1} \cdot h \cdot g \in G_x$ so isotropy groups are isomorphic by conjugation [12]. Therefore the choice of $x \in M$ is not important for the construction of the quotient. For a readable introduction to this type of construction, see [12], in particular Lecture 3.

A much encountered example is the hypersphere $M = S^{d-1}$ corresponding to the left action by $G = \mathrm{SO}(d)$, the Lie group of orthogonal $d \times d$ matrices with unit determinant. One has $S^{d-1} = \mathrm{SO}(d)/\mathrm{SO}(d-1)$.

The Stiefel manifold $\mathrm{St}(d, k)$ can be represented by the set of $d \times k$-matrices with orthonormal columns. An action on this set is obtained by left multiplication by $G = \mathrm{SO}(d)$. Lie group integrators for Stiefel manifolds are extensively studied in the literature, see e.g. [22, 37]. An important subclass of the homogeneous spaces is the symmetric spaces, also obtained through a transitive action by a Lie group G, where $M = G/G_x$, but here one requires in addition that the isotropy subgroup is an open subgroup of the fixed point set of an involution of G [61]. A prominent example of a symmetric space in applications is the Grassmann manifold, obtained as $\mathrm{SO}(d)/(\mathrm{SO}(k) \times \mathrm{SO}(d-k))$, which can alternatively be interpreted as the set of k-dimensional subspaces of a d-dimensional linear space.

4.5 Isospectral Flows

In isospectral integration one considers dynamical systems evolving on the manifold of $d \times d$-matrices sharing the same Jordan canonical form. Considering the case of symmetric matrices, one can use the transitive group action by $\mathrm{SO}(d)$ given as

$$g \cdot m = gmg^T.$$

This action is transitive, since any symmetric matrix can be diagonalised by an appropriately chosen orthogonal matrix. If the eigenvalues are distinct, then the isotropy group is discrete and consists of all matrices in $\mathrm{SO}(d)$ which are diagonal.

Lie group integrators for isospectral flows have been extensively studied, see for example [13, 14]. See also [16] for an application to the KdV equation.

4.6 Tangent and Cotangent Bundles

For mechanical systems the natural phase space will often be the tangent bundle TM as in the Lagrangian framework or the cotangent bundle T^*M in the Hamiltonian framework. The application of Lie group integrators in this setting is often done by using a prolongation of a group action on M to an action on TM or T^*M. The seminal paper by Lewis and Simo [41] discusses several Lie group integrators for mechanical systems on cotangent bundles, deriving methods which are symplectic, energy and momentum preserving. Engø [28] suggested a way to generalise the Runge–Kutta–Munthe-Kaas methods into a partitioned version when M is a Lie group. Marsden and collaborators have developed the theory of Lie group integrators from the variational viewpoint over the last two decades. See [47] for an overview. For more recent work pertaining to Lie groups in particular, see [8, 39, 70].

5 Isotropy

In Sect. 2 it was pointed out that when using frames to express the vector field $F \in \mathscr{X}(M)$, the functions $f_i : M \to M$ in the expression $F|_x = \sum_i f_i(x) E_i|_x$ were not necessarily unique. Similarly, using a map $f : M \to \mathfrak{g}$ in the group action framework to write $F = \rho \circ f$ where ρ is the infinitesimal action, we also remarked that f is not generally unique. In fact, taking any other map $z : M \to \mathfrak{g}$ satisfying $\rho(z(x))|_x = 0$, then $F|_x = \rho(f(x))|_x = \rho(f(x) + z(x))|_x$ hence $f + z$ represents the same vector field as f. So even though adding the map z to f does not alter the vector field F, it *does* generally alter any numerical Lie group integrator. A typical situation with isotropy was described in Sect. 4.4 with homogeneous manifolds.

An interesting example is the two-sphere $S^2 \simeq SO(3)/SO(2)$. If we represent elements of $\mathfrak{so}(3)$ as vectors in $\mathbb{R}^3$ and points on the sphere as three-vectors of unit length, we may express (3) by using the cross-product in $\mathbb{R}^3$

$$F|_y = f(y) \times y = (f(y) + \alpha(y)y) \times y$$

where $\alpha : S^2 \to \mathbb{R}$ is any smooth function. The Lie–Euler method (8) with initial value $y(0) = y$ would read for the first step

$$y_1 = \exp(h(f(y) + \alpha(y)y))y$$

We may assume that $(f(y), y) = 0$ for all $y \in S^2$ with the Euclidean inner product. Then, by using Rodrigues formula [69] for the exponential and simplifying, we can write

$$y_1 = \left(1 - h^2 \frac{1 - \cos\theta}{\theta^2}\right) y + h \frac{\sin\theta}{\theta} f(y) \times y + h^2 \frac{1 - \cos\theta}{\theta^2} \alpha(y) f(y)$$

where $\theta = h\sqrt{\|f(y)\|^2 + |\alpha(y)|^2}$. For small values of the step size h this can be approximated by

$$y_1 \approx (1 - \frac{1}{2}h^2)y + hf(y) \times y + \frac{1}{2}h^2\alpha(y)f(y)$$

so we see that by choosing $\alpha = 0$ one moves from y in the direction of the vector field $f(y) \times y$ whereas any nonzero α will give a contribution to the increment in the tangent plane orthogonal to the vector field. Lewis and Olver [40] pursue this analysis much further and show how the isotropy parameter $\alpha(y)$ can be used to minimise the orbital error by requiring that the curvatures of the exact and numerical solution agree.

Another situation in which special care should be taken in choosing the freedom left by isotropy is the case when M is a homogeneous manifold and the acting group G has much higher dimension than that of M as may be the case for instance with Stiefel and Grassmann manifolds discussed in Sect. 4.4. Typically, a naive implementation of a Lie group integrators has a computational cost of order d^3 whenever G is represented as a subgroup of $GL(d, \mathbb{R})$. But it is then useful to choose the map $f : M \to \mathfrak{g} \subset \mathfrak{gl}(d, \mathbb{R})$ in such a way that the resulting Lie algebra element has the lowest possible rank as a matrix. In [22] this idea was applied to Stiefel manifolds $\mathrm{St}(d, k) \simeq SO(d)/SO(d - k)$ and it was shown that a clever choice of isotropy component in $f(y)$ results in $\mathrm{rank} f(y) = 2k$, and then it was shown that the complexity of the Lie group integrator could be reduced from $\mathcal{O}(d^3)$ to $\mathcal{O}(dk^2)$. Krogstad [37] suggested a similar approach which also leads to $\mathcal{O}(dk^2)$ complexity algorithms. Recently Munthe-Kaas and Verdier [58] considered isotropy in the context of equivariance.

6 Order Theory for Lie Group Integrators

Order theory is concerned with the convergence order in terms of the step size for numerical integrators. In the setting of manifolds one can define the concept precisely as follows: Let $\Phi_{h,F} : M \to M$ be a numerical flow map applied to $F \in \mathscr{X}(M)$ in the sense that the numerical method is defined through $y_{n+1} = \Phi_{h,F}(y_n)$. We shall say that this method has *order p* if, for any $F \in \mathscr{X}(M)$, $\psi \in \mathscr{F}(M)$, and $y \in M$ it holds that

$$\psi(\exp(hF)y) - \psi(\Phi_h(y)) = \mathscr{O}(h^{p+1})$$

where here and in what follows write simply Φ_h for $\Phi_{h,F}$. The order theory amounts to writing the exact and numerical solutions to the problem in powers of the step-size h and comparing these term by term. A systematic development of the order theory for Crouch–Grossman and commutator-free Lie group was done in [64, 65] respectively, but the theory is now well established and founded on an advanced algebraic machinery much thanks to Munthe-Kaas and coauthors [27, 42, 56, 59] and Murua [62, 63]. We skip most of the details here, and remind the reader that what we refer to as "Munthe-Kaas like" schemes in Sect. 3.1 are relatively easy to deal with in practice since they can be seen as the application of a standard "vector space" Runge–Kutta schemes under a local change of variables. The schemes based on compositions of flows are less straightforward, one here needs to use the theory of ordered rooted trees. A key formula is the series giving the pullback of an arbitrary function $\psi \in \mathscr{F}(M)$ along the flow of a vector field

$$\exp(hF)^*\psi = \psi \circ \exp(hF) = \psi + hF(\psi) + \frac{h^2}{2!}F^2(\psi) + \cdots$$

Note also that by Leibniz' rule

$$F^2 = \mathrm{Fr}(\cdot, F)^2 + \nabla_F(F) = \sum_{i,j}(f_j f_i E_j E_i + f_j E_j(f_i) E_i)$$

The two terms can be associated to ordered rooted trees, each having three nodes.

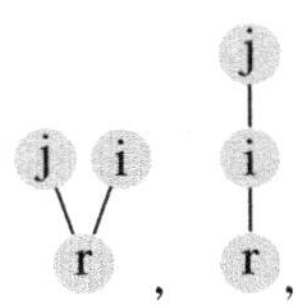

The reader may find it unnatural to include the fictitious root node r, the alternative is to represent the terms by means of ordered forests of rooted trees, but we shall stick with this notation, used for instance by Grossman and Larson [31], since it simplifies parts of the presentation.

Similarly, one can compute F^3

$$F^3 = \sum_{i,j,k} (f_k f_j f_i E_k E_j E_i + f_k f_j E_k(f_i) E_j E_i + f_k E_k(f_j) f_i E_j E_i$$
$$+ f_k f_j E_j(f_i) E_k E_i + f_k f_j E_k E_j(f_i) E_i + f_k E_k(f_j) E_j(f_i) E_i \quad (23)$$

where each term is associated to ordered trees as follows

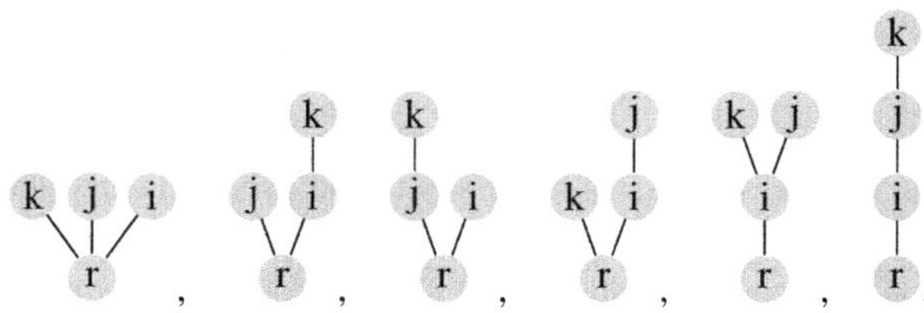

(24)

Going from F^2 to F^3 amounts to the following in our drawing convention: For every tree in F^2 run through all of its nodes and add a new node with a new label (k) as a new 'leftmost' child. Assuming the labels are ordered, say $i < j < k$, each labelled, monotonically ordered tree appears exactly once. In the convention used above the children of any node will have increasing labels from right to left. Also they must necessarily increase from parent to child. With this convention, it is transparent that the only difference between the second and the fourth term is that the summation indices j and k have been interchanged so these two terms are the same. In general, we have an equivalence relation on the set of labelled ordered trees. Let the set of $q + 1$ labelled nodes with a total order on the labels be denoted $\mathscr{A}_{q+1}$. To an ordered tree with $q + 1$ nodes, we associate a child-to-parent map $\mathbf{t} : \mathscr{A}_{q+1}\backslash\{r\} \to \mathscr{A}_{q+1}$. Now we define two labelled forests with child-to-parent maps $\mathbf{u}$ and $\mathbf{v}$ to be equivalent if there exists a permutation $\sigma : \mathscr{A}_{q+1} \to \mathscr{A}_{q+1}$ such that

1. $\sigma(r) = r$
2. $\mathbf{u} \circ \sigma = \sigma \circ \mathbf{v}$ on $\mathscr{A}_{q+1}\backslash\{r\}$
3. For all $z_1, z_2 \in \mathscr{A}_{q+1}\backslash\{r\}$ such that $\mathbf{u}(z_1) = \mathbf{u}(z_2)$ one has

$$\sigma(z_1) < \sigma(z_2) \Rightarrow z_1 < z_2$$

The ordered rooted trees are from now on taken to be equivalence classes of their labelled counterparts.

Notation summary.

T_O is the set of ordered rooted trees.

F_O is the set of ordered forests, i.e. an ordered set of trees, $t_1, t_2, \ldots, t_\mu$, each $t_i \in T_O$.

$|t|$ is the total number of nodes in a tree (resp. forest).

$\bar{T}_O$ is the set T_O augmented by the 'empty tree', $\emptyset$ where $|\emptyset| = 0$, similarly $\bar{F}_O = F_O \cup \emptyset$, i.e. F_O augmented with the empty forest.

B_+ is a map that takes any ordered forest into a tree by joining the roots of each tree to a new common node, $|B_+(t_1, \ldots, t_\mu)| = 1 + |t_1, \ldots, t_\mu|$, we also let $B_+(\emptyset) = \bullet$.

B_- is a map $B_- : T_O \to \bar{F}_O$ that removes the root of a tree and leaves the forest of subtrees. $B_+ \circ B_-$ is the identity map on T_O.

$\mathscr{T}_O$ The $\mathbb{R}$-linear space having the elements of T_O as basis.

We may now define $\alpha(t)$ to be the number of elements in the equivalence class which contains the labelled ordered tree t. A formula for $\alpha(t)$ for ordered rooted trees was found in [65].

Proposition 2 *Set* $\alpha(\bullet) = 1$. *For any* $t = B_+(t_1, t_2, \ldots, t_\mu) \in T_O$

$$\alpha(t) = \prod_{\ell=1}^{\mu} \binom{\sum_{i=1}^{\ell} |t_i| - 1}{|t_\ell| - 1} \alpha(t_\ell)$$

where $|t|$ *is the total number of nodes in the tree* t.

Example 2 We can work out the coefficient for a couple of examples. For instance

$\alpha($$) = \alpha(B_+(\bullet)) = \binom{0}{0} \alpha(\bullet) = 1$

For the tree  we get

$\alpha($$) = \binom{0}{0} \binom{2}{1} \alpha(\bullet)\alpha($$) = 1 \cdot 2 \cdot 1 \cdot 1 = 2,$

whereas

$\alpha($$) = \binom{1}{1} \binom{2}{0} \alpha($$)\alpha(\bullet) = 1 \cdot 1 \cdot 1 \cdot 1 = 1.$

For any tree $t = B_+(t_1, \ldots, t_\mu)$, $|t| > 1$, and vector field F we can associate operators $\mathbf{F}(t)$ acting on functions of M, defined in a recursive manner, setting

$$\mathbf{F}(\bullet) = 1,$$
$$\mathbf{F}(t) = \sum_{i_1 \ldots i_\mu} \mathbf{F}(t_1)(f_{i_1}) \cdots \mathbf{F}(t_\mu)(f_{i_\mu}) E_{i_1} \ldots E_{i_\mu}$$

Note in particular that according to the definition

$$\mathbf{F}(\bullet) = \sum_i f_i E_i = F.$$

Similarly, there is a counterpart to the frozen vector fields relative to a point $x \in M$

$$\mathbf{F}_x(t) = \sum_{i_1 \dots i_\mu} \mathbf{F}(t_1)(f_{i_1})|_x \cdots \mathbf{F}(t_\mu)(f_{i_\mu})|_x E_{i_1} \dots E_{i_\mu}$$

The formal Taylor expansion of the flow of the vector field F is

$$\psi \circ \exp(hF)y = \sum_{k=0}^{\infty} \frac{h^k}{k!} F^k(\psi)\big|_y, \quad \psi \in \mathscr{F}(M)$$

and in view of the above discussion we can therefore write the formal expansion

$$\psi \circ \exp(hF)y = \sum_{t \in T_O} \frac{h^{|t|-1}}{(|t|-1)!} \alpha(t)\, \mathbf{F}(t)(\psi)|_y \tag{25}$$

This infinite series is in fact a special instance of what we may call a Lie-Butcher series, in which the coefficient $\alpha(t)/|t|!$ is replaced by a general map $\mathbf{a} : F_O \to \mathbb{R}$. We define the operator series

$$B(\mathbf{a}, x) = \sum_{t \in T_O} h^{|t|-1} \mathbf{a}(t) \mathbf{F}_x(t) \tag{26}$$

At this point it is convenient to consider the free $\mathbb{R}$-vector space $\mathscr{T}_O$ over the set of ordered rooted trees T_O and to extend the map $\mathbf{F}$ to a linear map between $\mathscr{T}_O$ and the space of higher order derivations indicated above. The algebraic structures on $\mathscr{T}_O$ are by now well-known from the literature, see e.g. [2, 42, 56], and they originate from the algebra on higher order derivations under composition by requiring $\mathbf{F}$ to be a algebra homomorphism. We will not pursue further these issues here, but instead consider briefly how the commutator-free methods of Sect. 3.2.2 can be expanded in a series with the same type of terms as the exact flow, the details can be found in [64]. We introduce the concatenation product on $\mathscr{T}_O$ which is linear in both factors and on single trees u and v defined as

$$u \cdot v = B_+(u_1 u_2 \dots u_\mu v_1 \dots v_\nu), \quad u = B_+(u_1 u_2 \dots u_\mu),\ v = B_+(v_1 v_2 \dots v_\nu) \tag{27}$$

This is a morphism under $\mathbf{F}_x$ since obviously $\mathbf{F}_x(u) \circ \mathbf{F}_x(v) = \mathbf{F}_x(u \cdot v)$.

6.1 Order Conditions for Commutator-Free Lie Group Integrators

We now provide a few simple tools which we formulate through some lemmas with easy proofs which we omit, consult [64] for details.

Lemma 3 *Suppose $\phi_{\mathbf{a}}$ and $\phi_{\mathbf{b}}$ are maps of M with B-series $B(\mathbf{a},\cdot)$ and $B(\mathbf{b},\cdot)$ respectively, such that $\mathbf{a}(\bullet)=\mathbf{b}(\bullet)=1$. This means that formally, for any smooth function ψ*

$$\psi(\phi_{\mathbf{a}}(y)) = B(\mathbf{a}, y)(\psi)|_y, \qquad \psi(\phi_{\mathbf{b}}(y)) = B(\mathbf{b}, y)(\psi)|_y$$

The composition of maps $\phi_a \circ \phi_b$ has a B-series $B(\mathbf{ab}, y)$ with coefficients

$$\mathbf{ab}(\bullet) = 1,$$

and for $t = t_1 t_2 \dots t_\mu$

$$\mathbf{ab}(t) = \sum_{u\cdot v=t} \mathbf{a}(v)\mathbf{b}(u) = \sum_{k=0}^{\mu} \mathbf{a}(t_{k+1}\dots t_\mu)\mathbf{b}(t_1\dots t_k).$$

Lemma 4 *Suppose $a = \phi_{\mathbf{a}}(x)$ has a B-series $B(\mathbf{a}, x)$. Then the frozen vector field $F_a = \sum_i f_i(a)E_i \in V$ has the B-series $hF_a = B(\mathbf{F}_a, x)$ where*

$$\begin{aligned} \mathbf{F}_a(\bullet) &= 0 \\ \mathbf{F}_a(B_+(t_1\dots t_\mu)) &= 0, \quad \mu \geq 2 \\ \mathbf{F}_a(B_+(t)) &= \mathbf{a}(t), \ \forall t \in T_O \end{aligned}$$

Lemma 5 *Let $G \in V$ be any vector field with B-series of the form*

$$hG = \sum_{t\in T_O} h^{|t|-1}\mathbf{G}(t)\mathbf{F}_x(B_+(t)), \qquad \mathbf{G}(\bullet) = 0.$$

Then its h-flow $\exp(hG)$ *has again a B-series $B(\mathbf{g}, x)$ where*

$$\begin{aligned} g(\emptyset) &= 1, \\ \mathbf{g}(B_+(t_1\dots t_\mu)) &= \frac{1}{\mu!}\mathbf{G}(t_1)\cdots\mathbf{G}(t_\mu) \end{aligned}$$

To obtain a systematic development of the order conditions for the commutator-free schemes of Sect. 3.2.2, consider the definition (17)–(18), and define

$$Y_{r,0} = x \in M \quad \text{and} \quad Y_{r,j} = \exp\left(\sum_k \alpha_{r,j}^k F_k\right) Y_{r,j-1}, \quad j = 1, \ldots, J_r,$$

in this way $Y_{r,J_r} = Y_r$ in (17). For the occasion, we unify the notation by setting $\alpha_{s+1,j}^k := \beta_j^k$ for $1 \le k \le s$ and $1 \le j \le J$. Then we may also write Y_{s+1} for y_1 in (18) and we can write down the order conditions for the commutator free Lie group methods.

Theorem 2 *With the above definitions, the quantities $Y_{r,j}$, $1 \le r \le s+1$, $1 \le j \le J_r$ all have B-series $B(\mathbf{Y}_{r,j}, x)$ defined through the following formulas*

$$\begin{aligned}
\mathbf{Y}_{r,j}(\bullet) &= 1, \\
\mathbf{Y}_{r,0}(t) &= 0, \quad \forall t : |t| > 1, \\
\mathbf{Y}_{r,j}(B_+(t_1 \ldots t_\mu)) &= \sum_{k=0}^{\mu} \mathbf{Y}_{r,j-1}(B_+(t_1 \ldots t_k)) \cdot \mathbf{b}_{r,j}(B_+(t_{k+1} \ldots t_\mu)), \\
\mathbf{b}_{r,j}(\bullet) &= 1, \quad 1 \le r \le s+1, \ 1 \le j \le J_r, \\
\mathbf{b}_{r,j}(B_+(t_1 \ldots t_\mu)) &= \frac{1}{\mu!} \mathbf{G}_{r,j}(t_1) \cdots \mathbf{G}_{r,j}(t_\mu), \\
\mathbf{G}_{r,j}(t) &= \sum_{k=1}^{s} \alpha_{r,j}^k \mathbf{Y}_{k,J_r}(t)
\end{aligned}$$

We now have a B-series for the numerical solution, and the B-series of the exact solution is given by (25), so it follows that

Corollary 1 *A commutator-free Lie group method has order of consistency q if and only if*

$$\mathbf{Y}_{s+1,J}(t) = \frac{\alpha(t)}{(|t|-1)!}, \quad \forall t : |t| \le q+1.$$

Here $\mathbf{Y}_{s+1,J}$ is found from Theorem 2 and $\alpha(t)$ is given in Proposition 2.

A usual strategy for deriving methods of a given order of consistency is to first consider the classical order conditions for Runge–Kutta methods since these of course must be satisfied for the classical Runge–Kutta coefficients defined in (19). Since the computation of flows (exponential) is usually the most expensive operation, one next seeks the smallest possible number of exponentials per step, i.e. let each J_r be as small as possible while leaving enough free parameters to solve the remaining non-classical conditions $\alpha_{r,j}^k$ and β_j^k.

6.2 Selecting a Minimal Set of Conditions

The conditions arising from each ordered rooted tree are not independent. The B-series (26) we consider here are representing objects of different kinds, such as maps and vector fields. This means that the series representing these objects are subsets of $\mathcal{T}_O$ and we briefly characterise these subsets here in a purely algebraic fashion. Then we can better understand how to select a minimal set of conditions to solve.

Let $A_- = \{t \in T_O : t = B_+(u), u \in T_O\}$ and set $A = A_- \cup \{\bullet\}$. Any tree $t = B_+(t_1 \dots t_\mu)$ can be considered as a word of the alphabet A in the sense that it is formed by the finite sequence $B_+(t_1), \dots, B_+(t_\mu)$. With the concatenation product (27) we get all of T_O as the free monoid over the set A with identity element $\bullet$. This structure is then extended to $\mathcal{T}_O$ as an associative $\mathbb{R}$-algebra. The elements of $\mathcal{T}_O$ are the formal series on T_O and we denote by $(P, t) \in \mathbb{R}$ the coefficient of the tree t in the series P. The product of two series S and T is defined to be the series with coefficient

$$(ST, t) = \sum_{t=u\cdot v} (S, u)(T, v).$$

Next, notice that the commutator-free methods considered here are derived by composing flows of linear combinations of frozen vector fields. From Lemma 4 we see that these linear combinations (scaled by h) have expansions whose coefficients vanish on trees not belonging to A_-. On the other hand, the composition of exponentials can be written as a single exponential of a series via the Baker–Campbell–Hausdorff formula, and the resulting series belongs to the free Lie algebra on the set $A_- \subset T_O$.

There are three important subsets of $\mathcal{T}_O$

- the subspace $\mathfrak{g} \in \mathcal{T}_O$ which is the free Lie algebra on the set A
- $V \subset \mathfrak{g}$ is the subspace of $\mathfrak{g}$ consisting of series S such that

$$(S, t) = 0 \quad \text{whenever } t \notin A_-$$

- G is the group of formal exponential series $T = \exp(S),\ \ S \in \mathfrak{g}$. These series are such that $(T, \bullet) = 1$.

The combinatorial properties of ordered rooted trees and free Lie algebras are by now well understood, and many results hinge on the Poincare-Birkhoff-Witt theorem, see e.g. [33, Chap. 5]. The space $\mathcal{T}_O$ has a natural grading arising from the number of nodes in each ordered rooted tree, and we can define

$$\mathcal{T}_O^q = \text{span}\{t : |t| = q + 1\} \quad \text{thus } \mathcal{T}_O = \coprod_{n\geq 0} \mathcal{T}_O^n,$$

and similarly

$$\mathfrak{g} = \coprod_{n\geq 0} \mathfrak{g}^n, \quad \mathfrak{g}^n = \mathfrak{g} \cup \mathcal{T}_O^n.$$

The dimension of $\mathcal{T}_O^n$ is the Catalan number

$$\dim \mathcal{T}_O^n = \frac{1}{n+1}\binom{2n}{n}.$$

The next result is well-known, and its proof can be found for instance in [64].

Theorem 3

$$\dim \mathfrak{g}_n = \nu_n = \frac{1}{2n}\sum_{d|n}\mu(d)\binom{2n/d}{n/d}$$

where $\mu(d)$ is the Möbius function defined for any positive integer as $\mu(1) = 1$, $\mu(d) = (-1)^p$ when d is the product of p distinct primes, and $\mu(d) = 0$ otherwise. The sum is over all positive integers which divide n, including 1 and n.

We present a table over the numbers $\mathcal{T}_O^n$, $\mathfrak{g}^n$ and c^n, the last one being the dimensions of the graded components of the unordered trees, that count the number of order conditions for classical RK methods

n	1	2	3	4	5	6	7
$\mathcal{T}_O^n$	1	2	5	14	42	132	429
g^n	1	1	3	8	25	75	245
c^n	1	1	2	4	9	20	48

So in fact, the numbers $\mathfrak{g}^n$ give the number of order conditions to be considered for each order for commutator-free methods and a possible strategy would be to pick $\mathfrak{g}^n$ independent conditions out of the $\mathcal{T}_O^n$ found from Theorem 2. It should be observed that the dependency between conditions corresponding to ordered rooted trees arise amongst trees that share the same (unordered) set of subtrees, such as  and , in fact the condition corresponding to precisely one of these two trees can be discarded given that conditions of lower order are included. Using classical theory of free Lie algebras, one may characterize this dependency by a generalized Witt formula counting, for a given tree t, the dimension of the subspace of $\mathfrak{g}$ spanned by the set of trees obtained from permuting the subtrees of t. Consider the equivalence class $[t]$ characterized by a set of ν distinct subtrees, $t_i \in T_O,\ i = 1, \dots, \nu$, where there are exactly κ_i occurrences of t_i. The dimension of the subspace spanned by trees in $[t]$ can be derived from a formula in Bourbaki [9].

$$c(\kappa) = \frac{1}{|\kappa|}\sum_{d|\kappa}\mu(d)\frac{(|\kappa|/d)!}{(\kappa/d)!},$$

in other words, the dimensions depends only on the number of copies of each subtree and not on the subtrees themselves. For convenience, we give a few examples

$$
\begin{array}{lll}
c(n) & = 0, & n > 1 \\
c(n,1) & = 1, & n > 0, \\
c(n,1,1) & = n+1, & n > 0, \\
c(n,2) & = \lfloor \frac{n+1}{2} \rfloor & n > 0. \\
c(1,\ldots,1) & = (r-1)!, & (r \text{ distinct subtrees})
\end{array}
\tag{28}
$$

A detailed analysis of how methods of orders up to four are constructed can be found in [64]. Schemes of order 4 with 4 stages can be constructed with two exponentials in the fourth stage and the update stage as in (20). In this example, there is also a clever reuse of an exponential such that the total number of flow calculation is effectively 5. Yet another result from [64] shows that in the case of two exponentials, i.e. $J_r = 2$ the coefficients of stage r is only involved linearly in the order conditions.

As far as we know, no explicit commutator-free method of order five or higher has been derived at this point. A complication is then that one needs to have stages (including the final update) with at least three exponentials, and no simplification of the order conditions similar to the $J_r = 2$-case has been found.

7 Symplectic Lie Group Integrators

It is not clear whether there exist Lie group integrators of Munthe-Kaas or commutator-free type which are symplectic for an arbitrary symplectic manifold. Recently, McLachlan et al. [48] found an elegant adaptation to the classical mid-point rule to make it a symplectic integrator on product of 2-spheres. It is however relatively easy to find symplectic integrators on cotangent bundles of manifolds, an by looking at the special case where $M = T^*G$ for a Lie group G one can obtain symplectic integrators which are rather similar in form to partitioned Munthe-Kaas methods as defined by Engø in [28]. The approach we described here was introduced by Celledoni et al. [20] and by using ideas from Bou-Rabee and Marsden [8], Bogfjellmo and Marthinsen [7] extended it to high order partitioned symplectic Lie group integrators both in the Munthe-Kaas and Crouch–Grossman formats. We briefly describe the setting for these symplectic integrators.

The first step is to replace T^*G by $\mathbf{G} := G \ltimes \mathfrak{g}^*$ via right trivialisation, meaning that any $p_g \in T_g^*G$ is represented as the tuple (g, μ) where $\mu = R_g^* p_g$. We use the notation $R_{g*}v = TR_g v$ for any $v \in TG$ and similarly R_g^* for the adjoint operator such that $\langle R_g^* p, v\rangle = \langle p, R_{g*}v\rangle$ for any $p \in T^*G$, $v \in TG$. Next, we lift the group structure from G to $\mathbf{G}$ through

$$
(g_1, \mu_1) \cdot (g_2, \mu_2) = (g_1 \cdot g_2, \mu_1 + \mathrm{Ad}^*_{g_1^{-1}} \mu_2), \qquad 1_{\mathbf{G}} = (1_G, 0_{\mathfrak{g}^*}),
$$

where Ad_g^* is defined in (22). Similarly, the tangent map of right multiplication extends as

$$TR_{(g,\mu)}(R_{h*}\,\zeta,\nu) = (R_{hg*}\,\zeta, \nu - \mathrm{ad}^*_\zeta\,\mathrm{Ad}^*_{h^{-1}}\,\mu), \quad g,h \in G,\ \zeta \in \mathfrak{g},\ \mu,\nu \in \mathfrak{g}^*.$$

Of particular interest is the restriction of $TR_{(g,\mu)}$ to $T_1\mathbf{G} \cong \mathfrak{g} \times \mathfrak{g}^*$,

$$T_1 R_{(g,\mu)}(\zeta,\nu) = (R_{g*}\zeta, \nu - \mathrm{ad}^*_\zeta\,\mu).$$

The natural symplectic form on T^*G is defined as

$$\Omega_{(g,p_g)}((\delta v_1,\delta\pi_1),(\delta v_2,\delta\pi_2)) = \langle\delta\pi_2,\delta v_1\rangle - \langle\delta\pi_1,\delta v_2\rangle,$$

and by right trivialisation it may be pulled back to $\mathbf{G}$ and then takes the form

$$\omega_{(g,\mu)}((R_{g*}\xi_1,\delta\nu_1),(R_{g*}\xi_2,\delta\nu_2)) = \langle\delta\nu_2,\xi_1\rangle - \langle\delta\nu_1,\xi_2\rangle - \langle\mu,[\xi_1,\xi_2]\rangle, \tag{29}$$

where $\xi_1, \xi_2 \in \mathfrak{g}$. The presentation of differential equations on T^*G as in (3) is now achieved via the action by left multiplication, meaning that any vector field $F \in \mathcal{X}(\mathbf{G})$ is expressed by means of a map $f\colon \mathbf{G} \to T_1\mathbf{G}$,

$$F(g,\mu) = T_1 R_{(g,\mu)} f(g,\mu) = (R_{g*}f_1, f_2 - \mathrm{ad}^*_{f_1}\,\mu), \tag{30}$$

where $f_1 = f_1(g,\mu) \in \mathfrak{g}$, $f_2 = f_2(g,\mu) \in \mathfrak{g}^*$ are the two components of f. We are particularly interested in the case that F is a Hamiltonian vector field which means that F satisfies the relation

$$F \lrcorner\, \omega = \mathrm{d}H, \tag{31}$$

for some Hamiltonian function $H\colon T^*G \to \mathbb{R}$ and $\lrcorner$ is the interior product defined as $F \lrcorner\, \omega(X) := \omega(F,X)$ for any vector field X. From now on we let $H\colon \mathbf{G} \to \mathbb{R}$ denote the trivialised Hamiltonian. A simple calculation using (29), (30) and (31) shows that the corresponding map f for such a Hamiltonian vector field is

$$f(g,\mu) = \left(\frac{\partial H}{\partial \mu}(g,\mu), -R_g^*\frac{\partial H}{\partial g}(g,\mu)\right). \tag{32}$$

7.1 Variational Integrators on Lie Groups

A popular way to derive symplectic integrators in general is through the discretisation of a variational principle, this was first done by Veselov in [72] and Moser and Veselov [52], and further developed for Lie groups in [5, 6, 44, 45] For an extensive early review of variational integrators, see Marsden and West [47]. The exact solution to the mechanical system is the function $g(t)$ which minimises the action

$$S[g] = \int_{t_0}^{t_1} L(g,\dot{g})\,\mathrm{d}t. \tag{33}$$

The function $L : TM \to \mathbb{R}$ is called the Lagrangian. Taking the variation of this integral yields the Euler–Lagrange equations

$$\frac{\mathrm{d}}{\mathrm{d}t}\frac{\partial L}{\partial \dot{g}} = \frac{\partial L}{\partial g}.$$

It is usually more conventient to trivialise L by defining $\ell(g,\xi) = L(g, \mathbb{R}_{g*}\xi)$, $\ell : G \ltimes \mathfrak{g} \to \mathbb{R}$. In this case it is an easy exercise to derive the corresponding version of the Euler–Lagrange equations as

$$\frac{\mathrm{d}}{\mathrm{d}t}\frac{\partial \ell}{\partial \xi} = R_g^* \frac{\partial \ell}{\partial g} - \mathrm{ad}_\xi^* \frac{\partial \ell}{\partial \xi}.$$

At this point it is customary to introduce the Legendre transformation, defining

$$\mu = \frac{\partial \ell}{\partial \xi}(g,\xi) \in \mathfrak{g}^* \quad \Rightarrow \quad \xi = \iota(g,\mu) \text{ where } \iota : G \times \mathfrak{g}^* \to \mathfrak{g}, \tag{34}$$

assuming that the left equation can be solved for ξ. The corresponding differential equations for g and μ would be

$$\dot{g} = R_{g*}\iota(g,\mu), \qquad \dot{\mu} = R_g^* \frac{\partial \ell}{\partial g}(g, \iota(g,\mu)) - \mathrm{ad}^*_{\iota(g,\mu)}\,\mu,$$

which agrees with the formulation (30) with

$$f_1(g,\mu) = \iota(g,\mu) \quad \text{and} \quad f_2(g,\mu) = R_g^* \frac{\partial \ell}{\partial g}(g, \iota(g,\mu)). \tag{35}$$

Variational integrators are derived by extremising an approximation to (33),

$$S_d\left(\{g_k\}_{k=0}^{N-1}\right) = h \sum_{k=0}^{N-1} L_d(g_k, g_{k+1}),$$

with respect to the discrete trajectory of points $g_k \approx g(t_k)$. The function $L_d(x,y)$ is called *the discrete Lagrangian.* We compute the variation

$$\delta S_d = h \sum_{k=1}^{N-1} \langle D_1 L_d(g_k, g_{k+1}) + D_2 L_d(g_{k-1}, g_k), \delta g_k\rangle$$
$$+ h\langle D_1 L_d(g_0, g_1), \delta g_0\rangle + h\langle D_2 L_d(g_{N-1}, g_N), \delta g_N\rangle.$$

Leaving the end points fixed amounts to setting $\delta g_0 = \delta g_N = 0$ and this leads to the discrete version of the Euler–Lagrange equations (DEL)

$$D_1 L_d(g_k, g_{k+1}) + D_2 L_d(g_{k-1}, g_k) = 0, \quad k = 1, \dots, N-1.$$

We now present an example of how $L_d(g_k, g_{k+1})$ can be defined. Let $\xi^{-1} : \mathfrak{g} \to G$ be a map which satisfies

1. ξ^{-1} is a local diffeomorphism
2. $\xi^{-1}(0) = 1$
3. $T_0\xi^{-1} = \mathrm{Id}_{\mathfrak{g}}$.

For any $\eta \in \mathfrak{g}$ define the curve $g(t) = \xi^{-1}(t\eta)g_k$ on the Lie group. Clearly, we get $g(0) = g_k$ and by choosing

$$\eta = \xi(g_{k+1}g_k^{-1}) = \xi(\Delta_k), \quad \Delta_k := g_{k+1}g_k^{-1},$$

we get $g(1) = g_{k+1}$. As for the second argument of the continuous Lagrangian, we compute $\dot{g}(t) = R_{g_k *} \circ T_{t\eta}\xi^{-1} \circ \eta$. Taking $\dot{g}(0)$ as an approximation to the argument $\dot{g}$ of L, we find by 3. the approximation $\dot{g}(0) = R_{g_k*}\xi(g_{k+1}g_k^{-1}) = R_{g_k*}\xi(\Delta_k) := R_{g_k*}\xi_k$. So a possible approximation $L_d(g_k, g_{k+1})$ could be

$$L_d(g_k, g_{k+1}) = L(g_k, R_{g_k*}\xi_k) := \ell(g_k, \xi_k).$$

We can compute the variation of the action sum

$$\delta S_d = h \sum_{k=0}^{N-1} \Bigg(\langle \frac{\partial \ell}{\partial g}(g_k, \xi_k) - R^*_{g_k^{-1}} \mathrm{Ad}^*_{\Delta_k} \mathrm{d}\xi^*_{\Delta_k} \frac{\partial \ell}{\partial \xi}(g_k, \xi_k), \delta g_k \rangle + \langle R^*_{g_{k+1}^{-1}} \mathrm{d}\xi^*_{\Delta_k} \frac{\partial \ell}{\partial \xi}(g_k, \xi_k), \delta g_{k+1} \rangle \Bigg).$$

The DEL equations are thus

$$R^*_{g_k} \frac{\partial \ell}{\partial g}(g_k, \xi_k) - \mathrm{Ad}^*_{\Delta_k} \mathrm{d}\xi^*_{\Delta_k} \frac{\partial \ell}{\partial \xi}(g_k, \xi_k) + \mathrm{d}\xi^*_{\Delta_{k-1}} \frac{\partial \ell}{\partial \xi}(g_{k-1}, \xi_{k-1}) = 0, \quad 1 \le k \le N-1.$$

The trivialised discrete Legendre transformations can be seen as maps $\mathbb{F}L_d^{\pm} : G \times G \to G \times \mathfrak{g}^*$.

$$\mathbb{F}L_d^+(g, g') = (g', R^*_{g'} D_2 L_d(g, g')) = (g', \mathrm{d}\xi^*_{\Delta} \frac{\partial \ell}{\partial \xi}(g, \xi(\Delta))), \tag{36}$$

$$\mathbb{F}L_d^-(g, g') = (g, -R^*_g D_1 L_d(g, g')) = (g, -R^*_g \frac{\partial \ell}{\partial g}(g, \xi(\Delta)) + \mathrm{Ad}^*_{\Delta} \mathrm{d}\xi^*_{\Delta} \frac{\partial \ell}{\partial \xi}(g, \xi(\Delta))), \tag{37}$$

$$\Delta = g' g^{-1}. \tag{38}$$

One way of interpreting a method on the trivialised cotangent bundle $G \rtimes \mathfrak{g}^*$ is the following

1. Assume that (g_k, μ_k) is known. Then compute g_{k+1} by solving the equation
$$\mathbb{F}L_d^-(g_k, g_{k+1}) = \mu_k.$$
2. Next solve for μ_{k+1} explicitly by
$$\mu_{k+1} = \mathbb{F}L_d^+(g_k, g_{k+1}).$$

What we would like is to replace the occurrences of the Lagrangian $\ell(g, \xi)$ by the functions used in the RKMK formulation (30), (32). A plausible start is to define the variable $\bar{\mu}_k$ by
$$\bar{\mu}_k := \frac{\partial \ell}{\partial \xi}(g_k, \xi(g_{k+1}g_k^{-1})).$$

The continuous Legendre transformation (34) yields
$$\xi(g_{k+1}g_k^{-1}) = \xi(\Delta_k) = \iota(g_k, \bar{\mu}_k) = f_1(g_k, \bar{\mu}_k),$$
thus
$$g_{k+1} = \xi^{-1}(\iota(g_k, \bar{\mu}_k)) \cdot g_k.$$

To find an equation for $\bar{\mu}_k$ we need to consider $\mathbb{F}L_d^-(g_k, g_{k+1})$ as described in point 1 above..Note, again from (35) that
$$R_{g_k}^* \frac{\partial \ell}{\partial g}(g_k, \xi(\Delta_k)) = f_2(g_k, \bar{\mu}_k).$$

From (37) with $g = g_k$ and $g' = g_{k+1}$ we get
$$\mu_k = -f_2(g_k, \bar{\mu}_k) + \mathrm{Ad}_{\Delta_k}^* \, \mathrm{d}\xi_{\Delta_k}^* \bar{\mu}_k.$$

Finally, we need only use (36) to get μ_{k+1}. All equations for one step can be summarized as
$$\begin{aligned} g_{k+1} &= \xi^{-1}(f_1(g_k, \bar{\mu}_k)) \cdot g_k, \\ \mu_k &= -f_2(g_k, \bar{\mu}_k) + \mathrm{Ad}_{\Delta_k}^* \, \mathrm{d}\xi_{\Delta_k}^* \bar{\mu}_k, \\ \mu_{k+1} &= \mathrm{d}\xi_{\Delta_k}^* \bar{\mu}_k, \end{aligned}$$
where as before $\Delta_k = g_{k+1}g_k^{-1}$. We could use the map $\tau = \xi^{-1}$ instead, recalling that for $u = \xi(g)$ one has $\mathrm{d}\xi_g = (\mathrm{d}\xi_u^{-1})^{-1}$. One way of writing the resulting method would be
$$\begin{aligned} g_{k+1} &= \tau(f_1(g_k, \mathrm{d}\tau_{\xi_k}^* \mu_{k+1})) \cdot g_k, \\ \mu_{k+1} &= \mathrm{Ad}_{\Delta_k^{-1}}^*(\mu_k + f_2(g_k, \mathrm{d}\tau_{\xi_k}^* \mu_{k+1})), \end{aligned} \tag{39}$$

where $\xi_k = \xi(\Delta_k)$, i.e. $\Delta_k = \tau(\xi_k)$. It is already well-known that a scheme derived from such a variational principle leads to a symplectic method, see e.g. Marsden and West [47]. By replacing the discrete Lagrangian and action sum by other more advanced approximations, one can obtain various different variants of symplectic integrators on Lie groups, see e.g. [7, 20].

8 Preservation of First Integrals

There has been a significant interest over the last decades in constructing integrators which preserve one or more first integrals, such as energy or momentum. The reader who is interested in this topic should consult the pioneering paper by Gonzalez [30], but also McLachlan et al. [49] and the more recent work [67] and for preserving multiple first integrals simultaneously, see [26]. A key concept in integral preserving methods is that of discrete gradients, and in [23] these concepts were extended to Lie groups and retraction manifolds.

We shall begin by considering the case of a Lie group G and define a first integral to be any differentiable function $H : G \to \mathbb{R}$ which is invariant on solutions

$$\frac{\mathrm{d}}{\mathrm{d}t} H(y(t)) = \langle \mathrm{d}H, F\rangle = 0,$$

where we have introduced a duality pairing $\langle\cdot,\cdot\rangle$ between vector fields and one-forms. To any differential equation on a Lie group having a first integral H, there exists a bivector (dual two-form) ω such that

$$\dot{y} = F(y) = \omega(\mathrm{d}H, \cdot) = \mathrm{d}H \lrcorner\, \omega. \tag{40}$$

An explicit formula for ω can be given in the case when M is a Riemannian manifold. The gradient vector field is defined at the point x through the relation $\langle \mathrm{d}H, v\rangle_x = (\mathrm{grad}\, H|_x, v)_x$ for every $v \in T_xM$ where $(\cdot,\cdot)$ is the Riemannian inner product. An example of a bivector to be used in (40) is then given by

$$\omega = \frac{\mathrm{grad}\, H \wedge F}{\|\,\mathrm{grad}\, H\,\|^2}.$$

One should note that the bivector ω used to express the differential equation is not unique for a given ODE vector field F, but the choice of bivector will affect the resulting numerical method. The formulation (40) can easily be generalised to the case of k invariants, $H_1, \ldots, H_k$. In this case we replace the bivector by a $(k+1)$-vector and write the equation as

$$\dot{y} = F(y) = \omega(\mathrm{d}H_1, \ldots, \mathrm{d}H_k, \cdot),$$

and again, for Riemannian manifolds, we can define ω as

$$\omega = \frac{\omega_0 \wedge F}{\omega_0(\mathrm{d}H_1, \ldots, \mathrm{d}H_k)} \quad \text{where} \quad \omega_0 = \operatorname{grad} H_1 \wedge \cdots \wedge \operatorname{grad} H_k.$$

Integral preserving schemes on Lie groups

Let G be a Lie group with Lie algebra $\mathfrak{g}$, and define for each $g \in G$, the right multiplication operator $R_g : G \to G$ by $R_g(h) = h \cdot g$.

Definition 2 Let $H \in \mathscr{F}(G)$. We define the trivialised discrete differential $\bar{\mathrm{d}}H : G \times G \to \mathfrak{g}^*$ as any map that satisfies the conditions

$$\begin{aligned} H(v) - H(u) &= \langle \bar{\mathrm{d}}H(u, v), \log(v \cdot u^{-1}) \rangle, \\ \bar{\mathrm{d}}H(g, g) &= R_g^* \mathrm{d}H_g, \qquad \forall g \in G. \end{aligned}$$

We also need a trivialised approximation to the bivector ω in (40). For every pair of points $(u, v) \in G \times G$, we define an exterior 2-form on the linear space $\mathfrak{g}^*$, $\bar{\omega} : G \times G \to \Lambda^2(\mathfrak{g}^*)$, satisfying the consistency condition

$$\bar{\omega}(g, g)(R_g^*\alpha, R_g^*\beta) = \omega_g(\alpha, \beta), \qquad \forall g \in G, \quad \forall \alpha, \beta \in T_g^*G.$$

For practical purposes, $\bar{\omega}$ needs only to be defined in some suitable neighborhood of the diagonal subset $\{(g, g) : g \in G\}$. We can now write the numerical integral preserving method as

$$g^{i+1} = \exp(h\zeta(g^i, g^{i+1})) \cdot g^i, \qquad \zeta(g^i, g^{i+1}) = \bar{\mathrm{d}}H(g^i, g^{i+1}) \lrcorner \bar{\omega}(g^i, g^{i+1}).$$

That the integral H is exactly preserved is seen through the simple calculation

$$\begin{aligned} H(g^{i+1}) - H(g^i) &= \langle \bar{\mathrm{d}}H(g^i, g^{i+1}), \log(g^{i+1}(g^i)^{-1}) \rangle = h\langle \bar{\mathrm{d}}H(g^i, g^{i+1}), \zeta(g^i, g^{i+1}) \rangle \\ &= \bar{\omega}(g^i, g^{i+1})(\bar{\mathrm{d}}H(g^i, g^{i+1}), \bar{\mathrm{d}}H(g^i, g^{i+1})) = 0. \end{aligned}$$

Examples of trivialised discrete differentials

The first example has a counterpart on Euclidean space sometimes referred to as the *Averaged Vector Field* (AVF) gradient. It is defined as

$$\bar{\mathrm{d}}H(u, v) = \int_0^1 R_{\ell(\xi)}^* \mathrm{d}H_{\ell(\xi)} \, \mathrm{d}\xi, \quad \ell(\xi) = \exp(\xi \log(v \cdot u^{-1})). \tag{41}$$

Note that $\bar{\mathrm{d}}H(u, v) = \bar{\mathrm{d}}H(v, u)$. This trivialised discrete differential has the disadvantage that it can rarely be computed exactly a priori for general groups, although when G is Euclidean space it reduces to the standard AVF discrete gradient which has a wide range of applications.

Gonzalez [30] introduced a discrete gradient for Euclidean spaces which is often referred to as the midpoint discrete gradient. In the setting we use here, we need to introduce an inner product $(\cdot,\cdot)$ on the Lie algebra to generalise it to arbitrary Lie groups. We apply "index lowering" to any element $\eta \in \mathfrak{g}$ by defining $\eta^\flat \in \mathfrak{g}^*$ to be the unique element satisfying $\langle \eta^\flat, \zeta\rangle = (\eta, \zeta)$ for all $\zeta \in \mathfrak{g}$. We let

$$\bar{\mathrm{d}}H(u, v) = R_c^*\mathrm{d}H|_c + \frac{H(v) - H(u) - \langle R_c^*\mathrm{d}H|_c, \eta\rangle}{(\eta, \eta)}\eta^\flat, \quad \eta = \log(v \cdot u^{-1}), \tag{42}$$

where $c \in G$, is some point typically near u and v. One may for instance choose $c = \exp(\eta/2) \cdot u$, which implies symmetry, i.e. $\bar{\mathrm{d}}H(u, v) = \bar{\mathrm{d}}H(v, u)$.

Integral preserving schemes on a manifold with a retraction

What we present here is a basic and straightforward approach introduced in [23], but clearly there are other strategies that can be used. We use retractions as introduced on page xxx. Recall that the retraction restricted to the tangent space T_xM is denoted ϕ_x and is a diffeomorphism from some neighborhood $\mathscr{U}_x$ of $0_x \in T_xM$ into a subset $\mathscr{W}_x$ of M containing x. We tacitly assume the necessary restrictions on the integration step size h to ensure that both the initial and terminal points are contained in $\mathscr{W}_x$ for each time step. We also assume to have at our disposal a map c defined on some open subset of $M \times M$ containing all diagonal points (x, x), for which $c(x, y) \in M$. The differential equation is written in terms of a bivector ω and a first integral H as in (40). We introduce an approximate bivector $\bar{\omega}(x, y)$ such that

$$\bar{\omega}(x, x)(v, w) = \omega|_x, \quad \forall x \in M.$$

Contrary to the Lie group case we no longer assume a global trivialisation, so we introduce the *discrete differential* of a function H. To any pair of points $(x, y) \in M \times M$ we associate the covector $\bar{\mathrm{d}}H(x, y) \in T^*_{c(x,y)}M$ satisfying the relations

$$\begin{aligned} H(y) - H(x) &= \langle \bar{\mathrm{d}}H(x, y), \phi_c^{-1}(y) - \phi_c^{-1}(x)\rangle, \\ \bar{\mathrm{d}}H(x, x) &= \mathrm{d}H|_x, \quad \forall x \in M. \end{aligned}$$

where $c = c(x, y)$ is the map introduced above. The integrator on M is now defined as

$$y^{n+1} = \phi_c(W(y^n, y^{n+1})), \quad W(y^n, y^{n+1}) = \phi_c^{-1}(y^n) + h\,\bar{\mathrm{d}}H(y^n, y^{n+1}) \lrcorner\, \bar{\omega}(y^n, y^{n+1}).$$

One can easily see that this method is symmetric if the following three conditions are satisfied:

1. The map c is symmetric, i.e. $c(x, y) = c(y, x)$ for all x and y.
2. The discrete differential is symmetric in the sense that $\bar{\mathrm{d}}H(x, y) = \bar{\mathrm{d}}H(y, x)$.
3. The bivector $\bar{\omega}$ is symmetric in x and y: $\bar{\omega}(x, y) = \bar{\omega}(y, x)$.

The condition 1 can be achieved by solving the equation

$$\phi_c^{-1}(x) + \phi_c^{-1}(y) = 0, \tag{43}$$

with respect to c.

Both of the trivialised discrete differentials (41) and (42) have corresponding versions with retractions, in the former case, we write $\gamma_\xi = (1-\xi)v + \xi w$ where $x = \phi_c(v)$, $y = \phi_c(w)$. Then

$$\bar{\mathrm{d}}H(x, y) = \int_0^1 \phi_c^* \,\mathrm{d}H|_{\phi_c(\gamma_\xi)} \,\mathrm{d}\xi. \tag{44}$$

Similarly, assuming that M is Riemannian, we can define the following counterpart to the Gonzalez midpoint discrete gradient

$$\bar{\mathrm{d}}H(x, y) = \mathrm{d}H|_c + \frac{H(y) - H(x) - \langle \mathrm{d}H|_c, \eta\rangle}{(\eta, \eta)_c}\eta^\flat, \quad \eta = \phi_c^{-1}(y) - \phi_c^{-1}(x) \in T_cM, \tag{45}$$

where we may require that $c(x, y)$ satisfies (43) for the method to be symmetric. For clarity, we include an example taken from [23].

Example 3 We consider the sphere $M = S^{n-1}$ where we represent its points as vectors in $\mathbb{R}^n$ of unit length, $\|x\|_2 = 1$. The tangent space at x is then identified with the set of vectors in $\mathbb{R}^n$ orthogonal to x with respect to the Euclidean inner product $(\cdot, \cdot)$. A retraction is

$$\phi_x(v_x) = \frac{x + v_x}{\|x + v_x\|}, \tag{46}$$

its inverse is defined in the cone $\{y : (x, y) > 0\}$ where

$$\phi_x^{-1}(y) = \frac{y}{(x, y)} - x.$$

A symmetric map $c(x, y)$ satisfying (43) is simply

$$c(x, y) = \frac{x + y}{\|x + y\|_2}, \tag{47}$$

the geodesic midpoint between x and y in terms of the standard Riemannian metric on S^{n-1}. We compute the tangent map of the retraction to be

$$T_u\phi_c = \frac{1}{\|c + u\|_2}\left(I - \frac{(c + u) \otimes (c + u)}{\|c + u\|_2^2}\right).$$

As a toy problem, let us consider a mechanical system on S^2. Since the angular momentum in body coordinates for the free rigid body is of constant length, we may assume $(x, x) = 1$ for all x and we can model the problem as a dynamical system on the sphere. But in addition to this, the energy of the body is preserved, i.e.

$$H(x) = \frac{1}{2}(x, \mathbb{I}^{-1}x) = \frac{1}{2}\left(\frac{x_1^2}{\mathbb{I}_1} + \frac{x_2^2}{\mathbb{I}_2} + \frac{x_3^2}{\mathbb{I}_3}\right),$$

which we may take as the first integral to be preserved. Here the inertia tensor is $\mathbb{I} = \mathrm{diag}(\mathbb{I}_1, \mathbb{I}_2, \mathbb{I}_3)$. The system of differential equations can be written as follows

$$\frac{\mathrm{d}x}{\mathrm{d}t} = (\mathrm{d}H \lrcorner\, \omega)|_x = x \times \mathbb{I}^{-1}x,$$
$$\omega|_x(\alpha, \beta) = (x, \alpha \times \beta),$$

where the righthand side in both equations refers to the representation in $\mathbb{R}^3$. A symmetric consistent approximation to ω would be

$$\bar{\omega}(x, y)(\alpha, \beta) = \left(\frac{x+y}{2}, \alpha \times \beta\right).$$

We write $\ell_\xi = c + \gamma_\xi$ with the notation in (44), this is a linear function of the scalar argument ξ, and thus, $\phi_c(\gamma_\xi) = \ell_\xi / \|\ell_\xi\|$ from (46). We therefore derive for the AVF discrete gradient

$$\bar{\mathrm{d}}H(x, y) = \int_0^1 \frac{1}{\|\ell_\xi\|}\left(\mathbb{I}^{-1}\phi_c(\gamma_\xi) - (\phi_c(\gamma_\xi), \mathbb{I}^{-1}\phi_c(\gamma_\xi))\phi_c(\gamma_\xi)\right) \mathrm{d}\xi.$$

This integral is somewhat complicated to solve analytically. Instead, we may consider the discrete gradient (45) where we take as Riemannian metric the standard Euclidean inner product restricted to the tangent bundle of S^2. We obtain the following version of the discrete differential in the chosen representation

$$\bar{\mathrm{d}}H(x, y) = \frac{1}{\|m\|}\left(\mathbb{I}^{-1}m + \frac{\|m\|^2 - 1}{\|y - x\|^2}(H(y) - H(x))(y - x)\right), \quad m = \frac{x+y}{2}.$$

The corresponding method is symmetric, thus of second order.

References

1. Adler, R.L., Dedieu, J.P., Margulies, J.Y., Martens, M., Shub, M.: Newton's method on Riemannian manifolds and a geometric model for the human spine. IMA J. Numer. Anal. **22**(3), 359–390 (2002)
2. Berland, H., Owren, B.: Algebraic structures on ordered rooted trees and their significance to Lie group integrators. In: Group Theory and Numerical Analysis. CRM Proceedings and

Lecture Notes, vol. 39, pp. 49–63. American Mathematical Society, Providence, RI (2005)
3. Berland, H., Owren, B., Skaflestad, B.: B-series and order conditions for exponential integrators. SIAM J. Numer. Anal. **43**(4), 1715–1727 (2005) (electronic)
4. Blanes, S., Casas, F., Oteo, J.A., Ros, J.: The Magnus expansion and some of its applications. Phys. Rep. **470**(5–6), 151–238 (2009)
5. Bobenko, A.I., Suris, Yu.B.: Discrete Lagrangian reduction, discrete Euler-Poincaré equations, and semidirect products. Lett. Math. Phys. **49**(1), 79–93 (1999)
6. Bobenko, A.I., Suris, Yu.B.: Discrete time Lagrangian mechanics on Lie groups, with an application to the Lagrange top. Comm. Math. Phys. **204**(1), 147–188 (1999)
7. Bogfjellmo, G., Marthinsen, H.: High-Order Symplectic Partitioned Lie Group Methods. Foundations of Computational Mathematics, pp. 1–38 (2015)
8. Bou-Rabee, N., Marsden, J.E.: Hamilton-Pontryagin integrators on Lie groups. I. Introduction and structure-preserving properties. Found. Comput. Math. **9**(2), 197–219 (2009)
9. Bourbaki, N.: Lie Groups and Lie Algebras. Part I, Chapters 1–3, Addison-Wesley (1975)
10. Bras, S., Cunha, R., Silvestre, C.J., Oliveira, P.J.: Nonlinear attitude observer based on range and inertial measurements. IEEE Trans. Control Syst. Technol. **21**(5), 1889–1897 (2013)
11. Bruls, O., Cardona, A.: On the use of Lie group time integrators in multibody dynamics. J. Comput. Nonlinear Dyn. **5**(3), 031002 (2010)
12. Bryant, R.L.: An introduction to Lie groups and symplectic geometry. In: Freed, D.S., Uhlenbeck, K.K. (eds.) Geometry and Quantum Field Theory. IAS/Park City Mathematics Series, vol. 1, 2nd edn. American Mathematical Society (1995)
13. Calvo, M.P., Iserles, A., Zanna, A.: Runge-Kutta methods for orthogonal and isospectral flows. Appl. Numer. Math. **22**, 153–163 (1996)
14. Calvo, M.P., Iserles, A., Zanna, A.: Numerical solution of isospectral flows. Math. Comp. **66**(220), 1461–1486 (1997)
15. Casas, F., Owren, B.: Cost efficient Lie group integrators in the RKMK class. BIT Numer. Math. **43**(4), 723–742 (2003)
16. Celledoni, E.: A note on the numerical integration of the KdV equation via isospectral deformations. J. Phys. A: Math. Gener. **34**(11), 2205–2214 (2001)
17. Celledoni, E., Iserles, A.: Approximating the exponential from a Lie algebra to a Lie group. Math. Comp. (2000). Posted on March 15, PII S 0025-5718(00)01223-0 (to appear in print)
18. Celledoni, E., Iserles, A.: Methods for the approximation of the matrix exponential in a Lie-algebraic setting. IMA J. Numer. Anal. **21**(2), 463–488 (2001)
19. Celledoni, E., Marthinsen, A., Owren, B.: Commutator-free Lie group methods. Future Gener. Comput. Syst. **19**, 341–352 (2003)
20. Celledoni, E., Marthinsen, H., Owren, B.: An introduction to Lie group integrators—basics, new developments and applications. J. Comput. Phys. **257**(part B), 1040–1061 (2014)
21. Celledoni, E., Owren, B.: A class of intrinsic schemes for orthogonal integration. SIAM J. Numer. Anal. **40**(6), 2069–2084 (2002, 2003) (electronic)
22. Celledoni, E., Owren, B.: On the implementation of Lie group methods on the Stiefel manifold. Numer. Algorithms **32**(2–4), 163–183 (2003)
23. Celledoni, E., Owren, B.: Preserving first integrals with symmetric Lie group methods. Discrete Contin. Dyn. Syst. **34**(3), 977–990 (2014)
24. Chavel, I.: Riemannian Geometry: A Modern Introduction. Cambridge Tracts in Mathematics, vol. 108. Cambridge University Press (1993)
25. Crouch, P.E., Grossman, R.: Numerical integration of ordinary differential equations on manifolds. J. Nonlinear Sci. **3**, 1–33 (1993)
26. Dahlby, M., Owren, B., Yaguchi, T.: Preserving multiple first integrals by discrete gradients. J. Phys. A, **44**(30), 305205, 14 (2011)
27. Ebrahimi-Fard, K., Lundervold, A., Munthe-Kaas, H.: On the Lie enveloping algebra of a post-Lie algebra. J. Lie Theory **25**(4), 1139–1165 (2015)
28. Engø, K.: Partitioned Runge-Kutta methods in Lie-group setting. BIT **43**(1), 21–39 (2003)
29. Engø, K., Faltinsen, S.: Numerical integration of Lie-Poisson systems while preserving coadjoint orbits and energy. SIAM J. Numer. Anal. **39**(1), 128–145 (2001)

30. Gonzalez, O.: Time integration and discrete Hamiltonian systems. J. Nonlinear Sci. **6**, 449–467 (1996)
31. Grossman, R.L., Larson, R.G.: Differential algebra structures on families of trees. Adv. Appl. Math. **35**(1), 97–119 (2005)
32. Hochbruck, M., Ostermann, A.: Exponential integrators. Acta Numer. **19**, 209–286 (2010)
33. Humphreys, J.E.: Introduction to Lie Algebras and Representation Theory. Graduate Texts in Mathematics. Springer-Verlag (1972)
34. Iserles, A.: Magnus expansions and beyond. In: Combinatorics and Physics. Contemporary Mathematics, vol. 539, pp. 171–186. American Mathematical Society, Providence, RI (2011)
35. Iserles, A., Munthe-Kaas, H.Z., Nørsett, S.P., Zanna, A.: Lie-group methods. Acta Numerica **9**, 215–365 (2000)
36. Kobilarov, M., Crane, K., Desbrun, M.: Lie group integrators for animation and control of vehicles. ACM Trans. Graph. **28**(2) (2009)
37. Krogstad, S.: A low complexity Lie group method on the Stiefel manifold. BIT **43**(1), 107–122 (2003)
38. Krogstad, S., Munthe-Kaas, H., Zanna, A.: Generalized polar coordinates on Lie groups and numerical integrators. Numer. Math. **114**(1), 161–187 (2009)
39. Lee, T., Leok, M., McClamroch, N.H.: Lie group variational integrators for the full body problem. Comput. Methods Appl. Mech. Eng. **196**(29–30), 2907–2924 (2007)
40. Lewis, D., Olver, P.J.: Geometric integration algorithms on homogeneous manifolds. Found. Comput. Math. **2**(4), 363–392 (2002)
41. Lewis, D., Simo, J.C.: Conserving algorithms for the dynamics of Hamiltonian systems of Lie groups. J. Nonlinear Sci. **4**, 253–299 (1994)
42. Lundervold, A., Munthe-Kaas, H.: Hopf algebras of formal diffeomorphisms and numerical integration on manifolds. In: Combinatorics and Physics. Contemporary Mathematics, vol. 539, pp. 295–324. American Mathematical Society, Providence, RI (2011)
43. Malham, S.J.A., Wiese, A.: Stochastic Lie group integrators. SIAM J. Sci. Comput. **30**(2), 597–617 (2008)
44. Marsden, J.E., Pekarsky, S., Shkoller, S.: Discrete Euler-Poincaré and Lie-Poisson equations. Nonlinearity **12**(6), 1647–1662 (1999)
45. Marsden, J.E., Pekarsky, S., Shkoller, S.: Symmetry reduction of discrete Lagrangian mechanics on Lie groups. J. Geom. Phys. **36**(1–2), 140–151 (2000)
46. Marsden, J.E., Ratiu, T.S.: Introduction to Mechanics and Symmetry. Texts in Applied Mathematics, vol. 17, 2nd edn. Springer-Verlag (1999)
47. Marsden, J.E., West, M.: Discrete mechanics and variational integrators. Acta Numerica **10**, 357–514 (2001)
48. McLachlan, R., Modin, K., Verdier, O.: A minimal-variable symplectic integrator on spheres. Math. Comp. **86**(307), 2325–2344 (2017)
49. McLachlan, R.I., Quispel, G.R.W., Robidoux, N.: Geometric integration using discrete gradients. Phil. Trans. Royal Soc. A **357**, 1021–1046 (1999)
50. Moler, C., Van Loan, C.: Nineteen dubious ways to compute the exponential of a matrix. SIAM Rev. **20**(4), 801–836 (1978)
51. Moler, C., Van Loan, C.: Nineteen dubious ways to compute the exponential of a matrix, twenty-five years later. SIAM Rev. **45**(1), 3–49 (2003) (electronic)
52. Moser, J., Veselov, A.P.: Discrete versions of some classical integrable systems and factorization of matrix polynomials. Comm. Math. Phys. **139**(2), 217–243 (1991)
53. Munthe-Kaas, H.: Lie-Butcher theory for Runge-Kutta methods. BIT **35**(4), 572–587 (1995)
54. Munthe-Kaas, H.: Runge-Kutta methods on Lie groups. BIT **38**(1), 92–111 (1998)
55. Munthe-Kaas, H.: High order Runge-Kutta methods on manifolds. Appl. Numer. Math. **29**, 115–127 (1999)
56. Munthe-Kaas, H., Lundervold, A.: On post-Lie algebras, Lie-Butcher series and moving frames. Found. Comput. Math. **13**(4), 583–613 (2013)
57. Munthe-Kaas, H., Owren, B.: Computations in a free Lie algebra. Phil. Trans. Royal Soc. A **357**, 957–981 (1999)

58. Munthe-Kaas, H., Verdier, O.: Integrators on homogeneous spaces: isotropy choice and connections. Found. Comput. Math. **16**(4), 899–939 (2016)
59. Munthe-Kaas, H., Wright, W.M.: On the Hopf algebraic structure of Lie group integrators. Found. Comput. Math. **8**(2), 227–257 (2008)
60. Munthe-Kaas, H., Zanna, A.: Numerical integration of differential equations on homogeneous manifolds. In: Cucker, F., Shub, M. (eds.) Foundations of Computational Mathematics, pp. 305–315. Springer-Verlag (1997)
61. Munthe-Kaas, H.Z., Quispel, G.R.W., Zanna, A.: Symmetric spaces and Lie triple systems in numerical analysis of differential equations. BIT **54**(1), 257–282 (2014)
62. Murua, A.: Formal series and numerical integrators. I. Systems of ODEs and symplectic integrators. Appl. Numer. Math. **29**(2), 221–251 (1999)
63. Murua, A.: The Hopf algebra of rooted trees, free Lie algebras, and Lie series. Found. Comput. Math. **6**(4), 387–426 (2006)
64. Owren, B.: Order conditions for commutator-free Lie group methods. J. Phys. A **39**(19), 5585–5599 (2006)
65. Owren, B., Marthinsen, A.: Runge-Kutta methods adapted to manifolds and based on rigid frames. BIT **39**(1), 116–142 (1999)
66. Owren, B., Marthinsen, A.: Integration methods based on canonical coordinates of the second kind. Numer. Math. **87**(4), 763–790 (2001)
67. Quispel, G.R.W., McLaren, D.I.: A new class of energy-preserving numerical integration methods. J. Phys. A, **41**(4), 045206, 7 (2008)
68. Reutenauer, C.: Free Lie Algebras. London Mathematical Society Monographs. New Series, vol. 7. The Clarendon Press, Oxford University Press, Oxford Science Publications, New York (1993)
69. Rodrigues, O.: Des lois géometriques qui regissent les déplacements d' un systéme solide dans l' espace, et de la variation des coordonnées provenant de ces déplacement considérées indépendent des causes qui peuvent les produire. J. Math. Pures Appl. **5**, 380–440 (1840)
70. Saccon, A.: Midpoint rule for variational integrators on Lie groups. Int. J. Numer. Methods Eng. **78**(11), 1345–1364 (2009)
71. Terze, Z., Muller, A., Zlatar, D.: Lie-group integration method for constrained multibody systems in state space. Multibody Syst. Dyn. **34**(3), 275–305 (2015)
72. Veselov, A.P.: Integrable discrete-time systems and difference operators. Funct. Anal. Appl. **22**, 83–93 (1988)
73. Wensch, J., Knoth, O., Galant, A.: Multirate infinitesimal step methods for atmospheric flow simulation. BIT **49**(2), 449–473 (2009)

Lie–Butcher Series, Geometry, Algebra and Computation

Hans Z. Munthe-Kaas and Kristoffer K. Føllesdal

Abstract Lie–Butcher (LB) series are formal power series expressed in terms of trees and forests. On the geometric side LB-series generalizes classical B-series from Euclidean spaces to Lie groups and homogeneous manifolds. On the algebraic side, B-series are based on pre-Lie algebras and the Butcher-Connes-Kreimer Hopf algebra. The LB-series are instead based on post-Lie algebras and their enveloping algebras. Over the last decade the algebraic theory of LB-series has matured. The purpose of this paper is twofold. First, we aim at presenting the algebraic structures underlying LB series in a concise and self contained manner. Secondly, we review a number of algebraic operations on LB-series found in the literature, and reformulate these as recursive formulae. This is part of an ongoing effort to create an extensive software library for computations in LB-series and B-series in the programming language Haskell.

Keywords B-series · Lie–Butcher series · Post-Lie algebra · Pre-Lie algebra

MSC 16T05 · 17B99 · 17D99 · 65D30

1 Introduction

Classical B-series are formal power series expressed in terms of rooted trees (connected graphs without any cycle and a designated node called the root). The theory has its origins back to the work of Arthur Cayley [5] in the 1850s, where he realized

H. Z. Munthe-Kaas (✉) · K. K. Føllesdal
Department of Mathematics, University of Bergen, P.O. Box 7803, 5020 Bergen, Norway
e-mail: hans.munthe-kaas@uib.no

K. K. Føllesdal
e-mail: kristoffer.follesdal@uib.no

K. Ebrahimi-Fard and M. Barbero Liñán (eds.), *Discrete Mechanics, Geometric Integration and Lie–Butcher Series*, Springer Proceedings in Mathematics & Statistics 267, https://doi.org/10.1007/978-3-030-01397-4_3

that trees could be used to encode information about differential operators. Being forgotten for a century, the theory was revived through the efforts of understanding numerical integration algorithms by John Butcher in the 1960s and '70s [2, 3]. Ernst Hairer and Gerhard Wanner [14] coined the term *B-series* for an infinite formal series of the form

$$B_f(\alpha, y, t) := y + \sum_{\tau \in T} \frac{t^{|\tau|}}{\sigma(\tau)} \langle a, \tau \rangle \mathscr{F}_f(\tau)(y),$$

where $y \in \mathbb{R}^n$ is a 'base' point, $f\colon \mathbb{R}^n \to \mathbb{R}^n$ is a given vector field, $T = \{\bullet,$  $\ldots\}$ is the set of rooted trees, $|\tau|$ is the number of nodes in the tree, $\alpha\colon T \to \mathbb{R}$ are the coefficients of a given series and $\langle \alpha, \tau \rangle \in \mathbb{R}$ denotes evaluation of α at τ. The bracket hints that we later want to consider $\langle \alpha, \cdot \rangle$ as a linear functional on the vector space spanned by T. The animal $\mathscr{F}_f(\tau)\colon \mathbb{R}^n \to \mathbb{R}^n$ denotes special vector fields, called *elementary differentials*, which can be expressed in terms of partial derivatives of f. The coefficient $\sigma(\tau) \in \mathbb{N}$ is counting the number of symmetries in a given tree. This symmetry factor could have been subsumed into α, but is explicitly taken into the series due to the underlying algebraic structures, where this factor comes naturally. The B-series $t \mapsto B_f(\alpha, y, t)$ can be interpreted as a curve starting in y. By choosing different functions α, one may encode both the analytical solution of a differential equation $y'(t) = f(y(t))$ and also various numerical approximations of the solution.

During the 1980s and 1990s B-series evolved into an indispensable tool in analysis of numerical integration for differential equations evolving on $\mathbb{R}^n$. In the mid-1990s interest rose in the construction of numerical integration on Lie groups and manifolds [15, 17], and from this a need to interpret B-series type expansions in a differential geometric context, giving birth to *Lie–Butcher series* (LB-series), which combines B-series with Lie series on manifolds. It is necessary to make some modifications to the definition of the series to be interpreted geometrically on manifolds:

- We cannot add a point and a tangent vector as in $y + \mathscr{F}_f(\tau)$. Furthermore, it turns out to be very useful to regard the series as a Taylor-type series for the mapping $f \mapsto B_f$, rather than a series development of a curve $t \mapsto B_f(a, y, t)$. The target space of $f \mapsto B_f$ is differential operators, and we can remove explicit reference to the base point y from the series.
- The mapping $f \mapsto B_f$ inputs a vector field and outputs a series which may represent either a vector field or a solution operator (flow map). Flow maps are expressed as a series in higher order differential operators. We will see that trees encode first order differential operators. Higher order differential operators are encoded by products of trees, called *forests*. We want to also consider series in linear combinations of forests.
- We will in the sequel see that the elementary differential map $\tau \mapsto \mathscr{F}_f(\tau)$ is a *universal arrow* in a particular type of algebras. The existence of such a uniquely defined map expresses the fact that the vector space spanned by trees (with certain

algebraic operations) is a universal object in this category of algebras. Thus the trees encode faithfully the given algebraic structure. We will see that the algebra comes naturally from the geometric properties of a given *connection* (covariant derivation) on the manifold. For Lie groups the algebra of the natural connection is encoded by *ordered* rooted trees, where the ordering of the branches is important. The ordering is related to a non-vanishing torsion of the connection.

- The symmetry factor $\sigma(\tau)$ in the classical B-series is related to the fact that several different ordered trees correspond to the same unordered tree. This factor is absent in the Lie–Butcher series.
- The time parameter t is not essential for the algebraic properties of the series. Since $\mathscr{F}_{tf}(\tau) = t^{|\tau|}\mathscr{F}_f(\tau)$, we can recover the time factor through the substitution $f \mapsto tf$.

We arrive at the definition of an abstract Lie–Butcher series simply as

$$\sum_{\omega \in \mathrm{OF}} \langle \alpha, \omega \rangle \omega, \tag{1}$$

where

$\mathrm{OF} = \{\mathbb{I},$  $\}$

denotes the set of all ordered forests of ordered trees, $\mathbb{I}$ is the empty forest, and $\alpha\colon \mathrm{OF} \to \mathbb{R}$ are the coefficients of the series. This abstract series can be mapped down to a concrete algebra (e.g. an algebra of differential operators on a manifold) by a universal mapping $\omega \mapsto \mathscr{F}_f(\omega)$.

We can identify the function $\alpha\colon \mathrm{OF} \to \mathbb{R}$ with its series (1) and say that a Lie–Butcher series α is an element of the completion of the free vector space spanned by the forests of ordered rooted trees. However, to make sense of this statement, we have to attach algebraic and geometric meaning to the vector space of ordered forests. This is precisely explained in the sequel, where we see that the fundamental algebraic structures of this space arise because it is the universal enveloping algebra of a free post-Lie algebra. Hence we arrive at the precise definition:

An abstract Lie–Butcher series is an element of the completion of the enveloping algebra of the free post-Lie algebra.

We will in this paper present the basic geometric and algebraic structures behind LB-series in a self contained manner. Furthermore, an important goal for this work is to prepare a software package for computations on these structures. For this purpose we have chosen to present all the algebraic operations by recursive formulae, ideally suited for implementation in a functional programming language. We are in the process of implementing this package in the Haskell programming language. The implementation is still at a quite early stage, so a detailed presentation of the implementation will be reported later.

2 Geometry of Lie–Butcher Series

B-series and LB-series can both be viewed as series expansions in a connection on a fibre bundle, where B-series are derived from the canonical (flat and torsion free) connection on $\mathbb{R}^n$ and LB-series from a flat connection with constant torsion on a fibre bundle. Rather than pursuing this idea in an abstract general form, we will provide insight through the discussion of concrete and important examples.

2.1 *Parallel Transport*

Let M be a manifold, $\mathscr{F}(M)$ the set of smooth $\mathbb{R}$-valued scalar functions and $\mathfrak{X}(M)$ the set of real vector fields on M. For $t \in \mathbb{R}$ and $f \in \mathfrak{X}(M)$ let $\Psi_{t,f}\colon M \to M$ denote the solution operator such that the differential equation $\gamma'(t) = f(\gamma(t))$, $\gamma(0) = p \in M$ has solution $\gamma(t) = \Psi_{t,f}(p)$. For $\phi \in \mathscr{F}(M)$ we define *pullback along the flow* $\Psi^*_{t,f}\colon \mathscr{F}(M) \to \mathscr{F}(M)$ as

$$\Psi^*_{t,f}\phi = \phi \circ \Psi_{t,f}.$$

The directional derivative $f(\phi) \in \mathscr{F}(M)$ is defined as

$$f(\phi) = \left.\frac{d}{dt}\right|_{t=0} \Psi^*_{t,f}\phi.$$

Through this, we identify $\mathfrak{X}(M)$ with the first order derivations of $\mathscr{F}(M)$, and we obtain higher order derivations by iterating, i.e. $f * f$ is the second order derivation $f * f(\phi) := f(f(\phi))$. With $\mathbb{I}\phi = \phi$ being the 0-order identity operator, the set of all higher order differential operators on $\mathscr{F}(M)$ is called the *universal enveloping algebra* $U(\mathfrak{X}(M))$. This is an algebra with an associative product $*$. The pullback satisfies

$$\frac{d}{dt}\Psi^*_{t,f}\phi = \Psi^*_{t,f} f(\phi).$$

By iteration we find that $\left.\frac{d^n}{dt^n}\right|_{t=0} \Psi^*_{t,f}\phi = f(f(\cdots f(\phi))) = f^{*n}(\phi)$ and hence the Taylor expansion of the pullback is

$$\Psi^*_{t,f}\phi = \phi + tf(\phi) + \frac{t^2}{2!} f * f(\phi) + \cdots = \exp^*(tf)(\phi), \tag{2}$$

where we define the exponential as

$$\exp^*(tf) := \sum_{j=0}^{\infty} \frac{t^j}{j!} f^{*j}.$$

This exponential is an element of $U(\mathfrak{X}(M))$, or more correctly, since it is an infinite series, in the completion of this algebra. We can recover the flow $\Psi_{t,f}$ from $\exp^*(tf)$ by letting ϕ be the coordinate maps. However, some caution must be exercised, since pullbacks compose contravariantly $\left(\Psi_{t,f} \circ \Psi_{s,g}\right)^* = \Psi^*_{s,g} \circ \Psi^*_{t,f}$, we have that $\exp^*(sg) * \exp^*(tf)$ corresponds to the diffeomorphism $\Psi_{t,f} \circ \Psi_{s,g}$.

Numerical integrators are constructed by sampling a vector field in points near a base point. To understand this process, we need to transport vector fields. Pullback of vector fields is, however, less canonical than of scalar functions. The differential geometric concept of parallel transport of vectors is defined in terms of a *connection*. An affine connection is a $\mathbb{Z}$-bilinear mapping $\rhd \colon \mathfrak{X}(M) \times \mathfrak{X}(M) \to \mathfrak{X}(M)$ such that

$$\begin{aligned}(\phi f) \rhd g &= \phi(f \rhd g)\\ f \rhd (\phi g) &= f(\phi)g + \phi(f \rhd g)\end{aligned}$$

for all $f, g \in \mathfrak{X}(M)$ and $\phi \in \mathscr{F}(M)$. Note that the standard notation for a connection in differential geometry is is $\nabla_f g \equiv f \rhd g$. Our notation is chosen to emphasise the operation as a binary product on the set of vector fields. The triangle notation looks nicer when we iterate, such as in (3) below. Furthermore, the triangle notation is also standard in much of the algebraic literature on pre-Lie algebras, as well as in several recent works on post-Lie algebras.

There is an intimate relationship between connections and the concept of parallel transport. For a curve $\gamma(t) \in M$, let $\Gamma(\gamma)_s^t$ denote *parallel transport along* $\gamma(t)$, meaning that

- $\Gamma(\gamma)_s^t \colon TM_{\gamma(s)} \to TM_{\gamma(t)}$ is a linear isomorphism of the tangent spaces.
- $\Gamma(\gamma)_s^s = \mathrm{Id}$, the identity map.
- $\Gamma(\gamma)_t^u \circ \Gamma(\gamma)_s^t = \Gamma(\gamma)_s^u$.
- Γ depends smoothly on s, t and γ.

From Γ, let us consider the action of *parallel transport pullback* of vector fields, for $t \in \mathbb{R}$ and $f \in \mathfrak{X}(M)$ we denote $\Psi^*_{t,f} \colon \mathfrak{X}(M) \to \mathfrak{X}(M)$ the operation

$$\Psi^*_{t,f}g(p) := \Gamma(\gamma)_t^0 g(\gamma(t)), \quad \text{for the curve } \gamma(t) = \Psi_{t,f}(p).$$

Any connection can be obtained from a parallel transport as the rate of change of the parallel transport pullback. For a given Γ we can define a corresponding connection as

$$f \rhd g := \left.\frac{d}{dt}\right|_{t=0} \Psi^*_{t,f}g.$$

Conversely, we can recover Γ from $\rhd$ by solving a differential equation. We seek a power series expansion of the parallel transport pullback. Just like the case of scalars, it holds also for pullback of vector fields that

$$\frac{\partial}{\partial t}\Psi^*_{t,f}g = \Psi^*_{t,f} f \rhd g,$$

hence we obtain the following Taylor series of the pullback

$$\Psi_{t,f}^{*}g = g + tf \rhd g + \frac{t^2}{2} f \rhd (f \rhd g) + \frac{t^3}{3!} f \rhd (f \rhd (f \rhd g)) + \cdots. \quad (3)$$

Recall that in the case of pullback of a scalar function, we used $f(g(\phi)) = (f * g)(\phi)$ to express the pull-back in terms of $\exp^*(tf)$. Whether or not we can do similarly for vector fields depends on geometric properties of the connection. We would like to extend $\rhd$ from $\mathfrak{X}(M)$ to $U(\mathfrak{X}(M))$ such that $f \rhd (g \rhd h) = (f * g) \rhd h$ and hence (3) becomes $\Psi_{t,f}^{*}g = \exp^*(tf) \rhd g$. However, this requires that $f \rhd (g \rhd h) - g \rhd (f \rhd h) = [\![f, g]\!] \rhd h$, where $[\![f, g]\!] := f * g - g * f$ is the Jacobi bracket of vector fields. The *curvature tensor* of the connection $R\colon \mathfrak{X}(M) \wedge \mathfrak{X}(M) \to \mathrm{End}(\mathfrak{X}(M))$ is defined as

$$R(f, g)h := f \rhd (g \rhd h) - g \rhd (f \rhd h) - [\![f, g]\!] \rhd h.$$

Thus, we only expect to find a suitable extension of $\rhd$ to $U(\mathfrak{X}(M))$ if $\rhd$ is *flat*, i.e. when $R = 0$.

In addition to the curvature, the other important tensor related to a connection is the torsion. Given $\rhd$, we define an $\mathscr{F}(M)$-bilinear mapping $\cdot\colon \mathfrak{X}(M) \times \mathfrak{X}(M) \to U(\mathfrak{X}(M))$ as

$$f \cdot g := f * g - f \rhd g. \quad (4)$$

The skew-symmetrisation of this product called the *torsion*

$$T(f, g) := g \cdot f - f \cdot g \in \mathfrak{X}(M),$$

and if $f \cdot g = g \cdot f$ we say that $\rhd$ is *torsion free*.

The standard connection on $\mathbb{R}^n$ is flat and torsion free. In this case the algebra $\{\mathfrak{X}(M), \rhd\}$ forms a *pre-Lie* algebra (defined below). This gives rise to classical B-series. More generally, transport by left or right multiplication on a Lie group yields a flat connection where the product $\cdot$ is associative, but not commutative. The resulting algebra is called *post-Lie* and the series are called *Lie–Butcher series*. A third important example is the Levi–Civita connection on a symmetric space, where $\cdot$ is a Jordan product, $T = 0$ and R is constant, non-zero. This third case is the subject of forthcoming papers, but will not be discussed here.

2.2 The Flat Cartan Connection on a Lie Group

Let G be a Lie group with Lie algebra $\mathfrak{g}$. For $V \in \mathfrak{g}$ and $p \in G$ we let $Vp := TR_pV \in T_pG$. There is a 1–1 correspondence between functions $f \in C^\infty(G, \mathfrak{g})$ and vector fields $\xi_f \in \mathfrak{X}(G)$ given as $\xi_f(p) = f(p)p$. Left multiplication with $q \in G$ gives rise to a parallel transport

$$\Gamma_q\colon T_pG \to T_{qp}G\colon Vp \mapsto Vqp.$$

This transport is independent of the path between p and qp and hence gives rise to a flat connection. We express the corresponding parallel transport pullback on the space $C^\infty(G, \mathfrak{g})$ as

$$(\Gamma_q^* f)(p) = f(qp)$$

which yields the flat connection

$$(f \rhd g)(q) = \left.\frac{d}{dt}\right|_{t=0} g(\exp(tf(q))q).$$

The torsion is given as [21]

$$T(f, g)(p) = -[f(p), g(p)]_{\mathfrak{g}}.$$

The two operations $f \rhd g$ and $[f, g] := -T(f, g)$ turn $C^\infty(G, \mathfrak{g})$ into a *post-Lie algebra*, see Definition 3 below. This is the foundation of Lie–Butcher series.

We can alternatively express the connection and torsion on $\mathfrak{X}(G)$ via a basis $\{E_j\}$ for $\mathfrak{g}$. Let $\partial_j \in \mathfrak{X}(G)$ be the right invariant vector field $\partial_j(p) = E_j p$. For $F, G \in \mathfrak{X}(G)$, where $F = f^i\partial_i$, $G = g^j\partial_j$ [1] and $f^i, g^j \in \mathscr{F}(G)$, we have

$$\begin{aligned} F \rhd G &= f^i\partial_i(g^j)\partial_j \\ F \cdot G &= f^i g^j \partial_i\partial_j \\ T(F, G) &= f^i g^j(\partial_i\partial_j - \partial_j\partial_i). \end{aligned}$$

We return to $\rhd$ defined on $C^\infty(G, \mathfrak{g})$. Let $U(\mathfrak{g})$ be the span of the basis $\{E_{j_1}E_{j_2}\cdots E_{j_k}\}$, where $E_{j_1}E_{j_2}\cdots E_{j_k} \in U(\mathfrak{g})$ corresponds to the right invariant k-th order differential operator $\partial_{j_k}\cdots\partial_{j_2}\partial_{j_1} \in U(\mathfrak{X}(G))$. On $U(\mathfrak{g})$ we have two different associative products, the composition of differential operators $f * g$ and the 'concatenation product' $f \cdot g = f * g - f \rhd g$ which is computed as the concatenation of the basis, $f^i E_i \cdot g^j E_j = f^i g^j E_i E_j$. The general relationship between these two products and $\rhd$ extended to $U(\mathfrak{g})$ is given in (28)–(31) below. In particular we have

$$f \rhd (g \rhd h) = (f * g) \rhd h,$$

which yields the exponential form of the parallel transport

$$\Psi_{t,f}^* g = \exp^*(tf) \rhd g,$$

where $\exp^*(tf)$ is giving us the exact flow of f.

---

[1] Einstein summation convention.

We can also form the exponential with respect to the other product,

$$\exp^{\cdot}(tf) = I + tf + \frac{t^2}{2} f \cdot f + \frac{t^3}{3!} f \cdot f \cdot f + \cdots .$$

What is the geometric meaning of this? We say that a vector field g is *parallel along* f if the parallel transport pullback of g along the flow of f is constant, and we say that g is *absolutely parallel* if it is constant under *any* parallel transport. Infinitesimally we have that g is parallel along f if $f \rhd g = 0$ and g is absolutely parallel if $f \rhd g = 0$ for all f. In $C^\infty(G, \mathfrak{g})$ the absolutely parallel functions are constants $g(p) = V$, which correspond to right invariant vector fields $\xi_g \in \mathfrak{X}(G)$ given as $\xi_g(p) = Vp$. The flow of parallel vector fields are the geodesics of the connection. If g is absolutely parallel, we have $g * g = g \cdot g + g \rhd g = g \cdot g$, and more generally $g^{n*} = g^{n\cdot}$, hence $\exp^*(g) = \exp^{\cdot}(g)$. If $f(p) = g(p)$ at a point $p \in G$, then they define the same tangent at the point. Hence $f^{n\cdot}(p) = g^{n\cdot}(p)$ for all n, and we conclude that $\exp^{\cdot}(f)(p) = \exp^{\cdot}(g)(p) = \exp^*(g)(p)$. Thus, the concatenation exponential $\exp^{\cdot}(f)$ of a general vector field f produces the flow which in a given point follows the geodesic tangent to f at the given point.

On a Lie group, we have for two arbitrary vector fields represented by general functions $f, g \in C^\infty(G, \mathfrak{g})$ that

$$(\exp^{\cdot}(tf) \rhd g)(p) = g\left(\exp(tf(p))p\right). \tag{5}$$

2.3 Numerical Integration

Lie–Butcher series and its cousins are general mathematical tools with applications in numerics, stochastics and renormalisation. The problem of numerical integration on manifolds is a particular application which has been an important source of inspiration. We discuss a simple illustrative example.

Example 1 (*Lie–trapezoidal method*) Consider the classical trapezoidal method. For a differential equation $y'(t) = f(y(t))$, $y(0) = y_0$ on $\mathbb{R}^n$ a step from $t = 0$ to $t = h$ is given as

$$\begin{aligned} K &= \frac{h}{2}\left(f(y_0) + f(y_1)\right) \\ y_1 &= y_0 + K. \end{aligned}$$

Consider a curve $y(t) \in G$ evolving on a Lie group such that $y'(t) = f(y(t))y(t)$, where $f \in C^\infty(G, \mathfrak{g})$ and $y(0) = y_0$. In the Lie-trapezoidal integrator a step from y_0 to $y_1 \approx y(h)$ is given as

$$K = \frac{h}{2}\left(f(y_0) + f(y_1)\right)$$
$$y_1 = \exp_{\mathfrak{g}}(K)y_0,$$

where $\exp_{\mathfrak{g}} : \mathfrak{g} \to G$ is the classical Lie group exponential. We can write the method as a mapping $\Phi_{\text{trap}} : \mathfrak{X}(M) \to \text{Diff}(G)$ from vector fields to diffeomorphisms on G, given in terms of parallel transport on $\mathfrak{X}(M)$ as

$$K = \frac{1}{2}\left(f + \exp^{\cdot}(K) \rhd f\right) \tag{6}$$
$$\Phi_{\text{trap}}(f) := \exp^{\cdot}(K). \tag{7}$$

To simplify, we have removed the timestep h, but this can be recovered by the substitution $f \mapsto hf$. Note that we present this as a process in $U(\mathfrak{X}(M))$, without a reference to a given base point y_0. The method computes a diffeomorphism $\Phi_{\text{trap}}(f)$, which can be evaluated on a given base point y_0. This absence of an explicit base point facilitates an interpretation of the method as a process in the enveloping algebra of a free post-Lie algebra, an abstract model of $U(\mathfrak{X}(M))$ to be discussed in the sequel.

A basic problem of numerical integration is to understand in what sense a numerical method $\Phi(tf)$ approximates the exact flow $\exp^*(tf)$. The *order* of the approximation is computed by comparing the LB-series expansion of $\Phi(tf)$ and $\exp^*(tf)$, and comparing to which order in t the two series agree.

The *backward error* of the method is defined as a modified vector field $\widetilde{f}_h$ such that the exact flow of $\widetilde{f}_h$ interpolates the numerical solution at integer times.[2] The combinatorial definition of the backward error is

$$\exp^*(\widetilde{f}_h) = \Phi(hf).$$

The backward error is an important tool which yields important structural information of the numerical flow operator $f \mapsto \Phi(hf)$. The backward error analysis is fundamental in the study of geometric properties of numerical integration algorithms [8, 13].

Yet another problem is the numerical technique of *processing* a vector field, i.e. we seek a modified vector field $\widetilde{f}_h$ such that $\Phi(\widetilde{f}_h) = \exp^*(f)$. An important tool in the analysis of this technique is the characterization of a *substitution law*. What happens to the series expansion of $\Phi(hf)$ if f is replaced by a modified vector field $\widetilde{f}_h$ expressed in terms of a series expansion involving f?

The purpose of this essay is not to pursue a detailed discussion of numerical analysis of integration schemes. Instead we want to introduce the algebraic structures needed to formalize the structure of the series expansions. In particular we will present recursive formulas for the basic algebraic operations suitable for computer implementations.

---

[2]Technical issues about divergence of the backward error vector field is discussed in [1].

We finally remark that numerical integrators are typically defined as *families of mappings*, given in terms of unspecified coefficients. For example the Runge–Kutta family of integrators can be defined in terms of real coefficients $\{a_{i,j}\}_{i,j=1}^s$ and $\{b_j\}_{j=1}^s$ as

$$K_i = \exp^{\cdot}(\sum_{j=1}^{s} a_{i,j}K_j) \rhd f, \quad \text{for } i = 1, \ldots, s$$

$$\Phi_{\mathrm{RK}}(f) = \exp^{\cdot}(\sum_{j=1}^{s} b_j K_j).$$

In a computer package for computing with LB-series we want the possibility of computing series expansions of such parametrized families without specifying the coefficients. This is accomplished by defining the algebraic structures not over the concrete field of real numbers $\mathbb{R}$, but instead allowing this to be replaced by an abstract commutativ ring with unit, such as e.g. the ring of all real polynomials in the indeterminates $\{a_{i,j}\}_{i,j=1}^s$ and $\{b_j\}_{j=1}^s$.

3 Algebraic Structures of Lie–Butcher Theory

We give a concise summary of the basic algebraic structures behind Lie–Butcher series.

3.1 Algebras

All vector spaces we consider are over a field[3] k of characteristic 0, e.g. $\mathrm{k} \in \{\mathbb{R}, \mathbb{C}\}$.

Definition 1 (*Algebra*) An algebra $\{\mathscr{A}, *\}$ is a vector space $\mathscr{A}$ with a k-bilinear operation $*\colon \mathscr{A} \times \mathscr{A} \to \mathscr{A}$. $\mathscr{A}$ is called *unital* if it has a unit $\mathbb{I}$ such that $x * \mathbb{I} = \mathbb{I} * x$ for all $x \in \mathscr{A}$. The (minus-)*associator* of the product is defined as

$$a_*(x, y, z) := x * (y * z) - (x * y) * z.$$

If the associator is 0, the algebra is called *associative*.

[3]In the computer implementations we are relaxing this to allow k more generally to be a commutative ring, such as e.g. polynomials in a set of indeterminates. In this latter case the k-vector space should instead be called a free k-module. We will not pursue this detail in this exposition.

Definition 2 (*Lie algebra*) A *Lie-algebra* is an algebra $\{\mathfrak{g}, [\cdot, \cdot]\}$ such that

$$[x, y] = -[y, x]$$
$$[[x, y], z] + [[y, z], x] + [[z, x], y] = 0.$$

The bracket $[\cdot, \cdot]$ is called the *commutator* or *Lie bracket*. An associative algebra $\{\mathscr{A}, *\}$ give rise to a Lie algebra Lie($\mathscr{A}$), where $[x, y] = x * y - y * x$.

A connection on a fibre bundle which is flat and with constant torsion satisfies the algebraic conditions of a *post-Lie algebra* [21]. This algebraic structure first appeared in a purely operadic setting in [27].

Definition 3 (*Post-Lie algebra*) A *post-Lie algebra* $\{\mathscr{P}, [\cdot, \cdot], \triangleright\}$ is a Lie algebra $\{\mathscr{P}, [\cdot, \cdot]\}$ together with a bilinear operation $\triangleright : \mathscr{P} \times \mathscr{P} \to \mathscr{P}$ such that

$$x \triangleright [y, z] = [x \triangleright y, z] + [x, y \triangleright z] \tag{8}$$

$$[x, y] \triangleright z = a_{\triangleright}(x, y, z) - a_{\triangleright}(y, x, z). \tag{9}$$

A post-Lie algebra defines a relationship between *two* Lie algebras [21].

Lemma 1 *For a post-Lie algebra* $\mathscr{P}$ *the bi-linear operation*

$$[\![x, y]\!] = x \triangleright y - y \triangleright x + [x, y] \tag{10}$$

defines another Lie bracket.

Thus, we have two Lie algebras $\mathfrak{g} = \{\mathscr{P}, [\cdot, \cdot]\}$ and $\overline{\mathfrak{g}} = \{\mathscr{P}, [\![\cdot, \cdot]\!]\}$ related by $\triangleright$.

Definition 4 (*Pre-Lie algebra*) A *pre-Lie algebra* $\{\mathscr{L}, \triangleright\}$ is a post-Lie algebra where $[\cdot, \cdot] \equiv 0$, in other words an algebra such that

$$a_{\triangleright}(x, y, z) = a_{\triangleright}(y, x, z).$$

Pre- and post-Lie algebras appear naturally in differential geometry where post-Lie algebras are intimately linked with the differential geometry of Lie groups and pre-Lie algebras with Abelian Lie groups (Euclidean spaces).

3.2 *Morphisms and Free Objects*

All algebras of a given type form a *category*, which can be thought of as a directed graph where each node (object) represents an algebra of the given type and the arrows (edges) represent morphisms. Any composition of morphisms is again a morphism. Morphisms are mappings preserving the given algebraic structure. E.g. an algebra morphism $\phi\colon \mathscr{A} \to \mathscr{A}'$ is a k-linear map satisfying $\phi(x * y) = \phi(x) * \phi(y)$. A post-Lie morphism is, similarly, a linear mapping $\phi\colon \mathscr{P} \to \mathscr{P}'$ satisfying both $\phi([x, y]) = [\phi(x), \phi(y)]$ and $\phi(x \rhd y) = \phi(x) \rhd \phi(y)$.

In a given category a *free object over a set C* can informally be thought of as a generic algebraic structure. The only equations that hold between elements of the free object are those that follow from the defining axioms of the algebraic structure. Furthermore the free object is not larger than strictly necessary to be generic. Each of the elements of C correspond to generators of the free object. In software a free object can be thought of as a symbolic computing engine; formulas, identities and algebraic simplifications derived within the free object can be applied to any other object in the category. Thus, a detailed understanding of the free objects is crucial for the computer implementation of a given algebraic structure.

Definition 5 (*Free object over a set C*) In a given category we define[4] the free object over a set C as an object $\mathrm{Free}(C)$ together with a map $\mathrm{inj}\colon C \hookrightarrow \mathrm{Free}(C)$, called the canonical injection, such that for any object B in the category and any mapping $\phi\colon C \to B$ there exists a unique morphism $!\colon \mathrm{Free}(C) \to B$ such that the diagram commutes

$$\begin{array}{ccc} C & \xrightarrow{\mathrm{inj}} & \mathrm{Free}(C) \\ & \underset{\phi}{\searrow} & \downarrow{!} \\ & & B \end{array} \quad . \tag{11}$$

We will often consider $C \subset \mathrm{Free}(C)$ without mentioning the map inj.

Note 1 In category theory a free functor is intimately related to a *monad*, a concept which is central in the programming language Haskell. In Haskell the function "inj" is called "return" and the application of ! on $x \in \mathrm{Free}(C)$ is written $x >== \phi$.

A free object can be implemented in different ways, but different implementations are always algebraically isomorphic.

Example 2 **Free k-vector space** $\mathrm{k}^{(C)}$: Consider $C = \{1, 2, 3, \ldots\}$ and let $\mathrm{inj}(j) = \mathbf{e_j}$ represent a basis for $\mathrm{k}^{(C)}$. Then $\mathrm{k}^{(C)}$ consists of all *finite* $\mathbb{R}$-linear combinations of the basis vectors. Equivalently, we can consider $\mathrm{k}^{(C)}$ as the set of all functions $C \to \mathrm{k}$ with finite support. The unique morphism property states that a linear map is uniquely specified from its values on a set of basis vectors in its domain.

Example 3 **Free (associative and unital) algebra** $\mathrm{k}\langle C\rangle$: Think of C as an alphabet (collection of letters) $C = \{a, b, c, \ldots\}$. Let C^* denote all *words* over the alphabet, including the empty word $\mathbb{I}$,

---

[4]This definition is not strictly categorical, since the mappings inj and ϕ are not morphisms inside a category, but mappings from a set to an object of another category. A proper categorical definition of a free object, found in any book on category theory, is based on a *forgetful functor* mapping the given category into the category of sets. The *free functor* is the left adjoint of the forgetful functor.

$$C^* = \{\mathbb{I}, a, b, c, \dots, aa, ab, ac, \dots ba, bb, bc, \dots\}.$$

Then $\mathrm{k}\langle C\rangle = \{\mathrm{k}^{(C^*)}, \cdot\}$, is the vector space containing finite linear combinations of empty and non-empty words, equipped with a product $\cdot$ which on words is concatenation. Example $aba \cdot cus = abacus, \mathbb{I} \cdot abba = abba \cdot \mathbb{I} = abba$. This extends by linearity to $\mathrm{k}^{(C^*)}$ and yields an associative unital algebra. This is also called the non-commutative polynomial ring over C.

Example 4 **Free Lie algebra** $\mathrm{Lie}(C)$: Again, think of $C = \{a, b, c, d, \dots\}$ as an alphabet. $\mathrm{Lie}(C) \subset \mathrm{k}\langle C\rangle$ is the linear sub space generated by C under the Lie bracket $[w_1, w_2] = w_1 \cdot w_2 - w_2 \cdot w_1$ induced from the product in $\mathrm{k}\langle C\rangle$, thus $c \in C \Rightarrow c \in \mathrm{Lie}(C)$ and $x, y \in \mathrm{Lie}(C) \Rightarrow x \cdot y - y \cdot x \in \mathrm{Lie}(C)$. A basis for $\mathrm{Lie}(C)$ is given by the set of *Lyndon words* [26]. E.g. for $C = \{a, b\}$ the first Lyndon words a, b, ab, aab, abb (up to length 3) represent the commutators

$$\{a, b, [a, b], [a, [a, b]], [[a, b], b], \dots\}.$$

Computations in a free Lie algebra are important in many applications [20]. Relations such as $[[a, b], c] + [[b, c], a] = [[a, c], b]$ can be computed in $\mathrm{Lie}(C)$ and applied (evaluated) on concrete data in any Lie algebra $\mathfrak{g}$ via the Lie algebra morphism $\mathscr{F}_\phi\colon \mathrm{Lie}(C) \to \mathfrak{g}$, whenever an association of the letters with data in the concrete Lie algebra is provided through a map $\phi\colon C \to \mathfrak{g}$.

Example 5 **Free pre-Lie algebra** $\mathrm{preLie}(C)$: Consider $C = \{\bullet, \circ, \dots\}$ as a set of coloured nodes. In many applications $C = \{\bullet\}$, just a single color, and in that case we omit mentioning C. A *coloured rooted tree* is a finite connected directed graph where each node (from C) has exactly one outgoing edge, except the 'root' node which has no edge out. We illustrate a tree with the root on the bottom and the direction of the edges being down towards the root. Let T_C denote the set of all coloured rooted trees, e.g.

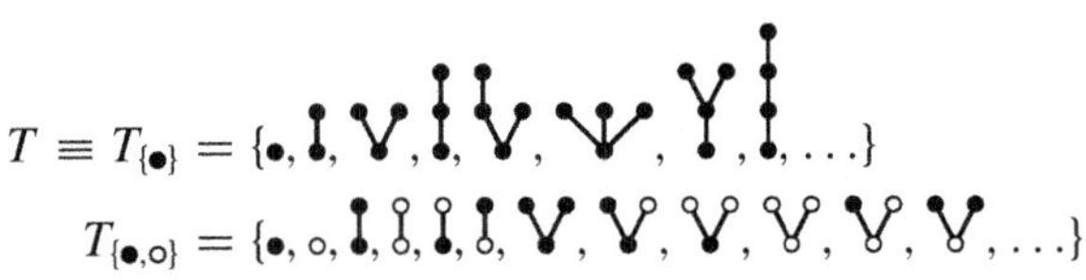

The trees are just graphs without considering an ordering of the branches, so

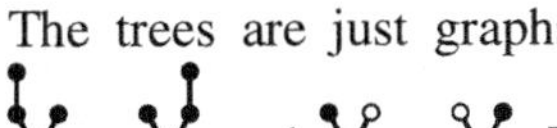

Let $\mathscr{T}_C = \mathrm{k}^{(T_C)}$. The free pre-Lie algebra over C is [6, 9] $\mathrm{preLie}(C) = \{\mathscr{T}_C, \rhd\}$, where $\rhd\colon \mathscr{T}_C \times \mathscr{T}_C$ denotes the *grafting product*. For $\tau_1, \tau_2 \in T_C$, the product $\tau_1 \rhd \tau_2$ is the sum of all possible attachments of the root of τ_1 to one of the nodes of τ_2 as shown in this example:

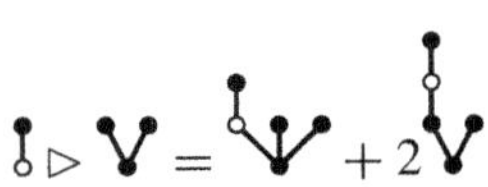

The grafting extends by linearity to all of $\mathscr{T}_C$.

Example 6 **Free magma** $\mathrm{Magma}(C) \cong \mathrm{OT}_C$: The algebraic definition of a *magma* is a set $C = \{\bullet, \circ, \ldots\}$ with a binary operation $\times$ without any algebraic relations imposed. The free magma over C consists of all possible ways to parenthesize binary operations on C, such as $(\bullet \times (\bullet \times \bullet)) \times (\circ \times \bullet)$. There are many isomorphic ways to represent the free magma. For our purpose it is convenient to represent the free magma as ordered (planar[5]) trees with coloured nodes. We let C denote a set of coloured nodes and let OT_C be the set of all ordered rooted trees with nodes chosen from C. On the trees we interpret $\times$ as the *Butcher product* [3]: $\tau_1 \times \tau_2 = \tau$ is a tree where the root of the tree τ_2 is attached to the right part of the root of the tree τ_1, e.g.:

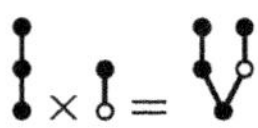

$$= (\bullet \times (\bullet \times \bullet)) \times (\circ \times \bullet).$$

If $C = \{\bullet\}$ has only one element, we write $\mathrm{OT} := \mathrm{OT}_{\{\bullet\}}$. The first few elements of OT are:

OT =

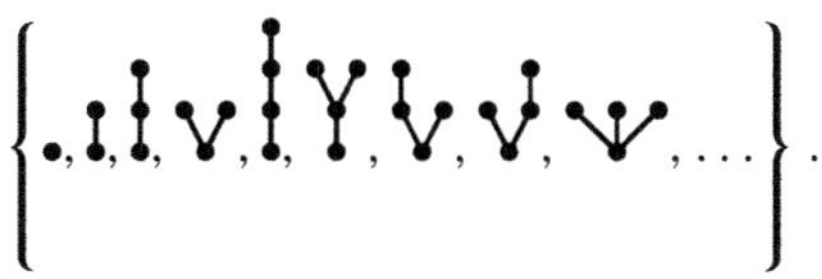

Example 7 **The free post-Lie algebra**, $\mathrm{postLie}(C)$, is given as

$$\mathrm{postLie}(C) = \{\mathrm{Lie}(\mathrm{Magma}(C)), \rhd\}, \tag{12}$$

where the product $\rhd$ is defined on $\mathrm{k}^{(\mathrm{Magma}(C))}$ as a derivation of the magmatic product

$$\tau \rhd c = c \times \tau \quad \text{for } c \in C, \tag{13}$$

$$\tau \rhd (\tau_1 \times \tau_2) = (\tau \rhd \tau_1) \times \tau_2 + \tau_1 \times (\tau \rhd \tau_2), \tag{14}$$

and it is extended by linearity and Eqs. (8)–(9) to all of $\mathrm{Lie}(\mathrm{Magma}(C))$.

Under the identification $\mathrm{Magma}(C) \cong \mathrm{OT}_C$, the product $\rhd\colon \mathrm{k}^{(\mathrm{OT}_C)} \times \mathrm{k}^{(\mathrm{OT}_C)} \to \mathrm{k}^{(\mathrm{OT}_C)}$ is given by *left* grafting. For $\tau_1, \tau_2 \in \mathrm{OT}_C$, the product $\tau_1 \rhd \tau_2$ is the sum of all possible attachments of the root of τ_1 to the left side of each node of τ_2 as shown in this example:

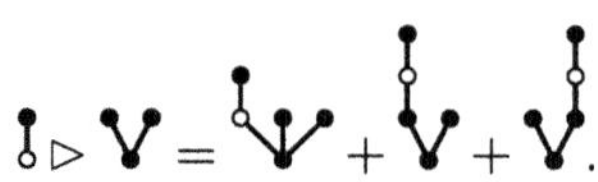

A Lyndon basis for $\mathrm{postLie}(C)$ is given in [19].

[5]Trees with different orderings of the branches are considered different, as embedded in the plane.

3.3 Enveloping Algebras

Lie algebras, pre- and post-Lie algebras are associated with algebras of first order differential operators (vector fields). Differential operators of higher order are obtained by compositions of these. Algebraically this is described through enveloping algebras.

3.3.1 Lie Enveloping Algebras

Recall that Lie$(\cdot)$ is a *functor* sending an associative algebra $\mathscr{A}$ to a Lie algebra Lie$(\mathscr{A})$, where $[x, y] = x \cdot y - y \cdot x$, and it sends associative algebra homomorphisms to Lie algebra homomorphisms. The universal enveloping algebra of a Lie algebra is defined via a functor U from Lie algebras to associative algebras being the left adjoint of Lie. This means the following:

Definition 6 (*Lie universal enveloping algebra* $U(\mathfrak{g})$) The universal enveloping algebra of a Lie algebra $\mathfrak{g}$ is a unital associative algebra $\{U(\mathfrak{g}), \cdot, \mathbb{I}\}$ together with a Lie algebra morphism $\mathrm{inj}\colon \mathfrak{g} \to \mathrm{Lie}(U(\mathfrak{g}))$ such that for any associative algebra $\mathscr{A}$ and any Lie algebra morphism $\phi\colon \mathfrak{g} \to \mathrm{Lie}(\mathscr{A})$ there exists a unique associative algebra morphism $!\colon U(\mathfrak{g}) \to \mathscr{A}$ such that $\phi = \mathrm{Lie}(!) \circ \mathrm{inj}$.

$$\begin{array}{ccc} \mathfrak{g} \xrightarrow{\mathrm{inj}} & \mathrm{Lie}(U(\mathfrak{g})) & \qquad U(\mathfrak{g}) \\ \quad\searrow^{\phi} & \big\downarrow \mathrm{Lie}(!) & \qquad \big\downarrow ! \\ & \mathrm{Lie}(\mathscr{A}) & \qquad \mathscr{A} \end{array} \tag{15}$$

The *Poincaré–Birkhoff–Witt Theorem* states that for any Lie algebra $\mathfrak{g}$ with a basis $\{e_j\}$, with some total ordering $e_j < e_k$, one gets a basis for $U(\mathfrak{g})$ by taking the set of all *canonical monomials* defined as the non-decreasing products of the basis elements $\{e_j\}$

$$\mathrm{PBWbasis}(U(\mathfrak{g})) = \{e_{j_1} \cdot e_{j_2} \cdots e_{j_r} \;:\; e_{j_1} \leq e_{j_2} \leq \cdots \leq e_{j_r}, r \in \mathbb{N}\},$$

where we have identified $\mathfrak{g} \subset U(\mathfrak{g})$ using inj. From this it follows that $U(\mathfrak{g})$ is a *filtered* algebra, splitting in a direct sum

$$U(\mathfrak{g}) = \oplus_{j=0}^{\infty} U_j(\mathfrak{g}),$$

where $U_j(\mathfrak{g})$ is the span of the canonical monomials of length j, $U_0 = \mathrm{span}(\mathbb{I})$ and $U_1(\mathfrak{g}) \cong \mathfrak{g}$. Furthermore, $U(\mathfrak{g})$ is *connected*, meaning that $U_0 \cong k$, and it is generated by U_1, meaning that $U(\mathfrak{g})$ has no proper subalgebra containing U_1.

3.3.2 Hopf Algebras

Recall that a bi-algebra is a unital associative algebra $\{B, \cdot, \mathbb{I}\}$ together with a co-associative co-algebra structure[6] $\{H, \Delta, \varepsilon\}$, where $\Delta \colon B \to B \otimes B$ is the coproduct and $\varepsilon \colon B \to \mathrm{k}$ is the co-unit. The product and coproduct must satisfy the compatibility condition

$$\Delta(x \cdot y) = \Delta(x) \cdot \Delta(y), \tag{16}$$

where the product on the right is componentwise in the tensor product.

Definition 7 (*Hopf algebra*) A *Hopf algebra* $\{H, \cdot, \mathbb{I}, \Delta, \varepsilon, S\}$ is a bi-algebra with an *antipode* $S \colon H \to H$ such that the diagram below commutes.

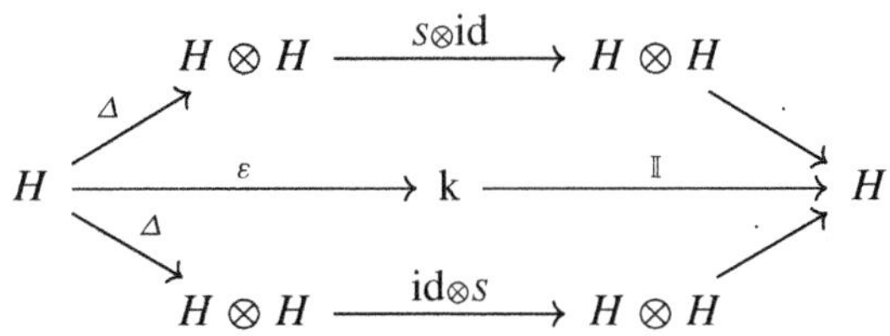

(17)

Example 8 **The concatenation de-shuffle Hopf algebra** $U(\mathfrak{g})$: The enveloping algebra $U(\mathfrak{g})$ has the structure of a Hopf algebra, where the coproduct $\Delta_{\sqcup\!\sqcup} \colon U(\mathfrak{g}) \to U(\mathfrak{g}) \otimes U(\mathfrak{g})$ is defined as

$$\Delta_{\sqcup\!\sqcup}(\mathbb{I}) = \mathbb{I} \otimes \mathbb{I} \tag{18}$$

$$\Delta_{\sqcup\!\sqcup}(x) = \mathbb{I} \otimes x + x \otimes \mathbb{I}, \quad \text{for all } x \in \mathfrak{g} \tag{19}$$

$$\Delta_{\sqcup\!\sqcup}(x \cdot y) = \Delta_{\sqcup\!\sqcup}(x) \cdot \Delta_{\sqcup\!\sqcup}(y), \quad \text{for all } x, y \in U(\mathfrak{g}). \tag{20}$$

We call this the de-shuffle coproduct, since it is the dual of the shuffle product. The co-unit is defined as

$$\varepsilon(\mathbb{I}) = 1 \tag{21}$$

$$\varepsilon(x) = 0, \quad x \in U_j(\mathfrak{g}), \quad j > 0, \tag{22}$$

and the antipode $S \colon U(\mathfrak{g}) \to U(\mathfrak{g})$ as

$$S(x_1 \cdot x_2 \cdots x_j) = (-1)^j x_j \cdots x_2 \cdot x_1 \quad \text{for all } x_1, \ldots, x_j \in \mathfrak{g}. \tag{23}$$

This turns $U(\mathfrak{g})$ into a filtered, connected, co-commutative Hopf algebra. Connected means that $U_0 \cong \mathrm{k}$ and co-commutative that $\Delta_{\sqcup\!\sqcup}$ satisfies the diagrams of a com-

[6] An associative algebra can be defined by commutative diagrams. The co-algebra structure is obtained by reversing all arrows.

mutative product, with the arrows reversed. The dual of a commutative product is co-commutative.

The *primitive elements* of a Hopf algebra H, defined as

$$\mathrm{Prim}(H) = \{x \in H \colon \Delta(x) = x \otimes \mathbb{I} + \mathbb{I} \otimes x\}$$

form a Lie algebra with $[x, y] = x \cdot y - y \cdot x$. The *Cartier–Milnor–Moore theorem* (CMM) states that if H is a connected, filtered, co-commutative Hopf algebra, then $U(\mathrm{Prim}(H))$ is isomorphic to H as a Hopf algebra. A consequence of CMM is that the enveloping algebra of a free Lie algebra over a set C is given as

$$U(\mathrm{Lie}(C)) = \mathrm{k}\langle C\rangle, \tag{24}$$

the non-commutative polynomials in C. Thus, a basis for $U(\mathrm{Lie}(C))$ is given by non-commutative monomials (the empty and non-empty words in C^*).

3.3.3 Post-Lie Enveloping Algebras

Enveloping algebras of pre- and post-Lie algebras are discussed by several authors [12, 21–23]. In our opinion the algebraic structure of the enveloping algebras are easiest to motivate by discussing the post-Lie case, and obtaining the pre-Lie enveloping algebra as a special case. For Lie algebras the enveloping algebras are associative algebras. The corresponding algebraic structure of a post-Lie enveloping algebra is called a D-algebra (D for derivation) [21, 22]:

Definition 8 (*D-algebra*) Let A be a unital associative algebra with a bilinear operation $\rhd \colon A \otimes A \to A$. Write $\mathrm{Der}(A)$ for the set of all $u \in A$ such that $v \mapsto u \rhd v$ is a derivation: $\mathrm{Der}(A) = \{u \in A \colon u \rhd (vw) = (u \rhd v)w + v(u \rhd w) \text{ for all } v, w \in A\}$. We call A a *D-algebra* if for any $u \in \mathrm{Der}(A)$ and any $v, w \in A$ we have

$$\mathbb{I} \rhd v = v \tag{25}$$

$$v \rhd u \in \mathrm{Der}(A) \tag{26}$$

$$(uv) \rhd w = a_{\rhd}(u, v, w) \equiv u \rhd (v \rhd w) - (u \rhd v) \rhd w. \tag{27}$$

In [21] it is shown:

Proposition 1 *For any D-algebra A the set of derivations forms a post-Lie algebra*

$$\mathrm{postLie}(A) := \{\mathrm{Der}(A), [\cdot, \cdot], \rhd\},$$

where $[x, y] = xy - yx$.

Thus, postLie$(\cdot)$ is a functor from the category of D-algebras to the category of post-Lie algebras. There is a functor $U(\cdot)$ from post-Lie algebras to D-algebras, which is the left adjoint of postLie(·). We can define post-Lie enveloping algebras similarly to Definition 6. A direct construction of the post-Lie enveloping algebra is obtained by extending $\rhd$ to the Lie enveloping algebra of the post-Lie algebra [21]:

Definition 9 *(Post-Lie enveloping algebra $U(\mathscr{P})$)* Let $\{\mathscr{P}, [\cdot,\cdot], \rhd\}$ be post-Lie, let $\{U_L, \cdot\} = U(\{\mathscr{P}, [\cdot,\cdot]\})$ be the Lie enveloping algebra and identify $\mathscr{P} \subset U_L$. The post-Lie enveloping algebra $U(\mathscr{P}) = \{U_L, \cdot, \rhd\}$ is defined by extending $\rhd$ from $\mathscr{P}$ to U_L according to

$$\mathbb{I} \rhd v = v \tag{28}$$
$$v \rhd \mathbb{I} = 0 \tag{29}$$
$$u \rhd (vw) = (u \rhd v)w + v(u \rhd w) \tag{30}$$
$$(uv) \rhd w = a_{\rhd}(u, v, w) := u \rhd (v \rhd w) - (u \rhd v) \rhd w \tag{31}$$

for all $u \in \mathscr{P}$ and $v, w \in U_L$. This construction yields $U(\cdot)$: postLie $\to$ D-algebra as a left adjoint functor of postLie(·).

A more detailed understanding of $U(\mathscr{P})$ is obtained by considering its Hopf algebra structures. A Lie enveloping algebra is naturally also a Hopf algebra with the de-shuffle coproduct $\Delta_{\sqcup\!\sqcup}$. With this coproduct $U(\mathscr{P})$ becomes a graded, connected, co-commutative Hopf algebra where $\mathrm{Der}(U(\mathscr{P})) = \mathrm{Prim}(U(\mathscr{P})) = \mathscr{P}$. Furthermore, the coproduct is compatible with $\rhd$ in the following sense [12]:

$$A \rhd \mathbb{I} = \varepsilon(A)$$
$$\varepsilon(A \rhd B) = \varepsilon(A)\varepsilon(B)$$
$$\Delta_{\sqcup\!\sqcup}(A \rhd B) = \sum_{\Delta_{\sqcup\!\sqcup}(A), \Delta_{\sqcup\!\sqcup}(B)} (A_{(1)} \rhd B_{(1)}) \otimes (A_{(2)} \rhd B_{(2)})$$

for all $A, B \in U(\mathscr{P})$. Here and in the sequel we employ Sweedler's notation for coproducts,

$$\Delta(A) =: \sum_{\Delta(A)} A_{(1)} \otimes A_{(2)}.$$

[7]Sometimes we need a repeated use of a coproduct. Let $\Delta\omega = \sum \omega_{(1)} \otimes \omega_{(2)}$. We continue by using Δ to split either $\omega_{(1)}$ or $\omega_{(2)}$. Since the coproduct is co-associative this yields the same result $\Delta^2\omega = \sum \omega_{(1)} \otimes \omega_{(2)} \otimes \omega_{(3)}$, and n applications are denoted

$$\Delta^n(A) =: \sum_{\Delta^n(A)} A_{(1)} \otimes A_{(2)} \otimes \cdots \otimes A_{(n+1)}.$$

Just as a post-Lie algebra always has two Lie algebras $\mathfrak{g}$ and $\overline{\mathfrak{g}}$, the post-Lie enveloping algebra $U(\mathscr{P})$ has two associative products $x, y \mapsto xy$ from the enveloping algebra $U(\mathfrak{g})$ and $x, y \mapsto x * y$ from $U(\overline{\mathfrak{g}})$. Both of these products define Hopf

[7]Splitting with regard to the coproduct Δ.

algebras with the same unit $\mathbb{I}$, co-unit ε and de-shuffle coproduct $\Delta_{\sqcup\!\sqcup}$, but with different antipodes.

Proposition 2 *[12] On $U(\mathscr{P})$ the product*

$$A * B := \sum_{\Delta_{\sqcup\!\sqcup}(A)} A_{(1)}(A_{(2)} \rhd B) \tag{32}$$

is associative. Furthermore $\{U(\mathscr{P}), *, \Delta_{\sqcup\!\sqcup}\} \cong U(\overline{\mathfrak{g}})$ are isomorphic as Hopf algebras.

The following result is crucial for handling the non-commutativity and non-associativity of $\rhd$:

Proposition 3 *[12, 22] For all $A, B, C \in U(\mathscr{P})$ we have*

$$A \rhd (B \rhd C) = (A * B) \rhd C. \tag{33}$$

The free enveloping post-Lie algebra.

Finally we introduce the enveloping algebra of the free post-Lie algebra $U(\text{postLie}(C))$. Due to CMM, we know that it is constructed from the Hopf algebra

$$U(\text{postLie}(C)) = U(Lie(\text{OT}_C)) = k\langle \text{OT}_C \rangle,$$

i.e. finite linear combinations of *words of ordered trees*, henceforth called *(ordered) forests* OF_C. If C contains only one element, we call the forests OF:

$\text{OF} = \{\mathbb{I},$  $\ldots\}$

The Hopf algebra has concatenation of forests as product and coproduct $\Delta_{\sqcup\!\sqcup}$ being de-shuffle of forests. Upon this we define $\rhd$ as left grafting on ordered trees, extended to forests by (28)–(31), where $\mathbb{I}$ is the empty forest, u is an ordered tree and v, w are ordered forests. The left grafting of a forest on another is combinatorially the sum of all possible left attachments of the roots of trees in the left forest to the nodes of the right forest, maintaining order when attaching to the same node, as in this example

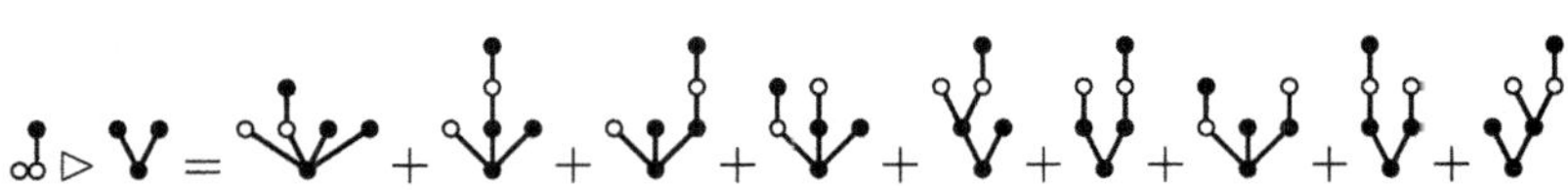

Four Hopf algebras on ordered forests.

On $k\langle \text{OF}_C \rangle$ we have two associative products $*$ and the concatenation product, denoted $\cdot$. Both these form Hopf algebras with the de-shuffle coproduct $\Delta_{\sqcup\!\sqcup}$ and antipodes $S_\cdot$ and S_*, where

$$S_{\cdot}(\tau_1 \cdot \tau_2 \cdots \tau_k) = (-1)^k \tau_k \cdots \tau_2 \cdot \tau_1, \quad \text{for } \tau_1 \cdot \tau_2 \cdots \tau_k \in \mathrm{OT}_C$$

and S_* given in (71). With their duals, we have the following four Hopf algebras:

$$\begin{aligned}
\mathscr{H}_{\cdot} &= \{\mathrm{k}\langle \mathrm{OF}_C\rangle, \Delta_{\sqcup\!\sqcup}, \cdot, S_{\cdot}\}\\
\mathscr{H}_{*} &= \{\mathrm{k}\langle \mathrm{OF}_C\rangle, \Delta_{\sqcup\!\sqcup}, *, S_{*}\}\\
\mathscr{H}_{\cdot}' &= \{\mathrm{k}\langle \mathrm{OF}_C\rangle, \Delta_{\cdot}, \sqcup\!\sqcup, S_{\cdot}\}\\
\mathscr{H}_{*}' &= \{\mathrm{k}\langle \mathrm{OF}_C\rangle, \Delta_{*}, \sqcup\!\sqcup, S_{*}\}.
\end{aligned}$$

The four share the same unit $\mathbb{I}\colon \mathrm{k} \to \mathscr{H}\colon 1 \mapsto \mathbb{I}$ and the same co-unit $\varepsilon\colon \mathscr{H} \to \mathrm{k}$, where $\varepsilon(\mathbb{I}) = 1$ and $\varepsilon(\omega) = 0$ for all $\omega \in \mathrm{OF}_C\backslash\{\mathbb{I}\}$. All four Hopf algebras are connected and graded with $|\omega|$ counting the number of nodes in a forest. $\mathscr{H}_{\cdot}$ and $\mathscr{H}_{\cdot}'$ are also connected and graded with the word length as a grading, although this grading is of less importance for our applications.

3.3.4 Lie–Butcher Series

The vector space $\mathrm{k}\langle \mathrm{OT}_C\rangle$ consists of *finite* linear combinations of forests. In order to be able to symbolically represent flow maps and backward error analysis, we do, however, need to extend the space to infinite sums. For a (non-commutative) polynomial ring $\mathrm{k}\langle C\rangle$, we denote $\mathrm{k}\langle\langle C\rangle\rangle$ the set of infinite (formal) power series. Let $\langle\cdot,\cdot\rangle\colon \mathrm{k}\langle C\rangle \times \mathrm{k}\langle C\rangle \to \mathrm{k}$ denote the inner product where the monomials (words in C^*) form an orthonormal basis. This extends to a dual pairing

$$\langle\cdot,\cdot\rangle\colon \mathrm{k}\langle\langle C\rangle\rangle \times \mathrm{k}\langle C\rangle \to \mathrm{k}, \tag{34}$$

which identifies $\mathrm{k}\langle\langle C\rangle\rangle = \mathrm{k}\langle C\rangle^*$ as the linar dual space. Any $\alpha \in \mathrm{k}\langle\langle C\rangle\rangle$ is uniquely determined by its evaluation on the finite polynomials, and we may write α as a formal infinite sum

$$\alpha = \sum_{w\in C^*} \langle \alpha, w\rangle w.$$

Any k-linear map $f\colon \mathrm{k}\langle\langle C\rangle\rangle \to \mathrm{k}\langle\langle C\rangle\rangle$ can be computed from its dual $f^*\colon \mathrm{k}\langle C\rangle \to \mathrm{k}\langle C\rangle$ as $\langle f(\alpha), w\rangle = \langle \alpha, f^*(w)\rangle$for all $w \in C^*$.

Definition 10 (*Lie–Butcher series* $\mathrm{LB}(C)$) The Lie–Butcher series over a set C is defined as the completion

$$\mathrm{LB}(C) := U(\mathrm{postLie}(C))^*.$$

This is the vector space $\mathrm{k}\langle\langle \mathrm{OT}_C\rangle\rangle$ (infinite linear combinations of ordered forests). All the operations we consider on this space are defined by their duals acting upon $\mathrm{k}\langle \mathrm{OT}_C\rangle$, see Sect. 4.1.

The space $\mathrm{LB}(C)$ has two important subsets, the *primitive elements* and the *group like elements*.

Definition 11 (*Primitive elements* $\mathfrak{g}_{\text{LB}}$) The primitive elements of $\text{LB}(C)$, denoted $\mathfrak{g}_{\text{LB}}$ are given as

$$\mathfrak{g}_{\text{LB}} = \{\alpha \in \text{LB}(C) \colon \Delta_{\sqcup\!\sqcup}(\alpha) = \alpha \otimes \mathbb{I} + \mathbb{I} \otimes \alpha\}, \tag{35}$$

where $\Delta_{\sqcup\!\sqcup}$ is the graded completion of the de-shuffle coproduct. This forms a post-Lie algebra which is the graded completion of the free post-Lie algebra $\text{postLie}(C)$.

Definition 12 (*The Lie–Butcher group* G_{LB}) The group like elements of $\text{LB}(C)$, denoted G_{LB} are given as

$$G_{\text{LB}} = \{\alpha \in \text{LB}(C) \colon \Delta_{\sqcup\!\sqcup}(\alpha) = \alpha \otimes \alpha\}, \tag{36}$$

where $\Delta_{\sqcup\!\sqcup}$ is the graded completion of the de-shuffle coproduct.

The Lie–Butcher group is a group both with respect to the concatenation product and the product $*$ in (32). There are also two exponential maps with respect to the two associative products sending primitive elements to group-like elements

$$\exp, \exp^* \colon \mathfrak{g}_{\text{LB}} \to G_{\text{LB}}.$$

Both these are 1–1 mappings with inverses given by the corresponding logarithms

$$\log, \log^* \colon G_{\text{LB}} \to \mathfrak{g}_{\text{LB}}.$$

4 Computing with Lie–Butcher Series

In this section, we will list important operations on Lie–Butcher series. A focus will be given on recursive formulations which are suited for computer implementations.

4.1 Operations on Infinite Series Computed by Dualisation

Lie–Butcher series are infinite series, and in principle the only computation we consider on an infinite series is the evaluation of the dual pairing (34). All operations on infinite Lie–Butcher series, $\alpha \in \text{LB}(C)$, are computed by dualisation, throwing the operation over to the finite right hand part of the dual pairing. By recursions, the dual computation on the right hand side is moving towards terms with a lower grade, and finally terminates. Some modern programming languages, such as Haskell, allow for *lazy evaluation*, meaning that terms are not computed before they are needed to produce a result. This way it is possible to implement proper infinite series.

Example 9 The computation of the de-shuffle coproduct of infinite series can be computed as

$$\langle \Delta_{\sqcup\!\sqcup}(\alpha), \omega_1 \otimes \omega_2 \rangle = \langle \alpha, \omega_1 \sqcup\!\sqcup\, \omega_2 \rangle, \tag{37}$$

where the pairing on the left is defined componentwise in the tensor product,

$$\langle \alpha_1 \otimes \alpha_2, \omega_1 \otimes \omega_2 \rangle = \langle \alpha_1, \omega_1 \rangle \cdot \langle \alpha_2, \omega_2 \rangle$$

and shuffle product $\omega \sqcup\!\sqcup\, \widetilde{\omega}$ of two words in an alphabet is the sum over all permutations of $\omega\widetilde{\omega}$ which are not changing the internal order of the letters coming from each part, e.g.

$$ab \sqcup\!\sqcup\, cd = abcd + acbd + cabd + acdb + cadb + cdab.$$

A recursive formula for the shuffle product is given below.

Any linear operation whose dual sends polynomials in $\mathrm{k}\langle \mathrm{OT}_C \rangle$ to polynomials (or tensor products of these) is well defined on infinite LB-series by such dualisation.

Linear algebraic operations.

$$+\colon \mathrm{LB}(C) \times \mathrm{LB}(C) \to \mathrm{LB}(C) \quad \text{(addition)}$$
$$\cdot\colon\ \mathrm{k} \times \mathrm{LB}(C) \to \mathrm{LB}(C) \quad \text{(scalar multiplication)}.$$

These are computed as $\langle \alpha + \beta, w \rangle = \langle \alpha, w \rangle + \langle \beta, w \rangle$ and $\langle c \cdot \alpha, w \rangle = c \cdot \langle \alpha, w \rangle$. Note that $\mathfrak{g}_{\mathrm{LB}} \subset \mathrm{LB}(C)$ is a linear subspace closed under these operations, $G_{\mathrm{LB}} \subset \mathrm{LB}(C)$ is *not* a linear subspace.

4.2 Operations on Forests Computed by Recursions in a Magma

Similar to the case of trees, Sect. 3.2, many recursion formulas for forests are suitably formulated in terms of magmatic products on forests. Let $B^-\colon \mathrm{OT}_C \to \mathrm{OF}_C$ denote the removal of the root, sending a tree to the forest containing the branches of the root, and for every $c \in C$ define $B_c^+\colon \mathrm{OF}_C \to \mathrm{OT}_C$ as the addition of a root of colour c to a forest, producing a tree, example

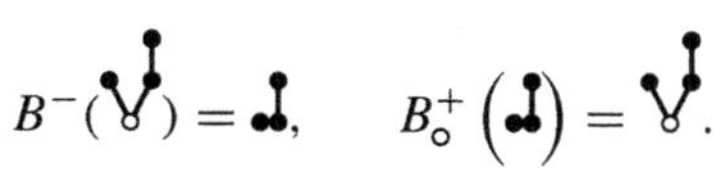

Definition 13 (*Magmatic products on* OF_C) For every $c \in C$, define a product $\times_c\colon \mathrm{OF}_C \times \mathrm{OF}_C \to \mathrm{OF}_C$ as

$$\omega_1 \times_c \omega_2 := \omega_1 B_c^+(\omega_2). \tag{38}$$

In the special case where $C = \{\bullet\}$ contains just one element, then $B^+ \colon \mathrm{OF} \to \mathrm{OT}$ is 1–1, sending the above product on forests to the Butcher product on trees; $B^+(\omega_1 \times_\bullet \omega_2) = B^+(\omega_1) \times B^+(\omega_2)$. Thus, in this case $\{\mathrm{OF}, \times_\bullet\} \cong \{\mathrm{OT}, \times\} \cong \mathrm{Magma}(\{\bullet\})$.

For a general C we have that any $\omega \in \mathrm{OF}_C \backslash \mathbb{I}$ has a unique decomposition

$$\omega = \omega_L \times_c \omega_R, \qquad c \in C, \quad \omega_L, \omega_R \in \mathrm{OF}_C. \tag{39}$$

The set of forests OF_C is freely generated from $\mathbb{I}$ by these products, e.g.

 $= (\mathbb{I} \times_\circ ((\mathbb{I} \times_\bullet \mathbb{I}) \times_\bullet \mathbb{I})) \times_\circ (\mathbb{I} \times_\bullet \mathbb{I}).$

Thus, there is a 1–1 correspondence between OF_C and binary trees where the internal nodes are coloured with C. We may take the binary tree representation as the *definition* of OF_C and express any computation in terms of this.

Definition 14 (*Magmatic definition of* OF_C) Given a set C, the ordered forests OF_C are defined recursively as

$$\mathbb{I} \in \mathrm{OF}_C \tag{40}$$

$$\omega = \omega_L \times_C \omega_R \in \mathrm{OF}_C \quad \text{for evey } \omega_L, \omega_R \in \mathrm{OF}_C \text{ and } c \in C. \tag{41}$$

OF_C has the following operations:

isEmpty: $\mathrm{OF}_C \to$ bool, defined by isEmpty$(\mathbb{I})$ = 'true', otherwise 'false'.
Left: $\mathrm{OF}_C \to \mathrm{OF}_C$, defined by $\mathrm{Left}(\omega_L \times_c \omega_R) = \omega_L$.
Right: $\mathrm{OF}_C \to \mathrm{OF}_C$, defined by $\mathrm{Right}(\omega_L \times_c \omega_R) = \omega_R$.
Root: $\mathrm{OF}_C \to C$, defined by $\mathrm{Root}(\omega_L \times_c \omega_R) = c$.

$\mathrm{Left}(\mathbb{I})$, $\mathrm{Right}(\mathbb{I})$ and $\mathrm{Root}(\mathbb{I})$ are undefined.

Any operation on forests can be expressed in terms of these. We can define ordered trees as the subset $\mathrm{OT}_C \subset \mathrm{OF}_C$

$$\mathrm{OT}_C := \{\tau \in \mathrm{OF}_C \colon \mathrm{Left}(\tau) = \mathbb{I}\},$$

and in particular the nodes $C \subset \mathrm{OF}_C$ are identified as $C \cong \{\mathbb{I} \times_c \mathbb{I}\}$. From this we define $B^- \colon \mathrm{OT}_C \to \mathrm{OF}_C$ and $B_c^+ \colon \mathrm{OF}_C \to \mathrm{OT}_C$ as

$$B^-(\tau) = \mathrm{Right}(\tau) \tag{42}$$

$$B_c^+(\omega) = \mathbb{I} \times_c \omega. \tag{43}$$

The Butcher product of two trees $\tau, \tau' \in \mathrm{OT}_C$, where $c = \mathrm{Root}(\tau)$, $c' = \mathrm{Root}(\tau')$ is

$$\tau \times \tau' := B_c^+(B^-(\tau) \times_{c'} B^-(\tau')).$$

4.3 Combinatorial Functions on Ordered Forests

The *order* of $\omega \in \mathrm{OF}_C$, denoted $|\omega| \in \mathbb{N}$, counts the number of nodes in the forest. It is computed by the recursion

$$|\mathbb{I}| = 0 \tag{44}$$

$$|\omega_L \times_\bullet \omega_R| = |\omega_L| + |\omega_R| + 1. \tag{45}$$

This counts the number of nodes in ω.

The *ordered forest factorial*, denoted $\omega¡ \in \mathbb{N}$ is defined by the recursion

$$\mathbb{I}¡ = 1 \tag{46}$$

$$\omega¡ = (\omega_L \times_\bullet \omega_R)¡ = |\omega| \cdot \omega_L¡ \cdot \omega_R¡. \tag{47}$$

We will see that the ordered factorial is important for characterising the flow map (exact solution) of a differential equation. This is a generalisation of the more well-known *tree factorial function for un-ordered trees*, which is denoted $\tau!$ and defined by the recursion

$$\bullet! = 1$$

$$\tau! = |\tau| \cdot \tau_1! \cdot \tau_2! \cdots \tau_p!$$

for $\tau = B^+(\tau_1\tau_2 \cdots \tau_p)$.

The relationship between the classical (unordered) and the ordered tree factorial functions is

$$\sigma(\tau) \sum_{\tau' \sim \tau} \frac{1}{\tau'¡} = \frac{1}{\tau!},$$

where the sum runs over all ordered trees that are equivalent under permutation of the branches and $\sigma(\tau)$ is the symmetry factor of the tree. This identity can be derived from the relationship between classical B-series and LB-series discussed in Sect. 4.1 of [22], by comparing the exact flow maps $\exp^*(\bullet)$ in the two cases. We omit details.

Example 10

$1/$  $¡ + 1/$  $¡ = \frac{1}{12} + \frac{1}{24} = \frac{1}{8} = 1/$  $!$

and

$2\Big(1/$  $¡ + 1/$  $¡ + 1/$  $¡\Big) = 2\left(\frac{1}{40} + \frac{1}{60} + \frac{1}{120}\right) = \frac{1}{10} = 1/$  $!.$

Table 1 Ordered forest factorial for all forest up to and including order 5

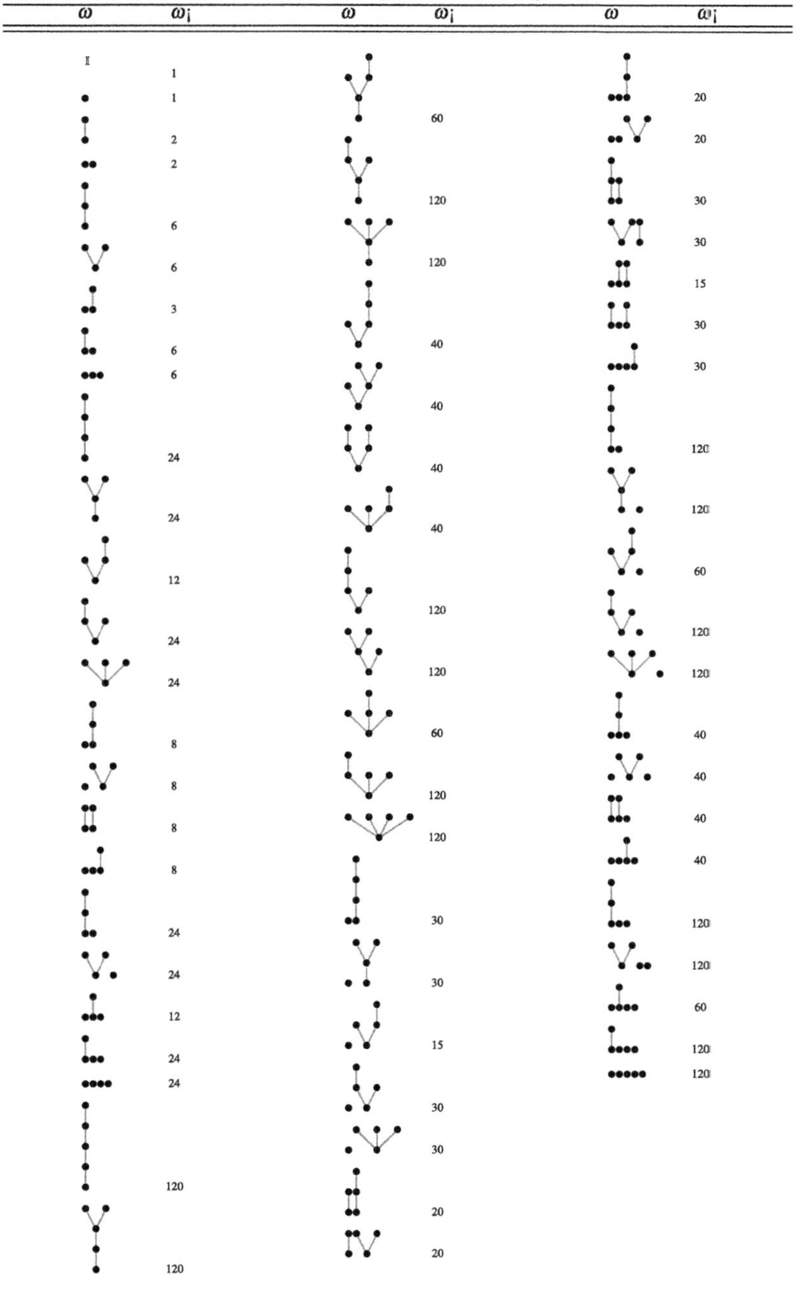

For the tall tree $\tau = \mathbb{I} \times_\bullet (\mathbb{I} \times_\bullet (\mathbb{I} \times_\bullet (\cdots \times_\bullet (\mathbb{I} \times_\bullet \mathbb{I}))))$ we have $\tau¡ = \tau! = |\tau|!$. Table 1 on p. 25 contains the ordered forest factorial for all ordered forests up to and including order 5.

4.4 Concatenation and De-concatenation

Concatenation and de-concatenation

$$\begin{aligned} \cdot &: \mathrm{k}\langle \mathrm{OT}_C\rangle \otimes \mathrm{k}\langle \mathrm{OT}_C\rangle \to \mathrm{k}\langle \mathrm{OT}_C\rangle \\ \Delta_\cdot &: \mathrm{k}\langle \mathrm{OT}_C\rangle \to \mathrm{k}\langle \mathrm{OT}_C\rangle \otimes \mathrm{k}\langle \mathrm{OT}_C\rangle \end{aligned}$$

form a pair of dual operations, just like $\sqcup\!\sqcup$ and $\Delta_{\sqcup\!\sqcup}$ in (37). On monomials $\omega \in \mathrm{OF}_C$ these are given by

$$\begin{aligned} \omega \cdot \omega' &= \omega\omega' \\ \Delta_\cdot(\omega) &= \sum_{\substack{\omega_1,\omega_2 \in \mathrm{OF}_C \\ \omega_1 \cdot \omega_2 = \omega}} \omega_1 \otimes \omega_2, \end{aligned}$$

thus for $\omega = \tau_1\tau_2\cdots\tau_k$, $\tau_1,\ldots,\tau_k \in \mathrm{OT}_C$ we have

$$\Delta_\cdot(\omega) = \omega \otimes \mathbb{I} + \mathbb{I} \otimes \omega + \sum_{j=1}^{k} \tau_1 \cdots \tau_j \otimes \tau_{j+1} \cdots \tau_k.$$

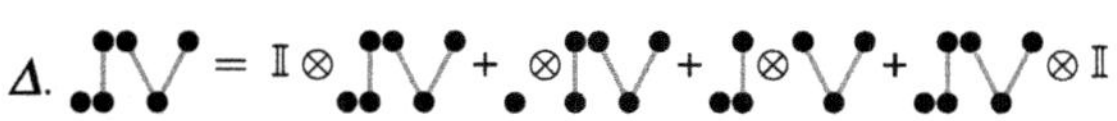

Recursive formulas, where $\widetilde{\omega} \in \mathrm{OF}_C$, $\omega = \omega_L \times_c \omega_R$ are

$$\widetilde{\omega} \cdot \mathbb{I} = \widetilde{\omega} \tag{48}$$

$$\widetilde{\omega} \cdot \omega = (\widetilde{\omega} \cdot \omega_L) \times_c \omega_R \tag{49}$$

and

$$\Delta_\cdot(\mathbb{I}) = \mathbb{I} \otimes \mathbb{I} \tag{50}$$

$$\Delta_\cdot(\omega) = \Delta_\cdot(\omega_L) \cdot (\mathbb{I} \otimes (\mathbb{I} \times_c \omega_R)) + \omega \otimes \mathbb{I}. \tag{51}$$

See Table 2[8] on p. 27 for deconcatenation of all ordered forests up to and including order 4.

[8]Note that the number under the terms are the coefficients to the terms.

Table 2 Deconcatenation and deshuffle for ordered forest up to and including order 4. Note that the numbers under the terms are the coefficients to the terms

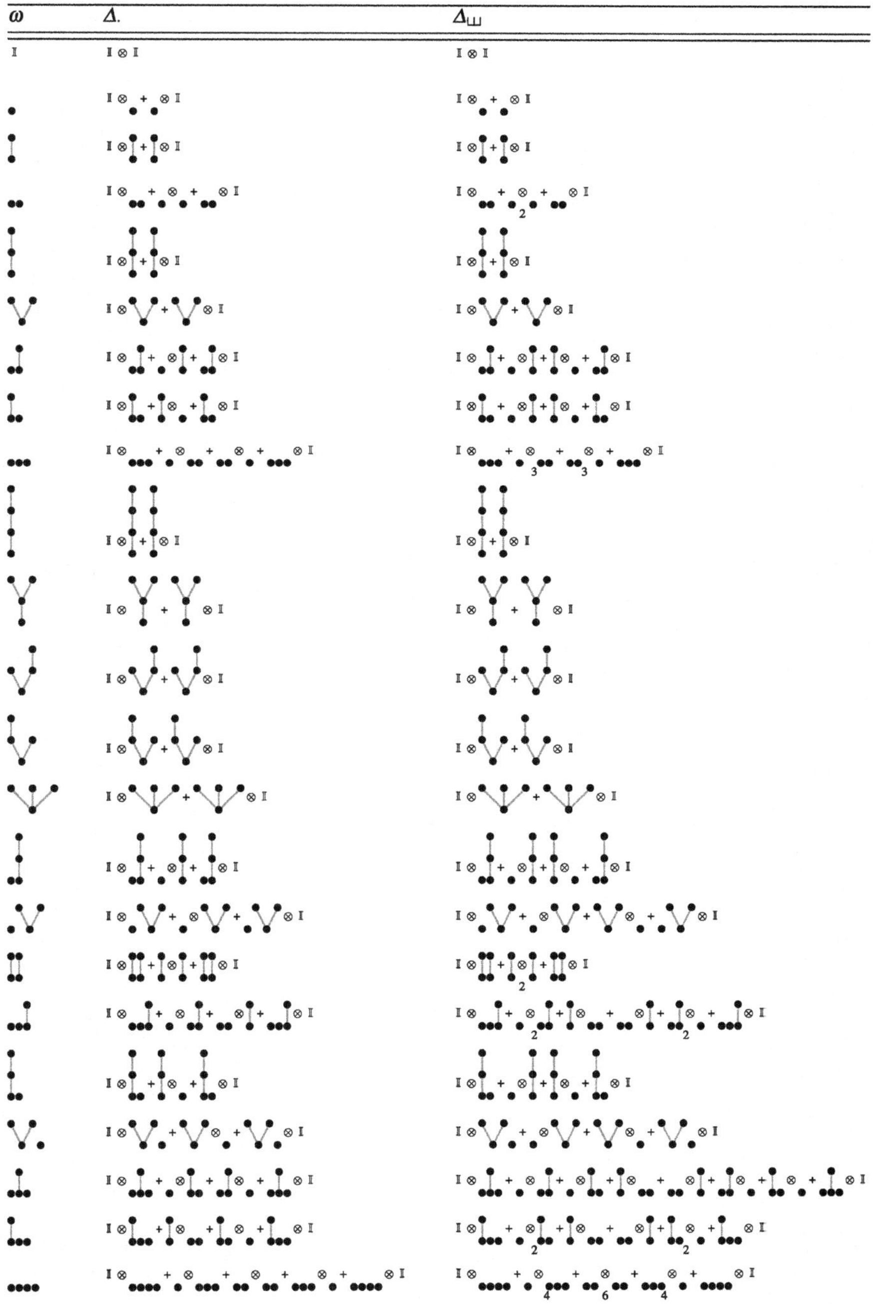

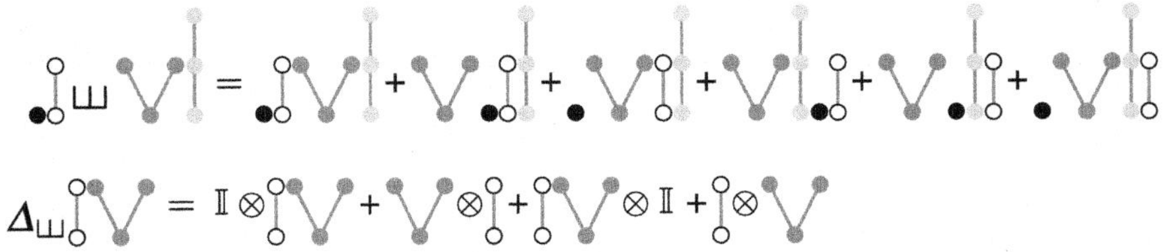

Fig. 1 See Table 2 on p. 27 for more examples on deshuffle

The concatenation antipode $S_.$, defined in (23), is computed by the recursion

$$S_.(\mathbb{I}) = \mathbb{I} \tag{52}$$

$$S_.(\omega_L \times_c \omega_R) = -B_c^+(\omega_R) \cdot S_.(\omega_L). \tag{53}$$

$S_.$ reverse the order of the trees in the forest and negate if there is a odd number of trees in the the forest. See Table 2 on p. 27.

4.5 Shuffle and De-shuffle

The duality of $\Delta_{\sqcup\!\sqcup}$ and $\sqcup\!\sqcup$ is given in (37). A recursive formula for $\omega \sqcup\!\sqcup \widetilde{\omega}$ where $\omega, \widetilde{\omega} \in \mathrm{OF}_C$ is obtained from the decomposition $\omega = \omega_L \times_c \omega_R$, $\widetilde{\omega} = \widetilde{\omega}_L \times_{\widetilde{c}} \widetilde{\omega}_R$ as

$$\mathbb{I} \sqcup\!\sqcup \omega = \omega \sqcup\!\sqcup \mathbb{I} = \omega \tag{54}$$

$$\omega \sqcup\!\sqcup \widetilde{\omega} = (\omega_L \sqcup\!\sqcup \widetilde{\omega}) \times_c \omega_R + (\omega \sqcup\!\sqcup \widetilde{\omega}_L) \times_{\widetilde{c}} \widetilde{\omega}_R, \tag{55}$$

while (18)–(20) yields the recursion

$$\Delta_{\sqcup\!\sqcup}(\mathbb{I}) = \mathbb{I} \otimes \mathbb{I} \tag{56}$$

$$\Delta_{\sqcup\!\sqcup}(\omega) = \Delta_{\sqcup\!\sqcup}(\omega_L) \cdot \left((\mathbb{I} \times_c \omega_R) \otimes \mathbb{I} + \mathbb{I} \otimes (\mathbb{I} \times_c \omega_R)\right). \tag{57}$$

The shuffle product $\sqcup\!\sqcup$ of two forests is the summation over all permutations of the trees in the forests while preserving the ordering of the trees in each of the initial forests (Fig. 1).

4.6 Grafting, Pruning, GL Product and GL Coproduct

These are four closely related operations. Grafting is defined in (13)–(14) for trees and (28)–(31) for forests (here u is a tree). Grafting can also be expressed directly through the magmatic definition of OF_C. First we need to decompose $\omega \in \mathrm{OF}_C \backslash \mathbb{I}$ as a concatenation of a tree on the left with a forest on the right, $\omega = \tau' \cdot \omega'$. We define

Fig. 2 See Table 3 on p. 30 and Table 4 on p. 31 for more examples

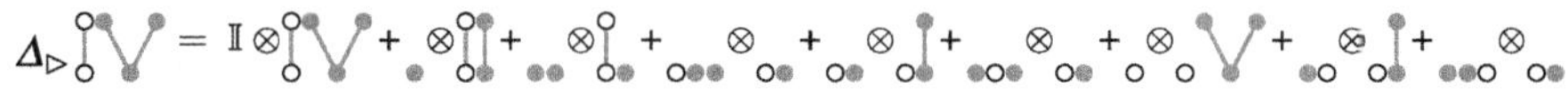

Fig. 3 See also Table 4 on p. 31

the decomposition $\tau' = \text{LeftTree}(\omega)$, $\omega' = \text{RightForest}(\omega)$ through the following recursions, where $\tau \in \text{OT}_C$ and $\omega = \omega_L \times_c \omega_R$:

$$\text{LeftTree}(\tau) = \tau \tag{58}$$
$$\text{LeftTree}(\omega) = \text{LeftTree}(\omega_L) \tag{59}$$
$$\text{RightForest}(\tau) = \mathbb{I} \tag{60}$$
$$\text{RightForest}(\omega) = \text{RightForest}(\omega_L) \times_c \omega_R. \tag{61}$$

The general recursion for grafting of forests becomes

$$\mathbb{I} \rhd \omega = \omega \tag{62}$$
$$\tau \rhd \mathbb{I} = 0 \tag{63}$$
$$\tau \rhd (\omega_L \times_c \omega_R) = (\tau \rhd \omega_L) \times_c \omega_R + \omega_L \times_c (\tau \cdot \omega_R + \tau \rhd \omega_R) \tag{64}$$
$$(\tau \cdot \omega) \rhd \widetilde{\omega} = \tau \rhd (\omega \rhd \widetilde{\omega}) - (\tau \rhd \omega) \rhd \widetilde{\omega}, \tag{65}$$

for all $\tau \in \text{OT}_C$, $\omega, \widetilde{\omega}, \omega_L, \omega_R \in \text{OF}_C$, $c \in C$. See Table 3 on p. 30 for examples.

The associative product $*$ defined in (32) is, in the context of polynomials of ordered trees $\text{k}\langle\text{OT}_C\rangle$, called the (ordered) Grossman–Larsson product [22], *GL product* for short. On $\text{k}\langle\text{OT}_C\rangle$ (and even on $\text{LB}(C)$), we can compute $*$ from grafting as

$$\omega_1 * \omega_2 = B^-(\omega_1 \rhd B^+(\omega_2)).$$

The colour of the added root is irrelevant, since this root is later removed by B^-. See Table 3 on p. 30 for examples (Figs. 2 and 3).

The dual of $*$, the GL coproduct $\Delta_* \colon \text{k}\langle\text{OF}_C\rangle \to \text{k}\langle\text{OF}_C\rangle \otimes \text{k}\langle\text{OF}_C\rangle$ has several different characterisations, in terms of left admissible cuts of trees and by recursion [22]. For $\omega = \omega_L \times_c \omega_R$ the recursion is

$$\Delta_*(\mathbb{I}) = \mathbb{I} \otimes \mathbb{I} \tag{66}$$
$$\Delta_*(\omega) = \omega \otimes \mathbb{I} + \Delta_*(\omega_L) \sqcup\!\sqcup \times_c \Delta_*(\omega_R), \tag{67}$$

Table 3 Grafting and Grossman-Larsson product for all combinations of non-empty trees with total order up to and including order 4. Note that the numbers under the terms are the coefficients to the terms

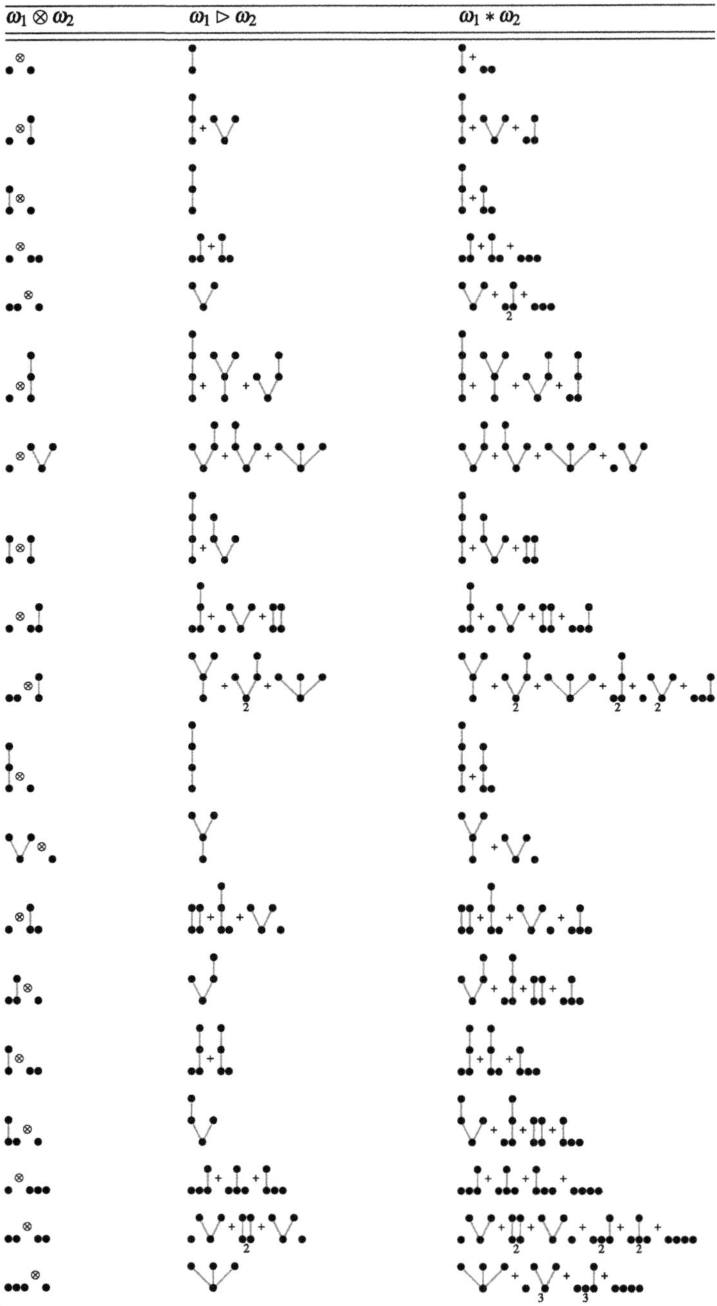

Table 4 Pruning and dual Grossman-Larsson coproduct for all forests up to and including order 4. Note that the numbers under the terms are the coefficients to the terms

ω	$\Delta_{\rhd}(\omega)$	$\Delta_{*}(\omega)$

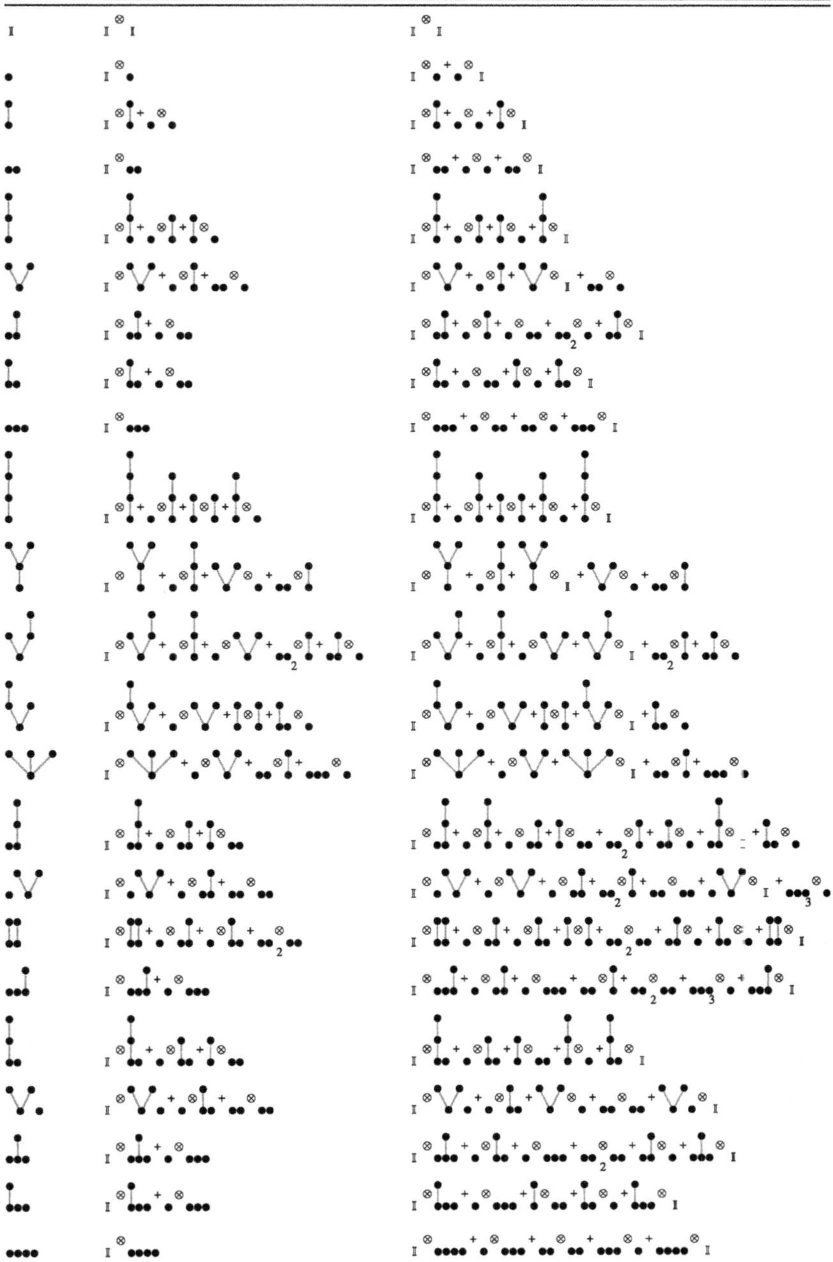

where $\sqcup\!\sqcup \times_c\colon \; \mathrm{k}\langle \mathrm{OT}_C\rangle \otimes \mathrm{k}\langle \mathrm{OT}_C\rangle \otimes \mathrm{k}\langle \mathrm{OT}_C\rangle \otimes \mathrm{k}\langle \mathrm{OT}_C\rangle \to \mathrm{k}\langle \mathrm{OT}_C\rangle \otimes \mathrm{k}\langle \mathrm{OT}_C\rangle$ denotes

$$(\alpha \otimes \widetilde{\alpha}) \sqcup\!\sqcup \times_c (\omega \otimes \widetilde{\omega}) := (\alpha \sqcup\!\sqcup \omega) \otimes (\widetilde{\alpha} \times_c \widetilde{\omega}).$$

The grafting operation $\rhd\colon \mathrm{k}\langle \mathrm{OT}_C\rangle \times \mathrm{k}\langle \mathrm{OT}_C\rangle \to \mathrm{k}\langle \mathrm{OT}_C\rangle$ has a right sided dual we call *pruning*, $\Delta_\rhd\colon \mathrm{k}\langle \mathrm{OT}_C\rangle \to \mathrm{k}\langle \mathrm{OT}_C\rangle \times \mathrm{k}\langle \mathrm{OT}_C\rangle$, dual in the usual sense

$$\langle \alpha \rhd \beta, \omega\rangle = \langle \alpha \otimes \beta, \Delta_\rhd(\omega)\rangle.$$

The pruning is characterised by admissible cuts in [16], or it can be computed by the following recursion involving both itself and the GL coproduct,

$$\Delta_\rhd(\mathbb{I}) = \mathbb{I} \otimes \mathbb{I} \tag{68}$$

$$\Delta_\rhd(\omega_L \times_c \omega_R) = \Delta_\rhd(\omega_L) \sqcup\!\sqcup \times_c \Delta_*(\omega_R). \tag{69}$$

The Lie–Butcher group and the antipode S_*.
The product in the Lie-Butcher group G_{LB} is the GL product $\alpha, \beta \mapsto \alpha * \beta$. The inverse is given by the *antipode* (with respect to $*$-product), an endomorphism $S_* \in \mathrm{End}(\mathrm{k}\langle \mathrm{OT}_C\rangle)$ such that

$$\langle \alpha^{*-1}, \omega\rangle = \langle \alpha, S_*(\omega)\rangle. \tag{70}$$

A recursive formula for S_* is found in [22]. In our magmatic representation of forests we have

$$S_*(\omega_L \times_c \omega_R) = -\sqcup\!\sqcup\left((S_* \otimes I)(\Delta_*(\omega_L) \sqcup\!\sqcup \times_c \Delta_*(\omega_R))\right). \tag{71}$$

Table 5 on p. 33 contain the the result of applying S_* to all ordered forests up to and including order 4.

4.7 Substitution, Co-substitution, Scaling and Derivation

A LB-series is an infinite series of forests built from nodes. The substitution law [4, 7, 16, 25] expresses the operation of replacing each node with an entire LB series. Since a node represents a primitive element, it is necessary to require that the LB-series in the substitution must be an element of $\mathfrak{g}_{\mathrm{LB}}$. The universal property of the free enveloping algebra $U(\mathrm{postLie}(C))$ implies that for any mapping $a\colon C \to \mathscr{P}$ from C into a post-Lie algebra $\mathscr{P}$, there exists a unique D-algebra morphism $!\colon U(\mathrm{postLie}(C)) \to U(\mathscr{P})$ such that the diagram commutes

Table 5 Concatenation and Grossman-Larsson antipode map for all forests up to and including order 4. Note that the numbers under the terms are the coefficients to the terms

$$\begin{array}{ccc} C & \xrightarrow{\text{inj}} & U(\text{postLie}(C)) \\ \downarrow{a} & & \downarrow{!} \\ \mathscr{P} & \xrightarrow{\text{inj}} & U(\mathscr{P}) \end{array} \tag{72}$$

In particular this holds if $\mathscr{P} = \text{postLie}(C)$, and it also holds if $U(\text{postLie}(C))$ is replaced with its graded completion $\text{LB}(C)$. From this we obtain the algebraic definition of substitution:

Definition 15 (*Substitution*) Given a mapping $a\colon C \to \mathfrak{g}_{\text{LB}}$ there exists a unique D-algebra automorphism $a\star\colon \text{LB}(C) \to \text{LB}(C)$ such that the diagram commutes

$$\begin{array}{ccc} C & \xrightarrow{\text{inj}} & \text{LB}(C) \\ \downarrow{a} & & \downarrow{a\star} \\ \mathfrak{g}_{\text{LB}} & \xrightarrow{\text{inj}} & \text{LB}(C). \end{array} \tag{73}$$

This morphism is called *substitution*.

The automorphism property implies that it enjoys many identities such as

$$a \star \mathbb{I} = \mathbb{I} \tag{74}$$

$$a \star (\omega\omega') = (a \star \omega)(a \star \omega') \tag{75}$$

$$a \star (\omega \rhd \omega') = (a \star \omega) \rhd (a \star \omega') \tag{76}$$

$$a \star (\omega * \omega') = (a \star \omega) * (a \star \omega') \tag{77}$$

$$(a \star \otimes a\star)(\Delta_{\sqcup\!\sqcup}(\omega)) = \Delta_{\sqcup\!\sqcup}(a \star \omega). \tag{78}$$

For more details, see [16].

As explained earlier, computations with LB-series are done by considering the series together with a pairing on the space of finite series and computations are performed by deriving how the given operation is expressed as an operation on finite series, via the dual. Thus, to compute substitution of infinite series, we need to characterise the dual map, called co-substitution.

Definition 16 (*Co-substitution*) Given a substitution $a\star\colon \text{LB}(C) \to \text{LB}(C)$, the *co-substitution* $a_\star^T$ is a k-linear map $a_\star^T\colon\ \text{k}\langle \text{OT}_C\rangle \to \text{k}\langle \text{OT}_C\rangle$ such that

$$\langle a \star \beta, x\rangle = \langle \beta, a_\star^T(x)\rangle$$

for all $\beta \in \text{LB}(C)$ and $x \in \text{k}\langle \text{OT}_C\rangle$.

A recursive formula for the co-substitution is derived in [16] in the case where $C = \{\bullet\}$. A general formula for arbitrary finite C is given here, the proof of this formula is similar to the proof in [16] but we omit it. The general formula for $a_\star^T(\omega)$ is based on decomposing ω with the de-concatenation coproduct $\Delta_.$ and thereafter decomposing the second component with the pruning coproduct $\Delta_{\triangleright}$. To clarify the notation, the decomposition is as follows

$$(I \otimes \Delta_{\triangleright}) \circ \Delta_{.}(\omega) = \sum_{\Delta_{.}(\omega)} \sum_{\Delta_{\triangleright}(\omega_{(2)})} \omega_{(1)} \otimes \omega_{(2)(1)} \otimes \omega_{(2)(2)}.$$

With this decomposition, a recursion for $a_{\star}^{T}$ is given as $a_{\star}^{T}(\mathbb{I}) = \mathbb{I}$ and for $\omega \in \mathrm{OF}_C \backslash \mathbb{I}$

$$a_{\star}^{T}(\omega) = \sum_{c \in C} \sum_{\Delta_{.}(\omega)} \sum_{\Delta_{\triangleright}(\omega_{(2)})} \left(a_{\star}^{T}(\omega_{(1)}) \times_c a_{\star}^{T}(\omega_{(2)(1)})\right) \langle a(c), \omega_{(2)(2)} \rangle. \tag{79}$$

The recursion is written more compactly as

$$a_{\star}^{T} = \sum_{c \in C} \mu_{.} \circ (\mu_{\times_c} \otimes I) \circ (a_{\star}^{T} \otimes a_{\star}^{T} \otimes a(c)) \circ (I \otimes \Delta_{\triangleright}) \circ \Delta_{.}\,,$$

where $\mu_{.}(\omega \otimes \omega') := \omega \cdot \omega'$, $\mu_{\times_c}(\omega \otimes \omega') := \omega \times_c \omega'$ and $a(c)\colon \mathrm{k}\langle \mathrm{OT}_C \rangle \to \mathrm{k}$ denotes $\omega \mapsto \langle a(c), \omega \rangle$.

See Table 6 on p. 36 where cosubstitution is calculated for all forests up to and including order 4, assuming a is a infinitesimal character.

Since $a\star$ is compatible with $\Delta_{\sqcup\!\sqcup}$ in the sense of (78), it follows that $a_{\star}^{T}$ is a shuffle homomorphism (a character) satisfying

$$a_{\star}^{T}(\omega \sqcup\!\sqcup \omega') = a_{\star}^{T}(\omega) \sqcup\!\sqcup a_{\star}^{T}(\omega').$$

Definition 17 (*Scaling*) For $t \in \mathrm{k}$ define the map $t(c) = tc\colon C \to \mathfrak{g}_{\mathrm{LB}}$. The corresponding substitution $\alpha \mapsto t \star \alpha$ is called *scaling by* t. For a fixed alpha $t \mapsto t \star \alpha$ defines a curve in $\mathrm{LB}(C)$

Note that $t \star \omega = t^{|\omega|}\omega$ and hence $\langle t \star \alpha, \omega \rangle = t^{|\omega|}\langle \alpha, \omega \rangle$ for all $\omega \in \mathrm{OF}_C$.

Definition 18 (*Derivation*) The derivative of a LB-series α, denoted $D\alpha$ is defined as

$$\langle D\alpha, \omega \rangle = |\omega| \langle \alpha, \omega \rangle.$$

Note that if $\mathrm{k} = \mathbb{R}$ we have $D\alpha = \frac{d}{dt}\big|_{t=1} (t \star \alpha)$.

4.8 Exponentials and Logarithms

We have three types of exponential type mappings $\exp^{\cdot}, \exp^{*}, \mathrm{evol}\colon \mathfrak{g}_{\mathrm{LB}} \to G_{\mathrm{LB}}$. These are all 1–1 mappings with an inverse being a kind of logarithm. In the interpretation of vector fields on Lie groups, $\exp^{\cdot}$ defines the geodesics of the connection and $\exp^{*}$ computes the exact flow of a vector field. The third of these, evol, computes

Table 6 Cosubstitution for an infinitesimal character α for all forest up to and including order 4

ω	$\alpha_\star^T(\omega)$

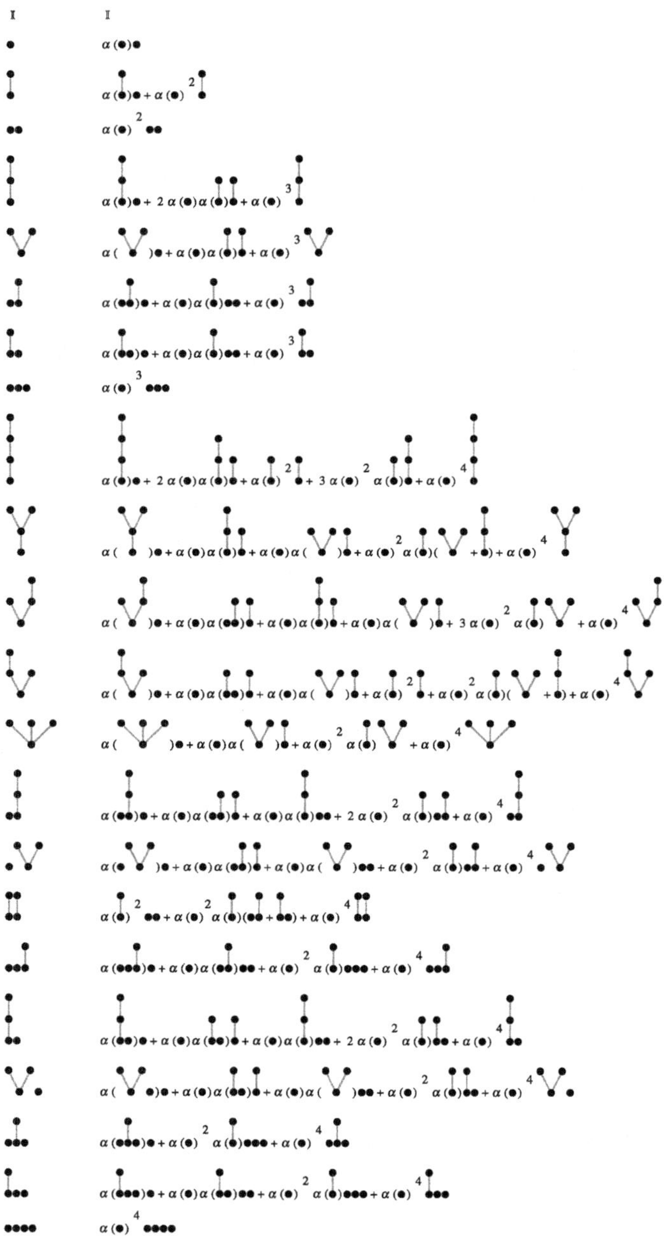

a curve in a Lie group from its development in the Lie algebra i.e. solves an equation of Lie type $y'(t) = y(t)\gamma(t)$ where $\gamma(t) = y^{-1}(t)y'(t)$ is the development of $y(t)$ (left logarithmic derivative). We will have a closer look at these three maps and their inverses.

Definition 19 (*Concatenation exponential*) The *concatenation exponential* $\exp^{\cdot}\colon \mathfrak{g}_{\mathrm{LB}} \to G_{\mathrm{LB}}$ is defined as

$$\exp^{\cdot}(\alpha) = \mathbb{I} + \alpha + \frac{1}{2}\alpha\alpha + \frac{1}{6}\alpha\alpha\alpha + \cdots = \sum_{j=0}^{\infty} \frac{1}{j!}\alpha^{\cdot j}. \tag{80}$$

In the algebra $U(\mathrm{postLie}(C))$, with the grading given by PBW, $U_0 = \mathrm{k}\,\mathbb{I}$, $U_1 = \mathrm{postLie}(C)$ and U_ℓ is generated from U_1 by ℓ-fold shuffle products. Since $\langle \exp^{\cdot}(\alpha), x \sqcup\!\sqcup y\rangle = \langle \exp^{\cdot}(\alpha), x\rangle\langle \exp^{\cdot}(\alpha), y\rangle$ we have the following result.

Lemma 2 *For* $\alpha \in \mathfrak{g}_{\mathrm{LB}}$, *the concatenation exponential* $\exp^{\cdot}(\alpha)$ *is the unique element of* G_{LB} *such that* $\langle \exp^{\cdot}(\alpha), x\rangle = \langle \alpha, x\rangle$ *for all* $x \in \mathrm{postLie}(C)$.

The GL-exponential is similarly defined from the GL product $*$:

Definition 20 (*GL-exponential*) The *GL-exponential* $\exp^*\colon \mathfrak{g}_{\mathrm{LB}} \to G_{\mathrm{LB}}$ is defined as

$$\exp^*(\alpha) = \mathbb{I} + \alpha + \frac{1}{2}\alpha * \alpha + \frac{1}{6}\alpha * \alpha * \alpha + \cdots = \sum_{j=0}^{\infty} \frac{1}{j!}\alpha^{*j}. \tag{81}$$

Recursive formulas for the coefficients of $\exp^*(\bullet)$ are found in [18, 24]. Here we derive a remarkably simple recursion formula based on the magmatic decomposition of OF, to our knowledge not found elsewhere:

Lemma 3 *For* $\omega = \omega_L \times_\bullet \omega_R$ *we have*

$$\langle \exp^*(\bullet), \mathbb{I}\rangle = 1 \tag{82}$$

$$\langle \exp^*(\bullet), \omega\rangle = \frac{1}{|\omega|} \cdot \langle \exp^*(\bullet), \omega_L\rangle \cdot \langle \exp^*(\bullet), \omega_R\rangle, \tag{83}$$

or equivalently

$$\langle \exp^*(\bullet), \omega\rangle = \frac{1}{\omega_{¡}}, \tag{84}$$

where $\omega_{¡}$ *denotes the ordered forest exponential.*

Proof The derivation $D\exp^*(\bullet)$ satisfies $\langle D\exp^*(\bullet), \omega\rangle = |\omega|\langle \exp^*(\bullet), \omega\rangle$. On the other hand, since the t-scaling of the exponential is $t \star \exp^*(\bullet) = \exp^*(t\bullet)$ we find

$$D\exp^*(\bullet) = \left.\frac{d}{dt}\right|_{t=1} \exp^*(t\bullet) = \exp^*(t\bullet) * \bullet\big|_{t=1} = \exp^*(\bullet) * \bullet = \exp^*(\bullet)(\exp^*(\bullet) \rhd \bullet),$$

where we in the rightmost equality use (32) and $\Delta_{\sqcup\sqcup}(\exp^*(\bullet)) = \exp^*(\bullet) \otimes \exp^*(\bullet)$, since $\exp^*(\bullet) \in G_{\mathrm{LB}}$. Since $\omega_L \times_\bullet \omega_R = \omega_L(\omega_R \rhd \bullet)$ we find

$$\begin{aligned}\langle \exp^*(\bullet), \omega\rangle &= \frac{1}{|\omega|}\cdot \langle D\exp^*(\bullet), \omega\rangle = \frac{1}{|\omega|}\cdot \langle \exp^*(\bullet)(\exp^*(\bullet) \rhd \bullet), \omega_L(\omega_R \rhd \bullet)\rangle \\ &= \frac{1}{|\omega|}\cdot \langle \exp^*(\bullet), \omega_L\rangle \cdot \langle \exp^*(\bullet), \omega_R\rangle.\end{aligned}$$

□

The exponential is thus given as

$$\exp^*(\bullet) = \sum_{\omega\in \mathrm{OF}} \frac{\omega}{\omega¡}, \tag{85}$$

which justifies the naming of ¡ as a factorial function.

The computation of $\exp^*(\alpha)$ for an arbitrary $\alpha \in \mathfrak{g}_{\mathrm{LB}}$ can be done by the substitution: If $a(\bullet) = \alpha$ then

$$\begin{aligned}\langle \exp^* \alpha, \omega\rangle &= \langle \exp^* a(\bullet), \omega\rangle = \langle \exp^*(a \star \bullet), \omega\rangle \\ &= \langle a \star \exp^*(\bullet), \omega\rangle = \langle \exp^*(\bullet), a^t_\star(\omega)\rangle = \frac{1}{a^t_\star(\omega)¡},\end{aligned}$$

where the forest exponential ¡ is extended to polynomials by linearity.

Backward error.
Whereas $\exp^*\colon \mathfrak{g}_{\mathrm{LB}} \to G_{\mathrm{LB}}$ computes the exact flow operator, the inverse $\log^*\colon G_{\mathrm{LB}} \to \mathfrak{g}_{\mathrm{LB}}$ inputs a flow map, and computes the vector field generating this flow. In numerical analysis this is called the backward error analysis operator and is an important tool for analysing numerical integrators. The GL-logarithm $\log^*$ is defined for $\alpha \in G_{\mathrm{LB}}$ as

$$\log^* \alpha = \sum_{n=1}^{\infty} \frac{(-1)^{n-1}}{n}(\alpha - \delta)^{*n},$$

where $\delta \in G_{\mathrm{LB}}$ is the identity in the Lie–Butcher group, given as $\langle \delta, \mathbb{I}\rangle = 1$ and $\langle \delta, \omega\rangle = 0$ for $\omega \in \mathrm{OF}_C\backslash\{\mathbb{I}\}$. The GL-logarithm can be computed via its dual operation, the *eulerian idempotent* $e \in \mathrm{End}(\mathrm{k}\langle \mathrm{OF}_C\rangle)$ such that

$$\langle \log^*(\alpha), \omega\rangle = \langle \alpha, e(\omega)\rangle.$$

To compute e, we introduce the augmented GL-coproduct defined as

$$\overline{\Delta}_*(\omega) := \Delta_*(\omega) - \omega \otimes \mathbb{I} - \mathbb{I} \otimes \omega.$$

The recursion for $\Delta_*(\omega)$ (66)–(67) yields the following recursion for $\overline{\Delta}_*(\omega)$:

$$\overline{\Delta}_*(\mathbb{I}) = -\mathbb{I} \otimes \mathbb{I} \tag{86}$$

$$\overline{\Delta}_*(\omega_L \times_c \omega_R) = (\overline{\Delta}_*(\omega_L) + \omega_L \otimes \mathbb{I}) ⧢ \times_c (\overline{\Delta}_*(\omega_R) + \omega_R \otimes \mathbb{I}). \tag{87}$$

The eulerian idempotent is computed as

$$e(\omega) = \sum_{n\geq 1} \frac{(-1)^{n-1}}{n} ⧢_n \overline{\Delta}_*^{n-1}(\omega),$$

where $⧢_n$ is the shuffle of n arguments and $\overline{\Delta}_*^n$ is the n-fold repeated application of the augmented GL coproduct. See Table 7 on p. 40 for calculations of the eulerian idempotent for all forests up to and including order 4.

Since α is a character, we obtain the following formula for the backward error

$$\langle \log^*(\alpha), \omega\rangle = \sum_{n\geq 1} \frac{(-1)^{n-1}}{n} \sum_{\overline{\Delta}_*^{n-1}(\omega)} \langle \alpha, \omega_{(1)}\rangle \cdot \langle \alpha, \omega_{(2)}\rangle \cdots \langle \alpha, \omega_{(n)}\rangle. \tag{88}$$

The development.
For a curve $y(t)$ on a Lie group G, the *development* is a curve $\gamma(t) \in \mathfrak{g}$ such that $y'(t) = \gamma(t)y(t)$, thus $\gamma(t) = y'(t)y(t)^{-1}$ is given by the logarithmic derivative. There is a corresponding[9] combinatorial operation on G_{LB}, given by a linear map $L\colon \mathrm{k}\langle \mathrm{OT}_C\rangle \to \mathrm{k}\langle \mathrm{OT}_C\rangle$ called the *Dynkin operator*, such that

$$\langle \alpha^{\cdot -1} \cdot D\alpha, \omega\rangle = \langle \alpha, L(\omega)\rangle \qquad \text{for every } \alpha \in G_{\mathrm{LB}}. \tag{89}$$

Lemma 4 *The Dynkin operator L is computed as a convolution of endomorphisms, $L, S_., D \in \mathrm{End}(\mathscr{H}'_.)$,*

$$L = S_. * D := ⧢(S_. \otimes D)\Delta_.,$$

where $\mathscr{H}'_.$ is the Hopf algebra on $\mathrm{k}\langle \mathrm{OT}_C\rangle$ with shuffle $⧢$ as product, de-concatenation $\Delta_.$ coproduct and antipode $S_.$, and with grading $|\omega|$ counting nodes in the forest. Explicitly we have

$$L(\omega) = \sum_{\Delta_.(\omega)} S_.(\omega_{(1)}) ⧢ \omega_{(2)} |\omega_{(2)}|. \tag{90}$$

[9]Since the action of differentiation operators composes contravariantly, the order of right and left is swapped in the mapping from LB-series to differential equations on manifolds.

Table 7 Dynkin map L and Eulerian idempotent e for all forest up to and including order 4. Note that the numbers under the terms are the coefficients to the terms

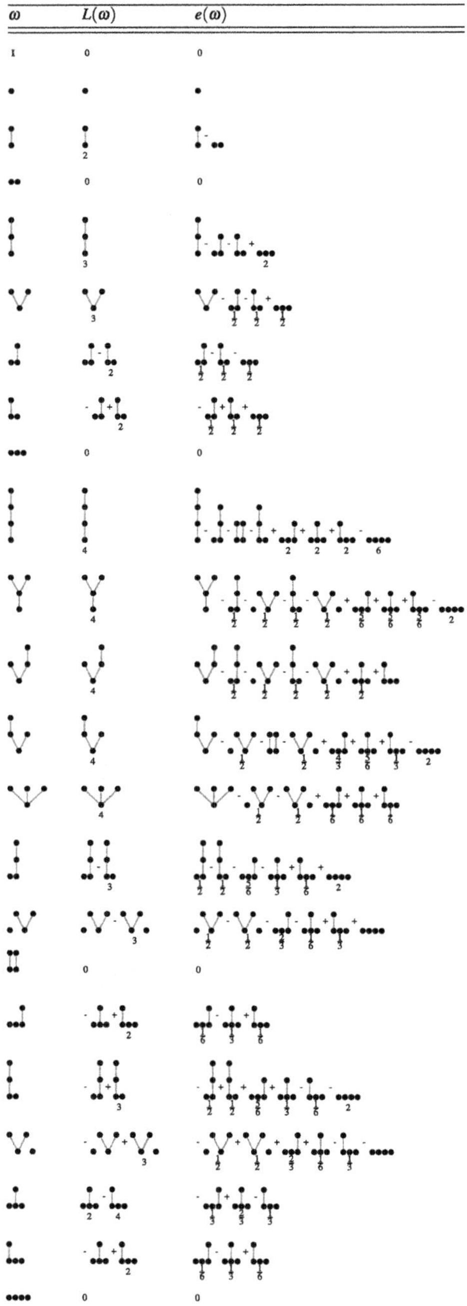

Proof

$$\langle \alpha^{\cdot -1} \cdot D\alpha, \omega \rangle = \langle \alpha^{\cdot -1} \otimes D\alpha, \Delta_{\cdot}\omega \rangle = \sum_{\Delta_{\cdot}(\omega)} \langle \alpha, S_{\cdot}(\omega_{(1)}) \rangle \langle \alpha, D(\omega_{(2)}) \rangle$$
$$= \langle \alpha, S_{\cdot}(\omega_{(1)}) \sqcup\!\sqcup D\omega_{(2)} \rangle = \langle \alpha, (S_{\cdot} * D)(\omega) \rangle.$$

□

Table 7 on p. 40 contain the Dynkin map applied to all ordered forests up to and including order 4.

The inverse of the Dynkin map, denoted evol: $\mathfrak{g}_{\mathrm{LB}} \to G_{\mathrm{LB}}$, yields a formal LB-series solution to equations of Lie type, $y'(t) = \gamma(t)y(t)$, for $y(t) \in G$, where $\gamma(t) \in \mathfrak{g}$ is given by a LB-series. In [10] it is proven that

$$\mathrm{evol}(\alpha) = \mathbb{I} + \sum_{n\geq 1} \sum_{\substack{n_1+\cdots+n_k=n \\ n_j>0}} \frac{\alpha_{n_1} * \alpha_{n_2} * \cdots * \alpha_{n_k}}{n_1(n_1+n_2)\cdots(n_1+n_2+\cdots+n_k)},$$

where $\alpha = \sum_{k\geq 1} \alpha_k$ and $|\alpha_k| = k$ and $*$ is the convolution in $\mathscr{H}'_{\cdot}$. For $\omega \in \mathrm{OF}_C \backslash \{\mathbb{I}\}$ this yields

$$\langle \mathrm{evol}(\alpha), \omega \rangle = \sum_{n\geq 1} \sum_{\Delta_{\cdot}^{n-1}(\omega)} \frac{\langle \alpha, \omega_{(1)} \rangle \cdot \langle \alpha, \omega_{(2)} \rangle \cdots \langle \alpha, \omega_{(n)} \rangle}{|\omega_{(1)}| \cdot \big(|\omega_{(1)}| + |\omega_{(2)}|\big) \cdots \big(|\omega_{(1)}| + |\omega_{(2)}| + \cdots + |\omega_{(n)}|\big)},$$

and from this we find the recursion formulae

$$\langle \mathrm{evol}(\alpha), \mathbb{I} \rangle = 1 \tag{91}$$
$$\langle \mathrm{evol}(\alpha), \omega \rangle = \frac{1}{|\omega|} \sum_{\Delta_{\cdot}(\omega)} \langle \mathrm{evol}(\alpha), \omega_{(1)} \rangle \cdot \langle \alpha, \omega_{(2)} \rangle \quad \text{for } \omega \in \mathrm{OF}_C \backslash \{\mathbb{I}\}. \tag{92}$$

5 Concluding Remarks

In this paper we have summarized the algebraic structures behind Lie–Butcher series. For the purpose of computer implementations, we have derived recursive formulae for all the basic operations on Lie–Butcher series that have appeared in the literature over the last decade. The simplicity of the recursive formulae are surprising to us. The GL-coproduct, the GL-exponential, the backward error and the inverse Dynkin map are in our opinion significantly simpler in their recursive formulations than the direct.

5.1 *Programming in Haskell*

We are in the process of making a software library for computations with post-Lie algebras and Lie-Butcher series. As we have seen in this paper, many of the structures and operations have nice recursive definitions. Functional programming languages are well suited for this type of implementation. Haskell is one of the most popular functional programming languages, it is named after the logician Haskell B. Curry. The development of Haskell started in 1987 after a meeting at the conference on *Functional Programming Languages and Computer Architecture* (FPCA 87), where the need for common language for research in functional programming languages was recognized. Haskell has since grown into a mature programming language, not only used in functional programming research but also in the industry.

Not only do Haskell encourage recursive definitions of functions, it also has algebraic data types which give us the opportunity to define recursive data types.

Functional programming language will usually result in shorter and more precise code compared to imperative languages. Mathematical ideas are often straightforward to translate into a functional language.

A feature of Haskell that come in handy when working with infinite structures is lazy evaluation, meaning that an expression will not be computed before it is needed. This is an excellent feature for working with Lie-Butcher series, since these are infinite series. The infinite series can only be evaluated on finite data, and when such a computation is requested the system performs the necessary intermediate computations.

Mathematical ideas such as functors and monads are very important concept in Haskell, for example IO in Haskell is implemented as a monad. Another example is the vector space constructor in Haskell is a monad, which makes it very easy to linear extend a function on basis element to a linear function between vector spaces. Two other examples of monads are the free functor and the universal enveloping functor. The elementary differential map of B-series and Lie–Butcher series fits also nicely into this picture.

Finally, we remark that the proof assistant Coq can output Haskell code, so for critical parts of the software one can prove correctness of the implementation in Coq and then output this as verified Haskell code.

References

1. Benettin, G., Giorgilli, A.: On the hamiltonian interpolation of near-to-the identity symplectic mappings with application to symplectic integration algorithms. J. Stat. Phys. **74**(5–6), 1117–1143 (1994)
2. Butcher, J.C.: Coefficients for the study of Runge-Kutta integration processes. J. Aust. Math. Soc. **3**(02), 185–201 (1963)
3. Butcher, J.C.: An algebraic theory of integration methods. Math. Comput. **26**(117), 79–106 (1972)

4. Calaque, D., Ebrahimi-Fard, K., Manchon, D.: Two interacting Hopf algebras of trees: a Hopf-algebraic approach to composition and substitution of B-series. Adv. Appl. Math. **47**(2), 282–308 (2011)
5. Cayley, A.: On the theory of the analytical forms called trees. Philos. Mag. **13**(19), 4–9 (1857)
6. Chapoton, F., Livernet, M.: Pre-Lie algebras and the rooted trees operad. Int. Math. Res. Not. **2001**(8), 395–408 (2001)
7. Chartier, P., Hairer, E., Vilmart, G.: Numerical integrators based on modified differential equations. Math. Comput. **76**(260), 1941 (2007)
8. Chartier, P., Hairer, E., Vilmart, G.: Algebraic structures of B-series. Found. Comput. Math. **10**(4), 407–427 (2010)
9. Dzhumadil'daev, A., Löfwall, C.: Trees, free right-symmetric algebras, free Novikov algebras and identities. Homol. Homotopy Appl. **4**(2), 165–190 (2002)
10. Ebrahimi-Fard, K., Gracia-Bondía, J.M., Patras, F.: A Lie theoretic approach to renormalization. Commun. Math. Phys. **276**(2), 519–549 (2007)
11. Ebrahimi-Fard, K., Lundervold, A., Mencattini, I., Munthe-Kaas, H.Z.: Post-Lie algebras and isospectral flows. Symmetry Integr. Geom. Methods Appl. (SIGMA) **11**(93) (2015)
12. Ebrahimi-Fard, K., Lundervold, A., Munthe-Kaas, H.: On the Lie enveloping algebra of a post-Lie algebra. J. Lie Theory **25**(4), 1139–1165 (2015)
13. Hairer, E., Lubich, C., Wanner, G.: Geometric Numerical Integration: Structure-Preserving Algorithms for Ordinary Differential Equations, Vol. 31. Springer series in computational mathematics (2006)
14. Hairer, E., Wanner, G.: On the Butcher group and general multi-value methods. Computing **13**(1), 1–15 (1974)
15. Iserles, A., Munthe-Kaas, H.Z., Nørsett, S.P., Zanna, A.: Lie-group methods. Acta Numer. **2000**(9), 215–365 (2000)
16. Lundervold, A., Munthe-Kaas, H.: Backward error analysis and the substitution law for Lie group integrators. Found. Comput. Math. **13**(2), 161–186 (2013)
17. Munthe-Kaas, H.: Lie-Butcher theory for Runge-Kutta methods. BIT Numer. Math. **35**(4), 572–587 (1995)
18. Munthe-Kaas, H.: Runge-Kutta methods on Lie groups. BIT Numer. Math. **38**(1), 92–111 (1998)
19. Munthe-Kaas, H., Krogstad, S.: On enumeration problems in Lie-Butcher theory. Future Gen. Comput. Syst. **19**(7), 1197–1205 (2003)
20. Munthe-Kaas, H., Owren, B.: Computations in a free Lie algebra. Philos. Trans. R. Soc. Lond. A Math. Phys. Eng. Sci. **357**(1754), 957–981 (1999)
21. Munthe-Kaas, H.Z., Lundervold, A.: On post-Lie algebras, Lie-butcher series and moving frames. Found. Comput. Math. **13**(4), 583–613 (2013)
22. Munthe-Kaas, H.Z., Wright, W.M.: On the Hopf algebraic structure of Lie group integrators. Found. Comput. Math. **8**(2), 227–257 (2008)
23. Oudom, J.-M., Guin, D.: On the Lie enveloping algebra of a pre-Lie algebra. J. K-Theory K-Theory Appl. Algebra Geom. Topol. **2**(01), 147–167 (2008)
24. Owren, B., Marthinsen, A.: Runge-Kutta methods adapted to manifolds and based on rigid frames. BIT Numer. Math. **39**(1), 116–142 (1999)
25. Philippe C., Hairer, E., Vilmart, G.: A substitution law for B-series vector fields. Technical Report 5498, INRIA (2005)
26. Reutenauer, C.: Free Lie Algebras. Oxford University Press (1993)
27. Vallette, B.: Homology of generalized partition posets. J. Pure Appl. Algebra **208**(2), 699–725 (2007)

Averaging and Computing Normal Forms with Word Series Algorithms

Ander Murua and Jesús M. Sanz-Serna

Abstract In the first part of the present work we consider periodically or quasiperiodically forced systems of the form $(d/dt)x = \varepsilon f(x, t\omega)$, where $\varepsilon \ll 1$, $\omega \in \mathbb{R}^d$ is a nonresonant vector of frequencies and $f(x, \theta)$ is 2π-periodic in each of the d components of θ (i.e. $\theta \in \mathbb{T}^d$). We describe in detail a technique for explicitly finding a change of variables $x = u(X, \theta; \varepsilon)$ and an (autonomous) averaged system $(d/dt)X = \varepsilon F(X; \varepsilon)$ so that, formally, the solutions of the given system may be expressed in terms of the solutions of the averaged system by means of the relation $x(t) = u(X(t), t\omega; \varepsilon)$. Here u and F are found as series whose terms consist of vector-valued maps weighted by suitable scalar coefficients. The maps are easily written down by combining the Fourier coefficients of f and the coefficients are found with the help of simple recursions. Furthermore these coefficients are *universal* in the sense that they do not depend on the particular f under consideration. In the second part of the contribution, we study problems of the form $(d/dt)x = g(x) + f(x)$, where one knows how to integrate the 'unperturbed' problem $(d/dt)x = g(x)$ and f is a perturbation satisfying appropriate hypotheses. It is shown how to explicitly rewrite the system in the 'normal form' $(d/dt)x = \bar{g}(x) + \bar{f}(x)$, where $\bar{g}$ and $\bar{f}$ are *commuting* vector fields and the flow of $(d/dt)x = \bar{g}(x)$ is conjugate to that of the unperturbed $(d/dt)x = g(x)$. In Hamiltonian problems the normal form directly leads to the explicit construction of formal invariants of motion. Again, $\bar{g}$, $\bar{f}$ and the invariants are written as series consisting of known vector-valued maps and universal scalar coefficients that may be found recursively.

A. Murua (✉)
Konputazio Zientziak eta A. A. Saila, Informatika Fakultatea, UPV/EHU,
20018 Donostia–San Sebastián, Spain
e-mail: Ander.Murua@ehu.eus

J. M. Sanz-Serna
Departamento de Matemáticas, Universidad Carlos III de Madrid, 28911 Leganés,
Madrid, Spain
e-mail: jmsanzserna@gmail.com

K. Ebrahimi-Fard and M. Barbero Liñán (eds.), *Discrete Mechanics, Geometric Integration and Lie–Butcher Series*, Springer Proceedings in Mathematics & Statistics 267, https://doi.org/10.1007/978-3-030-01397-4_4

Keywords Averaging · High-order averaging · Quasi-stroboscopic averaging Hamiltonian problems · Near-integrable systems · Normal forms

Mathematics Subject Classification (2010) 34C20 · 34C29 · 70H05

1 Introduction

In this article we illustrate how to use word series to manipulate systems of differential equations. Specifically we deal with the questions of high-order averaging of periodically or quasiperiodically forced systems and reduction to normal form of perturbations of integrable systems. The manipulations require operations with complex numbers rather than with vector fields.

Word series are patterned after B-series [10], a well-known tool to analyse numerical integrators (see [18] for a summary of the uses of formal series in the numerical analysis of differential equations). While B-series are parameterized by rooted trees, word series [14] possess one term for each word w that may be composed with the letters of a suitable alphabet A [15]. Each term $\delta_w f_w$ of a word series is the product of a scalar coefficient δ_w and a vector field f_w. The vector fields f_w may be immediately constructed and depend on the differential system under consideration. The coefficients δ_w are *universal*, in the sense that they do not change with the particular differential system being studied. Series of *differential operators* parameterized by words (Chen–Fliess series) are very common, e.g. in control theory [9] and dynamical systems [7] and have also been used in numerical analysis (see [11] among others). As discussed in [14], word series are mathematically equivalent to Chen–Fliess series, but being series of functions they are handled in a way very similar to the way numerical analysts handle B-series. In the present work, as in [3–5, 13], the formal series techniques originally introduced to analyze numerical integrators are applied to the study of dynamical systems.

The structure of this article is as follows. The use of word series is briefly reviewed in Sect. 2. Section 3 addresses the problem of averaging periodically or quasiperiodically forced systems. We find a change of variables that formally reduces the system to time-independent (averaged) form. Both the change of variables and the averaged system are expressed by means of word series with universal coefficients that may be computed by means of simple recursions. The averaged system obtained in this way has favourable geometric properties. It is equivariant with respect to arbitrary changes of variables, i.e., the operations of changing variables and averaging commute. In addition averaging a Hamiltonian/divergence free/... system results in a system that is also Hamiltonian/divergence free/.... Sections 4 and 5 are devoted to the reduction to normal form of general classes of perturbed problems.

Let us discuss the relation between this article and our earlier contributions. The problems envisaged here have been considered in [4]. However the treatment in [4] makes heavy use of B-series; word series results are derived, by means of the Hopf algebra techniques of [11], as a byproduct of B-series results. Here the circuitous derivations of [4] are avoided by working throughout with word series, without any reference to B-series. One of our aims when writing this article has been to provide potential users of word series techniques in application problems with a more focused, brief and clear guide than [4] provides. In addition, the class of perturbed problems considered in Sects. 4 and 5 below is much wider than that considered in [4]. In [13] we have recently addressed the reduction of perturbed problems to normal forms. The treatment in [13] is based on the application of successive changes of variables; here the normal form is directly obtained in the originally given variables. An application of word series techniques to stochastic problems is provided in [1]. The article [12] presents an application of the high-order averaging described here to a problem arising in vibrational resonance.

All the developments in the article use formal series of smooth maps. To streamline the presentation the words 'formal' and 'smooth' are often omitted. By truncating the formal expansions obtained in this article it is possible to obtain nonformal results, as in [5] or [6], but we shall not be concerned with such a task.

2 Word Series

We begin by presenting the most important rules for handling word series. For proofs and additional properties of word series, the reader is referred to [14].

2.1 Defining Word Series

Assume that A is a finite or infinite countable set of indices (the alphabet) and that for each element (letter) $\ell \in A$, $f_\ell(y)$ is a map $f_\ell : \mathbb{C}^d \to \mathbb{C}^d$. Associated with each nonempty word $\ell_1 \cdots \ell_n$ constructed with letters from the alphabet, there is a *word basis function*. These are defined recursively by

$$f_{\ell_1\cdots\ell_n}(y) = f'_{\ell_2\cdots\ell_n}(y) f_{\ell_1}(y), \quad n > 1,$$

where $f'_{\ell_2\cdots\ell_n}(y)$ is the Jacobian matrix of $f_{\ell_2\cdots\ell_n}(y)$. For the empty word, the corresponding basis function is the identity map $y \mapsto y$. The set of all words (including the empty word $\emptyset$) will be denoted by $\mathcal{W}$ and the symbol $\mathbb{C}^{\mathcal{W}}$ will be used to refer to the vector space of all mappings $\delta : \mathcal{W} \to \mathbb{C}$. For $\delta \in \mathbb{C}^{\mathcal{W}}$ and $w \in \mathcal{W}$, δ_w is the complex number that δ associates with w. To each $\delta \in \mathbb{C}^{\mathcal{W}}$ there corresponds a *word series* (relative to the mappings f_ℓ); this is the formal series

$$W_\delta(y) = \sum_{\delta \in \mathcal{W}} \delta_w f_w(y).$$

The numbers δ_w, $w \in \mathcal{W}$, are the *coefficients* of the series.

Let us present an example. If for each letter $\ell \in A$, $\lambda_\ell(t)$ is a scalar-valued function of the real variable t, the solution of initial value problem

$$\frac{d}{dt}y = \sum_{\ell \in A} \lambda_\ell(t) f_\ell(y), \quad y(t_0) = y_0 \in \mathbb{C}^D \tag{1}$$

has a formal expansion in terms of word series given by

$$y(t) = W_{\alpha(t;t_0)}(y_0), \tag{2}$$

where, for each t, t_0, the coefficients $\alpha_w(t; t_0)$ are the iterated integrals

$$\alpha_{\ell_1\cdots\ell_n}(t; t_0) = \int_{t_0}^{t} dt_n\, \lambda_{\ell_n}(t_n) \int_{t_0}^{t_n} dt_{n-1}\, \lambda_{\ell_{n-1}}(t_{n-1}) \cdots \int_{t_0}^{t_2} dt_1\, \lambda_{\ell_1}(t_1). \tag{3}$$

This series representation, whose standard derivation may be seen in e.g. [5] or [14], is essentially the Chen series used in control theory. (An alternative derivation is presented below.) Of much importance in what follows is the fact that the coefficients $\alpha_w(t; t_0)$ depend only on the $\lambda_\ell(t)$ in (1) and do not change with the vector fields $f_\ell(y)$; on the contrary, the word basis functions $f_w(y)$ depend on the $f_\ell(y)$ and do not change with the $\lambda_\ell(t)$.

2.2 The Convolution Product

The *convolution product* $\delta \star \delta' \in \mathbb{C}^{\mathcal{W}}$ of two elements $\delta, \delta' \in \mathbb{C}^{\mathcal{W}}$ is defined by

$$(\delta \star \delta')_{\ell_1\cdots\ell_n} = \delta_\emptyset \delta'_{\ell_1\cdots\ell_n} + \sum_{j=1}^{n-1} \delta_{\ell_1\cdots\ell_j} \delta'_{\ell_{j+1}\cdots\ell_n} + \delta_{\ell_1\cdots\ell_n}\delta'_\emptyset, \quad n \geq 1$$

($(\delta \star \delta')_\emptyset = \delta_\emptyset \delta'_\emptyset$). The operation $\star$ is not commutative, but it is associative and has a unit (the element $1\!\!1 \in \mathbb{C}^{\mathcal{W}}$ with $1\!\!1_\emptyset = 1$ and $1\!\!1_w = 0$ for $w \neq \emptyset$).

If w and w' are words, their *shuffle product* will be denoted by $w \sqcup\!\sqcup w'$; this is the formal sum of all words that may be formed by interleaving the letters of w with those of w' without altering the order in which those letters appear within w or w' (e.g., $\ell m \sqcup\!\sqcup n = \ell mn + \ell nm + n\ell m$). The set $\mathcal{G}$ consists of those $\gamma \in \mathbb{C}^{\mathcal{W}}$ that satisfy the following *shuffle relations:* $\gamma_\emptyset = 1$ and, for each $w, w' \in \mathcal{W}$,

$$\gamma_w \gamma_{w'} = \sum_{j=1}^{N} \gamma_{w_j} \quad \text{if} \quad w \sqcup\!\sqcup w' = \sum_{j=1}^{N} w_j.$$

This set is a group for the operation $\star$. For each t and t_0 the element $\alpha(t; t_0) \in \mathbb{C}^{\mathcal{W}}$ in (3) belongs to the group $\mathcal{G}$.

For $\gamma \in \mathcal{G}$, $\delta \in \mathbb{C}^{\mathcal{W}}$,

$$W_\delta\big(W_\gamma(x)\big) = W_{\gamma\star\delta}(x). \tag{4}$$

In words: the substitution of $W_\gamma(x)$ in an arbitrary word series $W_\delta(x)$ gives rise to a new word series whose coefficients are given by the convolution product $\gamma \star \delta$. We emphasize that this result does not hold for arbitrary $\gamma \in \mathbb{C}^{\mathcal{W}}$, the hypothesis $\gamma \in \mathcal{G}$ is essential.

Another property of the word series $W_\gamma(y)$ with $\gamma \in \mathcal{G}$ is its *equivariance* [18] with respect to arbitrary changes of variables $y = C(\bar{y})$. If $\bar{f}_\ell(\bar{y})$ is the result (pull-back) of changing variables in the field $f_\ell(y)$, i.e.,

$$\bar{f}_\ell(\bar{y}) = C'(\bar{y})^{-1} f_\ell(C(\bar{y})),$$

and $\bar{W}_\gamma(\bar{y})$ denotes the word series with coefficients γ_w constructed from the fields $\bar{f}_\ell(\bar{y})$, then

$$C\big(\bar{W}_\gamma(\bar{y})\big) = W_\gamma(C(\bar{y})).$$

We denote by $\mathfrak{g}$ the vector subspace of $\mathbb{C}^{\mathcal{W}}$ consisting of those β that satisfy the following shuffle relations: $\beta_\emptyset = 0$ and for each pair of nonempty words w, w',

$$\sum_{j=1}^{N} \beta_{w_j} = 0 \quad \text{if} \quad w \sqcup\!\sqcup\, w' = \sum_{j=1}^{N} w_j.$$

It is easily proved that the elements $\beta \in \mathfrak{g}$ are precisely the velocities $(d/dt)\gamma(0)$ at $t = 0$ of the smooth curves $t \mapsto \gamma(t) \in \mathcal{G}$ with $\gamma(0) = \mathbb{1}$, i.e., if $\mathcal{G}$ is formally viewed as a Lie group, then $\mathfrak{g}$ is the corresponding Lie algebra. In fact, $\mathcal{G}$ and $\mathfrak{g}$ are the group of characters and the Lie algebra of infinitesimal characters of the shuffle Hopf algebra (see [14] for details).

2.3 Universal Formulations

Let us consider once more the differential system (1). Define, for fixed t, $\beta(t) \in \mathfrak{g} \subset \mathbb{C}^{\mathcal{W}}$ by $\beta_\ell(t) = \lambda_\ell(t)$, for each $\ell \in A$, and $\beta_w(t) = 0$ if the word w is empty or has ≥ 2 letters. Then the right hand-side of (1) is simply the word series $W_{\beta(t)}(y)$. We look for the solution $y(t)$ in the word series form (2), with undetermined coefficients $\alpha_w(t; t_0)$ that have to be determined and *have to belong* to the group $\mathcal{G}$. By using the formula (4), we may write

$$\frac{\partial}{\partial t} W_{\alpha(t;t_0)}(y_0) = W_{\beta(t)}\big(W_{\alpha(t;t_0)}(y_0)\big) = W_{\alpha(t;t_0)\star\beta(t)}(y_0),$$
$$W_{\alpha(t_0,t_0)}(y_0) = y_0 = W_{1\!\!1}(y_0),$$

and these equations will be satisfied if

$$\frac{\partial}{\partial t}\alpha(t;t_0) = \alpha(t;t_0)\star\beta(t), \qquad \alpha(t_0,t_0) = 1\!\!1\,. \tag{5}$$

This is an initial value problem for the curve $t \mapsto \alpha(t;t_0)$ in the group $\mathcal{G}$ (t_0 is a parameter) and may be uniquely solved by successively determining the values $\alpha_w(t;t_0)$ for words of increasing length. In fact, for the empty word, the requirement $\alpha(t;t_0) \in \mathcal{G}$ implies $\alpha_\emptyset(t;t_0) = 1$. For words with one letter $\ell \in A$, using $\beta_\emptyset(t) = 0$ and the definition of the convolution product $\star$, we have the conditions

$$\frac{\partial}{\partial t}\alpha_\ell(t;t_0) = \beta_\ell(t)\alpha_\emptyset(t;t_0) = \lambda_\ell(t), \quad \alpha_\ell(t_0,t_0) = 0,$$

that lead to

$$\alpha_\ell(t;t_0) = \int_{t_0}^{t} dt_1 \lambda_\ell(t_1).$$

This procedure may be continued (see [14] for details) to determine uniquely $\alpha_w(t;t_0)$ for all words $w \in \mathcal{W}$. In addition, for each t and t_0, the element $\alpha(t;t_0) \in \mathbb{C}^{\mathcal{W}}$ found in this way belongs to $\mathcal{G}$, as it was desired. Of course this element coincides with that defined in (3).

In going from (1) to (5) we move from an initial value problem for the vector-valued function $y(t)$ to a seemingly more complicated initial value problem for the function $\alpha(t)$ with values in $\mathcal{G}$. However the abstract problem in $\mathcal{G}$ is linear and easily solvable. Of equal importance to us is the fact that (5) is *universal* in the sense that, once it has been integrated, one readily writes, by changing the word basis functions, the solution (2) of *each* problem obtained by replacing in (1) the mappings $f_\ell(y)$ by other choices. In particular the universal character of the formulation implies that (5) is independent of the dimension D of (1).

3 Averaging of Quasiperiodically Forced Systems

In this section, we consider the oscillatory initial value problem

$$\frac{d}{dt}y = \varepsilon f(y, t\omega), \qquad y(t_0) = y_0 \in \mathbb{C}^D, \tag{6}$$

in a long interval $t_0 \leq t \leq t_0 + L/\varepsilon$. The vector field $f(y, \theta)$ is 2π-periodic in each of the scalar components (angles) $\theta_1, \ldots, \theta_d$ of θ (i.e., $\theta \in \mathbb{T}^d$) and ω is a constant vector of frequencies $\omega_1, \ldots, \omega_d$. These are assumed to be *nonresonant* i.e. $\mathbf{k} \cdot \omega \neq 0$ for each multiindex $\mathbf{k} \in \mathbb{Z}^d$, $\mathbf{k} \neq \mathbf{0}$; resonant problems may be rewritten in nonresonant form by reducing the number of frequencies. Thus the forcing in (6) is quasiperiodic if $d > 1$ and periodic if $d = 1$. Our aim is to find a time-dependent change of variables $y = U(Y, t\omega; \varepsilon)$ that formally brings the differential system (6) into autonomous form [16]. Our approach is based on a *universal* formulation, analogous to the one we used above to deal with (1).

3.1 The Solution of the Oscillatory Problem

After Fourier expanding

$$f(y, \theta) = \sum_{\mathbf{k} \in \mathbb{Z}^d} \exp(i\mathbf{k} \cdot \theta) \widehat{f_{\mathbf{k}}}(y),$$

the problem (6) becomes a particular case of (1); each letter ℓ is a multiindex $\mathbf{k} \in \mathbb{Z}^d$, $f_\ell(y) = f_{\mathbf{k}}(y) = \varepsilon \widehat{f_{\mathbf{k}}}(y)$, and

$$\lambda_\ell(t) = \exp(i\mathbf{k} \cdot \omega t). \tag{7}$$

Each word basis function $f_w(y)$ contains the factor ε^n if w has length n. The first few coefficients (iterated integrals) $\alpha_w(t; t_0)$ in (3) are easily computed; here are a few instances

$$\begin{aligned}
\alpha_\emptyset(t; t_0) &= 1, \\
\alpha_{\mathbf{0}}(t; t_0) &= t - t_0, \\
\alpha_{\mathbf{k}}(t; t_0) &= \frac{i\big(\exp(i\mathbf{k} \cdot \omega t_0) - \exp(i\mathbf{k} \cdot \omega t)\big)}{\mathbf{k} \cdot \omega}, \quad \mathbf{k} \neq \mathbf{0}, \\
\alpha_{\mathbf{00}}(t; t_0) &= \frac{(t - t_0)^2}{2}, \\
\alpha_{\mathbf{kl}}(t; t_0) &= \frac{i(t - t_0)}{\mathbf{k} \cdot \omega} + \frac{1 - \exp(i\mathbf{k} \cdot \omega t)\exp(-i\mathbf{k} \cdot \omega t_0)}{(\mathbf{k} \cdot \omega)^2}, \quad \mathbf{k} \neq \mathbf{0},\ \mathbf{l} = -\mathbf{k}.
\end{aligned} \tag{8}$$

Note the oscillatory components present in some of the coefficients.

The following result shows how the coefficients $\alpha_w(t; t_0)$ can be determined recursively without explicitly carrying out the integrations in (3).

Proposition 1 *The coefficients (3) with the $\lambda_\ell(t)$ given by (7) are uniquely determined by the recursive formulas*

$$
\begin{aligned}
\alpha_{\mathbf{k}}(t;t_0) &= \frac{i\big(\exp(i\mathbf{k}\cdot\omega t_0)-\exp(i\mathbf{k}\cdot\omega t)\big)}{\mathbf{k}\cdot\omega},\\
\alpha_{\mathbf{0}^r}(t;t_0) &= (t-t_0)^r/r!,\\
\alpha_{\mathbf{0}^r\mathbf{k}}(t;t_0) &= \frac{i}{\mathbf{k}\cdot\omega}(\alpha_{\mathbf{0}^{r-1}\mathbf{k}}(t;t_0)-\alpha_{\mathbf{0}^r}(t;t_0)e^{i\mathbf{k}\cdot\omega t}),\\
\alpha_{\mathbf{k}\mathbf{l}_1\cdots\mathbf{l}_s}(t;t_0) &= \frac{i}{\mathbf{k}\cdot\omega}(e^{i\mathbf{k}\cdot\omega t_0}\alpha_{\mathbf{l}_1\cdots\mathbf{l}_s}(t;t_0)-\alpha_{(\mathbf{k}+\mathbf{l}_1)\mathbf{l}_2\cdots\mathbf{l}_s}(t;t_0)),\\
\alpha_{\mathbf{0}^r\mathbf{k}\mathbf{l}_1\cdots\mathbf{l}_s}(t;t_0) &= \frac{i}{\mathbf{k}\cdot\omega}(\alpha_{\mathbf{0}^{r-1}\mathbf{k}\mathbf{l}_1\cdots\mathbf{l}_s}(t;t_0)-\alpha_{\mathbf{0}^r(\mathbf{k}+\mathbf{l}_1)\mathbf{l}_2\cdots\mathbf{l}_s}(t;t_0)),
\end{aligned}
\tag{9}
$$

where $r\geq 1$, $\mathbf{k}\in\mathbb{Z}^d\backslash\{\mathbf{0}\}$, *and* $\mathbf{l}_1,\dots,\mathbf{l}_s\in\mathbb{Z}^d$.

Proof It is useful to point out that the formulas (9) are found by evaluating the innermost integral in (3). To prove the proposition we show that the coefficients $\alpha_{\mathbf{k}_1\cdots\mathbf{k}_n}(t;t_0)$ uniquely determined by (9) coincide with those in (3). The latter satisfy, for all words $w=\mathbf{k}_1\cdots\mathbf{k}_n$,

$$
\frac{d}{dt}\alpha_{\mathbf{k}_1\cdots\mathbf{k}_n}(t;t_0)=\exp(i\mathbf{k}_n\cdot\omega t)\alpha_{\mathbf{k}_1\cdots\mathbf{k}_{n-1}}(t;t_0),\qquad \alpha_{\mathbf{k}_1\cdots\mathbf{k}_n}(t_0;t_0)=0. \tag{10}
$$

We prove by induction on n that the coefficients in (9) also satisfy (10). One can trivially check the case $n=1$. For each word $w=\mathbf{k}_1\cdots\mathbf{k}_n$ with $n>1$, one arrives at (10) by differentiating with respect to t both sides of the equality in (9) that determines $\alpha_{\mathbf{k}_1\cdots\mathbf{k}_n}(t;t_0)$ and applying the induction hypothesis. □

3.2 The Transport Equation

It follows from Proposition 1 that each $\alpha_w(t;t_0)$ is of the form

$$
\alpha_w(t;t_0)=\Gamma_w(t-t_0,\omega t;\omega t_0), \tag{11}
$$

where $\Gamma_w(\tau,\theta;\theta_0)$ is a suitable scalar-valued function, which is, as a function of $\tau\in\mathbb{R}$, a polynomial and as a function of $\theta\in\mathbb{T}^d$ (or of $\theta_0\in\mathbb{T}^d$) a trigonometric polynomial. For instance, for $\mathbf{k}\neq\mathbf{0},\mathbf{l}=-\mathbf{k}$, (see (8)),

$$
\Gamma_{\mathbf{kl}}(\tau,\theta;\theta_0)=\frac{i\tau}{\mathbf{k}\cdot\omega}+\frac{1-\exp(i\mathbf{k}\cdot\theta)\exp(-i\mathbf{k}\cdot\theta_0)}{(\mathbf{k}\cdot\omega)^2}.
$$

Of course the Γ_w can be found recursively by mimicking (9). The following result summarizes this discussion.

Theorem 1 *Define, for each* $w\in\mathcal{W}$, $\Gamma_w(\tau,\theta;\theta_0)$ *by means of the following recursions.* $\Gamma_\emptyset(\tau,\theta;\theta_0)=1$, *and given* $r\geq 1$, $\mathbf{k}\in\mathbb{Z}^d-\{\mathbf{0}\}$, *and* $\mathbf{l}_1,\dots,\mathbf{l}_s\in\mathbb{Z}^d$,

$$
\begin{aligned}
\Gamma_{\mathbf{k}}(\tau,\theta;\theta_0) &= \frac{i}{\mathbf{k}\cdot\omega}(e^{i\mathbf{k}\cdot\theta_0} - e^{i\mathbf{k}\cdot\theta}),\\
\Gamma_{\mathbf{0}^r}(\tau,\theta;\theta_0) &= \tau^r/r!,\\
\Gamma_{\mathbf{0}^r\mathbf{k}}(\tau,\theta;\theta_0) &= \frac{i}{\mathbf{k}\cdot\omega}(\Gamma_{\mathbf{0}^{r-1}\mathbf{k}}(\tau,\theta;\theta_0) - \Gamma_{\mathbf{0}^r}(\tau,\theta;\theta_0)e^{i\mathbf{k}\cdot\theta}),\\
\Gamma_{\mathbf{k}\mathbf{l}_1\cdots\mathbf{l}_s}(\tau,\theta;\theta_0) &= \frac{i}{\mathbf{k}\cdot\omega}(e^{i\mathbf{k}\cdot\theta_0}\Gamma_{\mathbf{l}_1\cdots\mathbf{l}_s}(\tau,\theta;\theta_0) - \Gamma_{(\mathbf{k}+\mathbf{l}_1)\mathbf{l}_2\cdots\mathbf{l}_s}(\tau,\theta;\theta_0)),\\
\Gamma_{\mathbf{0}^r\mathbf{k}\mathbf{l}_1\cdots\mathbf{l}_s}(\tau,\theta;\theta_0) &= \frac{i}{\mathbf{k}\cdot\omega}(\Gamma_{\mathbf{0}^{r-1}\mathbf{k}\mathbf{l}_1\cdots\mathbf{l}_s}(\tau,\theta;\theta_0) - \Gamma_{\mathbf{0}^r(\mathbf{k}+\mathbf{l}_1)\mathbf{l}_2\cdots\mathbf{l}_s}(\tau,\theta;\theta_0)). \qquad (12)
\end{aligned}
$$

Then, for each $w \in \mathcal{W}$, $\Gamma_w(\tau,\theta;\theta_0)$ is a polynomial in τ and a trigonometric polynomial in θ and in θ_0 and the coefficient $\alpha_w(t;t_0)$ of the oscillatory solution satisfies (11).

Substituting (11) in the initial value problem (5) that characterizes $\alpha(t;t_0)$, we find, after using the chain rule,

$$
\begin{aligned}
&\frac{\partial}{\partial\tau}\Gamma(t-t_0,t\omega;t_0\omega) + \omega\cdot\nabla_\theta\Gamma(t-t_0,t\omega;t_0\omega) = \Gamma(t-t_0,t\omega;t_0\omega)\star B(t\omega),\\
&\Gamma(0,t_0\omega;t_0\omega) = \mathbb{1},
\end{aligned}
$$

where $B(\theta)\in\mathfrak{g}$ is defined as $B_{\mathbf{k}}(\theta) = \exp(i\mathbf{k}\cdot\theta)$, $\mathbf{k}\in\mathbb{Z}^d$, and $B_w(\theta) = 0$ if the length of w is not 1. We thus have that for all $(\tau,\theta;\theta_0)$ of the form $(t-t_0,t\omega;t_0\omega)$, the following equation is valid:

$$
\frac{\partial}{\partial\tau}\Gamma(\tau,\theta;\theta_0) + \omega\cdot\nabla_\theta\Gamma(\tau,\theta;\theta_0) = \Gamma(\tau,\theta;\theta_0)\star B(\theta), \quad \Gamma(0,\theta_0;\theta_0) = \mathbb{1}. \qquad (13)
$$

Actually, it can be proved, by mimicking the proof of Proposition 1 (with d/dt replaced by the operator $\partial/\partial\tau + \omega\cdot\nabla_\theta$), that (13) holds for arbitrary $(\tau,\theta;\theta_0)\in\mathbb{R}\times\mathbb{T}^d\times\mathbb{T}^d$.

We have thus found a *transport equation* for Γ as a function of τ and θ (θ_0 plays the role of a parameter).[1] For this partial differential equation, a standard initial condition would prescribe the value of $\Gamma(0,\theta;\theta_0)$ as a function of $\theta\in\mathbb{T}^d$ (and of the parameter θ_0); in (13), $\Gamma(0,\theta;\theta_0)$ is only given at the single point $\theta=\theta_0$. Therefore (13) may be expected to have many solutions; only one of them is such that, for each $w\in\mathcal{W}$, $\Gamma_w(\tau,\theta;\theta_0)$ depends polynomially on τ as we shall establish in Proposition 2. We shall use an auxiliary result whose simple proof will be ommitted (cf. Lemma 2.4 in [4]):

[1]The presence of this parameter is linked to the fact that the transport equation is nonautonomous in the variable θ.

Lemma 1 *Let the vector $\omega \in \mathbb{R}^d$ be nonresonant. If a smooth function $z : \mathbb{R} \times \mathbb{T}^d \to \mathbb{C}$ satisfies*

$$\frac{\partial}{\partial \tau} z(\tau, \theta) + \omega \cdot \nabla_\theta z(\tau, \theta) = 0, \quad z(0, \theta_0) = 0,$$

and $z(\tau, \theta)$ is polynomial in τ, then $z(\tau, \theta)$ is identically zero.

Proposition 2 *The function $\Gamma(\tau, \theta; \theta_0)$ given in Theorem 1 is the unique solution of problem (13) such that each $\Gamma_w(\tau, \theta; \theta_0))$, $w \in \mathcal{W}$, is smooth in θ and polynomial in τ.*

Proof Let $\delta(\tau, \theta; \theta_0)$ denote the difference of two solutions of (13). Then, for each $w \in \mathcal{W}$, $\delta_w(0, \theta_0; \theta_0) = 0$ and

$$\frac{\partial}{\partial \tau} \delta_w(\tau, \theta; \theta_0) + \omega \cdot \nabla_\theta \delta_w(\tau, \theta; \theta_0)$$

vanishes provided that the value of $\delta(\tau, \theta; \theta_0)$ at words with less letters than w vanish identically. Lemma 1 then allows us to prove, by induction on the number of letters, that $\delta_w(\tau, \theta; \theta_0) \equiv 0$ for all $w \in \mathcal{W}$. See [4], Sect. 2.4 for a similar proof. □

The transport problem is used in the proof of the following two theorems, which in turn play an important role in averaging.

Theorem 2 *For each $\tau \in \mathbb{R}$, $\theta \in \mathbb{T}^d$, $\theta_0 \in \mathbb{T}^d$, the element $\Gamma(\tau, \theta; \theta_0) \in \mathbb{C}^{\mathcal{W}}$ belongs to $\mathcal{G}$.*

Proof The proof is very similar to the proof given in Sect. 6.1.4 of [14] for nonautonomous ordinary linear differential equations in $\mathcal{G}$. We have to prove that

$$\sum_j \Gamma_{w_j}(\tau, \theta; \theta_0) = \Gamma_w(\tau, \theta; \theta_0) \Gamma_{w'}(\tau, \theta; \theta_0) \qquad (14)$$

for $w, w' \in \mathcal{W}$, with $w \sqcup\!\sqcup w' = \sum_j w_j$. This is established by induction on the sum of the number of letters of w and w'. Proceeding as in [14], by application of the induction hypothesis one arrives at

$$(\frac{\partial}{\partial \tau} + \omega \cdot \nabla_\theta) \left(\sum_j \Gamma_{w_j}(\tau, \theta; \theta_0) - \Gamma_w(\tau, \theta; \theta_0) \Gamma_{w'}(\tau, \theta; \theta_0) \right) = 0.$$

Since (14) holds at $(\tau, \theta) = (0, \theta_0)$, Lemma 1 implies that it does so for each value of $(\tau, \theta) \in \mathbb{R} \times \mathbb{T}^d$. □

Theorem 3 *For arbitrary $\tau_1, \tau_2 \in \mathbb{R}$ and $\theta_0, \theta_1, \theta_2 \in \mathbb{T}^d$,*

$$\Gamma(\tau_1, \theta_1; \theta_0) \star \Gamma(\tau_2, \theta_2; \theta_1) = \Gamma(\tau_1 + \tau_2, \theta_2; \theta_0).$$

Proof One can check that both $\gamma(\tau, \theta) = \Gamma(\tau_1, \theta_1; \theta_0) \star \Gamma(\tau - \tau_1, \theta; \theta_1)$ and $\gamma(\tau, \theta) = \Gamma(\tau, \theta; \theta_0)$ satisfy the following three conditions:

- for each $w \in \mathscr{W}$, $\gamma_w(\tau, \theta)$ is smooth in θ and depends polynomially on τ,
- $\gamma(\tau_1, \theta_1) = \Gamma(\tau_1, \theta_1; \theta_0)$, and
- for all $\tau \in \mathbb{R}$ and all $\theta \in \mathbb{T}^d$,

$$\frac{\partial}{\partial \tau}\gamma(\tau, \theta) + \omega \cdot \nabla_\theta \gamma(\tau, \theta) = \gamma(\tau, \theta) \star B(\theta).$$

Proceeding as in the proof of Proposition 2, one concludes that there is a unique $\gamma(\tau, \theta)$ satisfying the three conditions above. The required result is thus obtained by setting $\tau = \tau_1 + \tau_2$ and $\theta = \theta_2$. □

In particular

$$\Gamma(\tau_1, \theta_0; \theta_0) \star \Gamma(\tau_2, \theta_0; \theta_0) = \Gamma(\tau_1 + \tau_2, \theta_0; \theta_0),$$
$$\Gamma(0, \theta_1; \theta_0) \star \Gamma(0, \theta_2; \theta_1) = \Gamma(0, \theta_2; \theta_0),$$

for arbitrary $\tau_1, \tau_2 \in \mathbb{R}$, $\theta_0, \theta_1, \theta_2 \in \mathbb{T}^d$. The first of these identities shows that, as τ varies with θ_0 fixed, the elements $\Gamma(\tau, \theta_0; \theta_0)$ form a one-parameter subgroup of $\mathscr{G}$. The second identity is similar to what is sometimes called two-parameter group property of the solution operator of nonautonomous differential equations. Of course these identities reflect the fact that the transport equation is autonomous in τ and nonautonomous in θ.

3.3 The Averaged System and the Change of Variables

Before we average the oscillatory problem (6), we shall do so with the corresponding universal problem (5) in $\mathscr{G}$, whose solution α has been represented in (11) by means of the auxiliary function $\Gamma(\tau, \theta; \theta_0)$. Note that the oscillatory nature of α is caused by the second argument in Γ (each Γ_w, $w \in \mathscr{W}$, is a polynomial τ). Consider then the $\mathscr{G}$-valued function of t defined by

$$\bar{\alpha}(t; t_0) = \Gamma(t - t_0, t_0\omega; t_0\omega), \tag{15}$$

where the second argument in Γ has been frozen at its initial value. This satisfies $\bar{\alpha}(t_0; t_0) = \Gamma(0, t_0\omega; t_0\omega)$, or, from Proposition 2, $\bar{\alpha}(t_0; t_0) = \mathbb{1}$ so that $\bar{\alpha}(t; t_0)$ coincides with $\alpha(t; t_0)$ at the initial time $t = t_0$. Furthermore, due to the trigonometric dependence on θ, if $d = 1$ (periodic case), $\bar{\alpha}(t; t_0) = \Gamma(t - t_0, t_0\omega; t_0\omega)$ coincides with $\alpha(t; t_0) = \Gamma(t - t_0, t\omega; t_0\omega)$ at all times of the form $t = t_0 + 2k\pi/\omega$, k integer. If $d > 1$ (quasiperiodic case), as t varies, the point $t\omega \in \mathbb{T}^d$ never returns to the initial position $t_0\omega$; however it returns infinitely often to the neighborhood of $t_0\omega$. To sum up, the nonoscillatory $\bar{\alpha}(t; t_0)$ is a good description of the long-term evolution of

$\alpha(t;t_0)$ and, in fact, we shall presently arrange things in such a way that $\bar{\alpha}(t;t_0)$ is the solution of the averaged version of the problem (5).

Having identified the averaged *solution,* let us find the averaged *problem.* From Theorem 3, for each τ_1 and τ_2,

$$\bar{\alpha}(t_0+\tau_1+\tau_2;t_0)=\bar{\alpha}(t_0+\tau_1;t_0)\star\bar{\alpha}(t_0+\tau_2;t_0)$$

so that, as τ varies, the elements $\bar{\alpha}(t_0+\tau;t_0)$ form a (commutative) one-parameter group $\subset\mathcal{G}$. Therefore $\bar{\alpha}(t;t_0)$ is the solution of the *autonomous problem*

$$\frac{d}{dt}\bar{\alpha}(t;t_0)=\bar{\alpha}(t;t_0)\star\bar{\beta}(t_0),\qquad \bar{\alpha}(t_0;t_0)=\mathbb{1}\,, \tag{16}$$

with

$$\bar{\beta}(t_0)=\left.\frac{d}{dt}\bar{\alpha}(t;t_0)\right|_{t=t_0}. \tag{17}$$

For completeness we include here a proof of this fact, which is well known in the theory of differential equations,

$$\frac{d}{dt'}\bar{\alpha}(t';t_0)=\left.\frac{d}{dt}\bar{\alpha}(t'+t-t_0;t_0)\right|_{t=t_0}=\left.\frac{d}{dt}\bar{\alpha}(t';t_0)\star\bar{\alpha}(t;t_0)\right|_{t=t_0}=\bar{\alpha}(t';t_0)\star\bar{\beta}(t_0).$$

Note that (17) implies that $\bar{\beta}(t_0)\in\mathfrak{g}$.

After having found the averaged problem (16), we invoke once more Theorem 3 and write

$$\begin{aligned}\alpha(t;t_0)&=\Gamma(t-t_0,t\omega;t_0\omega)\\&=\Gamma(t-t_0,t_0\omega;t_0\omega)\star\Gamma(0,t\omega;t_0\omega)\\&=\bar{\alpha}(t;t_0)\star\Gamma(0,t\omega;t_0\omega).\end{aligned}$$

Thus, if we define

$$\kappa(\theta;t_0)=\Gamma(0,\theta;t_0\omega), \tag{18}$$

then κ depends periodically on the components of θ and $\kappa(t\omega;t_0)$ relates the averaged solution $\bar{\alpha}$ and the oscillatory solution α in the following way:

$$\alpha(t;t_0)=\bar{\alpha}(t;t_0)\star\kappa(t\omega;t_0). \tag{19}$$

To sum up, we have proved:

Theorem 4 *For $\theta\in\mathbb{T}^d$ define $\kappa(\theta;t_0)\in\mathcal{G}$ by (18). Then the solution of the problem (5) has the representation (19), where $\bar{\alpha}(t;t_0)$ satisfies the autonomous (averaged) initial value problem (16)–(17) with $\bar{\beta}(t_0)\in\mathfrak{g}$. Furthermore $\bar{\alpha}(t;t_0)$ may be found by means of (15).*

By inserting the word basis functions to obtain the corresponding series and recalling that the operation $\star$ for the coefficients represents the composition of the series, we conclude:

Theorem 5 *With the notation of the preceding theorem, the solution of (6) may be represented as*

$$y(t) = W_{\kappa(t\omega;t_0)}(Y(t))$$

where $Y(t) = W_{\bar{\alpha}(t;t_0)}(y_0)$ *solves the autonomous (averaged) initial value problem*

$$\frac{d}{dt}Y = W_{\bar{\beta}(t_0)}(Y), \qquad Y(t_0) = y_0.$$

3.4 Geometric Properties

Since $\bar{\beta}(t_0)$ is in the Lie algebra $\mathfrak{g}$, the Dynkin–Specht–Wever theorem [8], implies that the word series for the averaged vector field may be rewritten in terms of iterated Lie-Jacobi brackets

$$W_{\bar{\beta}(t_0)}(y) = \sum_{r=1}^{\infty} \sum_{\mathbf{k}_1,\ldots,\mathbf{k}_r \in Z^d} \frac{\varepsilon^r}{r} \beta_{\mathbf{k}_1\cdots\mathbf{k}_r}(t_0)\,[[\cdots[[f_{\mathbf{k}_1}, f_{\mathbf{k}_2}], f_{\mathbf{k}_3}]\cdots], f_{\mathbf{k}_r}](y). \quad (20)$$

(The bracket is defined by $[f, g](y) = g'(x)f(y) - f'(y)g(y)$.)

It follows from (20) that if all the $f_{\mathbf{k}}$ belong to a given Lie subalgebra of the Lie algebra of all vector fields (e.g., if they are all Hamiltonian or all divergence free), then the averaged vector field will also lie in that subalgebra (i.e., will be Hamiltonian or divergence free).

Additionally, the averaging procedure described above is *equivariant* with respect to arbitrary changes of variables: changing variables $y = C(\bar{y})$ in the oscillatory problem, followed by averaging, yields the same result as changing variables in the averaged system. This is a consequence of the equivariance of word series with coefficients in $\mathcal{G}$.

3.5 Finding the Coefficients

From (15), (respectively (18)) the quantities $\bar{\alpha}_w(t; t_0)$ (respectively $\kappa_w(\theta; t_0)$), $w \in \mathcal{W}$, may be found recursively by setting $\tau = t - t_0$, $\theta = \theta_0 = t_0\omega$ (respectively $\tau = 0$, $\theta_0 = t_0\omega$) in the formulas for $\Gamma_w(\tau, \theta; \theta_0)$ provided in Theorem 1. The following recurrences for $\bar{\beta}(t_0)$ are easily found via (17).

Theorem 6 *Given $r \geq 1$, $\mathbf{k} \in \mathbb{Z}^d \backslash \{0\}$, and $\mathbf{l}_1, \dots, \mathbf{l}_s \in \mathbb{Z}^d$,*

$$\begin{aligned}
\bar{\beta}_{\mathbf{k}}(t_0) &= 0,\\
\bar{\beta}_{\mathbf{0}}(t_0) &= 1,\\
\bar{\beta}_{\mathbf{0}^{r+1}}(t_0) &= 0,\\
\bar{\beta}_{\mathbf{0}^r\mathbf{k}}(t_0) &= \frac{i}{\mathbf{k}\cdot\omega}(\bar{\beta}_{\mathbf{0}^{r-1}\mathbf{k}}(t_0) - \bar{\beta}_{\mathbf{0}^r}(t_0)e^{i\mathbf{k}\cdot\omega t_0}),\\
\bar{\beta}_{\mathbf{k}\mathbf{l}_1\cdots\mathbf{l}_s}(t_0) &= \frac{i}{\mathbf{k}\cdot\omega}(e^{i\mathbf{k}\cdot\omega t_0}\bar{\beta}_{\mathbf{l}_1\cdots\mathbf{l}_s}(t_0) - \bar{\beta}_{(\mathbf{k}+\mathbf{l}_1)\mathbf{l}_2\cdots\mathbf{l}_s}(t_0)),\\
\bar{\beta}_{\mathbf{0}^r\mathbf{k}\mathbf{l}_1\cdots\mathbf{l}_s}(t_0) &= \frac{i}{\mathbf{k}\cdot\omega}(\bar{\beta}_{\mathbf{0}^{r-1}\mathbf{k}\mathbf{l}_1\cdots\mathbf{l}_s}(t_0) - \bar{\beta}_{\mathbf{0}^r(\mathbf{k}+\mathbf{l}_1)\mathbf{l}_2\cdots\mathbf{l}_s}(t_0)).
\end{aligned}$$

In the particular case $t_0 = 0$, after computing the coefficients $\bar{\beta}_w(0)$ for words with ≤ 3 letters by means of the formulas in the theorem, we obtain, with the help of the Jacobi identity for the bracket and the shuffle relations, the following explicit formula for the averaged system:

$$\frac{d}{dt}Y = \varepsilon f_{\mathbf{0}} + \varepsilon^2 F_2 + \varepsilon^3 F_3 + \mathscr{O}(\varepsilon^4),$$

where

$$\begin{aligned}
F_2 &= \sum_{\mathbf{k}>-\mathbf{k}} \frac{i}{\mathbf{k}\cdot\omega}([f_{\mathbf{k}} - f_{-\mathbf{k}}, f_{\mathbf{0}}] + [f_{-\mathbf{k}}, f_{\mathbf{k}}]),\\
F_3 &= \sum_{\mathbf{k}\neq\mathbf{0}} \frac{1}{(\mathbf{k}\cdot\omega)^2}\Big([f_{\mathbf{0}}, [f_{\mathbf{0}}, f_{\mathbf{k}}]] + [f_{\mathbf{k}}, [f_{\mathbf{k}}, f_{-\mathbf{k}}]] - \frac{1}{2}[f_{\mathbf{k}}, [f_{\mathbf{k}}, f_{-2\mathbf{k}}]] + [f_{-\mathbf{k}}, [f_{\mathbf{k}}, f_{\mathbf{0}}]]\Big)\\
&\quad + \sum_{\mathbf{0}\neq\mathbf{m}\neq-\mathbf{l}\neq\mathbf{0}} \frac{-1}{(\mathbf{l}\cdot\omega)((\mathbf{m}+\mathbf{l})\cdot\omega)}[f_{\mathbf{m}}, [f_{\mathbf{l}}, f_{\mathbf{0}}]]\\
&\quad + \sum_{-\mathbf{l}>\mathbf{k}<\mathbf{l},\ \mathbf{k}\neq\mathbf{0}} \frac{1}{(\mathbf{k}\cdot\omega)(\mathbf{l}\cdot\omega)}[f_{-\mathbf{l}}, [f_{\mathbf{l}}, f_{\mathbf{k}}]]\\
&\quad + \sum_{\substack{\mathbf{m}>\mathbf{k}<-\mathbf{k}\\ \mathbf{m}+\mathbf{k}\neq\mathbf{0}}} \frac{-1}{(\mathbf{k}\cdot\omega)(\mathbf{m}\cdot\omega)}[f_{\mathbf{m}}, [f_{-\mathbf{k}}, f_{\mathbf{k}}]]\\
&\quad + \sum_{\substack{\mathbf{0}\neq\mathbf{m}\neq\pm\mathbf{l}\neq\mathbf{0}\\ \mathbf{m}>-\mathbf{m}-\mathbf{l}<\mathbf{l}}} \frac{-1}{(\mathbf{m}\cdot\omega)((\mathbf{m}+\mathbf{l})\cdot\omega)}[f_{\mathbf{m}}, [f_{\mathbf{l}}, f_{-\mathbf{m}-\mathbf{l}}]].
\end{aligned}$$

In these formulas $<$ is some total ordering in the set of multi-indices $\mathbb{Z}^d$ such that $\mathbf{k} > \mathbf{0}$ for $\mathbf{k} \neq \mathbf{0}$.

3.6 Changing the Initial Time

There would have been no loss of generality if in (6) we had taken the initial time t_0 to be 0, as the general case may be reduced to the case where $t_0 = 0$ by a change of variables $t \to t' + t_0$. Here we give formulas that express $\Gamma(\tau, \theta; \theta_0)$ in terms of $\Gamma(\tau, \theta - \theta_0; 0)$ and therefore allows one to express the coefficients $\alpha(t; t_0)$, $\bar{\beta}(t_0)$, …in terms of the coefficients $\alpha(t; 0)$, $\bar{\beta}(0)$, …

We introduce, for each $\theta \in \mathbb{T}^d$, the linear map $\Xi_\theta : \mathbb{C}^{\mathscr{W}} \to \mathbb{C}^{\mathscr{W}}$ defined as follows: given $\delta \in \mathbb{C}^{\mathscr{W}}$, $(\Xi_\theta \delta)_\emptyset = \delta_\emptyset$, and for each word $w = \mathbf{k}_1 \cdots \mathbf{k}_n$ with $n > 0$ letters,

$$(\Xi_\theta \delta)_{\mathbf{k}_1 \cdots \mathbf{k}_n} = e^{i(\mathbf{k}_1 + \cdots + \mathbf{k}_n)\cdot\theta} \delta_{\mathbf{k}_1 \cdots \mathbf{k}_n}.$$

Note that Ξ_θ is actually an algebra automorphism, as it preserves the convolution product: $\Xi_\theta(\delta \star \delta') = (\Xi_\theta \delta) \star (\Xi_\theta \delta')$, if $\delta, \delta' \in \mathbb{C}^{\mathscr{W}}$. In addition it maps $\mathscr{G}$ into $\mathscr{G}$.

The following result may be proved by induction on the number of letters using the recursive formulas (12), or, alternatively, by using the transport equation.

Proposition 3 *For each $\tau \in \mathbb{R}$ and $\theta, \theta_0 \in \mathbb{T}^d$,*

$$\Gamma(\tau, \theta; \theta_0) = \Xi_{\theta_0} \Gamma(\tau, \theta - \theta_0; 0).$$

As a consequence we have (cf. Theorem 3):

Corollary 1 *For arbitrary $\tau_1, \tau_2 \in \mathbb{R}$ and $\theta_1, \theta_2 \in \mathbb{T}^d$,*

$$\Gamma(\tau_1, \theta_1; 0) \star \Xi_{\theta_1} \Gamma(\tau_2, \theta_2; 0) = \Gamma(\tau_1 + \tau_2, \theta_1 + \theta_2; 0).$$

4 Autonomous Problems

In this section consider a general class of perturbed autonomous problems. By building on the foundations laid down above we provide a method for reducing them to normal form.

4.1 Perturbed Problems

We now study initial value problems

$$\frac{d}{dt} x = g(x) + f(x), \quad x(0) = x_0, \tag{21}$$

where $f, g : \mathbb{C}^D \to \mathbb{C}^D$. In the situations we have in mind, this system is seen as a perturbation of the system $(d/dt)x = g(x)$ whose solutions are known. In what follows we denote by g_j, $j = 1, \ldots, d$, a family of linearly independent vector fields that commute with each other (i.e. $[g_j, g_k] = 0$) and, for each $u = [u_1, \cdots, u_d] \in \mathbb{C}^d$, we set

$$g^u = \sum_{j=1}^{d} u_j\, g_j. \tag{22}$$

We always work under the following hypotheses:

- f may be decomposed as

$$f(x) = \sum_{\ell \in A} f_\ell(x) \tag{23}$$

  for a set of indices A, referred to as the alphabet as in the preceding sections.
- For each $j = 1, \ldots, d$ and each $\ell \in A$, there is $\nu_{j,\ell} \in \mathbb{C}$ such that

$$[g_j, f_\ell] = \nu_{j,\ell}\, f_\ell. \tag{24}$$

- There is $v \in \mathbb{C}^d$ such that $g = g^v$.
- The alphabet A is an additive monoid with neutral element $\mathbf{0}$,[2] such that, for each $j = 1, \ldots, d$ and $\ell, \ell' \in A$, $\nu_{j,\ell+\ell'} = \nu_{j,\ell} + \nu_{j,\ell'}$. In particular, $\nu_{j,\mathbf{0}} = 0$ for all j.
- The vector $v = (v_1, \ldots, v_d) \in \mathbb{C}^d$ is *non-resonant*, in the sense that, given $\ell \in A$, $v_1\nu_{1,\ell} + \cdots + v_d\nu_{d,\ell} = 0$ if and only if $\ell = \mathbf{0}$.

The following proposition, whose proof may be seen in [13], shows that (24) may be reformulated in terms of the flows φ_u at time $t = 1$ of the vector fields g^u, $u \in \mathbb{C}^d$. Here and later, we use the notation

$$\nu_\ell^u = u_1\nu_{1,\ell} + \cdots + u_d\nu_{d,\ell}.$$

Proposition 4 *Equation (24) is equivalent to the requirement that for each $x \in \mathbb{R}^D$, $u \in \mathbb{C}^d$, $\ell \in A$,*

$$\varphi_u'(x)^{-1} f_\ell(\varphi_u(x)) = \exp(\nu_\ell^u) f_\ell(x). \tag{25}$$

Before providing examples of systems that satisfy the hypotheses above, we shall obtain a word series representation of the solution of (21). Use the ansatz $x(t) = \varphi_{tv}(z(t))$ and invoke (25), to find that $z(t)$ must be the solution of

$$\frac{d}{dt} z = \sum_{\ell \in A} \exp(t\nu_\ell^v) f_\ell(z), \quad z(0) = x_0.$$

Since this problem is of the form (1) with

$$\lambda_\ell(t) = \exp(t\,\nu_\ell^v), \qquad \ell \in A, \tag{26}$$

---

[2] Recall that this means that A possesses a binary operation $+$ that is commutative and associative and such that $\mathbf{0} + \ell = \ell$ for each $\ell \in A$.

we find that $z(t) = W_{\alpha(t;0)}(x_0)$, where the coefficients $\alpha(t; 0) \in \mathcal{G}$ are given by (3). In what follows, we will simply write $\alpha(t) = \alpha(t; 0)$. Thus the solution of (21) has the representation

$$x(t) = \varphi_{tv}(W_{\alpha(t)}(x_0)). \tag{27}$$

Note that the coefficients $\alpha_w(t)$ *depend on* v *and the* $v_{j,\ell}$ *but are otherwise independent of* g *and* f_ℓ, $\ell \in A$.

Systems that satisfy the assumptions include the following (additional examples and further discussion may be seen in [13]):

Example 1 Consider systems of the form

$$\frac{d}{dt}x = Lx + f(x), \tag{28}$$

where L is a diagonalizable $D \times D$ matrix and each component of $f(x)$ is a power series in the components of x. Let $\mu_1, \ldots, \mu_d$ denote the distinct nonzero eigenvalues of L, so that L may be uniquely decomposed as

$$L = \mu_1 L_1 + \cdots + \mu_d L_d,$$

where the $D \times D$ matrices $L_1, \ldots, L_d$ are projectors ($L_j^2 = L_j$) with $L_j L_k = 0$ if $j \neq k$. Thus (22) holds for $g_j(x) = L_j x$, $v_j = \mu_j$. Furthermore (see [13] for details) f may be decomposed as $f = \sum_{\mathbf{k}} f_{\mathbf{k}}$, where the 'letters' $\mathbf{k}$ are elements of $\mathbb{Z}^d$, $\mathbf{k} = [k_1, \ldots, k_d]$, and, for each j and $\mathbf{k}$, $[L_j, f_{\mathbf{k}}] = k_j f_{\mathbf{k}}$. Analytic systems of differential equations having an equilibrium at the origin are of the form (28), provided that the linearization at the origin is diagonalizable; the perturbation f then contains terms of degree >1 in the components of x.

As we shall point out later, Theorem 11 addresses the well-known problem, which goes back to Poincaré and Birkhoff [2], of reducing (28) to normal form.

Example 2 Consider next real systems of the form

$$\frac{d}{dt}\begin{bmatrix} y \\ \theta \end{bmatrix} = \begin{bmatrix} 0 \\ \omega \end{bmatrix} + f(y, \theta), \tag{29}$$

where $y \in \mathbb{R}^{D-d}$, $0 < d \leq D$, $\omega \in \mathbb{R}^d$ is a vector of frequencies $\omega_j \neq 0$, $j = 1, \ldots, d$, and θ comprises d angles, so that $f(y, \theta)$ is 2π-periodic in each component of θ with Fourier expansion

$$f(y, \theta) = \sum_{\mathbf{k} \in \mathbb{Z}^d} \exp(i\mathbf{k} \cdot \theta)\, \hat{f}_{\mathbf{k}}(y).$$

After introducing the functions

$$f_{\mathbf{k}}(y,\theta) = \exp(i\mathbf{k}\cdot\theta)\,\hat{f}_{\mathbf{k}}(y), \qquad y \in \mathbb{R}^{D-d},\ \theta \in \mathbb{R}^d,$$

the system takes the form (21) with $x = (y,\theta)$ and

$$g(y,\theta) = \begin{bmatrix} 0 \\ \omega \end{bmatrix}.$$

The decomposition (23) holds for the monoid $A = \mathbb{Z}^d$, and, if each $g_j(x)$ is taken to be a constant unit vector, then $g = g^\nu$, $\nu_j = \omega_j$ $(j = 1,\dots,d)$. In addition, (24) is satisfied with $\nu_{j,\mathbf{k}} = i\,k_j$, for each $j = 1,\dots,d$ and each $\mathbf{k} = (k_1,\dots,k_d) \in A$. Thus the nonresonance condition above (i.e., $\nu_1\nu_{1,\ell} + \cdots + \nu_d\nu_{d,\ell} = 0$ if and only if $\ell = \mathbf{0}$) now becomes the well-known requirement that $k_1\omega_1 + \cdots + k_d\omega_d = 0$ with integer k_j, $j = 1,\dots,d$, only if all k_j vanish.

In the particular case where the last d components of f vanish identically, the differential equations for θ yield $\theta = \omega t + \theta_0$, and (29) becomes a nonautonomous system for y of the form (6). Thus, the format (21) is a wide generalization of the format studied in the preceding section.

4.2 The Transport Equation. Normal Forms

All the results obtained in Sect. 3.2 can be generalized to the case at hand. We shall omit the proofs of the results that follow when they may be obtained by adapting the corresponding proofs in Sect. 3.

We first provide recurrences to find the coefficients required in (27).

Theorem 7 *Given $\tau \in \mathbb{R}$, $u \in \mathbb{C}^d$, define, for each $w \in \mathcal{W}$, $\gamma_w(\tau,u) \in \mathbb{C}$ by means of the following recursions. $\gamma_\emptyset(\tau,u) = 1$, and for $r \geq 1$, $\ell_0 \in A\backslash\{\mathbf{0}\}$, and $\ell_1,\dots,\ell_n \in A$,*

$$\begin{aligned}
\gamma_{\ell_0}(\tau,u) &= \frac{1}{\nu^\nu_{\ell_0}}(\exp(\nu^u_{\ell_0}) - 1),\\
\gamma_{\mathbf{0}^r}(\tau,u) &= \tau^r/r!,\\
\gamma_{\mathbf{0}^r\ell_0}(\tau,u) &= \frac{\gamma_{\mathbf{0}^r}(\tau,u)\exp(\nu^u_{\ell_0}) - \gamma_{\mathbf{0}^{r-1}\ell_0}(\tau,u)}{\nu^\nu_{\ell_0}},\\
\gamma_{\ell_0\ell_1\cdots\ell_n}(\tau,u) &= \frac{\gamma_{(\ell_0+\ell_1)\ell_2\cdots\ell_n}(\tau,u) - \gamma_{\ell_1\cdots\ell_n}(\tau,u)}{\nu^\nu_{\ell_0}},\\
\gamma_{\mathbf{0}^r\ell_0\ell_1\cdots\ell_n}(\tau,u) &= \frac{\gamma_{\mathbf{0}^r(\ell_0+\ell_1)\ell_2\cdots\ell_n}(\tau,u) - \gamma_{\mathbf{0}^{r-1}\ell_0\ell_1\cdots\ell_n}(\tau,u)}{\nu^\nu_{\ell_0}}.
\end{aligned} \tag{30}$$

Then, for each $w \in \mathcal{W}$,

$$\alpha_w(t) = \gamma_w(t, t\nu).$$

The transport problem (cf. (13)) is:

$$\frac{\partial}{\partial \tau}\gamma(\tau,u)+v\cdot\nabla_u\gamma(\tau,u)=\gamma(\tau,u)\star B(u),\quad \gamma(0,0)=1\!\!1. \tag{31}$$

where $B(u)\in\mathfrak{g}$ is defined as $B_\ell(u)=\exp(v_\ell^u)$, $\ell\in A$, and $B_w(u)=0$ if the length of $w\in W$ is not 1.

Lemma 1 needs some adaptation to the present circumstances. We say that a complex-valued function is *polynomially smooth* if it is a linear combination of terms of the form $\tau^k\exp(v_\ell^u)$, $j=1,2,3,\ldots$, $\ell\in A$. For each $w\in\mathscr{W}$, the function $\gamma_w:\mathbb{R}\times\mathbb{C}^d\to\mathbb{C}$ in Theorem 7 is clearly *polynomially smooth.*

Lemma 2 *Let the vector* $v\in\mathbb{C}^d$ *be nonresonant. If a polynomially smooth function* $z:\mathbb{R}\times\mathbb{C}^d\to\mathbb{C}$ *satisfies*

$$\frac{\partial}{\partial \tau}z(\tau,u)+v\cdot\nabla_u z(\tau,u)=0,\quad z(0,0)=0,$$

then $z(\tau,u)$ *is identically zero.*

Instead of Proposition 2 and Theorem 2, we now have the following result.

Theorem 8 *The function* $\gamma(\tau,u)$ *given in Theorem 7 is the unique solution of problem (31) such that each* $\gamma_w:\mathbb{R}\times\mathbb{C}^d\to\mathbb{C}$, $w\in\mathscr{W}$, *is polynomially smooth. Furthermore, for each* $\tau\in\mathbb{R}$, $u\in\mathbb{C}^d$, *the element* $\gamma(\tau,u)\in\mathbb{C}^{\mathscr{W}}$ *belongs to* $\mathscr{G}$.

Our next aim is to derive a result similar to Theorem 3. We need to introduce, for each $u\in\mathbb{C}^d$, the algebra map $\Xi_u:\mathbb{C}^{\mathscr{W}}\to\mathbb{C}^{\mathscr{W}}$ defined as follows: Given $\delta\in\mathbb{C}^{\mathscr{W}}$, $(\Xi_u\delta)_\emptyset=\delta_\emptyset$, and for each word $w=\ell_1\cdots\ell_n$ with $n\geq 1$ letters,

$$(\Xi_u\delta)_{\ell_1\cdots\ell_n}=\exp(v^u_{\ell_1+\cdots+\ell_n})\delta_{\ell_1\cdots\ell_n}.$$

This generalizes the map Ξ_θ we used in Sect. 3.

Theorem 9 *For arbitrary* $\tau_1,\tau_2\in\mathbb{R}$ *and* $u_1,u_2\in\mathbb{C}^d$,

$$\gamma(\tau,u)\star(\Xi_u\gamma(\tau',u'))=\gamma(\tau+\tau',u+u').$$

Let us provide an interpretation of the last result in terms of maps in $\mathbb{C}^D$ (rather than in terms of elements of $\mathscr{G}$). In [13], it is proved that, for arbitrary $\delta\in\mathscr{G}$ and $u\in\mathbb{C}^d$

$$W_\delta(\varphi_u(x))=\varphi_u(W_{\Xi_u\delta}(x)). \tag{32}$$

If we denote, for each $(\tau,u)\in\mathbb{R}\times\mathbb{C}^d$,

$$\Phi_{\tau,u}(x)=\varphi_u(W_{\gamma(\tau,u)}(x)),$$

then, for arbitrary $\tau,\tau'\in\mathbb{R}$ and $u,u'\in\mathbb{C}^d$,

$$\begin{aligned}\Phi_{\tau,u}(\Phi_{\tau',u'}(x)) &= \varphi_u(W_{\gamma(\tau,u)}(\varphi_{u'}(W_{\gamma(\tau',u')}(x))))\\ &= \varphi_u(\varphi_{u'}(W_{\Xi_u\gamma(\tau,u)}(W_{\gamma(\tau',u')}(x))))\\ &= \varphi_{u+u'}(W_{\gamma(\tau+\tau',u+u')}(x))\\ &= \Phi_{\tau+\tau',u+u'}(x).\end{aligned}$$

We have successively used the definition of Φ, Eq. (32), Theorem 9, Eq. (4), and, again, the definition of Φ. To sum up, we have proved the following result, which generalizes Proposition 5.3 in [4].

Theorem 10 *For arbitrary* $\tau, \tau' \in \mathbb{R}$ *and* $u, u' \in \mathbb{C}^d$,

$$\Phi_{\tau,u} \circ \Phi_{\tau',u'} = \Phi_{\tau+\tau',u+u'}$$

Since, in view of (27), the solution $x(t)$ may be written as $x(t) = \Phi_{t,tv}(x_0)$, the theorem implies the representations

$$x(t) = \Phi_{0,tv}(\Phi_{t,0}(x_0)) = \Phi_{t,0}(\Phi_{0,tv}(x_0)). \tag{33}$$

The group property $\Phi_{t,0} \circ \Phi_{t',0} = \Phi_{t+t',0}$ implies that $\Phi_{t,0}(x) = W_{\gamma(t,0)}(x)$ is the t-flow of the autonomous system

$$\frac{d}{dt}X = W_{\bar\beta}(X),$$

where $\bar\beta \in \mathfrak{g}$ is given by

$$\bar\beta = \frac{d}{dt}\gamma(t,0)\bigg|_{t=0}. \tag{34}$$

Similarly, $\Phi_{0,tu} \circ \Phi_{0,t'u} = \Phi_{0,(t+t')u}$, implies that, for each fixed $u \in \mathbb{C}^d$, $\Phi_{0,tu}(x) = \varphi_{tu}(W_{\gamma(0,tu)}(x))$ is the t-flow of an autonomous system

$$\frac{d}{dt}x = \tilde g^u(x);$$

differentiation with respect to t of the flow at $t = 0$ reveals that

$$\tilde g^u(x) = g^u(x) + W_{\rho(u)}(x), \tag{35}$$

where $\rho(u) \in \mathfrak{g}$ is given by

$$\rho(u) = \frac{d}{dt}\gamma(0,tu)\bigg|_{t=0}.$$

Since, from Theorem 10, the flows $\Phi_{t,0}$, $\Phi_{0,tu}(x)$, $\Phi_{0,tu'}(x)$, $u, u' \in \mathbb{C}^d$, commute with one another, so do the corresponding vector fields $W_{\bar{\beta}}(X)$, $\tilde{g}^u(x)$, $\tilde{g}^{u'}(x)$. After invoking (33), we summarize our findings as follows.

Theorem 11 *The system in the initial value problem* (21) *may be rewritten in the form*

$$\frac{d}{dt}x = g(x) + f(x) = \tilde{g}^{\nu}(x) + W_{\bar{\beta}}(x),$$

where $\tilde{g}^{\nu}(x)$ and $\bar{\beta}$ are respectively given by (35) *and* (34). *The vector fields $\tilde{g}^{\nu}(x)$ and $W_{\bar{\beta}}(x)$ commute with each other, with $g(x) + f(x)$ and with $\tilde{g}^u(x)$ for arbitrary $u \in \mathbb{C}^d$.*

Note that the recursions defining $\gamma(t, u)$ in Theorem 7 give rise to similar recursions that allows us to conveniently compute the coefficients $\bar{\beta}$, $\rho(u) \in \mathfrak{g}$.

In the particular case where $A = \mathbb{Z}^d$ and the eigenvalues $\nu_{j,\ell}$ lie on the imaginary axis, Theorem 11 essentially coincides with Theorem 5.5 of [4]. The techniques in [4] are similar to those used here, but use B-series rather than word series. The general case of Theorem 11 was obtained by means of *extended words series* (see Sect. 5) in [13], where in addition it is shown that the vector fields $\tilde{g}^u(x)$ are conjugate to $g^u(x)$ by a map of the form $x \mapsto W_\delta(x)$, where $\delta \in \mathcal{G}$. The decomposition in Theorem 11 may be regarded as providing a *normal form*, where the original vector field is written as a vector field $\tilde{g}^{\nu}(x)$ that is conjugate to $g^{\nu}(x)$ perturbed by a vector field $W_{\bar{\beta}}(x)$ that commutes with $\tilde{g}^{\nu}(x)$.

Remark 1 For Hamiltonian problems, the commutation results in Theorem 11 allows us to write down explicitly integrals of motion of the given problem. Details may be seen in [4, 13].

5 Further Extensions

In this section we study generalizations of the perturbed system in (21). Extended word series, introduced in [14], are a convenient auxiliary tool to study those generalizations.

5.1 *Extended Word Series*

Just as the study of systems of the form (1) leads to the introduction of word series via the representation (2), the expression (27) suggests the introduction of *extended word series*. Given the commuting vector fields g_j, $j = 1, \ldots, d$ and the vector fields f_ℓ, $\ell \in A$ in the preceding section, to each $(v, \delta) \in \mathbb{C}^d \times \mathbb{C}^{\mathcal{W}}$ we associate its *extended word series* [13, 14]:

$$\overline{W}_{(v,\delta)}(x) = \varphi_v(W_\delta(x)).$$

With this terminology, the solution of (21) in (27) may be written as $x(t) = \overline{W}_{(tv,\alpha(t))}(x_0)$.

The symbol $\overline{\mathcal{G}}$ denotes the set $\mathbb{C}^d \times \mathcal{G}$. Thus, for each t, the solution coefficients $(tv, \alpha(t))$ provide an example of element of $\overline{\mathcal{G}}$. For $(u, \gamma) \in \overline{\mathcal{G}}$ and $(v, \delta) \in \mathbb{C}^d \times \mathbb{C}^{\mathcal{W}}$ we set

$$(u, \gamma) \bigstar (v, \delta) = (v + \delta_\emptyset u, \gamma \star \Xi_u \delta) \in \mathbb{C}^d \times \mathbb{C}^{\mathcal{W}}.$$

For this operation $\overline{\mathcal{G}}$ is a noncommutative group, with unit $\overline{\mathbb{1}} = (0, \mathbb{1})$; $(\mathbb{C}^d, \mathbb{1})$ and $(0, \mathcal{G})$ are subgroups of $\overline{\mathcal{G}}$. (In fact $\overline{\mathcal{G}}$ is an outer semidirect product of $\mathcal{G}$ and the additive group $\mathbb{C}^d$, as discussed in Sect. 3.2 of [14].)

By using (4) and (32), it is a simple exercise to check that the product $\bigstar$ has the following implication for the composition of the corresponding extended word series

$$\overline{W}_{(v,\delta)}\big(\overline{W}_{(u,\gamma)}(x)\big) = \overline{W}_{(u,\gamma)\bigstar(v,\delta)}(x), \qquad \gamma, \in \mathcal{G}, \ \ \delta \in \mathbb{C}^{\mathcal{W}}, \ \ u, v \in \mathbb{C}^d.$$

5.2 More General Perturbed Problems

We now generalize the problem (21), and allow a more general perturbation:

$$\frac{d}{dt}x = g(x) + W_\beta(x), \quad x(0) = x_0,$$

where $\beta \in \mathfrak{g}$. Clearly, the original problem (21) corresponds to the particular case where $\beta_\ell = 1$ for each $\ell \in A$, and $\beta_w = 0$ if the length of the word w is not 1. Other choices of β are of interest [14] when analyzing numerical integrators by means of the method of modified equations [17].

Proceeding as in the derivation of (27), we find that the flow of (21) is given by

$$x(t) = \overline{W}_{(tv,\alpha(t))}(x(0)),$$

where $\alpha(t) \in \mathcal{G}$ is the solution of

$$\frac{d}{dt}\alpha(t) = \alpha(t) \star \Xi_{tv}\beta, \quad \alpha(0) = \mathbb{1}.$$

Moreover, $\alpha(t) = \gamma(t, tv)$, where $\gamma(\tau, u)$ is the unique polynomially smooth solution of the transport problem

$$\frac{\partial}{\partial \tau}\gamma(\tau, u) + v \cdot \nabla\gamma(\tau, u) = \gamma(\tau, u) \star \Xi_u\beta, \quad \gamma(0, 0) = \mathbb{1},$$

which clearly generalizes (31). For each $(\tau, u) \in \mathbb{R} \times \mathbb{C}^d$, the element $\gamma(\tau, u)$ belongs to the group $\mathcal{G}$. Note that the recursions (30) are not valid for general β.

In analogy with Theorem 9, we have that, for arbitrary $\tau, \tau' \in \mathbb{R}$ and $u, u' \in C^d$,

$$(u, \gamma(\tau, u))\bigstar(\tau', \gamma(\tau', u')) = (u + u', \gamma(\tau + \tau', u + u')).$$

Theorems 10 and 11 hold true for general $\beta \in \mathfrak{g}$.

Acknowledgements A. Murua and J. M. Sanz-Serna have been supported by projects MTM2013-46553-C3-2-P and MTM2013-46553-C3-1-P from Ministerio de Economía y Comercio, Spain. Additionally A. Murua has been partially supported by the Basque Government (Consolidated Research Group IT649-13).

References

1. Alamo, A., Sanz-Serna, J.M.: A technique for studying strong and weak local errors of splitting stochastic integrators. SIAM J. Numer. Anal. **54**, 3239–3257 (2016)
2. Arnold, V.I.: Geometrical Methods in the Theory of Ordinary Differential Equations, 2nd edn. Springer, New York (1988)
3. Chartier, P., Murua, A., Sanz-Serna, J.M.: Higher-order averaging, formal series and numerical integration I: B-series. Found. Comput. Math. **10**, 695–727 (2010)
4. Chartier, P., Murua, A., Sanz-Serna, J.M.: Higher-order averaging, formal series and numerical integration II: the quasi-periodic case. Found. Comput. Math. **12**, 471–508 (2012)
5. Chartier, P., Murua, A., Sanz-Serna, J.M.: A formal series approach to averaging: exponentially small error estimates. DCDS A **32**, 3009–3027 (2012)
6. Chartier, P., Murua, A., Sanz-Serna, J.M.: Higher-order averaging, formal series and numerical integration III: error bounds. Found. Comput. Math. **15**, 591–612 (2015)
7. Fauvet, F., Menous, F.: Ecalle's arborification-coarborification transforms and Connes-Kreimer Hopf algebra. Ann. Sci. Ec. Nom. Sup. **50**, 39–83 (2017)
8. Jacobson, N.: Lie Algebras. Dover, New York (1979)
9. Kawski, M., Sussmann, H.J.: Nonommutative power series and formal Lie algebraic techniques in nonlinear control theory. In: Helmke, U., Pratzel-Wolters, D., Zerz, E. (eds.) Operators, Systems, and Linear Algebra, pp. 111–118. Teubner, Stuttgart (1997)
10. Hairer, E., Wanner, G.: On the Butcher group and general multi-value methods. Computing **13**, 1–15 (1974)
11. Murua, A.: The Hopf algebra of rooted trees, free Lie algebras and Lie series. Found. Comput. Math. **6**, 387–426 (2006)
12. Murua, A., Sanz-Serna, J.M.: Vibrational resonance: a study with high-order word-series averaging. Appl. Math. Nonlinear Sci. **1**, 146–239 (2016)
13. Murua, A., Sanz-Serna, J.M.: Computing normal forms and formal invariants of dynamical systems by means of word series. Nonlinear Anal. **138**, 326–345 (2016)
14. Murua, A., Sanz-Serna, J.M.: Word series for dynamical systems and their numerical integrators. Found. Comput. Math. **17**, 675–712 (2017)
15. Reutenauer, C.: Free Lie Algebras. Clarendon Press, Oxford (1993)
16. Sanders, J.A., Verhulst, F., Murdock, J.: Averaging Methods in Nonlinear Dynamical Systems, 2nd edn. Springer, New York (2007)
17. Sanz-Serna, J.M., Calvo, M.P.: Numerical Hamiltonian Problems. Chapman and Hall, London (1994)
18. Sanz-Serna, J.M., Murua, A.: Formal series and numerical integrators: some history and some new techniques. In: Lei, G., Zhi, M. (eds.) Proceedings of the 8th International Congress on Industrial and Applied Mathematics (ICIAM 2015). Higher Education Press, Beijing, pp. 311–331 (2015)

Combinatorial Hopf Algebras for Interconnected Nonlinear Input-Output Systems with a View Towards Discretization

Luis A. Duffaut Espinosa, Kurusch Ebrahimi-Fard and W. Steven Gray

Abstract A detailed expose of the Hopf algebra approach to interconnected input-output systems in nonlinear control theory is presented. The focus is on input-output systems that can be represented in terms of Chen–Fliess functional expansions or Fliess operators. This provides a starting point for a discrete-time version of this theory. In particular, the notion of a discrete-time Fliess operator is given and a class of parallel interconnections is described in terms of the quasi-shuffle algebra.

Keywords Nonlinear control systems · Chen–Fliess series
Combinatorial Hopf algebras

AMS Subject Classification 93C10 · 93B25 · 16T05 · 16T30

L. A. Duffaut Espinosa
Department of Electrical and Biomedical Engineering, University of Vermont,
Burlington, VT 05405, USA
e-mail: lduffaut@uvm.edu

K. Ebrahimi-Fard
Department of Mathematical Sciences, Norwegian University of Science
and Technology – NTNU, NO-7491 Trondheim, Norway
e-mail: kurusch.ebrahimi-fard@ntnu.noaaaa
URL: https://folk.ntnu.no/kurusche/

W. S. Gray (✉)
Department of Electrical and Computer Engineering, Old Dominion University,
Norfolk, VA 23529, USA
e-mail: sgray@odu.edu

K. Ebrahimi-Fard and M. Barbero Liñán (eds.), *Discrete Mechanics, Geometric Integration and Lie–Butcher Series*, Springer Proceedings in Mathematics & Statistics 267, https://doi.org/10.1007/978-3-030-01397-4_5

1 Introduction

A central problem in control theory is understanding how dynamical systems behave when they are interconnected. In a typical design problem, one is given a fixed system representing the *plant*, say a robotic manipulator or a spacecraft. The objective is to find a second system, usually called the *controller*, which when interconnected with the first will make the output track a pre-specified trajectory. When the component systems are nonlinear, these problems are difficult to address by purely analytical means. The prevailing methodologies are geometric in nature and based largely on state variable analysis [43, 47, 53, 60]. A complementary approach, however, has begun to emerge where the input-output map of each component system is represented in terms of a Chen–Fliess functional expansion or Fliess operator. In this setting, concepts from combinatorics and algebra are employed to produce an explicit description of the interconnected system. The genesis of this method can be found in the work of Fliess [24, 25] and Ferfera [19, 20], who described key elements of the underlying algebras in terms of noncommutative formal power series. In particular, they identified the central role played by the shuffle algebra in this theory. The method was further developed by Gray et al. [38–40, 65] and Wang [66] who addressed basic analytical questions such as what inputs guarantee convergence of the component series, when are the interconnections well defined, and what is the nature of the output functions? A fundamental open problem on the algebraic side until 2011 was how to explicitly compute the generating series of two feedback interconnected Fliess operators. Largely inspired by interactions at the 2010 Trimester in Combinatorics and Control (COCO2010) in Madrid with researchers in the area of quantum field theory (see, for example, [21]), the problem was eventually solved by identifying a new combinatorial Hopf algebra underlying the calculation. The evolution of this structure took several distinct steps. The single-input, single-output (SISO) case was first addressed by Gray and Duffaut Espinosa in [28] via a certain graded Hopf algebra of combinatorial type. Foissy then introduced a new grading in [26] which rendered a connected version of this combinatorial Hopf algebra. This naturally provided a fully recursive formula for the antipode, which is central to the feedback calculation [31]. The multivariable, i.e., multi-input, multi-output (MIMO) case, was then treated in [32]. Next, a full combinatorial treatment, including a Zimmermann type forest formula for the antipode [2], was presented in [14]. This last result, based on an equivalent combinatorial Hopf algebra of decorated rooted circle trees, greatly reduces the number of computations involved by eliminating the inter-term cancellations that are intrinsic in the usual antipode calculation. Practical problems would be largely intractable without this innovation. The final and most recent development method is a description of this Hopf algebra based entirely on (co)derivation(-type) maps applied to the (co)product. This method was first observed to be implicit in the work of Devlin on the classical Poincaré center problem [12, 16]. In a state space setting, it can be related to computing iterated Lie derivatives to determine series coefficients [43]. Control applications ranging from guidance and chemical engineering to systems biology can be found in [15, 30, 32–34, 37].

This article has two general goals. First, an introduction to the method of combinatorial Hopf algebras in the context of feedback control theory is given in its most complete and up-to-date form. The idea is to integrate all of the advances described above into a single uniform treatment. This results in a new and distinct presentation of these ideas. In particular, the full solution to the problem of computing the generating series for a multivariable continuous-time dynamic output feedback system will be described. The origins of the forest formula related to this calculation will also be outlined. Then the focus is shifted to the second objective, which is largely an open problem in this field, namely how to recast this theory in a discrete-time setting. One approach suggested by Fliess in [23] is to describe the generating series of a discrete-time input-output map in terms of a complete tensor algebra defined on an algebra of polynomials over a noncommutative alphabet. While this is a very general approach and can be related to the notion of a Volterra series as described by Sontag in [61], to date it has not lead to any particularly useful algebraic structures in the context of system interconnections. Another approach developed by the authors in the context of numerical approximation is to define the notion of a discrete-time Fliess operator in terms of a series of iterated sums over a noncommutative alphabet [35, 36]. While not the most general set up, it has been shown to be related to a class of state affine rational input discrete-time systems in the case where the generating series is rational. Furthermore, this class of so called rational discrete-time Fliess operators is guaranteed to always converge. So this will be the approach taken here. As the main interest in this paper is interconnection theory, the analysis begins with the simplest type of interconnections, the parallel sum and parallel product connections. The former is completely trivial, but the latter induces the quasi-shuffle algebra of Hoffman (see [41]) on the vector space of generating series. Of particular interest is whether rationality is preserved under the quasi-shuffle product. It is well known to be the case for the shuffle product [24].

The paper is organized as follows. In Sect. 2, some preliminaries on Fliess operators and graded connected Hopf algebras are given to set the notation and provide some background. The subsequent section is devoted to describing the combinatorial algebras that are naturally induced by the interconnection of Fliess operators. In Sect. 4, the Hopf algebra of coordinate functions for the output feedback Hopf algebra is described in detail. The final section addresses elements of the discrete-time version of this theory.

2 Preliminaries

A finite nonempty set of noncommuting symbols $X = \{x_0, x_1, \ldots, x_m\}$ is called an *alphabet*. Each element of X is called a *letter*, and any finite sequence of letters from X, $\eta = x_{i_1} \cdots x_{i_k}$, is called a *word* over X. The *length* of the word η, denoted $|\eta|$, is given by the number of letters it contains. The set of all words with length k is denoted by X^k. The set of all words including the empty word, $\emptyset$, is designated by X^*, while $X^+ := X^* - \{\emptyset\}$. The set X^* forms a monoid under catenation. The set

ηX^* is comprised of all words with the prefix $\eta \in X^*$. For any fixed integer $\ell \geq 1$, a mapping $c : X^* \to \mathbb{R}^\ell$ is called a *formal power series*. The value of c at $\eta \in X^*$ is written as $(c, \eta) \in \mathbb{R}^\ell$ and called the *coefficient* of the word η in c. Typically, c is represented as the formal sum $c = \sum_{\eta \in X^*}(c, \eta)\eta$. If the *constant term* $(c, \emptyset) = 0$ then c is said to be *proper*. The *support* of c, supp(c), is the set of all words having nonzero coefficients in c. The *order* of c, ord(c), is the length of the shortest word in its support (ord(0) := ∞).[1] The $\mathbb{R}$-vector space of all formal power series over X^* with coefficients in $\mathbb{R}^\ell$ is denoted by $\mathbb{R}^\ell\langle\langle X \rangle\rangle$.[2] It forms a unital associative $\mathbb{R}$-algebra under the catenation product. Specifically, for $c, d \in \mathbb{R}^\ell\langle\langle X \rangle\rangle$ the catenation product is given by $cd = \sum_{\eta \in X^*}(cd, \eta)\, \eta$, where

$$(cd, \eta) = \sum_{\eta = \xi\nu} (c, \xi)(d, \nu), \quad \forall \eta \in X^*,$$

and the product on $\mathbb{R}^\ell$ is defined componentwise. The unit in this case is $\mathbf{1} := 1\emptyset$. $\mathbb{R}^\ell\langle\langle X \rangle\rangle$ is also a unital, commutative and associative $\mathbb{R}$-algebra under the shuffle product, denoted here by the shuffle symbol $\sqcup\!\sqcup$. The shuffle product of two words is defined inductively by

$$(x_i\eta) \sqcup\!\sqcup (x_j\xi) = x_i(\eta \sqcup\!\sqcup (x_j\xi)) + x_j((x_i\eta) \sqcup\!\sqcup \xi) \tag{1}$$

with $\eta \sqcup\!\sqcup \emptyset = \emptyset \sqcup\!\sqcup \eta = \eta$ given any words $\eta, \xi \in X^*$ and letters $x_i, x_j \in X$ [24, 56, 57]. For instance, $x_i \sqcup\!\sqcup x_j = x_i x_j + x_j x_i$ and

$$x_{i_1}x_{i_2} \sqcup\!\sqcup x_{i_3}x_{i_4} = x_{i_1}x_{i_2}x_{i_3}x_{i_4} + x_{i_3}x_{i_4}x_{i_1}x_{i_2} + x_{i_1}x_{i_3}(x_{i_2} \sqcup\!\sqcup x_{i_4}) + x_{i_3}x_{i_1}(x_{i_2} \sqcup\!\sqcup x_{i_4}).$$

The definition of the shuffle product is extended linearly to any two series $c, d \in \mathbb{R}^\ell\langle\langle X \rangle\rangle$ by letting

$$c \sqcup\!\sqcup d = \sum_{\eta, \xi \in X^*} (c, \eta)(d, \xi)\, \eta \sqcup\!\sqcup \xi, \tag{2}$$

where again the product on $\mathbb{R}^\ell$ is defined componentwise. For a fixed word $\nu \in X^*$, the coefficient

$$(\eta \sqcup\!\sqcup \xi, \nu) = 0 \ \text{ if } \ |\eta| + |\xi| \neq |\nu|.$$

Hence, the infinite sum in (2) is always well defined since the family of polynomials $\{\eta \sqcup\!\sqcup \xi\}_{\eta, \xi \in X^*}$ is locally finite. The unit for this product is $\mathbf{1}$.

Some standard concepts regarding rational formal power series, which are used in Sect. 5, are provided next [3]. A series $c \in \mathbb{R}\langle\langle X \rangle\rangle$ is called *invertible* if there exists a series $c^{-1} \in \mathbb{R}\langle\langle X \rangle\rangle$ such that $cc^{-1} = c^{-1}c = \mathbf{1}$. In the event that c is not proper, it is always possible to write

---

[1]For notational convenience, $p = (p, \emptyset)\emptyset \in \mathbb{R}\langle X \rangle$ is often abbreviated as $p = (p, \emptyset)$.

[2]The superscript ℓ will be dropped when $\ell = 1$.

$$c = (c, \emptyset)(\mathbf{1} - c'),$$

where $(c, \emptyset)$ is nonzero, and $c' \in \mathbb{R}\langle\langle X \rangle\rangle$ is proper. It then follows that

$$c^{-1} = \frac{1}{(c, \emptyset)}(\mathbf{1} - c')^{-1} = \frac{1}{(c, \emptyset)}(c')^*, \tag{3}$$

where

$$(c')^* := \sum_{i=0}^{\infty} (c')^i.$$

In fact, c is invertible if and *only if* c is not proper. Now let S be a subalgebra of the $\mathbb{R}$-algebra $\mathbb{R}\langle\langle X \rangle\rangle$ with the catenation product. S is said to be *rationally closed* when every invertible $c \in S$ has $c^{-1} \in S$ (or equivalently, every proper $c' \in S$ has $(c')^* \in S$). The *rational closure* of any subset $E \subset \mathbb{R}\langle\langle X \rangle\rangle$ is the smallest rationally closed subalgebra of $\mathbb{R}\langle\langle X \rangle\rangle$ containing E.

Definition 1 [3] A series $c \in \mathbb{R}\langle\langle X \rangle\rangle$ is ***rational*** if it belongs to the rational closure of $\mathbb{R}\langle X \rangle$. The subset of all rational series in $\mathbb{R}\langle\langle X \rangle\rangle$ is denoted by $\mathbb{R}_{rat}\langle\langle X \rangle\rangle$.

From Definition 1 it is clear that any given series in $\mathbb{R}_{rat}\langle\langle X \rangle\rangle$ is generated by a finite number of rational operations (scalar multiplication, addition, catenation, and inversion) applied to a finite set of polynomials over X. In the case where the alphabet X has an infinite number of letters, such a series can only involve a finite subset of X. Therefore, it is rational in exactly the sense described above when restricted to this sub-alphabet. This will be the notion of rationality employed in this manuscript whenever X is infinite. But the reader is cautioned that other notions of rationality for infinite alphabets appear in the literature, see, for example, [54].

It turns out that an entirely different characterization of a rational series is possible using a monoid structure on the set of $n \times n$ matrices over $\mathbb{R}$, denoted $\mathbb{R}^{n \times n}$, where the product is conventional matrix multiplication, and the unit is the $n \times n$ identity matrix I.

Definition 2 [3] A ***linear representation*** of a series $c \in \mathbb{R}\langle\langle X \rangle\rangle$ is any triple (μ, γ, λ), where

$$\mu : X^* \rightarrow \mathbb{R}^{n \times n}$$

is a monoid morphism, and $\gamma, \lambda^T \in \mathbb{R}^{n \times 1}$ are such that

$$(c, \eta) = \lambda\mu(\eta)\gamma, \quad \forall \eta \in X^*. \tag{4}$$

The integer $n > 0$ is the dimension of the representation.

Definition 3 [3] A series $c \in \mathbb{R}\langle\langle X \rangle\rangle$ is called ***recognizable*** if it has a linear representation.

Theorem 1 [59] *A formal power series is rational if and only if it is recognizable.*

A third characterization of rationality is given by the notion of *stability*. Define for any letter $x_i \in X$ and word $\eta = x_j\eta' \in X^*$ the left-shift operator

$$x_i^{-1}(\eta) = \delta_{ij}\eta', \tag{5}$$

where δ_{ij} is the standard Kronecker delta. Higher order shifts are defined inductively via $(x_i\xi)^{-1}(\cdot) = \xi^{-1}x_i^{-1}(\cdot)$, where $\xi \in X^*$. The left-shift operator is assumed to act linearly on $\mathbb{R}\langle\langle X\rangle\rangle$.

Definition 4 [3] A subset $V \subset \mathbb{R}\langle\langle X\rangle\rangle$ is called ***stable*** when $\xi^{-1}(c) \in V$ for all $c \in V$ and $\xi \in X^*$.

Theorem 2 [3] *A series $c \in \mathbb{R}\langle\langle X\rangle\rangle$ is rational if and only if there exists a stable finite dimensional $\mathbb{R}$-vector subspace of $\mathbb{R}\langle\langle X\rangle\rangle$ containing c.*

2.1 Chen–Fliess Series and Fliess Operators

One can associate with any formal power series $c \in \mathbb{R}^\ell\langle\langle X\rangle\rangle$ a functional series, F_c, known a *Chen–Fliess series*. Let $\mathfrak{p} \geq 1$ and $t_0 < t_1$ be given. For a Lebesgue measurable function $u : [t_0, t_1] \to \mathbb{R}^m$, define $\|u\|_\mathfrak{p} = \max\{\|u_i\|_\mathfrak{p} : 1 \leq i \leq m\}$, where $\|u_i\|_\mathfrak{p}$ is the usual $L_\mathfrak{p}$-norm for a measurable real-valued function, u_i, defined on $[t_0, t_1]$. Let $L_\mathfrak{p}^m[t_0, t_1]$ denote the set of all measurable functions defined on $[t_0, t_1]$ having a finite $\|\cdot\|_\mathfrak{p}$ norm and $B_\mathfrak{p}^m(R)[t_0, t_1] := \{u \in L_\mathfrak{p}^m[t_0, t_1] : \|u\|_\mathfrak{p} \leq R\}$. Assume $C[t_0, t_1]$ is the subset of continuous functions in $L_1^m[t_0, t_1]$. Define inductively for each word $\eta \in X^*$ the map $E_\eta : L_1^m[t_0, t_1] \to C[t_0, t_1]$ by setting $E_\emptyset[u] = 1$ and letting

$$\begin{aligned} E_{x_{i_1}\bar{\eta}}[u](t, t_0) &:= \int_{t_0}^{t} u_{i_1}(\tau_1) E_{\bar{\eta}}[u](\tau_1, t_0)\, d\tau_1 \\ &= \int_{\Delta_{[t,t_0]}^n} u_{i_1}(\tau_1)u_{i_2}(\tau_2)\cdots u_{i_n}(\tau_n)d\tau_n \cdots d\tau_2 d\tau_1, \end{aligned}$$

where $\Delta_{[t,t_0]}^n := \{(\tau_1, \ldots, \tau_n),\ t \geq \tau_1 \geq \cdots \geq \tau_n \geq t_0\}$, $\eta = x_{i_1}\cdots x_{i_n} = x_{i_1}\bar{\eta} \in X^*$, and $u_0 := 1$. For instance, the words x_i and $x_{i_1}x_{i_2}$ correspond to the integrals

$$E_{x_i}[u](t, t_0) = \int_{t_0}^{t} u_i(\tau)d\tau, \quad E_{x_{i_1}x_{i_2}}[u](t, t_0) = \int_{t_0}^{t} u_{i_1}(\tau_1) \int_{t_0}^{\tau_1} u_{i_2}(\tau_2)d\tau_2 d\tau_1.$$

The *Chen–Fliess series* corresponding to $c \in \mathbb{R}^\ell\langle\langle X\rangle\rangle$ is defined to be

$$F_c[u](t) = \sum_{\eta \in X^*} (c, \eta)\, E_\eta[u](t, t_0). \tag{6}$$

In the event that there exist real numbers $K_c, M_c > 0$ such that

$$|(c, \eta)| \leq K_c M_c^{|\eta|} |\eta|!, \quad \forall \eta \in X^*, \tag{7}$$

then F_c constitutes a well defined causal operator from $B_{\mathfrak{p}}^m(R)[t_0, t_0 + T]$ into $B_{\mathfrak{q}}^\ell(S)[t_0, t_0 + T]$ for some $S > 0$ provided $\bar{R} := \max\{R, T\} < 1/M_c(m+1)$, and the numbers $\mathfrak{p}, \mathfrak{q} \in [1, \infty]$ are conjugate exponents, i.e., $1/\mathfrak{p} + 1/\mathfrak{q} = 1$ [40]. (Here, $|z| := \max_i |z_i|$ when $z \in \mathbb{R}^\ell$.) In this case, F_c is called a *Fliess operator* and said to be *locally convergent* (LC). The set of all series satisfying (7) is denoted by $\mathbb{R}_{LC}^\ell\langle\langle X\rangle\rangle$. When c satisfies the more stringent growth condition

$$|(c, \eta)| \leq K_c M_c^{|\eta|}, \quad \forall \eta \in X^*, \tag{8}$$

the series (6) defines a Fliess operator from the extended space $L_{\mathfrak{p},e}^m(t_0)$ into $C[t_0, \infty)$, where

$$L_{\mathfrak{p},e}^m(t_0) := \{u : [t_0, \infty) \to \mathbb{R}^m : u_{[t_0,t_1]} \in L_{\mathfrak{p}}^m[t_0, t_1], \quad \forall t_1 \in (t_0, \infty)\},$$

and $u_{[t_0,t_1]}$ denotes the restriction of u to the interval $[t_0, t_1]$ [40]. In this case, the operator is said to be *globally convergent* (GC), and the set of all series satisfying (8) is designated by $\mathbb{R}_{GC}^\ell\langle\langle X\rangle\rangle$.

Most of the work regarding nonlinear input-output systems in control theory prior to the work of Fliess was based on Volterra series, see, for example, [4, 48, 58]. In many ways this earlier work set the stage for the introduction of the noncommutative algebraic framework championed by Fliess. As Fliess operators are series of weighted iterated integrals of control functions, they are also related to the work of K. T. Chen, who revealed that iterated integrals come with a natural algebraic structure [9–11]. Indeed, products of iterated integrals can again be written as linear combinations of iterated integrals. This is implied by the classical integration by parts rule for indefinite Riemann integrals, which yields for instance that

$$E_{x_{i_1}}[u](t, t_0) E_{x_{i_2}}[u](t, t_0) = E_{x_{i_1} x_{i_2}}[u](t, t_0) + E_{x_{i_2} x_{i_1}}[u](t, t_0).$$

Linearity allows one to relate this to the shuffle product (1)

$$E_{x_{i_1}}[u](t, t_0) E_{x_{i_2}}[u](t, t_0) = F_{x_{i_1} x_{i_2} + x_{i_2} x_{i_1}}[u](t) = F_{x_{i_1} \sqcup\!\sqcup\, x_{i_2}}[u](t).$$

This generalizes naturally to the shuffle product for iterated integrals with respect to words $\eta, \nu \in X^*$

$$E_\eta[u](t, t_0) E_\nu[u](t, t_0) = F_{\eta \sqcup\!\sqcup\, \nu}[u](t), \tag{9}$$

which in turn implies for Fliess operators corresponding to $c, d \in \mathbb{R}^\ell\langle\langle X\rangle\rangle$ that

$$F_c[u](t)F_d[u](t) = F_{c \sqcup\!\sqcup d}[u](t). \tag{10}$$

A Fliess operator F_c defined on $B_{\mathrm{p}}^m(R)[t_0, t_0 + T]$ is said to be *realizable* when there exists a state space model consisting of n ordinary differential equations and ℓ output functions

$$\dot{z}(t) = g_0(z(t)) + \sum_{i=1}^{m} g_i(z(t))\, u_i(t), \quad z(t_0) = z_0 \tag{11a}$$

$$y_j(t) = h_j(z(t)), \quad j = 1, 2, \ldots, \ell, \tag{11b}$$

where each g_i is an analytic vector field expressed in local coordinates on some neighborhood $\mathcal{W}$ of z_0, and each output function h_j is an analytic function on $\mathcal{W}$ such that (11a) has a well defined solution $z(t)$, $t \in [t_0, t_0 + T]$ for any given input $u \in B_{\mathrm{p}}^m(R)[t_0, t_0 + T]$, and $y_j(t) = F_{c_j}[u](t) = h_j(z(t))$, $t \in [t_0, t_0 + T]$, $j = 1, 2, \ldots, \ell$. It can be shown that for any word $\eta = x_{i_k} \cdots x_{i_1} \in X^*$

$$(c_j, \eta) = L_{g_\eta} h_j(z_0) := L_{g_{i_1}} \cdots L_{g_{i_k}} h_j(z_0), \tag{12}$$

where $L_{g_i} h_j$ is the *Lie derivative* of h_j with respect to g_i. For any $c \in \mathbb{R}^\ell\langle\langle X\rangle\rangle$, the $\mathbb{R}$-linear mapping $\mathcal{H}_c : \mathbb{R}\langle X\rangle \to \mathbb{R}^\ell\langle\langle X\rangle\rangle$ uniquely specified by $(\mathcal{H}_c(\eta), \xi) = (c, \xi\eta)$, $\xi, \eta \in X^*$ is called the *Hankel mapping* of c. The series c is said to have finite *Lie rank* $\rho_L(c)$ when the range of $\mathcal{H}_c$ restricted to the $\mathbb{R}$-vector space of Lie polynomials over X, i.e., the free Lie algebra $\mathcal{L}(X) \subset \mathbb{R}\langle X\rangle$, has dimension $\rho_L(c)$. It is well known that F_c is realizable if and only if $c \in \mathbb{R}_{LC}^\ell\langle\langle X\rangle\rangle$ has finite Lie rank [24, 25, 43–46, 62, 63]. In which case, all minimal realization have dimension $\rho_L(c)$ and are unique up to a diffeomorphism. In the event that $\mathcal{H}_c$ has finite rank on the entire vector space $\mathbb{R}\langle X\rangle$, usually referred to as the *Hankel rank* $\rho_H(c)$ of c, and $c \in \mathbb{R}_{GC}^\ell\langle\langle X\rangle\rangle$ then F_c has a minimal bilinear state space realization

$$\dot{z}(t) = A_0 z(t) + \sum_{i=1}^{m} A_i z(t) u_i(t), \quad z(t_0) = z_0$$

$$y_j(t) = C_j z(t), \quad j = 1, 2, \ldots, \ell$$

of dimension $\rho_H(c) \geq \rho_L(c)$, where A_i and C_j are real matrices of appropriate dimensions [24, 25, 43]. Here the state $z(t)$ is well defined on any interval $[t_0, t_0 + T]$, $T > 0$, when $u \in L_{1,e}^m(t_0)$, and the operator F_c always converges globally. In addition, (12) simplifies to

$$(c_j, x_{i_k} \cdots x_{i_1}) = C_j A_{i_k} \cdots A_{i_1} z_0. \tag{13}$$

In light of (4), it is not hard to see that c is recognizable in the SISO case if and only if F_c has a bilinear realization with $A_i = \mu(x_i)$ for $i = 0, 1$ and $z_0 = \gamma$, and $C = \lambda$.

2.2 *Graded Connected Hopf Algebras*

All algebraic structures here are considered over the base field $\mathbb{K}$ of characteristic zero, for instance, $\mathbb{C}$ or $\mathbb{R}$. Multiplication in $\mathbb{K}$ is denoted by $m_{\mathbb{K}} : \mathbb{K} \otimes \mathbb{K} \to \mathbb{K}$.

2.2.1 Algebra

A $\mathbb{K}$-*algebra* is denoted by the triple (A, m_A, η_A) where A is a $\mathbb{K}$-vector space carrying an associative product $m_A : A \otimes A \to A$, i.e., $m_A \circ (m_A \otimes \mathrm{id}_A) = m_A \circ (\mathrm{id}_A \otimes m_A) : A \otimes A \otimes A \to A$, and a unit map $\eta_A : \mathbb{K} \to A$. The algebra unit corresponding to the latter is denote by 1_A. A $\mathbb{K}$-*subalgebra* of the $\mathbb{K}$-algebra A is a $\mathbb{K}$-vector subspace $B \subseteq A$ such that $m_A(b \otimes b') \in B$ for all $b, b' \in B$. A $\mathbb{K}$-subalgebra $I \subseteq A$ is called a *(right-) left-ideal* if for any elements $i \in I$ and $a \in A$ the product $(m_A(i \otimes a))\, m_A(a \otimes i)$ is in I. An *ideal* $I \subseteq A$ is both a left- and right-ideal.

In order to motivate the concept of a $\mathbb{K}$-coalgebra, the definition of a $\mathbb{K}$-algebra A is rephrased in terms of commutative diagrams. Associativity of the $\mathbb{K}$-vector space morphism $m_A : A \otimes A \to A$ translates into commutativity of the diagram

$$\begin{CD} A \otimes A \otimes A @>{m_A \otimes \mathrm{id}_A}>> A \otimes A \\ @V{\mathrm{id}_A \otimes m_A}VV @VV{m_A}V \\ A \otimes A @>{m_A}>> A \end{CD} \tag{14}$$

The $\mathbb{K}$-algebra A is unital if the $\mathbb{K}$-vector space map $\eta_A : \mathbb{K} \to A$ satisfies the commutative diagram

$$\begin{array}{ccccc} \mathbb{K} \otimes A & \xrightarrow{\eta_A \otimes \mathrm{id}_A} & A \otimes A & \xleftarrow{\mathrm{id}_A \otimes \eta_A} & A \otimes \mathbb{K} \\ & {\scriptstyle \alpha_l} \searrow & \downarrow {\scriptstyle m_A} & \swarrow {\scriptstyle \alpha_r} & \\ & & A & & \end{array} \tag{15}$$

Here α_l and α_r are the isomorphisms sending $k \otimes a$ respectively $a \otimes k$ to ka for $k \in \mathbb{K}$, $a \in A$. Let $\tau := \tau_{A,A} : A \otimes A \to A \otimes A$ be the flip map, $\tau_{A,A}(x \otimes y) := y \otimes x$. The $\mathbb{K}$-algebra A is *commutative* if the next diagram commutes

$$\begin{array}{ccc} A \otimes A & \xrightarrow{\tau} & A \otimes A \\ {\scriptstyle m_A} \downarrow & \swarrow {\scriptstyle m_A} & \\ A & & \end{array} \tag{16}$$

An important observation is that nonassociative $\mathbb{K}$-algebras play a key role in the context of Hopf algebras, in particular, for those of combinatorial type. Recall that the Lie algebra $\mathcal{L}(A)$ associated with a $\mathbb{K}$-algebra A follows from anti-symmetrization of the algebra product m_A. Algebras that give Lie algebras in this way are called Lie admissible. Another class of Lie admissible algebras are pre-Lie $\mathbb{K}$-algebras. A *left pre-Lie algebra* [5, 8, 51] is a vector space V equipped with a bilinear product $\rhd: V \otimes V \to V$ such that the *(left) pre-Lie identity*,

$$a \rhd (b \rhd c) - (a \rhd b) \rhd c = b \rhd (a \rhd c) - (b \rhd a) \rhd c, \tag{17}$$

holds for $a, b, c \in V$. This identity rewrites as $L_{[a,b]} = [L_a, L_b]$, where $L_a : V \to V$ is defined by $L_a b := a \rhd b$. The bracket on the left-hand side is defined by $[a, b] := a \rhd b - b \rhd a$ and satisfies the Jacobi identity. Right pre-Lie algebras are defined analogously. Note that the (left) pre-Lie identity (17) can be understood as a relation between associators, i.e., let $\alpha_{\rhd} : V \otimes V \otimes V \to V$ be defined by

$$\alpha_{\rhd}(a, b, c) := a \rhd (b \rhd c) - (a \rhd b) \rhd c,$$

then (17) simply says that $\alpha_{\rhd}(a, b, c) = \alpha_{\rhd}(b, a, c)$. From this it is easy to see that any associative algebra is pre-Lie.

Example 1 [5] Let A be a commutative $\mathbb{K}$-algebra endowed with commuting derivatives $D := \{\partial_1, \ldots, \partial_n\}$. For $a \in A$ define $a\partial_i : A \to A$ by $(a\partial_i)(b) := a\partial_i b$ and $V(n) := \left\{ \sum_{i=1}^n a_i \partial_i \ : \ a_i \in A, \partial_i \in D \right\}$. The algebra $(V(n), \lhd)$, where $\sum_{i=1}^n a_i \partial_i \lhd \sum_{j=1}^n a_j \partial_j := \sum_{j,i=1}^n a_j (\partial_j a_i) \partial_i$, is a right pre-Lie algebra called the pre-Lie Witt algebra.

Example 2 [5, 8, 51] The next example of a pre-Lie algebra is of a geometric nature and similar to the one above. Let M be a differentiable manifold endowed with a flat and torsion-free connection. The corresponding covariant derivation operator ∇ on the space $\chi(M)$ of vector fields on M provides it with a left pre-Lie algebra structure, which is defined via $a \rhd b := \nabla_a b$ by virtue of the two equalities $\nabla_a b - \nabla_b a = [a, b]$ and $\nabla_{[a,b]} = [\nabla_a, \nabla_b]$. They express the vanishing of torsion and curvature respectively. Let $M = \mathbb{R}^n$ with its standard flat connection. For $a = \sum_{i=1}^n a_i \partial_i$ and $b = \sum_{i=1}^n b_i \partial_i$ it follows that

$$a \rhd b = \sum_{i=1}^n \left(\sum_{j=1}^n a_j (\partial_j b_i) \right) \partial_i.$$

Example 3 [5] Denote by W the space of all words over the alphabet $\{a, b\}$. Define for words v and $w = w_1 \cdots w_n$ in W the product $w \circ v := \sum_{i=0}^n \varepsilon(i) w \lhd_i v$, where $w \lhd_i v$ denotes inserting the word v between letters w_i and w_{i+1} of w, i.e., $w \lhd_i v = w_1 \cdots w_i v w_{i+1} \cdots w_n$ and

$$\varepsilon(i) := \begin{cases} -1, & w_i = a, w_{i+1} = b \\ +1, & w_i = b, w_{i+1} = a \text{ or } \emptyset \\ +1, & w_i = \emptyset, w_{i+1} = a \\ 0, & \text{otherwise.} \end{cases}$$

The algebra $(W, \circ)$ is right pre-Lie. For example,

$$a \circ a = aa, \quad a \circ ab = aba, \quad ab \circ a = aab - aab + aba = aba, \quad ba \circ ab = baba.$$

2.2.2 Coalgebra

The definition of a $\mathbb{K}$-coalgebra is most easily obtained by reversing the arrows in diagrams (14) and (15). Thus, a $\mathbb{K}$-*coalgebra* is a triple $(C, \Delta_C, \varepsilon_C)$, where C is a $\mathbb{K}$-vector space carrying a *coassociative coproduct* map $\Delta_C : C \to C \otimes C$, i.e., $(\Delta_C \otimes \mathrm{id}_C) \circ \Delta_C = (\mathrm{id}_C \otimes \Delta_C) \circ \Delta_C : C \to C \otimes C \otimes C$, and $\varepsilon_C : C \to \mathbb{K}$ is the counit map which satisfies $(\varepsilon_C \otimes \mathrm{id}_C) \circ \Delta_C = \mathrm{id}_C = (\mathrm{id}_C \otimes \varepsilon_C) \circ \Delta_C$. Its kernel $\ker(\varepsilon_C) \subset C$ is called the *augmentation ideal.*

A simple example of a coalgebra is the field $\mathbb{K}$ itself with the coproduct $\Delta_{\mathbb{K}} : \mathbb{K} \to \mathbb{K} \otimes \mathbb{K}$, $c \mapsto c \otimes 1_{\mathbb{K}}$ and $\varepsilon_{\mathbb{K}} := \mathrm{id}_{\mathbb{K}} : \mathbb{K} \to \mathbb{K}$.

Using Sweedler's notation for the coproduct of an element $x \in C$, $\Delta_C(x) = \sum_{(x)} x^{(1)} \otimes x^{(2)}$, provides a simple description of coassociativity

$$\sum_{(x)} \Big(\sum_{(x^{(1)})} x^{(1)(1)} \otimes x^{(1)(2)} \Big) \otimes x_{(2)} = \sum_{(x)} x^{(1)} \otimes \Big(\sum_{(x^{(2)})} x^{(2)(1)} \otimes x^{(2)(2)} \Big).$$

It permits the use of a transparent notation for iterated coproducts: $\Delta_C^{(n)} : C \to C^{\otimes n+1}$, where $\Delta^{(0)} := \mathrm{id}_C$, $\Delta_C^{(1)} := \Delta_C$, and

$$\Delta_C^{(n)} := (\mathrm{id}_C \otimes \Delta_C^{(n-1)}) \circ \Delta_C = (\Delta_C^{(n-1)} \otimes \mathrm{id}_C) \circ \Delta_C.$$

For example,

$$\Delta_C^{(2)}(x) := (\mathrm{id}_C \otimes \Delta_C) \circ \Delta_C(x) = (\Delta_C \otimes \mathrm{id}_C) \circ \Delta_C(x) = \sum_{(x)} x^{(1)} \otimes x^{(2)} \otimes x^{(3)}.$$

A *cocommutative* coalgebra satisfies $\tau \circ \Delta_C = \Delta_C$, which amounts to reversing the arrows in diagram (16). An element x in a coalgebra $(C, \Delta_C, \varepsilon_C)$ is called *primitive* if $\Delta_C(x) = x \otimes 1_C + 1_C \otimes x$. It is called *group-like* if $\Delta_C(x) = x \otimes x$. The set of primitive elements in C is denoted by $P(C)$. A subspace $I \subseteq C$ of a $\mathbb{K}$-coalgebra $(C, \Delta_C, \varepsilon_C)$ is a *subcoalgebra* if $\Delta_C(I) \subseteq I \otimes I$. A subspace $I \subseteq C$ is called a (left-, right-) *coideal* if $(\Delta_C(I) \subseteq I \otimes C, \Delta_C(I) \subseteq C \otimes I)$ $\Delta_C(I) \subseteq I \otimes C + C \otimes I$.

2.2.3 Bialgebra

A $\mathbb{K}$-*bialgebra* consists of a $\mathbb{K}$-algebra and a $\mathbb{K}$-coalgebra which are compatible [1, 21, 50, 55, 64]. More precisely, a $\mathbb{K}$-bialgebra is a quintuple $(B, m_B, \eta_B, \Delta_B, \varepsilon_B)$, where (B, m_B, η_B) is a $\mathbb{K}$-algebra, and $(B, \Delta_B, \varepsilon_B)$ is a $\mathbb{K}$-coalgebra, such that m_B and η_B are morphisms of $\mathbb{K}$-coalgebras with the natural coalgebra structure on the space $B \otimes B$. Commutativity of the following diagrams encodes the compatibilities

$$\begin{array}{ccc} B \otimes B & \xrightarrow{\quad m_B \quad} & B \\ {\scriptstyle \tau_2(\Delta_B \otimes \Delta_B)}\downarrow & & \downarrow{\scriptstyle \Delta_B} \\ B \otimes B \otimes B \otimes B & \xrightarrow{\; m_B \otimes m_B \;} & B \otimes B \end{array} \qquad \begin{array}{ccc} B \otimes B & \xrightarrow{\; \varepsilon_B \otimes \varepsilon_B \;} & \mathbb{K} \otimes \mathbb{K} \\ {\scriptstyle m_B}\downarrow & & \downarrow{\scriptstyle m_{\mathbb{K}}} \\ B & \xrightarrow{\quad \varepsilon_B \quad} & \mathbb{K} \end{array} \tag{18}$$

$$\begin{array}{ccc} \mathbb{K} & \xrightarrow{\quad \eta_B \quad} & B \\ {\scriptstyle \Delta_{\mathbb{K}}}\downarrow & & \downarrow{\scriptstyle \Delta_B} \\ \mathbb{K} \otimes \mathbb{K} & \xrightarrow{\; \eta_B \otimes \eta_B \;} & B \otimes B \end{array} \qquad \begin{array}{ccc} \mathbb{K} & \xrightarrow{\quad \eta_B \quad} & B \\ & {\scriptstyle \mathrm{id}_{\mathbb{K}}}\searrow \quad \swarrow{\scriptstyle \varepsilon_B} & \\ & \mathbb{K} & \end{array} , \tag{19}$$

where $\tau_2 := (\mathrm{id}_B \otimes \tau \otimes \mathrm{id}_B)$. Equivalently, Δ_B and ε_B are morphisms of $\mathbb{K}$-algebras with the natural algebra structure on the space $B \otimes B$. By a slight abuse of notation one writes $\Delta_B(m_B(b \otimes b')) = \Delta_B(b)\Delta_B(b')$ for $b, b' \in B$, saying that the *coproduct of the product is the product of the coproduct*. The identity element in B will be denoted by $\mathbf{1}_B$, and all algebra morphisms are required to be unital. Note that if x_1, x_2 are primitive in B, then $[x_1, x_2] := m_B(x_1 \otimes x_2) - m_B(x_2 \otimes x_1)$ is primitive as well, i.e., the set $P(B)$ of primitive elements of a bialgebra B is a Lie subalgebra of the Lie algebra $\mathcal{L}(B)$.

A bialgebra B is called *graded* if there are $\mathbb{K}$-vector spaces B_n, $n \geq 0$, such that

1. $B = \bigoplus_{n \geq 0} B_n$,
2. $m_B(B_n \otimes B_m) \subseteq B_{n+m}$,
3. $\Delta_B(B_n) \subseteq \bigoplus_{p+q=n} B_p \otimes B_q$.

Elements $x \in B_n$ are given a degree $\deg(x) = n$. For a *connected* graded bialgebra B, the degree zero part is $B_0 = \mathbb{K}\mathbf{1}_B$. Note that $\mathbf{1}_B$ is group-like. A graded bialgebra $B = \bigoplus_{n \geq 0} B_n$ is said to be of *finite type* if its homogeneous components B_n are $\mathbb{K}$-vector spaces of finite dimension.

Let B be a connected graded $\mathbb{K}$-bialgebra. One can show [50] that the coproduct of any element $x \in B_n$ is given by

$$\Delta_B(x) = x \otimes \mathbf{1}_B + \mathbf{1}_B \otimes x + \sum_{(x)}{}' x' \otimes x'',$$

where

$$\Delta'(x) := \sum_{(x)}{}' x' \otimes x'' \in \bigoplus_{\substack{p+q=n \\ p>0,\ q>0}} B_p \otimes B_q$$

is the *reduced coproduct*, which is coassociative on the augmentation ideal $\ker(\varepsilon_B) := B^+ := \bigoplus_{n>0} B_n$. Elements in the kernel of Δ'_B are primitive elements of B.

Example 4 Divided powers $(D, m_D, \eta_D, \Delta_D, \varepsilon_D)$ are a graded bialgebra, where $D = \bigoplus_{n=0}^{\infty} D_n$, $D_n := \mathbb{K}d_n$. The product is given by $m_D(d_m \otimes d_n) = \binom{m+n}{m} d_{m+n}$, and the unital map is $\eta_D : \mathbb{K} \to D$, $1_{\mathbb{K}} \mapsto d_0 := \mathbf{1}_D$. The coproduct $\Delta_D : D \to D \otimes D$ maps $d_n \mapsto \sum_{k=0}^{n} d_k \otimes d_{n-k}$, and $\varepsilon_D : D \to \mathbb{K}$, $d_n \mapsto \delta_{0,n} 1_{\mathbb{K}}$, where $\delta_{0,n}$ is the usual Kronecker delta.

For a $\mathbb{K}$-algebra A and a $\mathbb{K}$-coalgebra C, the *convolution product* of two linear maps $f, g \in L(C, A) := \mathrm{Hom}_{\mathbb{K}}(C, A)$ is defined to be the linear map $f \star g \in L(C, A)$ given for $a \in C$ by

$$(f \star g)(a) := m_A \circ (f \otimes g) \circ \Delta_C(a) = \sum_{(a)} f(a^{(1)})\, g(a^{(2)}). \tag{20}$$

In other words

$$C \xrightarrow{\Delta_C} C \otimes C \xrightarrow{f \otimes g} A \otimes A \xrightarrow{m_A} A.$$

It is easy to see that associativity of A and coassociativity of C imply the following.

Theorem 3 [1, 21, 50, 55, 64] *$L(C, A)$ with the convolution product (20) is a unital associative $\mathbb{K}$-algebra with unit $\eta := \eta_A \circ \varepsilon_C$.*

The algebra A can be replaced by the base field $\mathbb{K}$. For a bialgebra B the theorem describes the convolution algebra structure on $L(B, B)$ with unit $\eta := \eta_B \circ \varepsilon_B$.

For the maps $f_i \in L(C, A)$, $i = 1, \ldots, n$, $n > 1$, multiple convolution products are defined by

$$f_1 \star f_2 \star \cdots \star f_n := m_A \circ (f_1 \otimes f_2 \otimes \cdots \otimes f_n) \circ \Delta_C^{(n-1)}. \tag{21}$$

Recall that $\Delta_C^{(0)} := \mathrm{id}_C$, and for $n > 0$, $\Delta_C^{(n)} := (\Delta_C^{(n-1)} \otimes \mathrm{id}_C) \circ \Delta_C$.

2.2.4 Hopf Algebra

Definition 5 [1, 55, 64] A ***Hopf algebra*** is a $\mathbb{K}$-bialgebra $(H, m_H, \eta_H, \Delta_H, \varepsilon_H, S)$ together with a particular $\mathbb{K}$-linear map $S : H \to H$ called the ***antipode***, which satisfies the Hopf algebra axioms [1, 55, 64]. The algebra unit in H is denoted by $\mathbf{1}_H$.

The antipode map has the property of being an antihomomorphism for both the algebra and the coalgebra structures, i.e., $S(m_H(x \otimes y)) = m_H(S(y) \otimes S(x))$ and $\Delta_H \circ S = (S \otimes S) \circ \tau \circ \Delta_H$. The necessarily unique antipode $S \in L(H, H)$ is the inverse of the identity map $\mathrm{id}_H : H \to H$ with respect to the convolution product

$$S \star \mathrm{id}_H = m_H \circ (S \otimes \mathrm{id}_H) \circ \Delta_H = \eta_H \circ \varepsilon_H = m_H \circ (\mathrm{id}_H \otimes S) \circ \Delta_H = \mathrm{id}_H \star S. \tag{22}$$

If the Hopf algebra H is commutative or cocommutative, then $S \circ S = \mathrm{id}_H$.

Recall that the universal enveloping algebra $\mathcal{U}(\mathfrak{L})$ of a Lie algebra $\mathfrak{L}$ has the structure of a Hopf algebra [57] and provides a natural example. An important observation is contained in the next result.

Proposition 1 [21] *Any connected graded bialgebra $H = \bigoplus_{n\geq 0} H_n$ is a connected graded Hopf algebra. The antipode is defined by the geometric series $S := \mathrm{id}_H^{\star(-1)} = \big(\eta_H \circ \varepsilon_H - (\eta_H \circ \varepsilon_H - \mathrm{id}_H)\big)^{\star(-1)}$.*

See [21] for a proof and more details. Hence, for any $x \in H_n$ the antipode is computed by

$$S(x) = \sum_{k\geq 0} \big(\eta_H \circ \varepsilon_H - \mathrm{id}_H\big)^{\star k}(x). \tag{23}$$

It is well-defined as the sum on the right-hand side terminates at $\deg(x) = n$ due to the fact that the projector $P := \mathrm{id}_H - \eta_H \circ \varepsilon_H$ maps H to its augmentation ideal $\ker(\varepsilon_H)$. Note that the antipode preserves the grading, i.e., $S(H_n) \subseteq H_n$.

Corollary 1 [21] *The antipode S for a connected graded Hopf algebra $H = \bigoplus_{n\geq 0} H_n$ may be defined recursively in terms of either of the two formulae*

$$S(x) = -S \star P(x) = -x - \sum_{(x)}' S(x')x'', \tag{24a}$$

$$S(x) = -P \star S(x) = -x - \sum_{(x)}' x' S(x'') \tag{24b}$$

for $x \in \ker(\varepsilon_H) = \bigoplus_{n>0} H_n$, which follow readily from (22) and $S(\mathbf{1}_H) = \mathbf{1}_H$.

These recursions make sense due to the fact that on the right-hand side the antipode is calculated on elements x' or x'', which are of strictly smaller degree than the element x. Let A be a $\mathbb{K}$-algebra and H a Hopf algebra. An element $\phi \in L(H, A)$ is called a *character* if ϕ is a unital algebra morphism, that is, $\phi(\mathbf{1}_H) = \mathbf{1}_A$ and

$$\phi\big(m_H(x \otimes y)\big) = m_A\big(\phi(x) \otimes \phi(y)\big). \tag{25}$$

An *infinitesimal character* with values in A is a linear map $\xi \in L(H, A)$ such that for $x, y \in \ker(\varepsilon_H)$, $\xi(m_H(x \otimes y)) = 0$, which implies $\xi(\mathbf{1}_H) = 0$. An equivalent way to characterize infinitesimal characters is as *derivations*, i.e.,

$$\xi\big(m_H(x \otimes y)\big) = m_A\big(\eta_A \circ \varepsilon_H(x) \otimes \xi(y)\big) + m_A\big(\xi(x) \otimes \eta_A \circ \varepsilon_H(y)\big) \tag{26}$$

for any $x, y \in H$. The set of characters (respectively infinitesimal characters) is denoted by $G_A \subset L(H, A)$ (respectively $g_A \subset L(H, A)$).

Let A be a commutative $\mathbb{K}$-algebra. The linear space of infinitesimal characters, g_A, forms a Lie algebra with respect to the Lie bracket defined on $L(H, A)$ in terms of the convolution product

$$[\alpha, \beta] := \alpha \star \beta - \beta \star \alpha.$$

Moreover, A-valued characters, G_A, form a group. The inverse is given by composition with the antipode S of H. Since $\alpha(\mathbf{1}_H) = 0$ for $\alpha \in g_A$, the exponential defined by its power series with respect to convolution, $\exp^\star(\alpha)(x) := \sum_{j \geq 0} \frac{1}{j!}\alpha^{\star j}(x)$, is a finite sum terminating at $j = n$ for any $x \in H_n$.

Proposition 2 *[21, 50]* $\exp^\star$ *restricts to a bijection from* g_A *onto* G_A.

The compositional inverse of $\exp^\star$ is given by the logarithm defined with respect to the convolution product, $\log^\star(\eta_A \circ \varepsilon_H + \gamma)(x) = \sum_{k \geq 1} \frac{(-1)^{k-1}}{k} \gamma^{\star k}(x)$, where $\gamma \in g_A$. Again the sum terminates at $k = n$ for any $x \in H_n$ as $\gamma(\mathbf{1}_H) = 0$. For details and proofs of these well-known facts the reader is referred to [21, 50].

Remark 1 An important result which is due to Milnor and Moore [52] concerns the structure of cocommutative connected graded Hopf algebras of finite type. It states that any such Hopf algebra H is isomorphic to the universal enveloping algebra of its primitive elements, i.e., $H \cong \mathcal{U}(P(H))$. See also [22].

Remark 2 An observation concerning the relationship between the group $G_A \subset L(H, A)$ and the Hopf algebra H will be important in the analysis which follows. The reader is referred to the paper of Frabetti and Manchon [27] for details and additional references. By definition, elements in G_A map all of H into the commutative unital algebra A. However, an element $x \in H$ can also be seen as an A-valued function on G_A. Indeed, let $\Phi \in G_A$, then $x(\Phi) := \Phi(x) \in A$, and the usual pointwise product of functions $(xy)(\Phi) = x(\Phi)y(\Phi)$ follows from (25) since $\Phi \in G_A$. The definition of the convolution product (20) in terms of the coproduct of H implies a natural coproduct on functions $x \in H$, that is, $\Delta(x)(\Phi, \Psi) := (\Phi \star \Psi)(x) \in A$. Similarly, the inverse of G_A as well as its unit correspond naturally to the antipode and counit map on H, respectively. This *reversed* perspective on the relationship between H and its group of characters G_A allows one to interpret H as the (Hopf) algebra of *coordinate functions* of the group G_A. More precisely, H contains the *representative functions* over G_A. The reader is directed to Cartier's work [7] for a comprehensive review of this topic. In the context of input-output systems in nonlinear control theory, a particular group of unital Fliess operators is central. Its product, unit and inverse are used to identify its Hopf algebra of coordinate functions.

Example 5 Three examples of Hopf algebra are presented: the unshuffle, shuffle [3] and quasi-shuffle Hopf algebras [41].

1. Let $X := \{x_1, x_2, x_3, \ldots, x_m\}$ be an alphabet with m letters. As before, X^* is the set of words with letters in X. The length of a word $\eta = x_{i_1} \cdots x_{i_n}$ in X^* is defined by the number of letters it contains, i.e., $|\eta| := n$. The empty word $\mathbf{1} \in X^*$ has length zero. The vector space $\mathbb{K}\langle X \rangle$, which is freely generated by X^* and graded by length, becomes a noncommutative, unital, connected, graded algebra by concatenating words, i.e., for $\eta = x_{i_1} \cdots x_{i_n}$ and $\eta' = x_{j_1} \cdots x_{j_l}$, $\eta \cdot \eta' := x_{i_1} \cdots x_{i_n} x_{j_1} \cdots x_{j_l}$, and $|\eta \cdot \eta'| = n + l$. The *unshuffle coproduct* is defined

by declaring the elements in X to be primitive, i.e., $\Delta^{\sqcup\!\sqcup}(x_i) := x_i \otimes \mathbf{1} + \mathbf{1} \otimes x_i$ for all $x_i \in X$, and by extending it multiplicatively. In this case, $\mathbb{K}\langle X\rangle$ is turned into the cocommutative *unshuffle Hopf algebra* H_{conc}. For instance, for the letters $x_{i_1}, x_{i_2} \in X$ the coproduct of the length two word $\eta = x_{i_1}x_{i_2}$ is

$$\Delta^{\sqcup\!\sqcup}(\eta) = \Delta^{\sqcup\!\sqcup}(x_{i_1})\Delta^{\sqcup\!\sqcup}(x_{i_2}) = x_{i_1}x_{i_2} \otimes \mathbf{1} + \mathbf{1} \otimes x_{i_1}x_{i_2} + x_{i_1} \otimes x_{i_2} + x_{i_2} \otimes x_{i_1}.$$

The general form of $\Delta^{\sqcup\!\sqcup}$ for an arbitrary word $\eta = x_{i_1} \cdots x_{i_l} \in X^*$ is given by

$$\Delta^{\sqcup\!\sqcup}(\eta) = \prod_{j=1}^{l} \Delta^{\sqcup\!\sqcup}(x_{i_j}) = \sum_{\alpha,\beta \in X^*} \langle \alpha \sqcup\!\sqcup \beta, \eta\rangle \alpha \otimes \beta. \tag{27}$$

The coefficient in the sum over words $\alpha, \beta \in X^*$ on the right-hand side is defined through the linearly extended bracket $\langle \alpha, \nu\rangle := 1$ if $\alpha = \nu$, and zero otherwise. The product $\sqcup\!\sqcup$ displayed in (27) is the shuffle product of words (1) introduced above. The antipode of H_{conc} turns out to be

$$S(x_{i_1} \cdots x_{i_l}) = (-1)^l x_{i_l} \cdots x_{i_1}. \tag{28}$$

It is interesting to check that (28) satisfies the recursions (24a, 24b).

2. The same space $\mathbb{K}\langle X\rangle$ can be turned into a unital, connected, graded, commutative, noncocommutative Hopf algebra, $H_{\sqcup\!\sqcup}$, known as *shuffle Hopf algebra*, by defining its algebra structure in terms of the shuffle product on words (1), and its coproduct by *deconcatenation*. The latter is defined on words $\eta = x_{i_1} \cdots x_{i_l} \in X^*$ as follows

$$\Delta(\eta) = \eta \otimes \mathbf{1} + \mathbf{1} \otimes \eta + \sum_{k=1}^{l-1} x_{i_1} \cdots x_{i_k} \otimes x_{i_{k+1}} \cdots x_{i_l}. \tag{29}$$

It is easy to show—and left to the reader—that this gives a coassociative coproduct which is compatible with the shuffle product. The antipode of $H_{\sqcup\!\sqcup}$ is the same as that of H_{conc}, i.e., $S(x_{i_1} \cdots x_{i_l}) := (-1)^l x_{i_l} \cdots x_{i_1}$. Again, it is interesting to verify that it satisfies both recursions (24a, 24b).

3. The last example of a connected graded Hopf algebra is defined on the countable alphabet A. Here A^* denotes the monoid of words $w = a_{i_1} \cdots a_{i_l}$ generated by the letters from A with concatenation as product. The degree of an element $a_i \in A$ is defined to be $\deg(a_i)$. It is extended to words additively, i.e., $\deg(a_{i_1} \cdots a_{i_l}) = \deg(a_{i_1}) + \cdots + \deg(a_{i_l})$. The empty word $\mathbf{1} \in A^*$ is of degree zero. Moreover, it is assumed that A itself is a graded commutative semigroup with bilinear product $[-\ -]\colon A \times A \to A$. The degree $\deg([a_i\ a_j]) := \deg(a_i) + \deg(a_j)$. Commutativity and associativity of $[-\ -]$ allow for a notational simplification, i.e., $[a_{i_1} \cdots a_{i_n}] := [a_{i_1}\ [\cdots [a_{i_{n-1}}\ a_{i_n}]] \cdots]$. The free noncommutative algebra of

words $w = a_{i_1} \cdots a_{i_l}$ over the alphabet A is denoted $\mathbb{K}\langle A\rangle$. The commutative and associative quasi-shuffle product on words $w, v \in \mathbb{K}\langle A\rangle$ is defined by

(i) $\mathbf{1} \star v := v \star \mathbf{1} := v$,
(ii) $a_i v \star a_j w := a_i(v \star a_j w) + a_j(a_i v \star w) + [a_i\ a_j](v \star w)$,

where a_i, a_j are letters in A. For instance,

$$\begin{aligned} a_{i_1} \star a_{i_2} &= a_{i_1}a_{i_2} + a_{i_2}a_{i_1} + [a_{i_1}\ a_{i_2}] \\ a_{i_1} \star a_{i_2}a_{i_3} &= a_{i_1}a_{i_2}a_{i_3} + a_{i_2}a_{i_1}a_{i_3} + a_{i_2}a_{i_3}a_{i_1} + [a_{i_1}\ a_{i_2}]a_{i_3} + a_{i_2}[a_{i_1}a_{i_3}]. \end{aligned}$$

Hoffman [41] showed that the quasi-shuffle algebra $H_\star$ is a Hopf algebra with respect to the deconcatenation coproduct (29). The antipode $S : H_\star \to H_\star$ is deduced from the recursion (24a), e.g.,

$$\begin{aligned} S(a_{i_1} \cdots a_{i_n}) &= -(S \star P)(a_{i_1} \cdots a_{i_n}) \\ &= \big(-\mathrm{id}_{H_\star} - m_\star \circ (S \otimes \mathrm{id}_{H_\star}) \circ \Delta'\big)(a_{i_1} \cdots a_{i_n}) \qquad (30) \\ &= -a_{i_1} \cdots a_{i_n} - \sum_{l=1}^{n-1} S(a_{i_1} \cdots a_{i_l}) \star a_{i_{l+1}} \cdots a_{i_n}, \qquad (31) \end{aligned}$$

where the projector $P := \mathrm{id}_{H_\star} - \eta_{H_\star} \circ \varepsilon_{H_\star}$ maps $H_\star$ to its augmentation ideal $\ker(\varepsilon_{H_\star})$. For example, the antipode for the letter a_i and the word $a_i a_j$ are respectively

$$S(a_i) = -a_i, \quad S(a_i a_j) = -a_i a_j + a_i \star a_j = a_j a_i + [a_i\ a_j].$$

If $[a_i\ a_j] = 0$ for any letters $a_i, a_j \in A$, then the quasi-shuffle product reduces to the ordinary shuffle product (1) on words, and $H_\star$ reduces to $H_{\sqcup\!\sqcup}$.

In Sect. 4 another example of a connected graded Hopf algebra is presented, one which plays a key role in the context of the output feedback interconnection. In Sect. 5 the quasi-shuffle product is employed in the context of products of discrete-time Fliess operators.

3 Algebras Induced by the Interconnection of Fliess Operators

In engineering applications, where input-output systems are represented in terms of Fliess operators, it is natural to interconnect systems to create models of more complex systems. It is known that all the basic interconnection types, such as the parallel, cascade and feedback connections, are well-posed. For example, under suitable assumptions, the output of a given Fliess operator generates an admissible input for another Fliess operator in the cascade connection. In each case it is also known that the aggregate system has a Fliess operator representation. Therefore, a family of formal power series products is naturally induced whereby the generating series of an aggregate system can be computed in terms of the generating series of its subsystems. Of particular interest here are the cascade and feedback interconnections, as their respective products define a semi-group and group that are central in control theory. In this section, these algebraic structures are described in detail.

3.1 Cascade Interconnections

Consider the cascade interconnections of two Fliess operators shown in Figure 1. The first is a simple cascade interconnection corresponding to a direct composition of the operators F_c and F_d, namely, $F_c \circ F_d$. The other involves a *direct feed term* passing from the input u to the input v so that $F_c \circ (I + F_d)$, where I denotes the identity operator. This direct feed term is a common feature found in some control systems and is particularly important in feedback systems as will be discussed shortly. The primary claim is that each cascade interconnection induces a locally finite product on the level of formal power series, which unambiguously describes the interconnected system as generating series of Fliess operators are known to be unique [25, 66]. Specifically, the *composition product* satisfies $F_c \circ F_d = F_{c \circ d}$, and the *modified composition product* satisfies $F_c \circ (I + F_d) = F_{c \tilde{\circ} d}$. Each product is defined in terms of a certain algebra homomorphism. The morphism for the modified composition product ultimately defines a pre-Lie product, which is at the root of all the underlying combinatorial structures at play.

For a fixed $d \in \mathbb{R}^m\langle\langle X\rangle\rangle$ and alphabet $X = \{x_0, x_1, \ldots, x_m\}$, let ψ_d be the continuous (in the ultrametric sense) algebra homomorphism mapping $\mathbb{R}\langle\langle X\rangle\rangle$ to the set of vector space endomorphisms $\mathrm{End}(\mathbb{R}\langle\langle X\rangle\rangle)$ uniquely specified by $\psi_d(x_i\eta) = \psi_d(x_i) \circ \psi_d(\eta)$ for $x_i \in X$, $\eta \in X^*$ with

$$\psi_d(x_i)(e) = x_0(d_i \sqcup\!\sqcup e),$$

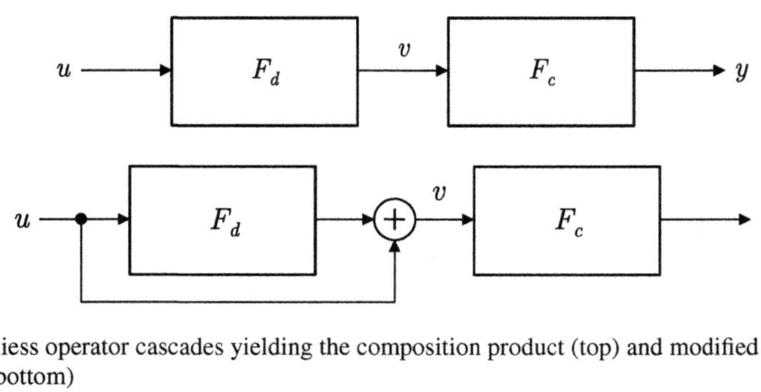

Fig. 1 Fliess operator cascades yielding the composition product (top) and modified composition product (bottom)

$i = 0, 1, \ldots, m$ and any $e \in \mathbb{R}\langle\langle X \rangle\rangle$, and where d_i is the i-th component series of $d \in \mathbb{R}^m\langle\langle X \rangle\rangle$ ($d_0 := \emptyset$). By definition, $\psi_d(\emptyset)$ is the identity map on $\mathbb{R}\langle\langle X \rangle\rangle$. The following theorem describes the generating series for the direct cascade connection of two Fliess operators.

Theorem 4 *[19, 20, 38, 65] Given any $c \in \mathbb{R}^{\ell}_{LC}\langle\langle X \rangle\rangle$ and $d \in \mathbb{R}^m_{LC}\langle\langle X \rangle\rangle$, the composition $F_c \circ F_d = F_{c \circ d}$, where the* ***composition product*** *of c and d is given by*

$$c \circ d = \sum_{\eta \in X^*} (c, \eta)\, \psi_d(\eta)(\mathbf{1}),$$

and $c \circ d \in \mathbb{R}^{\ell}_{LC}\langle\langle X \rangle\rangle$.

It is not difficult to show that this composition product is locally finite (and therefore summable), associative and $\mathbb{R}$-linear in its left argument. As this product lacks an identity, it defines a semi-group on $\mathbb{R}^m\langle\langle X \rangle\rangle$. In addition, this product distributes to the left over the shuffle product, which reflects the fact that in general

$$F_{(c \,\sqcup\!\sqcup\, d)\circ e} = (F_c F_d) \circ F_e = F_c[F_e] F_d[F_e] = F_{(c\circ e) \,\sqcup\!\sqcup\, (d \circ e)}.$$

Finally, it is known that the composition product defines an ultrametric contraction on $\mathbb{R}^m\langle\langle X \rangle\rangle$.

Example 6 In the case of two linear time-invariant systems with analytic kernels $h_c(t) = \sum_{i\geq 0}(c, x_0^i x_1) t^i / i!$ and $h_d(t) = \sum_{i \geq 0}(d, x_0^i x_1) t^i / i!$, respectively, a direct calculation gives the kernel for the composition

$$(h_c * h_d)(t) := \int_0^t h_c(t-\tau) h_d(\tau)\, d\tau = \sum_{k=1}^{\infty} \left[\sum_{j=0}^{k-1} (c, x_0^{k-j-1} x_1)(d, x_0^j x_1) \right] \frac{t^k}{k!}$$
$$= \sum_{k=1}^{\infty} (c \circ d, x_0^k x_1) \frac{t^k}{k!} =: h_{c\circ d}(t).$$

The introduction of a direct feed term in the composition interconnection requires a modification to the set up, namely, the new algebra homomorphism ϕ_d from $\mathbb{R}\langle\langle X \rangle\rangle$ to $\mathrm{End}(\mathbb{R}\langle\langle X \rangle\rangle)$ where $\phi_d(x_i \eta) = \phi_d(x_i) \circ \phi_d(\eta)$ for $x_i \in X$, $\eta \in X^*$ with

$$\phi_d(x_i)(e) = x_i e + x_0(d_i \,\sqcup\!\sqcup\, e),$$

$i = 0, 1, \ldots, m$ and any $e \in \mathbb{R}\langle\langle X \rangle\rangle$, and where $d_0 := 0$. Again, $\phi_d(\emptyset)$ is the identity map on $\mathbb{R}\langle\langle X \rangle\rangle$. The direct feed term is encoded in the new term $x_i e$ shown above. This yields the desired formal power series product as described next.

Theorem 5 *[38, 49] Given any $c \in \mathbb{R}^{\ell}_{LC}\langle\langle X \rangle\rangle$ and $d \in \mathbb{R}^m_{LC}\langle\langle X \rangle\rangle$, the composition $F_c \circ (I + F_d) = F_{c \,\tilde{\circ}\, d}$, where the* ***modified composition product*** *of c and d is given by*

Table 1 Composition products involving $c, d, c_\delta = \delta + c, d_\delta = \delta + d$ when $X = \{x_0, x_1, \ldots, x_m\}$

Name	Symbol	Map	Operator Identity	Remarks
Composition	$c \circ d$	$\mathbb{R}^\ell\langle\langle X\rangle\rangle \times \mathbb{R}^m\langle\langle X\rangle\rangle \to \mathbb{R}^\ell\langle\langle X\rangle\rangle$	$F_c \circ F_d = F_{c\circ d}$	Associative
Modified composition	$c \tilde{\circ} d$	$\mathbb{R}^\ell\langle\langle X\rangle\rangle \times \mathbb{R}^m\langle\langle X\rangle\rangle \to \mathbb{R}^\ell\langle\langle X\rangle\rangle$	$F_c \circ (I + F_d)$	Nonassociative
Mixed composition	$c \circ d_\delta$	$\mathbb{R}^\ell\langle\langle X\rangle\rangle \times \mathbb{R}^m\langle\langle X_\delta\rangle\rangle \to \mathbb{R}^\ell\langle\langle X\rangle\rangle$	$F_c \circ F_{d_\delta} = F_{c\circ d_\delta}$	$c \circ d_\delta = c \tilde{\circ} d$
Group composition	$c \circledcirc d$	$\mathbb{R}^m\langle\langle X\rangle\rangle \times \mathbb{R}^m\langle\langle X\rangle\rangle \to \mathbb{R}^m\langle\langle X\rangle\rangle$	$(I + F_c) \circ (I + F_d) = I + F_{c\circledcirc d}$	$c \circledcirc d = d + c \tilde{\circ} d$
Group product	$c_\delta \circ d_\delta$	$\mathbb{R}^m\langle\langle X_\delta\rangle\rangle \times \mathbb{R}^m\langle\langle X_\delta\rangle\rangle \to \mathbb{R}^m\langle\langle X_\delta\rangle\rangle$	$F_{c_\delta} \circ F_{d_\delta} = F_{c_\delta\circ d_\delta}$	$c_\delta \circ d_\delta = \delta + c \circledcirc d$

$$c \tilde{\circ} d = \sum_{\eta \in X^*} (c, \eta)\, \phi_d(\eta)(\mathbf{1}),$$

and $c \tilde{\circ} d \in \mathbb{R}^\ell_{LC}\langle\langle X\rangle\rangle$.

This product is also always summable, but it is not associative, in fact, in general

$$(c \tilde{\circ} d) \tilde{\circ} e = c \tilde{\circ} (d \tilde{\circ} e + e) \tag{32}$$

[49]. The following lemma describes some other elementary properties of the modified composition product.

Lemma 1 *[32, 38] The modified composition product*

(1) is left $\mathbb{R}$*-linear;*
(2) satisfies $c \tilde{\circ} 0 = c$*;*
(3) satisfies $c \tilde{\circ} d = k \in \mathbb{R}^\ell$ *for any fixed d if and only if* $c = k$*;*
(4) satisfies $(x_0 c) \tilde{\circ} d = x_0(c \tilde{\circ} d)$ *and* $(x_i c) \tilde{\circ} d = x_i(c \tilde{\circ} d) + x_0(d_i \sqcup\!\sqcup (c \tilde{\circ} d))$*;*
(5) distributes to the left over the shuffle product.

It is also known that the modified composition product is an ultrametric contraction on $\mathbb{R}^m\langle\langle X\rangle\rangle$. A summary description of the composition and modified composition product is given in Table 1 along with other types of composition products which will be presented shortly.

3.2 Output Feedback

A central object of study in control theory is the output feedback interconnection as shown in Figure 2. As with the cascade systems discussed above, this class of interconnections is also closed in the sense that the mapping $u \mapsto y$ always has a Fliess

Fig. 2 Output feedback connection

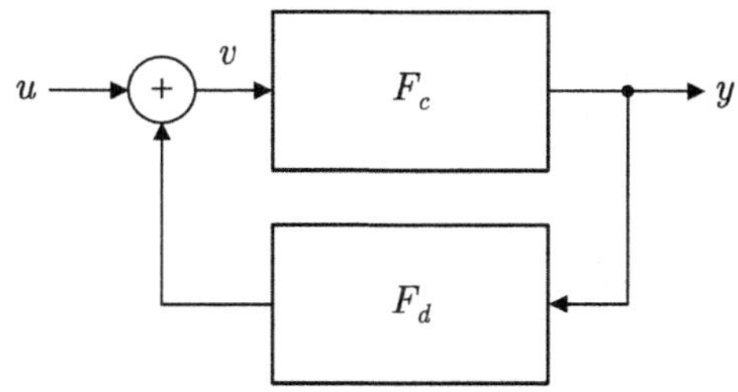

operator representation. The computation of the corresponding generating series, however, is much more involved. To see the source of the difficulty, consider the following calculation assuming $c, d \in \mathbb{R}^m_{LC}\langle\langle X\rangle\rangle$ (for the most general case, which requires two alphabets, see [32]). Clearly the function v in Figure 2 must satisfy the identity

$$v = u + F_{d\circ c}[v].$$

Therefore, from $(I + F_{-d\circ c})[v] = u$ one deduces that

$$v = \left(I + F_{(-d\circ c)}\right)^{-1}[u] =: \left(I + F_{(-d\circ c)^{-1}}\right)[u],$$

where I is the identity operator, and the superscript "-1" denotes the composition inverse in both the operator sense and in terms of formal power series. Thus, the generating series for the closed-loop system, denoted by the output feedback product $c@d$, is

$$F_{c@d}[u] = F_c[v] = F_{c\,\tilde{\circ}\,(-d\circ c)^{-1}}[u]. \tag{33}$$

The crux of the problem is how to compute the generating series $(-d \circ c)^{-1}$ of the inverse operator $(I + F_{(-d\circ c)})^{-1}$. The approach described next is based on identifying an underlying combinatorial Hopf algebra whose antipode acts on a certain character group in such a way as to explicitly produce this inverse generating series. This requires that one first identifies the relevant group structures.

Consider the set of *unital Fliess operators* $\mathscr{F}_\delta := \{I + F_c : c \in \mathbb{R}^m_{LC}\langle\langle X\rangle\rangle\}$. It is convenient to introduce the symbol δ as the (fictitious) generating series for the identity map. That is, $F_\delta := I$ such that $I + F_c := F_{\delta+c} = F_{c_\delta}$ with $c_\delta := \delta + c$. The set of all such generating series for $\mathscr{F}_\delta$ will be denoted by $\mathbb{R}^m_{LC}\langle\langle X_\delta\rangle\rangle$. The central idea is that $(\mathscr{F}_\delta, \circ, I)$ forms a group under operator composition

$$\begin{aligned} F_{c_\delta} \circ F_{d_\delta} &= (I + F_c) \circ (I + F_d) = I + F_d + F_c \circ (I + F_d) \\ &= I + F_d + F_{c\,\tilde{\circ}\,d} =: F_{c_\delta \circ d_\delta}, \end{aligned}$$

where

$$c_\delta \circ d_\delta := \delta + d + c\,\tilde{\circ}\,d =: \delta + c \circledcirc d. \tag{34}$$

(The series $c \circledcirc d$ is clearly locally convergent since all the operations employed in its definition preserve local convergence.) This group will be referred to as the *output feedback group* of unital Fliess operators. It is natural to think of this group as acting on an arbitrary Fliess operator, say F_c, to produce another Fliess operator, as is evident in (33), by defining the right action $F_c \circ F_{d_\delta} = F_{c \circ d_\delta}$ with $c \circ d_\delta := c \,\tilde{\circ}\, d$. This product will be referred to as the *mixed composition product* on $\mathbb{R}^\ell\langle\langle X\rangle\rangle \times \mathbb{R}^m\langle\langle X_\delta\rangle\rangle$.[3] The following lemma gives the basic properties of the composition product on $\mathbb{R}^m\langle\langle X_\delta\rangle\rangle$.

Lemma 2 *[29] The composition product (34) on* $\mathbb{R}^m\langle\langle X_\delta\rangle\rangle$

(1) satisfies $\delta \circ c_\delta = c_\delta \circ \delta = c_\delta$*;*
(2) satisfies $(c \circ d_\delta) \circ e_\delta = c \circ (d_\delta \circ e_\delta)$ *(mixed associativity);*
(3) is associative.

Proof (1) This claim follows from the definition of this composition product.

(2) In light of item (1) in Lemma 1 it is sufficient to prove the claim only for $c = \eta \in X^k$, $k \geq 0$. The cases $k = 0$ and $k = 1$ are trivial. Assume the claim holds up to some fixed $k \geq 0$. Then via item (4) in Lemma 1 and the induction hypothesis it follows that

$$\begin{aligned}((x_0\eta) \circ d_\delta) \circ e_\delta &= (x_0(\eta \circ d_\delta)) \circ e_\delta = x_0((\eta \circ d_\delta) \circ e_\delta) = x_0(\eta \circ (d_\delta \circ e_\delta)) \\ &= (x_0\eta) \circ (d_\delta \circ e_\delta).\end{aligned}$$

In a similar fashion, apply the properties (1), (4), and (5) in Lemma 1 to get

$$\begin{aligned}((x_i\eta) \circ d_\delta) \circ e_\delta &= [x_i(\eta \circ d_\delta) + x_0(d_i \sqcup\!\sqcup (\eta \circ d_\delta))] \circ e_\delta \\ &= [x_i(\eta \circ d_\delta)] \circ e_\delta + [x_0(d_i \sqcup\!\sqcup (\eta \circ d_\delta))] \circ e_\delta \\ &= x_i[(\eta \circ d_\delta) \circ e_\delta] + x_0[e_i \sqcup\!\sqcup ((\eta \circ d_\delta) \circ e_\delta)] + \\ &\quad\; x_0[(d_i \circ e_\delta) \sqcup\!\sqcup ((\eta \circ d_\delta) \circ e_\delta)].\end{aligned}$$

Now employ the induction hypothesis so that

$$\begin{aligned}((x_i\eta) \circ d_\delta) \circ e_\delta &= x_i[\eta \circ (d_\delta \circ e_\delta)] + x_0[e_i \sqcup\!\sqcup (\eta \circ (d_\delta \circ e_\delta))] + \\ &\quad\; x_0[(d_i \circ e_\delta) \sqcup\!\sqcup (\eta \circ (d_\delta \circ e_\delta))] \\ &= x_i[\eta \circ (d_\delta \circ e_\delta)] + x_0[(e_i + (d_i \circ e_\delta)) \sqcup\!\sqcup (\eta \circ (d_\delta \circ e_\delta))] \\ &= x_i[\eta \circ (d_\delta \circ e_\delta)] + x_0[(d_\delta \circ e_\delta)_i \sqcup\!\sqcup (\eta \circ (d_\delta \circ e_\delta))] \\ &= (x_i\eta) \circ (d_\delta \circ e_\delta).\end{aligned}$$

[3] The same symbol will be used for composition on $\mathbb{R}^m\langle\langle X\rangle\rangle$, $\mathbb{R}^m\langle\langle X_\delta\rangle\rangle$, and $\mathbb{R}^\ell\langle\langle X\rangle\rangle \times \mathbb{R}^m\langle\langle X_\delta\rangle\rangle$. It will always be clear which product is being used since the arguments of these products have a distinct notation, namely, c versus c_δ.

Therefore, the claim holds for all $\eta \in X^*$, and the identity is proved.
(3) Applying item (1) in Lemma 1 and the previous result it follows

$$\begin{aligned}(c_\delta \circ d_\delta) \circ e_\delta &= \delta + e + (c \circledcirc d) \circ e_\delta \\ &= \delta + e + (d + c \circ d_\delta) \circ e_\delta \\ &= \delta + e + d \circ e_\delta + (c \circ d_\delta) \circ e_\delta \\ &= \delta + d \circledcirc e + c \circ (d_\delta \circ e_\delta) \\ &= c_\delta \circ (d_\delta \circ e_\delta).\end{aligned}$$

Given the uniqueness of generating series of Fliess operators, their set of generating series forms the group $(\mathbb{R}^m_{LC}\langle\langle X_\delta \rangle\rangle, \circ, \delta)$. In fact, it is a subgroup of the group described next since it can be shown that the composition inverse preserves local convergence [32].

Theorem 6 *[32] The triple* $(\mathbb{R}^m\langle\langle X_\delta \rangle\rangle, \circ, \delta)$ *is a group.*

Proof From Lemma 2, item (1), δ is the identity element of the group. Associativity of the group product was also established in this lemma. For a fixed $c_\delta \in \mathbb{R}^m\langle\langle X_\delta \rangle\rangle$, the composition inverse, $c_\delta^{-1} = \delta + c^{-1}$, must satisfy $c_\delta \circ c_\delta^{-1} = \delta$ and $c_\delta^{-1} \circ c_\delta = \delta$, which reduce, respectively, to

$$c^{-1} = (-c)\,\tilde{\circ}\,c^{-1} \tag{35a}$$

$$c = (-c^{-1})\,\tilde{\circ}\,c. \tag{35b}$$

It was shown in [38] that $e \mapsto (-c)\,\tilde{\circ}\,e$ is always a contraction on $\mathbb{R}^m\langle\langle X \rangle\rangle$ when viewed as an (complete) ultrametric space and thus has a unique fixed point, c^{-1}. So it follows directly that c_δ^{-1} is a right inverse of c_δ, i.e., satisfies (35a). To see that this same series is also a left inverse, first observe that (35a) is equivalent to

$$c^{-1}\,\tilde{\circ}\,0 + c\,\tilde{\circ}\,c^{-1} = 0, \tag{36}$$

using Lemma 1, items (1) and (2). Substituting (36) back into itself where zero appears and applying (32) gives

$$\begin{aligned}c^{-1}\,\tilde{\circ}\,(c\,\tilde{\circ}\,c^{-1} + c^{-1}) + c\,\tilde{\circ}\,c^{-1} &= 0 \\ (c^{-1}\,\tilde{\circ}\,c)\,\tilde{\circ}\,c^{-1} + c\,\tilde{\circ}\,c^{-1} &= 0.\end{aligned}$$

Again from left linearity of the modified composition product it follows that

$$(c^{-1}\,\tilde{\circ}\,c + c)\,\tilde{\circ}\,c^{-1} = 0.$$

Finally, Lemma 1, item (3) implies that $c^{-1}\,\tilde{\circ}\,c + c = 0$, which is equivalent to (35b). This concludes the proof.

In light of the identities $c \circ \delta = c$ and Lemma 2, item (2), it is therefore established that $\mathbb{R}^m\langle\langle X_\delta\rangle\rangle$ acts as a right transformation group on $\mathbb{R}^m\langle\langle X\rangle\rangle$. In which case, the feedback product $c@d = c\,\tilde{\circ}\,(-d \circ c)^{-1} = c \circ (-d \circ c)_\delta^{-1}$ can be viewed as specific example of such a right action.

4 The Hopf Algebra of Coordinate Functions

In this section the Hopf algebra of coordinate functions for the group $\mathbb{R}^m\langle\langle X_\delta\rangle\rangle$ is described. This provides an explicit computational framework for computing group inverses, and thus, to calculate the feedback product as described in the previous section. The strategy is to first introduce a connected graded commutative algebra and then a compatible noncocommutative coalgebra, resulting in a connected graded commutative noncocommutative bialgebra. The connectedness property ensures then that this bialgebra is a connected graded Hopf algebra. The section ends by providing a purely inductive formula for the antipode of this Hopf algebra.

4.1 Multivariable Hopf Algebra of Output Feedback

Let $X = \{x_0, x_1, x_2, \ldots, x_m\}$ be a finite alphabet with $m + 1$ letters. As usual the monoid of words is denoted by X^* and includes the empty word $\mathrm{e} = \emptyset$. The degree of a word $\eta = x_{i_1} \cdots x_{i_n} \in X^*$ of length $|\eta| := n$, where $x_{i_l} \in X$, is defined by

$$\|\eta\| := 2|\eta|_0 + |\eta|_1. \tag{37}$$

Here $|\eta|_0$ denotes the number of times the letter $x_0 \in X$ appears in η, and $|\eta|_1$ is the number letters $x_{j\neq 0} \in X$ appearing in the word η. Note that $|\mathrm{e}| = 0 = \|\mathrm{e}\|$.

Recall Remark 2 above. For any word $\eta \in X^*$ and $i = 1, \ldots, m$, the *coordinate function* a^i_η is defined to be the element of the dual space $\mathbb{R}^*\langle\langle X_\delta\rangle\rangle$ giving the coefficient of the i-th component series for the word $\eta \in X^*$, namely,

$$a^i_\eta(c) := (c_i, \eta).$$

In this context, a^i_δ denotes the coordinate function with respect to δ, where $a^i_\delta(\delta) = 1$ and zero otherwise. Consider the vector space V generated by the coordinate functions a^i_η, where $\eta \in X^*$ and $1 \leq i \leq m$. It is turned into a polynomial algebra H with unit denoted by $\mathbf{1}$. By defining the degree of elements in H as $\deg(\mathbf{1}) := 0$ and for $k > 0$, $\eta \in X^*$

$$\deg(a^k_\eta) := 1 + \|\eta\|, \tag{38}$$

$\deg(a^k_\eta a^l_\kappa) := \deg(a^k_\eta) + \deg(a^l_\kappa)$, H becomes a graded connected algebra, $H := \bigoplus_{n\geq 0} H_n$. Note, in particular, that $\deg(a^k_{\mathrm{e}}) := 1$.

The left- and right-shift maps, $\theta_j : H \to H$ respectively $\tilde{\theta}_j : H \to H$ are defined by

$$\theta_j a^k_\eta := a^k_{x_j\eta}, \qquad \tilde{\theta}_j a^k_\eta := a^k_{\eta x_j}$$

for $x_j \in X$, and $\theta_j \mathbf{1} = \tilde{\theta}_j \mathbf{1} = 0$. On products these maps act by definition as derivations

$$\theta_j a^k_\eta a^l_\mu := (\theta_j a^k_\eta) a^l_\mu + a^k_\eta (\theta_j a^l_\mu),$$

and analogously for $\tilde{\theta}_j$. For a word $\eta = x_{i_1} \cdots x_{i_n} \in X^*$

$$\theta_\eta := \theta_{i_1} \circ \cdots \circ \theta_{i_n}, \qquad \tilde{\theta}_\eta := \tilde{\theta}_{i_n} \circ \cdots \circ \tilde{\theta}_{i_1}.$$

Hence, any element a^i_η, $\eta \in X^*$ can be written

$$a^i_\eta = \theta_\eta a^i_{\mathrm{e}} = \tilde{\theta}_\eta a^i_{\mathrm{e}}.$$

Both maps can be used to define a particular coproduct $\Delta : H \to H \otimes H$. In this approach, the right-shift map is considered first. Later it will be shown that the left-shift map gives rise to the same coproduct. The coordinate function with respect to the empty word, a^l_{e}, $1 \leq l \leq m$, is defined to be primitive

$$\Delta a^l_{\mathrm{e}} := a^l_{\mathrm{e}} \otimes \mathbf{1} + \mathbf{1} \otimes a^l_{\mathrm{e}}. \tag{39}$$

The next step is to define Δ inductively on any a^i_η, $|\eta| > 0$, by specifying intertwining relations between the map $\tilde{\theta}_\eta$ and the coproduct

$$\Delta \circ \tilde{\theta}_i := \Big(\tilde{\theta}_i \otimes \mathrm{id} + \mathrm{id} \otimes \tilde{\theta}_i + \delta_{0i} \sum_{j=1}^{m} \tilde{\theta}_j \otimes A^{(j)}_{\mathrm{e}}\Big) \circ \Delta, \tag{40}$$

where δ_{0i} is the usual Kronecker delta. The map $A^{(j)}_{\mathrm{e}}$ for $0 < j \leq m$ is defined by

$$A^{(j)}_{\mathrm{e}} a^i_\eta := a^i_\eta a^j_{\mathrm{e}}. \tag{41}$$

The following notation is used, $\Delta \circ \tilde{\theta}_i = \tilde{\Theta}_i \circ \Delta$, where

$$\tilde{\Theta}_i := \tilde{\theta}_i \otimes \mathrm{id} + \mathrm{id} \otimes \tilde{\theta}_i + \delta_{0i} \sum_{j=1}^{m} \tilde{\theta}_j \otimes A_{\mathrm{e}}^{(j)}, \tag{42}$$

and $\tilde{\Theta}_\eta := \tilde{\Theta}_{i_n} \circ \cdots \circ \tilde{\Theta}_{i_1}$ for $\eta = x_{i_1} \cdots x_{i_n} \in X^*$. The functions $a^l_{x_j}$ for $0 < l, j \leq m$ are primitive since

$$\begin{aligned}\Delta a^l_{x_j} = (\tilde{\theta}_j \otimes \mathrm{id} + \mathrm{id} \otimes \tilde{\theta}_j) \circ \Delta a^l_{\mathrm{e}} &= (\tilde{\theta}_j \otimes \mathrm{id} + \mathrm{id} \otimes \tilde{\theta}_j)(a^l_{\mathrm{e}} \otimes \mathbf{1} + \mathbf{1} \otimes a^l_{\mathrm{e}}) \\ &= a^l_{x_j} \otimes \mathbf{1} + \mathbf{1} \otimes a^l_{x_j},\end{aligned}$$

which follows from $\tilde{\theta}_j \mathbf{1} = 0$. However, for $a^l_{x_0}$ the coproduct is

$$\begin{aligned}\Delta a^l_{x_0} = \tilde{\Theta}_0 \circ \Delta a^l_{\mathrm{e}} &= \Big(\tilde{\theta}_0 \otimes \mathrm{id} + \mathrm{id} \otimes \tilde{\theta}_0 + \sum_{j=1}^{m} \tilde{\theta}_j \otimes A_{\mathrm{e}}^{(j)}\Big) \circ \Delta a^l_{\mathrm{e}} \\ &= a^l_{x_0} \otimes \mathbf{1} + \mathbf{1} \otimes a^l_{x_0} + \sum_{j=1}^{m} a^l_{x_j} \otimes a^j_{\mathrm{e}}.\end{aligned} \tag{43}$$

Observe that the coproduct is compatible with the grading. Indeed, $\deg(a^l_{x_0}) = 1 + 2$ and $\deg(a^l_{x_j} \otimes a^j_{\mathrm{e}}) = \deg(a^l_{x_j}) + \deg(a^j_{\mathrm{e}}) = 1 + 1 + 1$ for any $j > 0$. For the element $a^l_{x_i x_j}$, $i, j > 0$, one finds the following coproduct

$$\begin{aligned}\Delta a^l_{x_i x_j} = \tilde{\Theta}_j \circ \tilde{\Theta}_i \circ \Delta a^l_{\mathrm{e}} &= \big(\tilde{\theta}_j \otimes \mathrm{id} + \mathrm{id} \otimes \tilde{\theta}_j\big) \circ \big(\tilde{\theta}_i \otimes \mathrm{id} + \mathrm{id} \otimes \tilde{\theta}_i\big) \circ \Delta a^l_{\mathrm{e}} \\ &= \big(\tilde{\theta}_j \tilde{\theta}_i \otimes \mathrm{id} + \mathrm{id} \otimes \tilde{\theta}_j \tilde{\theta}_i + \tilde{\theta}_j \otimes \tilde{\theta}_i + \tilde{\theta}_i \otimes \tilde{\theta}_j\big) \circ \Delta a^l_{\mathrm{e}} \\ &= a^l_{x_i x_j} \otimes \mathbf{1} + \mathbf{1} \otimes a^l_{x_i x_j}.\end{aligned} \tag{44}$$

Again, this follows from $\tilde{\theta}_k \mathbf{1} = 0$ and generalizes to any word η in the monoid $\tilde{X}^*$ made from the reduced alphabet $\tilde{X} := X - \{x_0\} = \{x_1, \ldots, x_m\}$. For the coproduct of $a^l_{x_i x_0}$, $i > 0$, it follows that

$$\begin{aligned}\Delta a^l_{x_i x_0} &= \tilde{\Theta}_0 \circ \tilde{\Theta}_i \circ \Delta a^l_{\mathrm{e}} \\ &= \Big(\tilde{\theta}_0 \otimes \mathrm{id} + \mathrm{id} \otimes \tilde{\theta}_0 + \sum_{j=1}^{m} \tilde{\theta}_j \otimes A_{\mathrm{e}}^{(j)}\Big) \circ \big(\tilde{\theta}_i \otimes \mathrm{id} + \mathrm{id} \otimes \tilde{\theta}_i\big) \circ \Delta a^l_{\mathrm{e}} \\ &= \Big(\tilde{\theta}_0 \otimes \mathrm{id} + \mathrm{id} \otimes \tilde{\theta}_0 + \sum_{j=1}^{m} \tilde{\theta}_j \otimes A_{\mathrm{e}}^{(j)}\Big)(a^l_{x_i} \otimes \mathbf{1} + \mathbf{1} \otimes a^l_{x_i}) \\ &= a^l_{x_i x_0} \otimes \mathbf{1} + \mathbf{1} \otimes a^l_{x_i x_0} + \sum_{j=1}^{m} a^l_{x_i x_j} \otimes a^j_{\mathrm{e}}.\end{aligned} \tag{45}$$

This should be compared with the coproduct of $a^l_{x_0 x_i}$, $i > 0$

$$
\begin{aligned}
\Delta a^l_{x_0 x_i} &= \tilde{\Theta}_i \circ \tilde{\Theta}_0 \circ \Delta a^l_{\mathrm{e}} \\
&= \big(\tilde{\theta}_i \otimes \mathrm{id} + \mathrm{id} \otimes \tilde{\theta}_i\big) \circ \Big(\tilde{\theta}_0 \otimes \mathrm{id} + \mathrm{id} \otimes \tilde{\theta}_0 + \sum_{j=1}^{m} \tilde{\theta}_j \otimes A^{(j)}_{\mathrm{e}}\Big) \circ \Delta a^l_{\mathrm{e}} \\
&= \big(\tilde{\theta}_i \otimes \mathrm{id} + \mathrm{id} \otimes \tilde{\theta}_i\big)\Big(a^l_{x_0} \otimes \mathbf{1} + \mathbf{1} \otimes a^l_{x_0} + \sum_{j=1}^{m} a^l_{x_j} \otimes a^j_{\mathrm{e}}\Big) \\
&= a^l_{x_0 x_i} \otimes \mathbf{1} + \mathbf{1} \otimes a^l_{x_0 x_i} + \sum_{j=1}^{m} a^l_{x_j x_i} \otimes a^j_{\mathrm{e}} + \sum_{j=1}^{m} a^l_{x_j} \otimes a^j_{x_i}.
\end{aligned} \tag{46}
$$

Finally, the coproduct of $a^l_{x_0 x_0}$ is calculated

$$
\begin{aligned}
\Delta a^l_{x_0 x_0} &= \tilde{\Theta}_0 \circ \tilde{\Theta}_0 \circ \Delta a^l_{\mathrm{e}} \\
&= \Big(\tilde{\theta}_0 \otimes \mathrm{id} + \mathrm{id} \otimes \tilde{\theta}_0 + \sum_{n=1}^{m} \tilde{\theta}_n \otimes A^{(n)}_{\mathrm{e}}\Big) \circ \Big(\tilde{\theta}_0 \otimes \mathrm{id} + \mathrm{id} \otimes \tilde{\theta}_0 + \sum_{j=1}^{m} \tilde{\theta}_j \otimes A^{(j)}_{\mathrm{e}}\Big) \circ \Delta a^l_{\mathrm{e}} \\
&= \Big(\tilde{\theta}_0 \otimes \mathrm{id} + \mathrm{id} \otimes \tilde{\theta}_0 + \sum_{n=1}^{m} \tilde{\theta}_n \otimes A^{(n)}_{\mathrm{e}}\Big)\Big(a^l_{x_0} \otimes \mathbf{1} + \mathbf{1} \otimes a^l_{x_0} + \sum_{j=1}^{m} a^l_{x_j} \otimes a^j_{\mathrm{e}}\Big) \\
&= a^l_{x_0 x_0} \otimes \mathbf{1} + \mathbf{1} \otimes a^l_{x_0 x_0} + \sum_{j=1}^{m} a^l_{x_j x_0} \otimes a^j_{\mathrm{e}} + \sum_{n=1}^{m} a^l_{x_0 x_n} \otimes a^n_{\mathrm{e}} \\
&\quad + \sum_{j=1}^{m} a^l_{x_j} \otimes a^j_{x_0} + \sum_{n,j=1}^{m} a^l_{x_j x_n} \otimes a^j_{\mathrm{e}} a^n_{\mathrm{e}}.
\end{aligned} \tag{47}
$$

The coproduct Δ is extended multiplicatively to all of H, and the coproduct of the unit $\mathbf{1}$ is defined to be $\Delta(\mathbf{1}) := \mathbf{1} \otimes \mathbf{1}$. In particular, $\Delta a^l_\eta \in V \otimes H$ as opposed to simply $H \otimes H$ due to the left linearity of the modified composition product as shown in Lemma 1. This has interesting practical consequences for the antipode calculation as described in [14]. Namely, there is a significant difference between the computational efficiencies of the two antipode recursions in (24a, 24b).

Theorem 7 *The algebra H with the multiplicatively extended coproduct*

$$
\Delta a^l_\eta := \tilde{\Theta}_\eta (a^l_{\mathrm{e}} \otimes \mathbf{1} + \mathbf{1} \otimes a^l_{\mathrm{e}}) \in V \otimes H \tag{48}
$$

is a connected graded commutative noncocommutative Hopf algebra.

Proof Since H is connected, graded and commutative by construction, the antipode can be calculated recursively via one of the recursions in (24a, 24b). It is clear as well that the coproduct (48) is noncocommutative. Hence, coassociativity remains to be checked. This is done inductively with respect to the length of the word $\eta \in X^*$.

First observe that

$$\Delta(a^k_{\eta x_i}) = \Delta \circ \tilde{\theta}_i(a^k_\eta) = \tilde{\Theta}_i \circ \Delta(a^k_\eta).$$

This implies that

$$\begin{aligned}
&(\Delta \otimes \mathrm{id}) \circ \Delta(a^k_{\eta x_i}) = (\Delta \otimes \mathrm{id}) \circ \tilde{\Theta}_i \circ \Delta(a^k_\eta) \\
&= \Big(\Delta \circ \tilde{\theta}_i \otimes \mathrm{id} + (\mathrm{id} \otimes \mathrm{id}) \circ \Delta \otimes \tilde{\theta}_i + \delta_{0i} \sum_{j=1}^m \Delta \circ \tilde{\theta}_j \otimes A_{\mathrm{e}}^{(j)}\Big) \circ \Delta(a^k_\eta) \qquad (49) \\
&= \Big(\tilde{\Theta}_i \otimes \mathrm{id} + \mathrm{id} \otimes \mathrm{id} \otimes \tilde{\theta}_i + \delta_{0i} \sum_{j=1}^m \tilde{\Theta}_j \otimes A_{\mathrm{e}}^{(j)}\Big)(\Delta \otimes \mathrm{id}) \circ \Delta(a^k_\eta) \\
&= \Big(\tilde{\theta}_i \otimes \mathrm{id} \otimes \mathrm{id} + \mathrm{id} \otimes \tilde{\theta}_i \otimes \mathrm{id} + \mathrm{id} \otimes \mathrm{id} \otimes \tilde{\theta}_i \\
&\quad + \delta_{0i} \sum_{j=1}^m \tilde{\theta}_j \otimes A_{\mathrm{e}}^{(j)} \otimes \mathrm{id} + \delta_{0i} \sum_{j=1}^m \tilde{\theta}_j \otimes \mathrm{id} \otimes A_{\mathrm{e}}^{(j)} \\
&\quad + \delta_{0i} \sum_{j=1}^m \mathrm{id} \otimes \tilde{\theta}_j \otimes A_{\mathrm{e}}^{(j)}\Big)(\mathrm{id} \otimes \Delta) \circ \Delta(a^k_\eta) \\
&= \Big(\tilde{\theta}_i \otimes \mathrm{id} \otimes \mathrm{id} + \mathrm{id} \otimes \tilde{\Theta}_i + \delta_{0i} \sum_{j=1}^m \tilde{\theta}_j \otimes \big(A_{\mathrm{e}}^{(j)} \otimes \mathrm{id} + \mathrm{id} \otimes A_{\mathrm{e}}^{(j)}\big)\Big)(\mathrm{id} \otimes \Delta) \circ \Delta(a^k_\eta) \\
&= (\mathrm{id} \otimes \Delta) \circ \Big(\tilde{\theta}_i \otimes \mathrm{id} + \mathrm{id} \otimes \tilde{\theta}_i + \delta_{0i} \sum_{j=1}^m \tilde{\theta}_j \otimes A_{\mathrm{e}}^{(j)}\Big) \circ \Delta(a^k_\eta) \\
&= (\mathrm{id} \otimes \Delta) \circ \Delta(a^k_{\eta x_i}). \qquad (50)
\end{aligned}$$

The identity

$$\Delta \circ A_{\mathrm{e}}^{(i)} = \big(A_{\mathrm{e}}^{(i)} \otimes \mathrm{id} + \mathrm{id} \otimes A_{\mathrm{e}}^{(i)}\big) \circ \Delta$$

was used above, which follows from $A_{\mathrm{e}}^{(l)} a^k_\eta = a^l_{\mathrm{e}} a^k_\eta$ and a^l_{e} being primitive for all $0 < l \leq m$.

Remark 3 Note that the coproduct (48) can be simplified

$$\Delta a^l_\eta := \tilde{\Theta}_\eta(a^l_{\mathrm{e}} \otimes \mathbf{1}) + \mathbf{1} \otimes a^l_\eta, \qquad (51)$$

which follows from $\tilde{\theta}_k \mathbf{1} = 0$ and the form of $\tilde{\Theta}_i$.

A variant of Sweedler's notation is used for the reduced coproduct, i.e., $\Delta'(a^l_\eta) = \sum' a^l_{\eta'} \otimes a^{l'}_{\eta''}$, as well as for the full coproduct

$$\Delta(a^l_\eta) = \sum a^l_{\eta_{(1)}} \otimes a^{l'}_{\eta_{(2)}} = a^l_\eta \otimes \mathbf{1} + \mathbf{1} \otimes a^l_\eta + \Delta'(a^l_\eta).$$

Connectedness of H implies that its antipode $S : H \to H$ can be calculated using (24a, 24b), namely,

$$Sa^l_\eta = -a^l_\eta - \sum{}' S(a^l_{\eta'})a^{l'}_{\eta''} = -a^l_\eta - \sum{}' a^l_{\eta'} S(a^{l'}_{\eta''}). \tag{52}$$

A few examples are given next. The coproduct (39) implies for the elements a^k_{e} that $Sa^k_{\mathrm{e}} = -a^k_{\mathrm{e}}$. For $0 < j, k, l \leq m$

$$Sa^k_{x_j} = -a^k_{x_j}, \qquad Sa^l_{x_0} = -a^l_{x_0} + \sum_{i=1}^{m} a^l_{x_i} a^i_{\mathrm{e}}. \tag{53}$$

The next theorem uses the coproduct formula (48) to provide an alternative formula for the antipode of H.

Theorem 8 *For any nonempty word $\eta = x_{i_1} \cdots x_{i_l}$, the antipode $S : H \to H$ satisfies*

$$Sa^k_\eta = (-1)^{|\eta|+1} \tilde{\Theta}'_\eta(a^k_{\mathrm{e}}), \tag{54}$$

where

$$\tilde{\Theta}'_\eta := \tilde{\theta}'_{i_l} \circ \cdots \circ \tilde{\theta}'_{i_1} \tag{55}$$

and

$$\tilde{\theta}'_l := -\tilde{\theta}_l + \delta_{0l} \sum_{j=1}^{m} a^j_{\mathrm{e}} \tilde{\theta}_j. \tag{56}$$

Proof The claim is equivalent to saying that

$$S \circ \tilde{\theta}_\eta = -\tilde{\theta}'_{i_l} \circ S \circ \tilde{\theta}_{i_{l-1}} \circ \cdots \circ \tilde{\theta}_{i_1}$$

for the word $\eta = x_{i_1} \cdots x_{i_l} \in X^*$. The proof is via induction on the degree of a^k_η. The degree one case is excluded by assumption as it corresponds to $Sa^k_{\mathrm{e}} = -a^k_{\mathrm{e}}$. For degree two, three, four and five the claim is quickly verified. For $i > 0$ observe that

$$Sa^k_{x_i} = S \circ \tilde{\theta}_i a^k_{\mathrm{e}} = -\tilde{\theta}'_i \circ Sa^k_{\mathrm{e}} = \tilde{\theta}'_i a^k_{\mathrm{e}} = -\tilde{\theta}_i a^k_{\mathrm{e}} = -a^k_{x_i}$$

and

$$Sa^k_{x_0} = S \circ \tilde{\theta}_0 a^k_{\mathrm{e}} = \tilde{\theta}'_0 a^k_{\mathrm{e}} = -a^k_{x_0} + \sum_{i=1}^{m} a^k_{x_i} a^i_{\mathrm{e}},$$

which coincide with (53). For $j > 0$

$$Sa^k_{x_j x_0} = S \circ \tilde{\theta}_0 \circ \tilde{\theta}_j a^k_{\mathrm{e}} = (-1)\tilde{\theta}'_0 \circ \tilde{\theta}'_j (a^k_{\mathrm{e}}) = -a^k_{x_j x_0} + \sum_{i=1}^{m} a^k_{x_j x_i} a^i_{\mathrm{e}}$$

$$= \tilde{\theta}_0' \circ \tilde{\theta}_j' S(a_{\mathrm{e}}^k)$$
$$= -\tilde{\theta}_0' \circ S \circ \tilde{\theta}_j(a_{\mathrm{e}}^k).$$

For degree five

$$Sa_{x_0x_0}^k = S \circ \tilde{\theta}_0 \circ \tilde{\theta}_0 a_{\mathrm{e}}^k = -\tilde{\theta}_0' \circ \tilde{\theta}_0'(a_{\mathrm{e}}^k)$$
$$= -\Big(-\tilde{\theta}_0 + \sum_{j=1}^m a_{\mathrm{e}}^j \tilde{\theta}_j \Big)\Big(-\tilde{\theta}_0 + \sum_{n=1}^m a_{\mathrm{e}}^n \tilde{\theta}_n \Big)(a_{\mathrm{e}}^k)$$
$$= \Big(-\tilde{\theta}_0\tilde{\theta}_0 + \sum_{n=1}^m a_{x_0}^n \tilde{\theta}_n + \sum_{n=1}^m a_{\mathrm{e}}^n \tilde{\theta}_0\tilde{\theta}_n + \sum_{j=1}^m a_{\mathrm{e}}^j \tilde{\theta}_j\tilde{\theta}_0 - \sum_{j=1}^m\sum_{n=1}^m a_{\mathrm{e}}^j \tilde{\theta}_j a_{\mathrm{e}}^n \tilde{\theta}_n \Big)(a_{\mathrm{e}}^k)$$
$$= \Big(-\tilde{\theta}_0\tilde{\theta}_0 + \sum_{n=1}^m a_{x_0}^n \tilde{\theta}_n + \sum_{n=1}^m a_{\mathrm{e}}^n \tilde{\theta}_0\tilde{\theta}_n + \sum_{j=1}^m a_{\mathrm{e}}^j \tilde{\theta}_j\tilde{\theta}_0$$
$$- \sum_{j=1}^m\sum_{n=1}^m a_{\mathrm{e}}^j a_{x_j}^n \tilde{\theta}_n - \sum_{j=1}^m\sum_{n=1}^m a_{\mathrm{e}}^j a_{\mathrm{e}}^n \tilde{\theta}_j\tilde{\theta}_n \Big)(a_{\mathrm{e}}^k)$$
$$= -a_{x_0x_0}^k + \sum_{n=1}^m a_{x_n}^k a_{x_0}^n + \sum_{n=1}^m a_{x_nx_0}^k a_{\mathrm{e}}^n + \sum_{n=1}^m a_{x_0x_n}^k a_{\mathrm{e}}^n$$
$$- \sum_{j=1}^m\sum_{n=1}^m a_{x_n}^k a_{x_j}^n a_{\mathrm{e}}^j - \sum_{j=1}^m\sum_{n=1}^m a_{x_nx_j}^n a_{\mathrm{e}}^j a_{\mathrm{e}}^n.$$

Recall that Sweedler's notation for the reduced coproduct is in use. It is assumed that the theorem holds up to degree $n \geq 2$. Recall that $\tilde{\theta}_{i_l}'$ are derivations on H and that for the augmentation ideal projector P it holds that $P\mathbf{1} = 0$. Working with the second recursion in (52) one finds for $\deg(a_\eta^k) = n+1$, $\eta = x_{i_1} \cdots x_{i_l} = \bar{\eta}x_{i_l} \in X^*$ that

$$Sa_\eta^k = m_H \circ (P \otimes S) \circ \Delta a_\eta^k$$
$$= m_H \circ \Big(P \circ \tilde{\theta}_{i_l} \otimes S + P \otimes S \circ \tilde{\theta}_{i_l} + \delta_{0i_l} \sum_{n=1}^m P \circ \tilde{\theta}_n \otimes S \circ A_{\mathrm{e}}^{(n)} \Big) \circ \Delta a_{\bar{\eta}}^k$$
$$= m_H \circ \big(P \circ \tilde{\theta}_{i_l} \otimes S \big) \circ (a_{\bar{\eta}}^k \otimes \mathbf{1} + \Delta' a_{\bar{\eta}}^k) + m_H \circ \big(P \otimes S \circ \tilde{\theta}_{i_l} \big) \circ \Delta' a_{\bar{\eta}}^k$$
$$+ m_H \circ \Big(\delta_{0i_l} \sum_{n=1}^m P \circ \tilde{\theta}_n \otimes S \circ A_{\mathrm{e}}^{(n)} \Big) \circ (a_{\bar{\eta}}^k \otimes \mathbf{1} + \Delta' a_{\bar{\eta}}^k).$$

The critical term is

$$m_H \circ (P \otimes S \circ \tilde{\theta}_{i_l}) \circ \Delta' a_{\bar{\eta}}^k = m_H \circ (P \otimes S \circ \tilde{\theta}_{i_l}) \circ \sum{}' a_{\eta'}^l \otimes a_{\eta''}^{l'}.$$

Since $\deg(\tilde{\theta}_{i_l} a^{l'}_{\eta''}) < n+1$, it can be written as

$$m_H \circ (P \otimes S \circ \tilde{\theta}_{i_l}) \circ \Delta' a^k_{\bar{\eta}} = -m_H \circ (P \otimes \tilde{\theta}'_{i_l} \circ S) \circ \Delta' a^k_{\bar{\eta}}.$$

This yields

$$\begin{aligned}
Sa^k_\eta &= m_H \circ (P \otimes S) \circ \Delta a^k_\eta \\
&= m_H \circ \big(P \circ \tilde{\theta}_{i_l} \otimes S - P \otimes \tilde{\theta}'_{i_l} \circ S + \delta_{0 i_l} \sum_{n=1}^{m} P \circ \tilde{\theta}_n \otimes S \circ A^{(n)}_{\mathrm{e}}\big) \circ \Delta a^k_{\bar{\eta}} \\
&= m_H \circ \big(\tilde{\theta}_{i_l} \otimes \mathrm{id} + \mathrm{id} \otimes \tilde{\theta}_{i_l} - \mathrm{id} \otimes \delta_{0 i_l} \sum_{n=1}^{m} A^{(n)}_{\mathrm{e}} \tilde{\theta}_n - \delta_{0 i_l} \sum_{n=1}^{m} \tilde{\theta}_n \otimes A^{(n)}_{\mathrm{e}}\big) \circ (P \otimes S) \circ \Delta a^k_{\bar{\eta}} \\
&= -\big(-\tilde{\theta}_{i_l} + \delta_{0 i_l} \sum_{n=1}^{m} A^{(n)}_{\mathrm{e}} \tilde{\theta}_n\big) m_H \circ (P \otimes S) \circ \Delta a^k_{\bar{\eta}} \\
&= -\tilde{\theta}'_{i_l} S a^k_{\bar{\eta}},
\end{aligned}$$

which proves the theorem. Note that the next to the last equality used the fact that the $\tilde{\theta}_i$ are derivations on H.

Remark 4 Consider the case where $m = 1$ in Theorem 7. That is, the alphabet $X := \{x_0, x_1\}$, and the Hopf algebra H is generated by the coordinate functions a_η, $\eta \in X^*$. Note that the upper index on the coordinate functions can be omitted as $m = 1$. The element $a_\eta \in H$ has the coproduct defined in terms of $\Delta \circ \tilde{\theta}_i = \tilde{\Theta}_i \circ \Delta$, $i = 0, 1$, where $\tilde{\Theta}_1 = \tilde{\theta}_1 \otimes \mathrm{id} + \mathrm{id} \otimes \tilde{\theta}_1$ and $\tilde{\Theta}_0 := \tilde{\theta}_0 \otimes \mathrm{id} + \mathrm{id} \otimes \tilde{\theta}_0 + \tilde{\theta}_1 \otimes A_{\mathrm{e}}$. The antipode $S : H \to H$ for any nonempty word $\eta = x_{i_1} \cdots x_{i_l}$ is given by $Sa_\eta = (-1)^{|\eta|+1} \tilde{\Theta}'_\eta(a_{\mathrm{e}})$, where $\tilde{\Theta}'_\eta := \tilde{\theta}'_{i_p} \circ \cdots \circ \tilde{\theta}'_{i_1}$ and $\tilde{\theta}'_l := -\tilde{\theta}_l + \delta_{0l} a_{\mathrm{e}} \tilde{\theta}_1, l = 0, 1$. Here $|\eta| := 2|\eta|_0 + |\eta|_1$, where $|\eta|_0$ denotes the number of times the letter x_0 appears in η, and $|\eta|_1$ is the number of times the letter x_1 is appearing in the word η. One can verify directly that this Hopf algebra coincides with the single-input/single-output (SISO) feedback Hopf algebra described in [28]. The reader is also referred to [14, 16] for more details. This connection between Hopf algebras will be studied further in future work regarding the multivariable (MIMO) case as described in [32].

Finally, returning to the antipode recursions in (52) one realizes quickly the intricacies that result from the signs of the different terms. Surprisingly, the computational aspects of the two formulas are rather different. It turns out that the rightmost recursion is optimal in the sense that its expansion is free of cancellations [14]. This triggers immediately the question whether the antipode formula (54) shares similar properties, which is answered by the next result.

Proposition 3 *The antipode formula (54) is free of cancellations.*

Proof First recall that the algebra of coordinate functions is polynomially free by construction. Then the absence of cancellations follows from looking at (55) and (56) and noting that $\tilde{\theta}_i \tilde{\theta}_j \neq \tilde{\theta}_j \tilde{\theta}_i$.

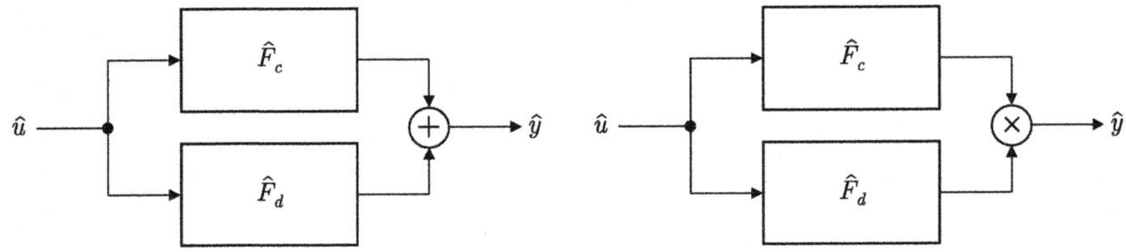

Fig. 3 Parallel sum (left) and parallel product (right) interconnections of two discrete-time Fliess operators

5 Towards Discretization

This section lays the foundation for a discrete-time analogue of the continuous-time Fliess operator theory described in the previous sections. The starting point is the introduction of a *discrete-time* Fliess operator, where the basic idea is to replace the iterated integrals in (6) with iterated sums. This concept was originally exploited in [35, 36] to provide numerical approximations of continuous-time Fliess operators. The main results were developed without any a priori assumption regarding the existence of a state space realization. It was shown, however, that discrete-time Fliess operators are realizable by the class of state space realizations which are rational in the input and affine in the state whenever the generating series is rational. Some specific examples of this will be given here.

The main focus of this section is on parallel interconnections of discrete-time Fliess operators as shown in Figure 3. In the continuous-time theory presented earlier, it was evident that virtually all the results about interconnections flow from the shuffle algebra, which is induced by the parallel product interconnection. The hypothesis here is that an analogous situation holds in the discrete-time case. So it will be shown that the parallel product of discrete-time Fliess operators induces a *quasi-*shuffle algebra on the set of generating series. Given the natural suitability of rational generating series for discrete-time realization theory, a natural question to pursue is whether rationality is preserved under the quasi-shuffle product. The question was affirmatively answered in [42] but without proof. So here a complete proof will be given.

5.1 Discrete-Time Fliess Operators

The set of admissible inputs for discrete-time Fliess operators will be drawn from the real sequence space

$$l_{\infty}^{m+1}[N_0] := \{\hat{u} = (\hat{u}(N_0), \hat{u}(N_0+1), \ldots) : \exists \hat{R}_u \text{ with } 0 \leq |\hat{u}(N)| < \hat{R}_{\hat{u}} < \infty, \forall N \geq N_0\},$$

where $\hat{u} = [\hat{u}_0, \hat{u}_1, \ldots, \hat{u}_m]^T$ and $|\hat{u}(N)| := \max_{i=0,1,\ldots,m} |\hat{u}_i(N)|$. In which case, $\|\hat{u}\|_\infty := \sup_{N \geq N_0} |\hat{u}(N)|$ is always finite. Define a ball of radius $\hat{R}$ in $l_\infty^{m+1}[N_0]$ as

$$B_\infty^{m+1}[N_0](\hat{R}) = \{\hat{u} \in l_\infty^{m+1}[N_0] : \|\hat{u}\|_\infty \leq \hat{R}\}.$$

The subset of finite sequences over $[N_0, N_f]$ is denoted by $B_\infty^{m+1}[N_0, N_f](\hat{R})$. That is, $\hat{u} \in B_\infty^{m+1}[N_0, N_f](\hat{R})$ if $\max_{N \in [N_0, N_f]} |\hat{u}(N)| \leq \hat{R}$. The following definition is of central importance.

Definition 6 [35, 36] For any $c \in \mathbb{R}^\ell\langle\langle X\rangle\rangle$, the corresponding **discrete-time Fliess operator** is

$$\hat{y}(N) = \hat{F}_c[\hat{u}](N) = \sum_{\eta \in X^*} (c, \eta) S_\eta[\hat{u}](N), \tag{57}$$

where $\hat{u} \in l_\infty^{m+1}[1]$, $N \geq 1$, and the iterated sum for any $x_i \in X$ and $\eta \in X^*$ is defined inductively by

$$S_{x_i \eta}[\hat{u}](N) = \sum_{k=1}^{N} \hat{u}_i(k) S_\eta[\hat{u}](k) \tag{58}$$

with $S_\emptyset[\hat{u}](N) := 1$.

The following lemma will be used for providing sufficient conditions for the convergence of such operators.

Lemma 3 *[35, 36] If $\hat{u} \in B_\infty^{m+1}[1](\hat{R})$ then for any $\eta \in X^*$ and $N \geq 1$*

$$|S_\eta[\hat{u}](N)| \leq \hat{R}^{|\eta|} \binom{N-1+|\eta|}{|\eta|} \leq 2^{N-1}(2\hat{R})^{|\eta|}.$$

Proof If $\eta = x_{i_j} \cdots x_{i_1} \in X^*$ then for any $N \geq 1$

$$\begin{aligned} |S_\eta[\hat{u}](N)| &= \left| \sum_{k_j=1}^{N} \hat{u}_{i_j}(k_j) \sum_{k_{j-1}=1}^{k_j} \hat{u}_{i_{j-1}}(k_{j-1}) \cdots \sum_{k_1=1}^{k_2} \hat{u}_{i_1}(k_1) \right| \\ &\leq \sum_{k_j=1}^{N} |\hat{u}_{i_j}(k_j)| \sum_{k_{j-1}=1}^{k_j} |\hat{u}_{i_{j-1}}(k_{j-1})| \cdots \sum_{k_1=1}^{k_2} |\hat{u}_{i_1}(k_1)| \\ &\leq \hat{R}^{|\eta|} \sum_{k_j=1}^{N} \sum_{k_{j-1}=1}^{k_j} \cdots \sum_{k_1=1}^{k_2} 1 = \hat{R}^{|\eta|} \binom{N-1+|\eta|}{|\eta|}, \end{aligned}$$

using the fact that the final nested sum above has $\binom{N-1+|\eta|}{|\eta|}$ terms [6]. The remaining inequality is standard.

Since the upper bound on $|S_\eta[\hat{u}](N)|$ in this lemma is achievable, it is not difficult to see that when the generating series c satisfies the growth bound (7), the series (57) defining $\hat{F}_c$ can diverge. For example, if $(c,\eta) = K_c M_c^{|\eta|}\,|\eta|!$ for all $\eta \in X^*$, and $\hat{u}_i(N) = \hat{R}$, $N \geq 1$, $i = 0, 1, \ldots, m$ then

$$F[\hat{u}](N) = \sum_{\eta \in X^*} K_c M_c^{|\eta|}\,|\eta|!\,\hat{R}^{|\eta|}\binom{N-1+|\eta|}{|\eta|}$$
$$= K_c \sum_{j=0}^{\infty} (M_c(m+1)\hat{R})^j\, j! \binom{N-1+j}{j}.$$

Since $\lim_{j\to\infty}\binom{N-1+j}{j} = 1$, this series diverges even when $\hat{R} < 1/M_c(m+1)$. The next theorem shows that this problem is averted when c satisfies the stronger growth condition (8).

Theorem 9 *[35, 36] Suppose $c \in \mathbb{R}^\ell\langle\langle X\rangle\rangle$ has coefficients which satisfy*

$$|(c,\eta)| \leq K_c M_c^{|\eta|}, \quad \forall \eta \in X^*.$$

Then there exists a real number $\hat{R} > 0$ such that for each $\hat{u} \in B_\infty^{m+1}[1](\hat{R})$, the series (57) converges absolutely for any $N \geq 1$.

Proof Fix $N \geq 1$. From the assumed coefficient bound and Lemma 3, it follows that

$$\left|\hat{F}_c(\hat{u})(N)\right| \leq \sum_{j=0}^{\infty}\sum_{\eta \in X^j} |(c,\eta)|\,\left|S_\eta[\hat{u}](N)\right| \leq \sum_{j=0}^{\infty} K_c (M_c(m+1))^j\, 2^{N-1}(2\hat{R})^j$$
$$= \frac{K_c 2^{N-1}}{1 - 2M_c(m+1)\hat{R}},$$

provided $\hat{R} < 1/2M_c(m+1)$.

The final convergence theorem shows that the restriction on the norm of $\hat{u}$ can be removed if an even more stringent growth condition is imposed on c.

Theorem 10 *[35, 36] Suppose $c \in \mathbb{R}^\ell\langle\langle X\rangle\rangle$ has coefficients which satisfy*

$$|(c,\eta)| \leq K_c M_c^{|\eta|}\frac{1}{|\eta|!}, \quad \forall \eta \in X^*$$

for some real numbers $K_c, M_c > 0$. Then for every $\hat{u} \in l_\infty^{m+1}[1]$, the series (57) converges absolutely for any $N \geq 1$.

Proof Following the same argument as in the proof of the previous theorem, it is clear for any $\hat{u} \in l_\infty^{m+1}[1]$ and $N \geq 1$ that

Table 2 Summary of convergence conditions for F_c and $\hat{F}_c$

Growth Rate	F_c	$\hat{F}_c$
$\lvert(c,\eta)\rvert \leq K_c M_c^{\lvert\eta\rvert} \lvert\eta\rvert!$	LC	Divergent
$\lvert(c,\eta)\rvert \leq K_c M_c^{\lvert\eta\rvert}$	GC	LC
$\lvert(c,\eta)\rvert \leq K_c M_c^{\lvert\eta\rvert} \frac{1}{\lvert\eta\rvert!}$	GC (at least)	GC

$$\left|\hat{F}_c(\hat{u})(N)\right| \leq \sum_{j=0}^{\infty} K_c (M_c(m+1))^j \frac{1}{j!} 2^{N-1} (2\|\hat{u}\|_\infty)^j = K_c 2^{N-1} \mathrm{e}^{2M_c(m+1)\|\hat{u}\|_\infty}.$$

Assuming the analogous definitions for local convergence (LC) and global convergence (GC) of the operator $\hat{F}_c$, note the incongruence between the convergence conditions for continuous-time and discrete-time Fliess operators as summarized in Table 2. In each case, for a fixed series c, the sense in which the discrete-time Fliess operator $\hat{F}_c$ converges is *weaker* than that for F_c. The source of this dichotomy is the observation in Lemma 3 that iterated sums of $\hat{u}$ do not grow as a function of word length like $\hat{R}^{|\eta|}/\,|\eta|!$, which is the case for iterated integrals, but at a potentially faster rate. On the other hand, it is well known that rational series have coefficients that grow as in (8). Thus, as indicated in Table 2, their corresponding discrete-time Fliess operators always converge locally. Therefore, in the following sections this case will be considered in further detail.

5.2 *Rational Discrete-Time Fliess Operators*

An example of a system that can be described by a discrete-time Fliess operator is

$$\begin{aligned} \begin{bmatrix} z_1(N+1) \\ z_2(N+1) \end{bmatrix} &= \begin{bmatrix} 1 & 0 \\ 0 & 1 \end{bmatrix} \begin{bmatrix} z_1(N) \\ z_2(N) \end{bmatrix} + \begin{bmatrix} 1 \\ 0 \end{bmatrix} \hat{u}_1(N+1) + \begin{bmatrix} 0 \\ z_1(N) \end{bmatrix} \hat{u}_2(N+1) \\ &\quad + \begin{bmatrix} 0 \\ 1 \end{bmatrix} \hat{u}_1(N+1)\hat{u}_2(N+1), \\ y(N) &= \begin{bmatrix} 0 & 1 \end{bmatrix} \begin{bmatrix} z_1(N) \\ z_2(N) \end{bmatrix}, \end{aligned} \tag{59}$$

where $\hat{u}_1$ and $\hat{u}_2$ are suitable input sequences. Letting $z_1(0) = z_2(0) = 0$. Observe that

$$z_1(N+1) = z_1(N) + \hat{u}_1(N+1)$$

implies that $z_1(N) = \sum_{k=1}^{N} \hat{u}_1(k)$. Thus, it follows that

$$\begin{aligned} z_2(N+1) &= z_2(N) + z_1(N)\hat{u}_2(N+1) + \hat{u}_2(N+1)\hat{u}_1(N+1) \\ &= z_2(N) + \hat{u}_2(N+1)z_1(N+1) \\ &= \sum_{k_2=1}^{N+1} \hat{u}_2(k_2) \sum_{k_1=1}^{k_2} \hat{u}_1(k_1). \end{aligned} \tag{60}$$

The corresponding output is then

$$y(N) = \sum_{k_2=1}^{N+1} \hat{u}_2(k_2) \sum_{k_1=1}^{k_2} \hat{u}_1(k_1) = S_{x_2x_1}[\hat{u}](N),$$

which has the form of (57). System (59) falls into the category of *polynomial input and state affine* systems [61]. A simple discretization procedure can also yield discrete-time systems that are rational functions of the inputs. Consider, for instance, the following continuous-time system

$$\dot{z}(t) = z(t)u(t), \quad z(0) = 0. \tag{61}$$

For small $\Delta > 0$, an Euler type approximation gives

$$\begin{aligned} \tilde{z}((N+1)\Delta) &= \tilde{z}(N\Delta) + \int_{N\Delta}^{(N+1)\Delta} \tilde{z}(t)u(t)\,dt \\ &\approx \tilde{z}(N\Delta) + \int_{N\Delta}^{(N+1)\Delta} u(t)\,dt\ \tilde{z}((N+1)\Delta) \\ &= \tilde{z}(N\Delta) + \hat{u}(N+1)\,\tilde{z}((N+1)\Delta), \end{aligned}$$

and therefore, letting $\hat{z}(N) = \tilde{z}(N\Delta)$, observe that

$$\hat{z}(N+1) = (1 - \hat{u}(N+1))^{-1}\hat{z}(N) \tag{62}$$

In this case, $(1 - \hat{u}(N+1))^{-1}$ is a rational function and fall into the following class of systems.

Definition 7 [36] A discrete-time state space realization is ***rational input*** and ***state affine*** if its transition map has the form

$$\hat{z}_i(N+1) = \sum_{j=1}^{n} r_{ij}(\hat{u}(N+1))\hat{z}_j(N) + s_i(\hat{u}(N+1)),$$

$i = 1, 2, \ldots, n$, where $\hat{z}(N) \in \mathbb{R}^n$, $\hat{u} = [\hat{u}_0, \hat{u}_1, \ldots, \hat{u}_m]^T$, r_{ij} and s_i are rational functions, and the output map $h : \hat{z} \mapsto \hat{y}$ is linear.

The general situation is described by the following realization theorem.

Theorem 11 *[36] Let $c \in \mathbb{R}\langle\langle X\rangle\rangle$ be a rational series over $X = \{x_0, x_1, \ldots, x_m\}$ with linear representation (μ, γ, λ). Then $\hat{y} = \hat{F}_c[\hat{u}]$ has a finite dimensional rational input and state affine realization on $B_\infty^{m+1}[0, N_f](\hat{R})$ for any $N_f > 0$ provided $\hat{R} < \left(\sum_{j=0}^m \|\mu(x_j)\|\right)^{-1}$, where $\|\cdot\|$ is any matrix norm.*

5.3 Parallel Interconnections and the Quasi-shuffle Algebra

Given two continuous-time Fliess operators F_c and F_d with $c, d \in \mathbb{R}_{LC}\langle\langle X\rangle\rangle$, the parallel interconnections as shown in Figure 3 satisfy $F_c + F_d = F_{c+d}$ and $F_c F_d = F_{c \sqcup\!\sqcup d}$ [24]. In the discrete-time case, the parallel sum interconnection is characterized trivially by the addition of generating series, i.e., $\hat{F}_c + \hat{F}_d = \hat{F}_{c+d}$ due to the vector space nature of $\mathbb{R}_{LC}\langle\langle X\rangle\rangle$. But the parallel product connection in this case is characterized by the so-called *quasi-shuffle* product introduced in Example 5. The main objective of this section is to give a description of the quasi-shuffle algebra $H_{qsh} = (\mathbb{R}\langle X\rangle, \circledast)$ in the context of discrete-time Fliess operators and show that rationality is preserved under the quasi-shuffle product.

5.3.1 Quasi-shuffle Algebra

The shuffle product (1) describes the product of iterated integrals. However, it cannot account for products of iterated sums. For instance, observe that the product

$$\sum_{i=1}^{N} \hat{u}_1(i) \sum_{j=1}^{N} \hat{u}_2(j) = \sum_{i=1}^{N}\sum_{j=1}^{i} \hat{u}_1(i)\hat{u}_2(j) + \sum_{i=1}^{N}\sum_{j=1}^{i} \hat{u}_1(j)\hat{u}_2(i) - \sum_{i=1}^{N} \hat{u}_1(i)\hat{u}_2(i), \tag{63}$$

where $\hat{u} \in B_\infty^{m+1}[0, N_f](R)$ for suitable R and N_f. If $X = \{x_0, x_1, x_2\}$, then using (58) it follows that (63) can be written as

$$S_{x_1}[\hat{u}](N) S_{x_2}[\hat{u}](N) = S_{x_1x_2}[\hat{u}](N) + S_{x_2x_1}[\hat{u}](N) - \sum_{i=1}^{N} \hat{u}_1(i)\hat{u}_2(i).$$

Note that the last term $\sum_{i=1}^{N} \hat{u}_1(i)\hat{u}_2(i)$ does not correspond to a letter in X nor to a word in X^*. Therefore, the alphabet X needs to be augmented to account for this fact. Associating the input $\hat{u}_1\hat{u}_2$ with the new letter $x_{1,2}$, one can now write

$$S_{x_1}[\hat{u}](N) S_{x_2}[\hat{u}](N) = S_{x_1x_2}[\hat{u}](N) + S_{x_2x_1}[\hat{u}](N) + S_{x_{1,2}}[\hat{u}](N).$$

Therefore, the general setting in which products of iterated sums are considered requires a countable alphabet. The extra letters, in addition to those in $X = \{x_0, x_1, \ldots, x_m\}$, account for all possible finite products of inputs. Recall item (3) in

Example 5, where the quasi-shuffle Hopf algebra is defined. Here the alphabet X is extended to a graded commutative semigroup by defining the commutative bracket operation of letters in X to be $[x_i x_j] = x_{i,j}$, which is assumed to be associative, i.e., $[[x_i x_j]x_l] = [x_i[x_j x_l]]$ for letters $x_i, x_j, x_l \in X$. Iterated brackets may therefore be denoted by $x_{i_1,\ldots,i_n} := [[[x_{i_1} x_{i_2}] \cdots]x_{i_n}]$. The augmented alphabet $\bar{X}$ contains X as well as all finitely iterated brackets $x_{i_1,\ldots,i_n}$. The monoid of words with letters from $\bar{X}$ is denoted $\bar{X}^*$. The definition (58) of iterated sums has to be extended to include the additional words in $\bar{X}^*$, for instance,

$$S_{x_k x_{i_1,i_2,\ldots,i_n}}[\hat{u}](N) := \sum_{i=1}^{N} \hat{u}_k(i) \sum_{j=1}^{i} \hat{u}_{i_1}(j)\hat{u}_{i_2}(j) \cdots \hat{u}_{i_n}(j).$$

It follows now that the product $S_{x_1}[\hat{u}](N)S_{x_2}[\hat{u}](N)$ is encoded symbolically in terms of a quasi-shuffle product on $\bar{X}^*$

$$x_1 \circledast x_2 = x_1 x_2 + x_2 x_1 - x_{1,2} \in \mathbb{R}\langle \bar{X} \rangle. \tag{64}$$

The foundation of discrete-time Fliess operator theory is the summation operator, which is used inductively in the construction of the iterated sums in (58). In general, the summation operator Z is defined as

$$Z(f)(x) := \sum_{k=1}^{[x/\theta]} \theta f(\theta k) \tag{65}$$

for a suitable class of functions f. It is known to satisfy the so-called Rota–Baxter relation of weight θ [18]

$$Z(f)(x)Z(g)(x) = Z\big(Z(f)g + fZ(g) - \theta fg\big)(x). \tag{66}$$

This relation generalizes the integration by parts rule for indefinite Riemann integrals and provides the corresponding formula for iterated sums. Specifically, (63) corresponds to (66) where $\theta = 1$, $f = \hat{u}_1$ and $g = \hat{u}_2$. The quasi-shuffle product, introduced in item (3) of Example 5, defined on $\bar{X}^*$ provides an extension of (64) and (66). For words $\eta = \eta_1 \cdots \eta_n$ and $\xi = \xi_1 \cdots \xi_m$, where $\eta_i, \xi_j \in \bar{X}$, the recursive definition of the quasi-shuffle product on $\bar{X}^*$ is given by

$$\eta \circledast \xi = \eta_1(\eta_1^{-1}(\eta) \circledast \xi) + \xi_1(\eta \circledast \xi_1^{-1}(\xi)) - [\eta_1 \xi_1]\big(\eta_1^{-1}(\eta) \circledast \xi_1^{-1}(\xi)\big) \tag{67}$$

with $\emptyset \circledast \eta = \eta \circledast \emptyset = \eta$ for $\eta \in \bar{X}^*$, and $\eta_1^{-1}(\cdot)$ is the left-shift operator defined in (5). This implies that

$$S_\eta[\hat{u}](N) \cdot S_\xi[\hat{u}](N) = S_{\eta \circledast \xi}[\hat{u}](N) \tag{68}$$

with $\eta \circledast \xi \in \mathbb{R}\langle\bar{X}\rangle$. Observe that since $|\eta|, |\xi| < \infty$, then $\mathrm{supp}\{\eta \circledast \xi\}$ is generated by a finite subset of $\bar{X}$. The quasi-shuffle product $\circledast$ is linearly extended to series $c, d \in \mathbb{R}\langle\langle\bar{X}\rangle\rangle$ so that

$$c \circledast d = \sum_{\eta,\xi\in\bar{X}^*} (c,\eta)(d,\xi)\eta \circledast \xi = \sum_{\nu\in\bar{X}^*} \underbrace{\sum_{\eta,\xi\in\bar{X}^*} (c,\eta)(d,\xi)(\eta \circledast \xi,\nu)}_{(c\circledast d,\nu)}\,\nu.$$

Note that the coefficient $(\eta\circledast\xi,\nu) \neq 0$ only when $\nu \in X^*$ is such that $|\eta| + |\xi| - \min(|\eta|, |\xi|) \leq |\nu| \leq |\eta| + |\xi|$. Therefore, $(c \circledast d, \nu)$ is finite since the set $I_{\circledast}(\nu) \triangleq \{(\eta,\xi) \in \bar{X}^* \times \bar{X}^* : (\eta \circledast \xi, \nu) \neq 0\}$ is finite. Hence, the summation defining $c \circledast d$ is locally finite, and therefore summable. It can be shown that the quasi-shuffle product is commutative, associative and distributes over addition [17, 41]. Thus, the vector space $\mathbb{R}\langle\langle\bar{X}\rangle\rangle$ endowed with the quasi-shuffle product forms a commutative $\mathbb{R}$-algebra, the so-called *quasi-shuffle algebra* with multiplicative identity element **1**.

5.3.2 Rationality of the Quasi-shuffle Product

In this section the question of whether the quasi-shuffle product of two rational series is again rational is addressed. In light of Definition 1 and the remark thereafter, it is clear that a rational series c over $\bar{X}$ is also rational over a finite sub-alphabet $X_c \subset \bar{X}$. In which case, Theorems 1 and 2 still apply in the present setting. Also note that in the context of the parallel product connection the underlying alphabets for the generating series of $\hat{F}_c$ and $\hat{F}_d$ are always the same since the inputs are identical. But there is no additional complexity introduced if the alphabets are allowed to be distinct. So let $X_c, X_d \subset \bar{X}$ be finite sub-alphabets of $\bar{X}$ corresponding to the generating series c and d and with cardinalities N_c and N_d, respectively. Define $[X_cX_d] = \{[x_i^c x_j^d] : x_i^c \in X_c, x_j^d \in X_d, i = 1,\ldots,N_c, j = 1\ldots,N_d\}$. The main theorem of the section is given first.

Theorem 12 *Let $c, d \in \mathbb{R}\langle\langle\bar{X}\rangle\rangle$ be rational series with underlying finite alphabets $X_c, X_d \subset \bar{X}$, then $e = c \circledast d \in \mathbb{R}\langle\langle\bar{X}\rangle\rangle$ is rational with underlying alphabet $X_e = X_c \cup X_d \cup [X_cX_d] \subset \bar{X}$.*

Proof In light of (67), the series $e = c \circledast d$ is clearly defined over the finite alphabet X_e. Therefore, a stable finite dimensional vector space V_e is constructed which contains e in order to apply Theorem 2. Since c and d are both rational, let V_c and V_d be stable finite dimensional vector subspaces of $\mathbb{R}\langle\langle X_c\rangle\rangle$ and $\mathbb{R}\langle\langle X_d\rangle\rangle$ containing c and d, respectively. Let $\{\bar{c}_i\}_{i=1}^{n_c}$ and $\{\bar{d}_j\}_{j=1}^{n_d}$ denote their corresponding bases. Define

$$V_e = \mathrm{span}_{\mathbb{R}}\{\bar{c}_i \circledast \bar{d}_j \;:\; i = 1,\ldots,n_c,\quad j = 1,\ldots,n_d\}.$$

Clearly, $V_e \subset \mathbb{R}\langle\langle X_e \rangle\rangle$ is finite dimensional. If one writes

$$c = \sum_{i=1}^{n_c} \alpha_i \bar{c}_i, \quad d = \sum_{j=1}^{n_d} \beta_j \bar{d}_j,$$

it then follows directly that

$$e = c \circledast d = \sum_{i,j=1}^{n_c,n_d} \alpha_i \beta_j \; \bar{c}_i \circledast \bar{d}_j \in V_e.$$

So it only remains to be shown that V_e is stable. Observe from (67) that for any $x \in X_e$ the left-shift operator acts on the quasi-shuffle product as

$$x^{-1}(\eta \circledast \xi) = x^{-1}(\eta) \circledast \xi + \eta \circledast x^{-1}(\xi) + \delta_{x,[x_i x_j]}(x_i^{-1}(\eta) \circledast x_j^{-1}(\xi)), \tag{69}$$

where $\eta = x_i \eta', \xi = x_j \xi' \in \bar{X}^*$ and $\delta_{x,[x_i x_j]} = 1$ if $x = [x_i x_j]$ and 0 otherwise. Writing $c = (c, \emptyset) + \sum_{i=0}^{N_c} x_i^c \; (x_i^c)^{-1}(c)$ and $d = (d, \emptyset) + \sum_{i=0}^{N_d} x_i^d \; (x_i^d)^{-1}(d)$ and using the bilinearity of the quasi-shuffle, it follows that

$$x^{-1}(e) = x^{-1}(c) \circledast d + c \circledast x^{-1}(d) + \sum_{i,j=0}^{N_c,N_d} \delta_{x,[x_i^c x_j^d]}((x_i^c)^{-1}(c) \circledast (x_j^d)^{-1}(d)).$$

But since $V_c, V_d \subset \mathbb{R}\langle\langle X_e \rangle\rangle$ are stable vector spaces by assumption, it is immediate that $(x_i^c)^{-1}(c) \in V_c$ and $(x_j^d)^{-1}(d) \in V_d$, and therefore $x^{-1}(e) \in V_e$ as well. It then follows that V_e is a stable vector space, and hence e is rational.

The following corollary describes the generating series for the parallel product connection in the context of rational series.

Corollary 2 *If $c, d \in \mathbb{R}\langle\langle \bar{X} \rangle\rangle$ are rational series with underlying finite alphabets $X_c, X_d \subset \bar{X}$, then $\hat{F}_c \hat{F}_d = \hat{F}_{c \circledast d}$ with $e = c \circledast d \in \mathbb{R}_{rat}\langle\langle X_e \rangle\rangle$, where $X_e = X_c \cup X_d \cup [X_c X_d]$.*

Proof From (68) the product connection of two operators as in (57) is

$$\begin{aligned}
\hat{F}_c[\hat{u}](N)\hat{F}_d[\hat{u}](N) &= \sum_{\eta \in X_c^*} (c, \eta) S_\eta[\hat{u}](N) \cdot \sum_{\xi \in X_d^*} (d, \xi) S_\xi[\hat{u}](N) \\
&= \sum_{\eta \in X_c^*, \xi \in X_d^*} (c, \eta)(d, \xi) S_{\eta \circledast \xi}[\hat{u}](N) \\
&= F_{c \circledast d}[\hat{u}](N) =: F_e[\hat{u}](N).
\end{aligned}$$

Here $e \in \mathbb{R}_{rat}\langle\langle X_e \rangle\rangle$ since by Theorem 12 the quasi-shuffle preserves rationality.

The following lemma will be used in the final example of this section.

Lemma 4 *For any* $i, j \geq 0$

$$x_1^i \circledast x_1^j = \sum_{k=0}^{\min\{i,j\}} \binom{i+j-2k}{\min\{i,j\}-k} x_{1,1}^k \sqcup\!\sqcup\, x_1^{i+j-2k}. \tag{70}$$

Proof Without loss of generality assume $i \leq j$. The identity is proved by induction over $i+j$. The cases for $i+j = 0, 1$ are trivial. Assume (70) holds up to some fixed $i+j$. Using (67) compute

$$x_1^i \circledast x_1^{j+1} = x_1\left(x_1^{i-1} \circledast x_1^{j+1}\right) + x_1\left(x_1^i \circledast x_1^j\right) + x_{1,1}\left(x_1^{i-1} \circledast x_1^j\right).$$

By the induction hypothesis and since $i \leq j$,

$$\begin{aligned}
x_1^i \circledast x_1^{j+1} &= \sum_{k=0}^{i-1} \binom{i+j-2k}{i-1-k} x_1(x_{1,1}^k \sqcup\!\sqcup\, x_1^{i+j-2k}) \\
&+ \sum_{k=0}^{i} \binom{i+j-2k}{i-k} x_1(x_{1,1}^k \sqcup\!\sqcup\, x_1^{i+j-2k}) + \sum_{k=0}^{i-1} \binom{i+j-1-2k}{i-1-k} x_{1,1}(x_{1,1}^k \sqcup\!\sqcup\, x_1^{i+j-1-2k}) \\
&= \sum_{k=0}^{i-1} \binom{i+j-2k}{i-1-k} x_1(x_{1,1}^k \sqcup\!\sqcup\, x_1^{i+j-2k}) \\
&+ \sum_{k=0}^{i-1} \binom{i+j-2k}{i-k} x_1(x_{1,1}^k \sqcup\!\sqcup\, x_1^{i+j-2k}) + \binom{j-i}{0} x_1(x_{1,1}^k \sqcup\!\sqcup\, x^{j-i}) \\
&+ \sum_{k=1}^{i-1} \binom{i+j-2k+1}{i-k} x_{1,1}(x_{1,1}^k \sqcup\!\sqcup\, x_1^{i+j-2k+1}) + \binom{j-i+1}{0} x_1(x_{1,1}^{k-1} \sqcup\!\sqcup\, x^{j-i+1}) \\
&= (x_{1,1}^k \sqcup\!\sqcup\, x_1^{i+j-2k}) + \binom{i+j+1}{i} x_1^{i+j+1} + \sum_{k=0}^{i-1} \binom{i+j-2k+1}{i-k} x_1(x_{1,1}^k \sqcup\!\sqcup\, x_1^{i+j-2k}) \\
&+ \sum_{k=1}^{i-1} \binom{i+j-2k+1}{i-k} x_{1,1}(x_{1,1}^k \sqcup\!\sqcup\, x_1^{i+j-2k+1}) \\
&= \sum_{k=0}^{i} \binom{i+j-2k+1}{i-k} x_{1,1}^k \sqcup\!\sqcup\, x_1^{i+j-2k+1}.
\end{aligned}$$

This complete the proof since it was assumed that $\min\{i, j\} = i$.

Example 7 Let $X = \{x_1\}$ and consider the rational series $c = x_1^* := \sum_{k \geq 0} x_1^k$. It can be shown directly that

$$x_1^* \sqcup\!\sqcup\, x_1^* = \sum_{n=0}^{\infty} \sum_{i=1}^{n} \binom{n}{i} x_1^n = \sum_{\eta \in X^*} 2^{|\eta|} \eta, \tag{71}$$

using the identity $x_1^i \sqcup\!\sqcup x_1^j = \binom{i+j}{i} x_1^{i+j}$ [66]. Since the shuffle product is known to preserve rationality, it follows from Theorem 1 that $x_1^* \sqcup\!\sqcup x_1^*$ must have a linear representation (μ, γ, λ), in this case $\mu(\eta) = 2^{|\eta|}$ and $\gamma = \lambda = 1$. This is easily verified by setting $z_i = F_c[u]$, which gives the bilinear state space realization $\dot{z}_i = z_i u$, $y_i = z_i$. Then the parallel product connection $y = y_1 y_2 = F_c^2[u] = z$ has the realization $\dot{z} = 2zu$, $y = z$. One can confirm using iterated Lie derivatives that the generating series for this system is exactly (71). □

Example 8 The goal now is to produce the quasi-shuffle analogue of (71). Note here that $X = \{x_1, x_{1,1}\}$. Using Lemma 4 it follows that

$$x_1^* \circledast x_1^* = \sum_{i,j=0}^{\infty} x_1^i \circledast x_1^j = \sum_{n=0} \sum_{i+j=n} \sum_{k=0}^{\min\{i,j\}} \binom{i+j-2k}{\min\{i,j\}-k} x_{1,1}^k \sqcup\!\sqcup x_1^{i+j-2k}.$$

For fixed $k', n' \in \mathbb{N} \cup \{0\}$, let $\eta \in X^{n'}$ be such that $|\eta|_{x_{1,1}} = k'$, where $|\eta|_{x_i}$ denotes the number of times the letter $x_i \in X$ appears in $\eta \in X^*$. It follows that the coefficient $(x_1^* \circledast x_1^*, \eta)$ is

$$(x_1^* \circledast x_1^*, \eta) = \sum_{n=0} \sum_{i+j=n} \sum_{k=0}^{\min\{i,j\}} \binom{i+j-2k}{\min\{i,j\}-k} (x_{1,1}^k \sqcup\!\sqcup x_1^{i+j-2k}, \eta). \tag{72}$$

Notice first that the number of elements in $\mathrm{supp}(x_{1,1}^k \sqcup\!\sqcup x_1^{i+j-2k})$ is $\binom{i+j-k}{k}$, and secondly that it implies that $(x_{1,1}^k \sqcup\!\sqcup x_1^{i+j-2k}, \eta) = 1$ when $k = k'$ and $i + j = n' + k'$, otherwise $(x_{1,1}^k \sqcup\!\sqcup x_1^{i+j-2k}, \eta) = 0$. The former statement is related to the fact that $Q^N = \sum_{i_1+\cdots+i_m=N} q_1^{i_1} \sqcup\!\sqcup \cdots \sqcup\!\sqcup q_m^{i_m}$ for an arbitrary alphabet $Q = \{q_1, \ldots, q_m\}$ [13]. The coefficient $(x_1^* \circledast x_1^*, \eta)$ can be further developed as

$$\begin{aligned}
(x_1^* \circledast x_1^*, \eta) &= \sum_{i+j=n'+k'} \binom{n'-k'}{\min\{i,j\}-k'} = \sum_{i+j=n'+k'} \binom{n'-k'}{\min\{i-k', j-k'\}} \\
&= \sum_{i=k'}^{n'} \binom{n'-k'}{\min\{i-k', n'-i\}} = \sum_{i=k'}^{n'} \binom{n'-k'}{i-k'} = \sum_{i=0}^{n'-k'} \binom{n'-k'}{i} = 2^{|\eta|_{x_1}}.
\end{aligned}$$

Thus, one can write

$$x_1^* \circledast x_1^* = \sum_{\eta \in X^*} (x_1^* \circledast x_1^*, \eta)\eta = \sum_{\eta \in X^*} 2^{|\eta|_{x_1}} \eta. \tag{73}$$

In light of Theorem 12, the series $x_1^* \circledast x_1^*$ must be rational. In particular, a straightforward linear representation for $x_1^* \circledast x_1^*$ can be obtained due to (73). That is, (μ, γ, λ) is identified as $\mu(\eta) = 2^{|\eta|_{x_1}}$ and $\lambda = \gamma = 1$. Finally, a direct computation confirms that

$$\begin{aligned} x_1^* \circledast x_1^* = \emptyset &+ 2x_1 + x_{1,1} + 4x_1^2 + 2x_1x_{1,1} + 2x_{1,1}x_1 + x_{1,1}^2 + 8x_1^3 + 4x_1^2x_{1,1} \\ &+ 4x_1x_{1,1}x_1 + 4x_{1,1}x_1^2 + 2x_1x_{1,1}^2 + 2x_{1,1}x_1x_{1,1} + 2x_{1,1}^2x_1 + x_{1,1}^3 + 16x_1^4 \\ &+ 8x_1^3x_{1,1} + 8x_1^2x_{1,1}x_1 + 8x_1x_{1,1}x_1^2 + 8x_{1,1}x_1^3 + 4x_1^2x_{1,1}^2 + 4x_1x_{1,1}x_1x_{1,1} \\ &+ 4x_1x_{1,1}^2x_1 + 4x_{1,1}x_1^2x_{1,1} + 4x_{1,1}x_1x_{1,1}x_1 + 4x_{1,1}^2x_1^2 + 2x_1x_{1,1}^3 + 2x_{1,1}x_1x_{1,1}^2 \\ &+ 2x_{1,1}^2x_1x_{1,1} + 2x_{1,1}^3x_1 + x_{1,1}^4 + \cdots, \end{aligned}$$

where it is clear that (73) holds. □

Acknowledgements The third author was supported by grant SEV-2011-0087 from the Severo Ochoa Excellence Program at the Instituto de Ciencias Matemáticas in Madrid, Spain. This research was also supported by a grant from the BBVA Foundation.

References

1. Abe, E.: Hopf Algebras. Cambridge University Press, Cambridge (1980)
2. Anshelevich, M., Effros, E.G., Popa, M.: Zimmerman type cancellation in the free Faà di Bruno algebra. J. Funct. Anal. **237**, 76–104 (2006)
3. Berstel, J., Reutenauer, C.: Rational Series and Their Languages. Springer, Berlin (1988)
4. Brockett, R.: Volterra series and geometric control theory. Automatica **12**, 167–176 (1976), (addendum with E. Gilbert, **12**, p. 635 (1976))
5. Burde, D.: Left-symmetric algebras, or pre-Lie algebras in geometry and physics. Cent. Eur. J. Math. **4**, 323–357 (2006)
6. Butler, S., Karasik, P.: A note on nested sums. J. Integer Seq. **13**, article 10.4.4 (2010)
7. Cartier, P.: A primer of Hopf algebras. In: Cartier, P., Moussa, P., Julia, B., Vanhove, P. (eds.) Frontiers in Number Theory, Physics and Geometry II, pp. 537–615. Springer, Berlin Heidelberg (2007)
8. Cartier, P.: Vinberg algebras, Lie groups and combinatorics. Clay Math. Proc. **11**, 107–126 (2011)
9. Chen, K.T.: Iterated integrals and exponential homomorphisms. Proc. Lond. Math. Soc. **4**, 502–512 (1954)
10. Chen, K.T.: Algebraization of iterated integration along paths. Bull. AMS **73**, 975–978 (1967)
11. Chen, K.T.: Integration of paths, geometric invariants and a generalized Baker–Hausdorff formula. Ann. Math., 2nd Ser. **65**, 163–178 (1975)
12. Devlin, J.: Word problems related to periodic solutions of a non-autonomous system. Math. Proc. Camb. Philos. Soc. **108**, 127–151 (1990)
13. Duffaut Espinosa, L.A.: Interconnections of nonlinear systems driven by L_2-Itô stochastic processes. Dissertation, Old Dominion University, Norfolk, Virginia (2009)
14. Duffaut Espinosa, L.A., Ebrahimi-Fard, K., Gray, W.S.: A combinatorial Hopf algebra for nonlinear output feedback control systems. J. Algebra **453**, 609–643 (2016). arXiv:1406.5396
15. Duffaut Espinosa, L.A., Gray, W.S.: Integration of output tracking and trajectory generation via analytic left inversion. In: Proceedings of the 21st International Conference on System Theory, Control and Computing, Sinaia, Romania, pp. 802–807 (2017)
16. Ebrahimi-Fard, K., Gray, W.S.: Center problem, Abel equation and the Faà di Bruno Hopf algebra for output feedback. Int. Math. Res. Not. **2017**, 5415–5450 (2017). arXiv:1507.06939
17. Ebrahimi-Fard, K., Guo, L.: Mixable shuffles, quasi-shuffles and Hopf algebras. J. Algebraic Comb. **24**, 83–101 (2006)
18. Ebrahimi-Fard, K., Patras, F.: La structure combinatoire du calcul intégral. Gaz. Math. **138**, 5–22 (2013)

19. Ferfera, A.: Combinatoire du monoïde libre appliquée à la composition et aux variations de certaines fonctionnelles issues de la théorie des systèmes. Dissertation, University of Bordeaux I, Talence (1979)
20. Ferfera, A.: Combinatoire du monoïde libre et composition de certains systèmes non linéaires. Astérisque **75–76**, 87–93 (1980)
21. Figueroa, H., Gracia-Bondía, J.M.: Combinatorial Hopf algebras in quantum field theory I. Rev. Math. Phys. **17**, 881–976 (2005)
22. Figueroa, H., Gracia-Bondía, J.M., Várilly, J.C.: Elements of Noncommutative Geometry, Birkhäuser, Boston (2001)
23. Fliess, M.: Generating series for discrete-time nonlinear systems. IEEE Trans. Autom. Control **AC-25**, 984–985 (1980)
24. Fliess, M.: Fonctionnelles causales non linéaires et indéterminées non commutatives. Bull. Soc. Math. Fr. **109**, 3–40 (1981)
25. Fliess, M.: Réalisation locale des systèmes non linéaires, algèbres de Lie filtrées transitives et séries génératrices non commutatives. Invent. Math. **71**, 521–537 (1983)
26. Foissy, L.: The Hopf algebra of Fliess operators and its dual pre-Lie algebra. Commun. Algebra **43**, 4528–4552 (2015). arXiv:0805.4385v2
27. Frabetti, A., Manchon, D.: Five interpretations of Faà di Bruno's formula. In: Ebrahimi-Fard, K., Fauvet, F. (eds.) Faà di Bruno Hopf Algebras, Dyson-Schwinger Equations, and Lie-Butcher Series. IRMA Lectures in Mathematics and Theoretical Physics, vol. 21, pp. 91–147. European Mathematical Society, Zürich, Switzerland (2015)
28. Gray, W.S., Duffaut Espinosa, L.A.: A Faà di Bruno Hopf algebra for a group of Fliess operators with applications to feedback. Syst. Control Lett. **60**, 441–449 (2011)
29. Gray, W.S., Duffaut Espinosa, L.A.: Feedback transformation groups for nonlinear input-output systems. In: Proceedings of the 52nd IEEE Conference on Decision and Control, Florence, Italy, pp. 2570–2575 (2013)
30. Gray, W.S., Duffaut Espinosa, L.A.: A Faà di Bruno Hopf algebra for analytic nonlinear feedback control systems. In: Ebrahimi-Fard, K., Fauvet, F. (eds.) Faà di Bruno Hopf Algebras, Dyson-Schwinger Equations, and Lie-Butcher Series. IRMA Lectures in Mathematics and Theoretical Physics, vol. 21, pp. 149–217. European Mathematical Society, Zürich, Switzerland (2015)
31. Gray, W.S., Duffaut Espinosa, L.A., Ebrahimi-Fard, K.: Recursive algorithm for the antipode in the SISO feedback product. In: Proceedings of the 21st International Symposium on Mathematical Theory of Networks and Systems, Groningen, The Netherlands, pp. 1088–1093 (2014)
32. Gray, W.S., Duffaut Espinosa, L.A., Ebrahimi-Fard, K.: Faà di Bruno Hopf algebra of the output feedback group for multivariable Fliess operators. Syst. Control Lett. **74**, 64–73 (2014)
33. Gray, W.S., Duffaut Espinosa, L.A., Ebrahimi-Fard, K.: Analytic left inversion of SISO Lotka-Volterra models. In: Proceedings of the 49th Conference on Information Sciences and Systems, Baltimore, Maryland (2015)
34. Gray, W.S., Duffaut Espinosa, L.A., Ebrahimi-Fard, K.: Analytic left inversion of multivariable Lotka-Volterra models. In: Proceedings of the 54th IEEE Conference on Decision and Control, Osaka, Japan, pp. 6472–6477 (2015)
35. Gray, W.S., Duffaut Espinosa, L.A., Ebrahimi-Fard, K.: Discrete-time approximations of Fliess operators. In: Proceedings of the American Control Conference. Boston, Massachusetts, pp. 2433–2439 (2016)
36. Gray, W.S., Duffaut Espinosa, L.A., Ebrahimi-Fard, K.: Discrete-time approximations of Fliess operators. Numerische Mathematik **137**, 35–62 (2017). arXiv:1510.07901
37. Gray, W.S., Duffaut Espinosa, L.A., Thitsa, M.: Left inversion of analytic nonlinear SISO systems via formal power series methods. Automatica **50**, 2381–2388 (2014)
38. Gray, W.S., Li, Y.: Generating series for interconnected analytic nonlinear systems. SIAM J. Control Optim. **44**, 646–672 (2005)
39. Gray, W.S., Thitsa, M.: A unified approach to generating series for cascaded nonlinear input-output systems. Int. J. Control **85**, 1737–1754 (2012)

40. Gray, W.S., Wang, Y.: Fliess operators on L_p spaces: convergence and continuity Syst. Control Lett. **46**, 67–74 (2002)
41. Hoffman, M.E.: Quasi-shuffle products. J. Algebraic Comb. **11**, 49–68 (2000)
42. Houseaux, V., Jacob, G., Oussous, N.E., Petitot, M.: A complete Maple package for noncommutative rational power series. In: Li, Z., Sit, W.Y. (eds.) Computer Mathematics: Proceeding of the Sixth Asian Symposium. Lecture Notes Series on Computing, vol. 10, pp. 174–188. World Scientific, Singapore (2003)
43. Isidori, A.: Nonlinear Control Systems, 3rd edn. Springer, London (1995)
44. Jakubczyk, B.: Existence and uniqueness of realizations of nonlinear systems. SIAM J. Control Optim. **18**, 445–471 (1980)
45. Jakubczyk, B.: Local realization of nonlinear causal operators. SIAM J. Control Optim. **24**, 230–242 (1986)
46. Jakubczyk, B.: Realization theory for nonlinear systems: three approaches. In: Fliess, M., Hazewinkel M. (eds.) Algebraic and Geometric Methods in Nonlinear Control Theory, pp. 3–31. D. Reidel Publishing Company, Dordrecht (1986)
47. Khalil, H.K.: Nonlinear Systems, 3rd edn. Prentice-Hall, Upper Saddle River, New Jersey (2002)
48. Lesiak, C., Krener, A.J.: The existence and uniqueness of Volterra series for nonlinear systems. IEEE Trans. Autom. Control **AC-23**, 1090–1095 (1978)
49. Li, Y.: Generating series of interconnected nonlinear systems and the formal Laplace-Borel transform. Dissertation, Old Dominion University, Norfolk, Virginia (2004)
50. Manchon, D.: Hopf algebras, from basics to applications to renormalization. Comptes-rendus des Rencontres mathématiques de Glanon 2001 (2001). arXiv:0408405
51. Manchon, D.: A short survey on pre-Lie algebras. In: Carey, A. (ed.) Noncommutative Geometry and Physics: Renormalisation, Motives, Index Theory, pp. 89–102. European Mathematical Society, Zürich, Switzerland (2011)
52. Milnor, J.W., Moore, J.C.: On the structure of Hopf algebras. Ann. Math. **81**, 211–264 (1965)
53. Nijmeijer, H., van der Schaft, A.J.: Nonlinear Dynamical Control Systems. Springer, New York (1990)
54. Pittou, M., Rahonis, G.: Weighted recognizability over infinite alphabets. Acta Cybern. **23**, 283–317 (2017)
55. Radford, D.E.: Hopf Algebras. World Scientific Publishing, Hackensack, New Jersey (2012)
56. Ree, R.: Lie elements and an algebra associated with shuffles. Ann. Math., 2nd Ser. **68**, 210–220 (1958)
57. Reutenauer, C.: Free Lie Algebras. Oxford University Press, New York (1993)
58. Rugh, W.J.: Nonlinear System Theory, The Volterra/Wiener Approach. The Johns Hopkins University Press, Baltimore, Maryland (1981)
59. Schützenberger, M.P.: On the definition of a family of automata. Inf. Control **4**, 245–270 (1961)
60. Slotine, J.-J.E., Li, W.: Applied Nonlinear Control. Prentice-Hall, Englewood Cliffs, New Jersey (1991)
61. Sontag, E.D.: Polynomial Response Maps. Springer, Berlin (1979)
62. Sussmann, H.J.: Existence and uniqueness of minimal realizations of nonlinear systems. Math. Syst. Theory **10**, 263–284 (1977)
63. Sussmann, H.J.: A proof of the realization theorem for convergent generating series of finite Lie rank, internal report SYCON-90-02, Rutgers Center for Systems and Control (1990)
64. Sweedler, M.E.: Hopf Algebras. W. A. Benjamin Inc., New York (1969)
65. Thitsa, M., Gray, W.S.: On the radius of convergence of interconnected analytic nonlinear input-output systems. SIAM J. Control Optim. **50**, 2786–2813 (2012)
66. Wang, Y.: Differential equations and nonlinear control systems. Dissertation, Rutgers University, New Brunswick, New Jersey (1990)

Computational Aspects of Some Exponential Identities

Fernando Casas

Abstract The notion of the exponential of a matrix is usually introduced in elementary textbooks on ordinary differential equations when solving a constant coefficients linear system, also providing some of its properties and in particular one that *does not* hold unless the involved matrices commute. Several problems arise indeed from this fundamental issue, and it is our purpose to review some of them in this work, namely: (i) is it possible to write the product of two exponential matrices as the exponential of a matrix? (ii) is it possible to "disentangle" the exponential of a sum of two matrices? (iii) how to write the solution of a time-dependent linear differential system as the exponential of a matrix? To address these problems the Baker–Campbell–Hausdorff series, the Zassenhaus formula and the Magnus expansion are formulated and efficiently computed, paying attention to their convergence. Finally, several applications are also considered.

Keywords Matrix exponential · Baker–Campbell–Hausdorff series
Zassenhaus formula · Magnus expansion

MSC codes 5A16 · 22E60 · 17B66 · 34L99

1 Introduction

The exponential of a linear operator $T : \mathbb{C}^n \longrightarrow \mathbb{C}^n$, defined by the absolutely convergent series

F. Casas (✉)
Institut de Matemàtiques i Aplicacions de Castelló and Departament de Matemàtiques, Universitat Jaume I, 12071 Castellón, Spain
e-mail: Fernando.Casas@uji.es

K. Ebrahimi-Fard and M. Barbero Liñán (eds.), *Discrete Mechanics, Geometric Integration and Lie–Butcher Series*, Springer Proceedings in Mathematics & Statistics 267, https://doi.org/10.1007/978-3-030-01397-4_6

$$e^T = \sum_{k=0}^{\infty} \frac{1}{k!} T^k,$$

appears in a natural way when solving linear systems of differential equations of the form

$$\frac{dy}{dt} = Ay, \qquad y(0) = y_0. \tag{1}$$

Assuming that the linear transformation T on $\mathbb{C}^n$ is represented by the $n \times n$ matrix A, then the unique solution of (1) is given by [54]

$$y(t) = e^{tA} y_0 = \sum_{k=0}^{\infty} \frac{t^k}{k!} A^k y_0.$$

The $n \times n$ matrix e^{tA} can be computed in several ways, not all of them feasible from a numerical point of view [48].

Closely associated with Eq. (1) is the matrix differential equation

$$\frac{dY}{dt} = A\,Y, \qquad Y(0) = I, \tag{2}$$

in the sense that $y(t) = Y(t)y_0 \in \mathbb{C}^n$ is the solution of (1) if and only if $Y(t)$ is the solution of (2) [22].

The exponential of a matrix satisfies some remarkable properties:

- $e^{0A} = I$;
- $e^{(t+s)A} = e^{tA}\, e^{sA}$;
- $(e^{tA})^{-1} = e^{-tA}$;
- if A and P are $n \times n$ matrices and $B = PAP^{-1}$, then $e^B = P\, e^A\, P^{-1}$;
- if A and B commute, i.e., $AB = BA$, then $e^{A+B} = e^A\, e^B = e^B\, e^A$.

It is less well known, however, that the converse of the last property is not true in general: there are simple examples of matrices A, B such that $AB \neq BA$, but $e^{A+B} = e^A\, e^B = e^B\, e^A$ [68, 69].

It turns out that the *commutator*

$$[A, B] = AB - BA$$

plays indeed a fundamental role when analyzing the exponential or a product of exponentials of matrices, as we will see in the sequel. More specifically, the issues we will address in this work can be summarized as follows.

- **Problem 1**. Since $e^A\, e^B \neq e^{A+B}$ in general, one could ask whether some additional term C exists such that $e^A\, e^B = e^{A+B+C}$ and, if the answer is in the affirmative, how C can be obtained from A and B. This, of course, leads to the much celebrated *Baker–Campbell–Hausdorff formula*.

- **Problem 2**. The dual of the previous problem is the following. Is it possible to get matrices $C_1, C_2, \ldots$ such that $\mathrm{e}^{A+B} = \mathrm{e}^A\,\mathrm{e}^B\,\mathrm{e}^{C_1}\,\mathrm{e}^{C_2}\ldots$? Such an expression is called the *Zassenhaus formula*.
- **Problem 3**. Suppose that the coefficient matrix A in the linear differential equation (2) depends explicitly on time, $Y' = A(t)Y$. As is well known [22], the solution in that case is

$$Y(t) = \exp\left(\int_0^t A(s)ds\right) \qquad \text{only if} \qquad \left[\int_0^t A(s)ds,\, A(t)\right] = 0.$$

The question is: can we still write $Y(t)$ as the exponential of a certain matrix $\Omega(t)$, where

$$\Omega(t) = \int_0^t A(s)ds + \Delta\Omega(t)$$

and the additional term $\Delta\Omega(t)$ stands for the necessary correction in the general case? As it turns out, the *Magnus expansion* (sometimes also called the continuous analogue of the Baker–Campbell–Hausdorff formula [43]) provides an algorithmic procedure to solve this problem.

Although these problems have been established in terms of matrices, they can be generalized to linear operators defined on a certain Hilbert space (this in fact corresponds to the original formulation of the Magnus expansion [43]) and elements in a Lie group $\mathcal{G}$ and its corresponding Lie algebra $\mathfrak{g}$ (the tangent space at the identity of $\mathcal{G}$). One should recall the fundamental role the exponential mapping $\exp : \mathfrak{g} \longrightarrow \mathcal{G}$ plays in this setting: given $\beta(t) \in \mathcal{G}$ the one-parameter group solution of the differential equation

$$\frac{d\beta(t)}{dt} = X\beta(t), \qquad \beta(0) = e,$$

where e is the identity of $\mathcal{G}$ and X is a smooth left-invariant vector field, the exponential transformation is defined as $\exp(X) = \beta(1)$ [33, 58]. This exponential map coincides with the usual exponential matrix function if $\mathcal{G}$ is a matrix Lie group. Given the ubiquitous nature of Lie groups and Lie algebras in many fields of science (classical and quantum mechanics, statistical mechanics, quantum computing, control theory, etc.), very often we will consider the general case where no particular algebraic structure is assumed beyond what is common to all Lie algebras, i.e., we will work in a *free Lie algebra*, especially when addressing Problems 1 and 2 above. For the sake of completeness, we have included an Appendix with some basic properties of free Lie algebras.

Before starting with our study, let us mention another well known result concerning exponentials of matrices and operators, namely the *Lie product formula* and the

Trotter product formula [56, 65]. The former is formulated in terms of matrices A and B and states that

$$e^{A+B} = \lim_{m\to\infty} \left(e^{\frac{A}{m}} e^{\frac{B}{m}} \right)^m, \tag{3}$$

whereas the latter establishes that (3) and its proof can indeed be extended to the case where A and B are unbounded self-adjoint operators and $A + B$ is also self-adjoint on the common domain of A and B. This important theorem has found many applications, in particular in the numerical treatment of partial differential equations.

2 The Baker–Campbell–Hausdorff Formula

2.1 General Considerations

Problem 1 can be established in general as follows. Let X and Y be two non commuting indeterminates. Then, clearly

$$e^X e^Y = \sum_{p,q=0}^{\infty} \frac{1}{p!q!} X^p Y^q. \tag{4}$$

When this series is substituted in the formal series defining the logarithm of the operator Z,

$$\log Z = \sum_{k=1}^{\infty} \frac{(-1)^{k-1}}{k} (Z - I)^k,$$

one gets, after some work,

$$Z = \log(e^X e^Y) = \sum_{k=1}^{\infty} \frac{(-1)^{k-1}}{k} \sum \frac{X^{p_1} Y^{q_1} \ldots X^{p_k} Y^{q_k}}{p_1! q_1! \ldots p_k! q_k!}, \tag{5}$$

where the inner summation extends over all non-negative integers $p_1, q_1, \ldots, p_k, q_k$ for which $p_i + q_i > 0$ $(i = 1, 2, \ldots, k)$. The first terms in the previous expression read explicitly

$$Z = (X + Y + XY + \frac{1}{2}X^2 + \frac{1}{2}Y^2 + \cdots) - \frac{1}{2}(XY + YX + X^2 + Y^2 + \cdots) + \cdots$$
$$= X + Y + \frac{1}{2}(XY - YX) + \cdots = X + Y + \frac{1}{2}[X, Y] + \cdots$$

Campbell [17], Baker [6] and Hausdorff [34], among others, addressed the question whether Z in (5) can be represented as a series of nested commutators of X and Y, concluding that this is indeed the case, although they were not able either to provide a rigorous proof of this feature or to give an explicit formula (or a method of construction). As Bourbaki states, "each considered that the proofs of his predecessors were not convincing" [15, p. 425]. It was only in 1947 that Dynkin [24, 25] finally obtained an explicit formula by considering from the outset a normed Lie algebra. Specifically, he obtained

$$Z = \sum_{k=1}^{\infty} \sum_{p_i, q_i} \frac{(-1)^{k-1}}{k} \frac{[X^{p_1} Y^{q_1} \dots X^{p_k} Y^{q_k}]}{(\sum_{i=1}^{k}(p_i + q_i))\, p_1! q_1! \dots p_k! q_k!}, \tag{6}$$

where the inner summation is taken over all non-negative integers $p_1, q_1, \dots, p_k, q_k$ such that $p_1 + q_1 > 0, \dots, p_k + q_k > 0$ and $[X^{p_1} Y^{q_1} \dots X^{p_k} Y^{q_k}]$ denotes the right nested commutator based on the *word* $X^{p_1} Y^{q_1} \dots X^{p_k} Y^{q_k}$, i.e.,

$$[XY^2X^2Y] \equiv [XYYXXY] \equiv [X, [Y, [Y, [X, [X, Y]]]]].$$

Expression (6) is known, for obvious reasons, as the *Baker–Campbell–Hausdorff series in the Dynkin form* and the reader is referred to [14] for a detailed account of the genesis, development and history of this important result.

Gathering together in (6) those terms for which $p_1 + q_1 + \dots + p_k + q_k = m$ one arrives at the following expressions up to $m = 5$:

$$\begin{aligned}
m = 1; \quad Z_1 &= X + Y \\
m = 2: \quad Z_2 &= \frac{1}{2}[X, Y] \\
m = 3: \quad Z_3 &= \frac{1}{12}[X, [X, Y]] - \frac{1}{12}[Y, [X, Y]] \\
m = 4: \quad Z_4 &= -\frac{1}{24}[Y, [X, [X, Y]]] \\
m = 5: \quad Z_5 &= -\frac{1}{720}[X, [X, [X, [X, Y]]]] - \frac{1}{120}[X, [Y, [X, [X, Y]]]] \\
&\quad -\frac{1}{360}[X, [Y, [Y, [X, Y]]]] + \frac{1}{360}[Y, [X, [X, [X, Y]]]] \\
&\quad +\frac{1}{120}[Y, [Y, [X, [X, Y]]]] + \frac{1}{720}[Y, [Y, [Y, [X, Y]]]].
\end{aligned}$$

In general, one has

$$Z = \log(e^X e^Y) = X + Y + \sum_{m=2}^{\infty} Z_m, \tag{7}$$

where $Z_m(X, Y)$ is a *homogeneous Lie polynomial* in X and Y of degree m, i.e., a linear combination of commutators of the form $[V_1, [V_2, \dots, [V_{m-1}, V_m] \dots]]$ with $V_i \in \{X, Y\}$ for $1 \le i \le m$, the coefficients being rational constants. This is the content of the Baker–Campbell–Hausdorff (BCH) theorem, whereas the expression $e^X e^Y = e^Z$ is called the Baker–Campbell–Hausdorff formula, although other different labels (e.g., Campbell–Baker–Hausdorff, Baker–Hausdorff, Campbell–Hausdorff) are also used in the literature [14].

Although (6) solves in principle the mathematical problem addressed in this section, it is barely useful from a practical point of view, due to its complexity and the existing redundancies. Notice in particular, that different choices of p_i, q_i, k in (6) may lead to several terms in the same commutator. Thus, for instance, $[X^3Y^1] = [X^1Y^0X^2Y^1] = [X,[X,[X,Y]]]$. An additional source of redundancies arises from the fact that not all the commutators are independent, due to the Jacobi identity [66]:

$$[X_1,[X_2,X_3]] + [X_2,[X_3,X_1]] + [X_3,[X_1,X_2]] = 0, \tag{8}$$

for any three variables X_1, X_2, X_3. From this perspective, it would be certainly preferable to have an explicit expression for Z formulated directly in terms of a basis of the free Lie algebra $\mathscr{L}(X,Y)$ generated by X and Y, or at least a systematic and efficient procedure to generate the coefficients in such an expression. In this way, different combinatorial properties of the series, such as the distribution of the coefficients, etc., could be analyzed in detail.

In addition to the Dynkin form (6) there are other presentations of the BCH series. In particular, the associative presentation (as a linear combination of words in X and Y) is also widely used:

$$Z = X + Y + \sum_{m=2}^{\infty} \sum_{w,|w|=m} g_w\, w. \tag{9}$$

Here g_w are rational coefficients and the inner sum is taken over all words w with length $|w| = m$ (the length of w is just the number of letters it contains). The values of g_w can be computed with a procedure based on a family of recursively computable polynomials due to Goldberg [31].

Although the presentation (9) is commutator-free, a direct application of the Dynkin–Specht–Wever theorem [37] allows one to write it as

$$Z = X + Y + \sum_{m=2}^{\infty} \frac{1}{m} \sum_{w,|w|=m} g_w\, [w], \tag{10}$$

i.e., the individual terms are the same as in (9) except that the word $w = a_1a_2\ldots a_m$ is replaced with the right nested commutator $[w] \equiv [a_1,[a_2,\ldots[a_{m-1},a_m]\ldots]]$ and the coefficient g_w is divided by the word length m [62].

The series Z can also be obtained by stating and solving iteratively differential equations. In particular, for sufficiently small $t \in \mathbb{R}$, if we write $\exp(tX)\exp(tY) = \exp(Z(t))$, then $Z(t)$ is an analytic function around $t = 0$ which verifies [66]

$$\frac{dZ}{dt} = X + Y + \frac{1}{2}[X - Y, Z] + \sum_{p=1}^{\infty} \frac{B_{2p}}{(2p)!} \mathrm{ad}_Z^{2p}(X + Y), \qquad Z(0) = 0. \tag{11}$$

in terms of the adjoint operator (90) and the Bernoulli numbers B_k [1]. By writing $Z(t) = \sum_{n=1}^{\infty} t^m Z_m(X, Y)$, with $Z_1 = X + Y$, one arrives at the following recursion for Z_m:

$$mZ_m = \frac{1}{2}[X - Y, Z_{m-1}] + \sum_{p=1}^{[(m-1)/2]} \frac{B_{2p}}{(2p)!} \left(\mathrm{ad}_Z^{2p}(X + Y)\right)_m, \qquad m \geq 1, \tag{12}$$

where

$$\left(\mathrm{ad}_Z^{2p}(X + Y)\right)_m \equiv \sum_{\substack{k_1+\cdots+k_{2p}=m-1 \\ k_1 \geq 1,\ldots,k_{2p} \geq 1}} [Z_{k_1}, [\cdots [Z_{k_{2p}}, X + Y] \cdots]].$$

Equivalently, if we denote by $\mathscr{L}(X, Y)_m$ ($m \geq 1$) the homogeneous subspace of $\mathscr{L}(X, Y)$ of degree m (the subspace of all nested commutators involving precisely m operators X, Y), then $\left(\mathrm{ad}_Z^{2p}(X + Y)\right)_m$ is nothing but the projection of $\mathrm{ad}_Z^{2p}(X + Y)$ onto $\mathscr{L}(X, Y)_m$.

Other differential equations can be considered instead. For instance, in [8] the function $Z(t)$ in $\exp(Z(t)) = \exp(tX)\exp(Y)$ is shown to verify

$$\frac{dZ}{dt} = \frac{\mathrm{ad}_Z}{e^{\mathrm{ad}_Z} - I}(X) \equiv \sum_{k=0}^{\infty} \frac{B_k}{k!} \mathrm{ad}_Z^k X, \qquad Z(0) = Y. \tag{13}$$

Then, it is possible to get the recurrence

$$Z_1(t) = Xt + Y, \qquad Z_m(t) = \sum_{j=1}^{m-1} \frac{B_j}{j!} \int_0^t (\mathrm{ad}_Z^j X)_m ds \tag{14}$$

or alternatively

$$Z_m(t) = \sum_{j=1}^{m-1} \frac{B_j}{j!} \sum_{\substack{k_1+\cdots+k_j=m-1 \\ k_1 \geq 1,\ldots,k_j \geq 1}} \int_0^t \mathrm{ad}_{Z_{k_1}(s)} \mathrm{ad}_{Z_{k_2}(s)} \cdots \mathrm{ad}_{Z_{k_j}(s)} X \, ds \qquad m \geq 2,$$

whence the BCH series is recovered by taking $t = 1$. Any of these procedures allow one to construct the BCH series up to arbitrary degree in terms of commutators, but, as in the case of the Dynkin presentation, not all of them are independent due to the Jacobi identity (and other identities involving nested commutators of higher degree which are originated by it [53]). Although it is always possible to express the resulting formulas in terms of a basis of $\mathscr{L}(X, Y)$ with the help of a symbolic algebra package, this rewriting process is very expensive both in terms of computational time and memory resources. As a matter of fact, the complexity of the problem grows exponentially with m: the number of terms involved in the series grows as the dimension c_m of the homogeneous subspace $\mathscr{L}(X, Y)_m$ and this number $c_m = \mathscr{O}(2^m/m)$ according to Witt's formula (96).

2.2 An Efficient Algorithm for Generating the Series

In reference [20], an optimized algorithm is presented for expressing the BCH series as

$$Z = \log(\exp(X)\exp(Y)) = \sum_{i\geq 1} z_i\, E_i, \tag{15}$$

where $z_i \in \mathbb{Q}\,(i \geq 1)$ and $\{E_i\ :\ i = 1, 2, 3, \ldots\}$ is a Hall–Viennot basis of $\mathscr{L}(X, Y)$ (see the Appendix). The elements E_i are of the form

$$E_1 = X, \quad E_2 = Y, \quad \text{and} \quad E_i = [E_{i'}, E_{i''}] \quad i \geq 3, \tag{16}$$

for positive integers $i', i'' < i$ $(i = 3, 4, \ldots)$. Each E_i in (16) is a homogeneous Lie polynomial of degree $|i|$, where

$$|1| = |2| = 1, \quad \text{and} \quad |i| = |i'| + |i''| \quad \text{for} \quad i \geq 3. \tag{17}$$

As reviewed in the Appendix, the classical Hall basis and the Lyndon basis are particular examples of Hall–Viennot bases [57, 67].

The algorithm for generating the BCH series is based on the treatment done by Murua [50] relating a certain Lie algebra structure $\mathfrak{g}$ on rooted trees with the description of a free Lie algebra in terms of a Hall–Viennot basis. Essentially, the idea is to construct algorithmically a sequence of labeled rooted trees in a one-to-one correspondence with a Hall–Viennot basis in such a way that each element in the basis of $\mathscr{L}(X, Y)$ can be characterized in terms of a tree in this sequence.

The procedure can be implemented in a computer algebra system (in particular, in *Mathematica*) and gives the BCH series up to a prescribed degree directly in terms of a Hall-Viennot basis of $\mathscr{L}(X, Y)$. This allowed the authors of [20] to provide for the first time the explicit expression of the term of degree 20, Z_{20}, in the series (7). Since a fully detailed treatment of the algorithm can be found in [20], we only summarize here its main features.

The starting point is the set $\mathscr{T}$ of rooted trees with black and white vertices

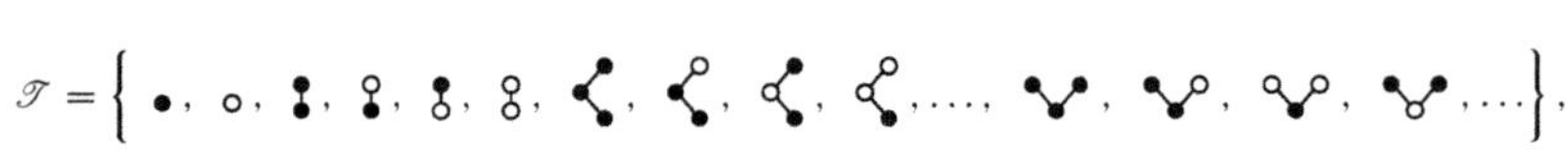

whose elements are referred to as bicoloured rooted trees. Here and in what follows all trees grow up.

Next one considers the vector space $\mathfrak{g}$ of real maps defined on $\mathscr{T}$, $\alpha : \mathscr{T} \to \mathbb{R}$. This set can be endowed with a Lie algebra structure by defining the Lie bracket $[\alpha, \beta] \in \mathfrak{g}$ of two arbitrary maps $\alpha, \beta \in \mathfrak{g}$ as follows. For each $u \in \mathscr{T}$ the action of the new map $[\alpha, \beta]$ is given by

$$[\alpha, \beta](u) = \sum_{j=1}^{|u|-1} \big(\alpha(u_{(j)})\beta(u^{(j)}) - \alpha(u^{(j)})\beta(u_{(j)})\big), \tag{18}$$

where $|u|$ denotes the number vertices of u, and each of the pairs of trees $(u_{(j)}, u^{(j)}) \in \mathscr{T} \times \mathscr{T}$, $j = 1, \ldots, |u| - 1$, is obtained from u by removing one of the $|u| - 1$ edges of the rooted tree u, the root of $u_{(j)}$ being the original root of u. Thus, in particular,

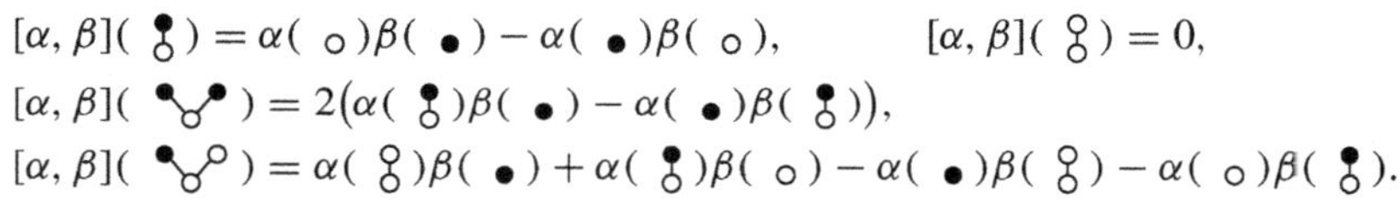

(19)

Consider now the maps $X, Y \in \mathfrak{g}$ defined as

$$X(u) = \begin{cases} 1 \text{ if } u = \bullet \\ 0 \text{ if } u \in \mathscr{T} \backslash \{\bullet\} \end{cases}, \qquad Y(u) = \begin{cases} 1 \text{ if } u = \circ \\ 0 \text{ if } u \in \mathscr{T} \backslash \{\circ\} \end{cases} \tag{20}$$

and the subalgebra of $\mathfrak{g}$ generated by them, which we denote by $\mathscr{L}(X, Y)$. It has been shown that $\mathscr{L}(X, Y)$ is a free Lie algebra over the set $\{X, Y\}$ [50]. Moreover, for each particular Hall–Viennot basis $\{E_i : i = 1, 2, 3, \ldots\}$ of this free Lie algebra $\mathscr{L}(X, Y)$ one can associate a bicoloured rooted tree u_i to each element E_i. For instance, in Table 1 we collect the bicoloured rooted trees u_i associated with the elements E_i of the Hall basis (94), whereas in Table 2 we depict the corresponding to the Lyndon basis (95). Then, for any map $\alpha \in \mathscr{L}(X, Y)$ it is true that

$$\alpha = \sum_{i \geq 1} \frac{\alpha(u_i)}{\sigma(u_i)} E_i, \tag{21}$$

were $\sigma(u_i)$ is a certain positive integer associated to the rooted tree u_i (the number of symmetries of u_i, also called the symmetry number of u_i). Again, the value of $\sigma(u_i)$ up to $i = 5$ is collected in Tables 1 and 2.

Denoting by α_m the projection on the map $\alpha \in \mathscr{L}(X, Y)$ onto the homogeneous subspace $\mathscr{L}(X, Y)_m$, then [50]

Table 1 First elements E_i of the Hall basis (94), their corresponding Hall words w_i and bicoloured rooted trees u_i, the values of $|i|$, i', i'', $\sigma(u_i)$, and the coefficients $z_i = Z(u_i)/\sigma(u_i)$ in the BCH series (15)

i	$\|i\|$	i'	i''	w_i	E_i	u_i	$\sigma(u_i)$	$z_i = \frac{Z(u_i)}{\sigma(u_i)}$
1	1	1	0	x	X	$\bullet$	1	1
2	1	2	0	y	Y	$\circ$	1	1
3	2	2	1	yx	$[Y, X]$		1	$-\frac{1}{2}$
4	3	3	1	yxx	$[[Y, X], X]$		2	$\frac{1}{12}$
5	3	3	2	yxy	$[[Y, X], Y]$		1	$-\frac{1}{12}$
6	4	4	1	$yxxx$	$[[[Y, X], X], X]$	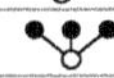	6	0
7	4	4	2	$yxxy$	$[[[Y, X], X], Y]$	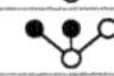	2	$\frac{1}{24}$
8	4	5	2	$yxyy$	$[[[Y, X], Y], Y]$	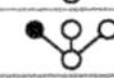	2	0
9	5	6	1	$yxxxx$	$[[[[Y, X], X], X], X]$	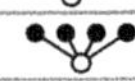	24	$-\frac{1}{720}$
10	5	6	2	$yxxxy$	$[[[[Y, X], X], X], Y]$	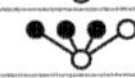	6	$-\frac{1}{180}$
11	5	7	2	$yxxyy$	$[[[[Y, X], X], Y], Y]$	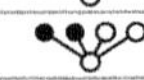	4	$\frac{1}{180}$
12	5	8	2	$yxyyy$	$[[[[Y, X], Y], Y], Y]$	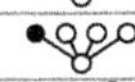	6	$\frac{1}{720}$
13	5	4	3	$yxxyx$	$[[[Y, X], X], [Y, X]]$		2	$-\frac{1}{120}$
14	5	5	3	$yxyyx$	$[[[Y, X], Y], [Y, X]]$	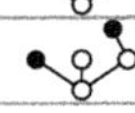	1	$-\frac{1}{360}$

$$\alpha_m(u) = \begin{cases} \alpha(u) & \text{if} \quad |u| = m \\ 0 & \text{otherwise} \end{cases} \tag{22}$$

for each $u \in \mathscr{T}$. A basis of $\mathscr{L}(X, Y)_m$ is given by $\{E_i : |i| = m\}$.

Consider now the Lie algebra of Lie series, i.e., series of the form

$$\alpha = \alpha_1 + \alpha_2 + \alpha_3 + \cdots, \quad \text{where} \quad \alpha_m \in \mathscr{L}(X, Y)_m.$$

A map $\alpha \in \mathfrak{g}$ is then a Lie series if and only if (21) holds. In particular, the BCH series given by (12) (or (14)) is a Lie series if X and Y are defined as in (20), and so it can be characterized by an expression of the form (21). Specifically, starting with (12) one has $Z(\bullet) = Z(\circ) = 1$, and for $m = 2, 3, 4, \ldots$

$$mZ(u) = \frac{1}{2}[X - Y, Z](u) + \sum_{p=1}^{[(m-1)/2]} \frac{B_{2p}}{(2p)!} \left(\mathrm{ad}_Z^{2p}(X + Y)\right)(u) \tag{23}$$

for each $u \in \mathscr{T}$ with $m = |u|$. Evaluating the corresponding brackets $[\alpha, \beta](u)$ according with the prescription (18), one can compute the value of $Z(u)$ for trees with arbitrarily high number of vertices. For the Hall basis considered in Table 1 we have

Table 2 First elements E_i of the Lyndon basis, their corresponding Lyndon words w_i and bicoloured rooted trees u_i, the values $|i|, i'', i', \sigma(u_i)$, and the coefficients $z_i = Z(u_i)/\sigma(u_i)$ in the BCH series (15)

i	$\lvert i\rvert$	i'	i''	w_i	E_i	u_i	$\sigma(u_i)$	$z_i = \frac{Z(u_i)}{\sigma(u_i)}$
1	1	1	0	x	X		1	1
2	1	2	0	y	Y		1	1
3	2	1	2	xy	$[X, Y]$		1	$\frac{1}{2}$
4	3	3	2	xyy	$[[X, Y], Y]$		2	$\frac{1}{12}$
5	3	1	3	xxy	$[X, [X, Y]]$		1	$\frac{1}{12}$
6	4	4	2	$xyyy$	$[[[X, Y], Y], Y]$	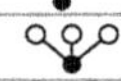	6	0
7	4	1	4	$xxyy$	$[X, [[X, Y], Y]]$	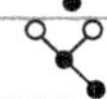	2	$\frac{1}{24}$
8	4	1	5	$xxxy$	$[X, [X, [X, Y]]]$		1	0
9	5	6	2	$xyyyy$	$[[[[X, Y], Y], Y], Y]$	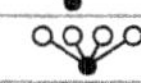	24	$\frac{1}{720}$
10	5	5	3	$xxyxy$	$[[X, [X, Y]], [X, Y]]$	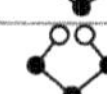	2	$\frac{1}{360}$
11	5	3	4	$xyxyy$	$[[X, Y], [[X, Y], Y]]$	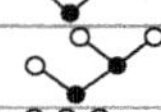	2	$\frac{1}{120}$
12	5	1	6	$xxyyy$	$[X, [[[X, Y], Y], Y]]$	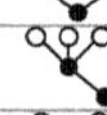	6	$\frac{1}{180}$
13	5	1	7	$xxxyy$	$[X, [X, [[X, Y], Y]]]$	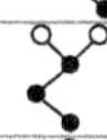	2	$\frac{1}{180}$
14	5	1	8	$xxxxy$	$[X, [X, [X, [X, Y]]]]$		1	$-\frac{1}{720}$

$$Z = \sum_{i\geq 1} z_i E_i = \sum_{i\geq 1} \frac{Z(u_i)}{\sigma(u_i)} E_i$$

$= Z(\bullet)X + Z(\circ)Y + Z($  $)[Y, X] + \frac{Z(}{}$  $\frac{)}{2}[[Y, X], X] + Z($  $)[[Y, X], Y] + \cdots,$

where the first coefficients $Z(u_i)$ are given by [20]

$Z(\bullet) = Z(\circ) = 1, \quad Z($  $) = -\frac{1}{2}, \quad Z($  $) = \frac{1}{6}, \quad Z($ 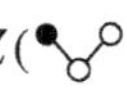 $) = -\frac{1}{12}.$

If one instead works with the Lyndon basis (95) of Table 2, then it results in

$$Z = \sum_{i\geq 1} z_i E_i = \sum_{i\geq 1} \frac{Z(u_i)}{\sigma(u_i)} E_i$$

$= Z(\bullet)X + Z(\circ)Y + Z($  $)[X, Y] + \frac{Z(}{}$  $\frac{)}{2}[[X, Y], Y] + Z($  $)[X, [X, Y]] + \cdots,$

with

$Z($  $) = \frac{1}{2}, \quad Z($ 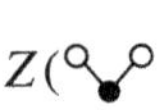 $) = \frac{1}{6}, \quad Z($  $) = \frac{1}{12}.$

This process can be carried out for arbitrarily large values of m in a fully automatic way once the bicoloured rooted trees corresponding to each Hall–Viennot basis in the free Lie algebra have been generated up to the prescribed degree. In this respect, the computational efficiency depends on the particular basis one chooses for $\mathscr{L}(X, Y)$ and the representation used for the BCH series. For instance, in the Hall basis of Table 1 one needs to generate 724018 bicoloured rooted trees up to $m = 20$, whereas in the Lyndon basis of Table 2 this number raises up to 1952325. In consequence, more memory and computation time is required in the later case. Nevertheless, in the Lyndon basis the number of non-vanishing coefficients z_i is greatly reduced in comparison with the Hall basis: 76760 versus 109697 (out of 111013 elements E_i) up to degree $m = 20$. In [20] an explanation can be found for this phenomenon. On the other hand, working with the Lie series defined by (14) is slightly more efficient in practice.

2.3 The BCH Series of a Given Degree with Respect to *Y*

The series (6) can in principle be reordered with respect to the increasing number of times the operator Y appears in the expression. We can then write Z as

$$Z = \sum_{n=0}^{\infty} Z_n^Y,$$

where Z_n^Y is the part of Z which is homogeneous of degree n with respect to Y, i.e.,

$$Z_n^Y = \sum_{k=1}^{\infty} \frac{(-1)^{k-1}}{k} \sum_{p_i, q_i} \frac{[X^{p_1} Y^{q_1} \dots X^{p_k} Y^{q_k}]}{(\sum_{i=1}^{k}(p_i + q_i))\, p_1! q_1! \dots p_k! q_k!},$$

where now $q_1 + \cdots + q_k = n$ in the inner sum. In particular, $Z_0^Y = X$, whereas the expression of Z_1^Y can be found in e.g. [14, 57]. A recursion for the homogeneous component Z_n^Y can be obtained as follows.

Let us introduce a parameter $\varepsilon > 0$ and consider the series

$$Z(\varepsilon) = Z_0^Y + \varepsilon Z_1^Y + \varepsilon^2 Z_2^Y + \cdots \tag{24}$$

in $\exp(Z(\varepsilon)) = \exp(X)\exp(\varepsilon Y)$. Then $Z(\varepsilon)$ verifies the initial value problem [8]

$$\frac{dZ(\varepsilon)}{d\varepsilon} = \sum_{j=0}^{\infty} (-1)^j \frac{B_j}{j!} \mathrm{ad}_Z^j(Y), \qquad Z(0) = X. \tag{25}$$

Notice the close similarity of this equation with (13). It is clear that

$$\frac{dZ(\varepsilon)}{d\varepsilon} = \sum_{j=0}^{n} (j+1)\varepsilon^j Z^Y_{j+1} + \mathcal{O}(\varepsilon^{n+1})$$

and

$$\mathrm{ad}_Z = \mathrm{ad}_{Z^Y_0} + \varepsilon\,\mathrm{ad}_{Z^Y_1} + \varepsilon^2\mathrm{ad}_{Z^Y_2} + \cdots + \varepsilon^n\,\mathrm{ad}_{Z^Y_n} + \mathcal{O}(\varepsilon^{n+1}).$$

In consequence,

$$\begin{aligned}\mathrm{ad}^2_Z &= \mathrm{ad}_{Z^Y_0}\mathrm{ad}_{Z^Y_0} + \varepsilon(\mathrm{ad}_{Z^Y_0}\mathrm{ad}_{Z^Y_1} + \mathrm{ad}_{Z^Y_1}\mathrm{ad}_{Z^Y_0}) + \cdots \\ &= \sum_{\ell=0}^{n} \varepsilon^\ell \sum_{\substack{k_1+k_2=\ell \\ k_1\geq 0, k_2\geq 0}} \mathrm{ad}_{Z^Y_{k_1}}\,\mathrm{ad}_{Z^Y_{k_2}} + \mathcal{O}(\varepsilon^{n+1})\end{aligned}$$

and in general

$$\mathrm{ad}^j_Z = \sum_{\ell=0}^{n} \varepsilon^\ell \sum_{\substack{k_1+\cdots+k_j=\ell \\ k_1\geq 0,\ldots,k_j\geq 0}} \mathrm{ad}_{Z^Y_{k_1}}\,\mathrm{ad}_{Z^Y_{k_2}} \cdots \mathrm{ad}_{Z^Y_{k_j}} + \mathcal{O}(\varepsilon^{n+1}).$$

In this way

$$\begin{aligned}&\sum_{j=0}^{\infty}(-1)^j \frac{B_j}{j!}\mathrm{ad}^j_Z(Y) = \\ &Y + \sum_{j=1}^{\infty}(-1)^j \frac{B_j}{j!}\left(\mathrm{ad}^j_{Z^Y_0} Y + \sum_{\ell=1}^{n}\varepsilon^\ell \sum_{\substack{k_1+\cdots+k_j=\ell \\ k_1\geq 0,\ldots,k_j\geq 0}} \mathrm{ad}_{Z^Y_{k_1}}\,\mathrm{ad}_{Z^Y_{k_2}}\cdots\mathrm{ad}_{Z^Y_{k_j}} Y + \mathcal{O}(\varepsilon^{n+1})\right) = \\ &Y + \sum_{j=1}^{\infty}(-1)^j \frac{B_j}{j!}\mathrm{ad}^j_{Z^Y_0} Y + \sum_{j=1}^{\infty}(-1)^j \frac{B_j}{j!}\sum_{\ell=1}^{n}\varepsilon^\ell \sum_{\substack{k_1+\cdots+k_j=\ell \\ k_1\geq 0,\ldots,k_j\geq 0}} \mathrm{ad}_{Z^Y_{k_1}}\,\mathrm{ad}_{Z^Y_{k_2}}\cdots\mathrm{ad}_{Z^Y_{k_j}} Y + \mathcal{O}(\varepsilon^{n+1})\end{aligned}$$

Substituting these expressions in (25) and identifying the coefficients of ε^ℓ on both sides we arrive at $Z^Y_0 = X$,

$$Z^Y_1 = \sum_{k=0}^{\infty} \frac{(-1)^k}{k!} B_k\,\mathrm{ad}^k_X(Y) \equiv \frac{\mathrm{ad}_X}{I - \mathrm{e}^{-\mathrm{ad}_X}}(Y) \tag{26}$$

and, for $n \geq 1$,

$$(n+1)Z^Y_{n+1} = \sum_{j=1}^{\infty}(-1)^j \frac{B_j}{j!} \sum_{\substack{k_1+\cdots+k_j=n \\ k_1\geq 0,\ldots,k_j\geq 0}} \mathrm{ad}_{Z^Y_{k_1}}\,\mathrm{ad}_{Z^Y_{k_2}}\cdots\mathrm{ad}_{Z^Y_{k_j}} Y. \tag{27}$$

This recursion can be written in a more compact form by introducing the operators $S_n^{(j)}$, $j = 0, 1, 2, \ldots$, as

$$\begin{aligned} S_1^{(j)} &= \mathrm{ad}^j_{Z_0^Y} Y \\ S_n^{(0)} &= 0, \qquad S_n^{(j)} = \sum_{\ell=0}^{n-1} \mathrm{ad}_{Z_\ell^Y} S_{n-\ell}^{(j-1)}, \qquad n \geq 2. \end{aligned} \tag{28}$$

Then we have

$$Z_n^Y = \frac{1}{n} \sum_{j=1}^{\infty} (-1)^j \frac{B_j}{j!} S_n^{(j)}. \tag{29}$$

By working out this recurrence it is possible in principle to obtain closed expressions for each homogeneous term Z_n^Y, although their structure is increasingly complex for $n \geq 2$. In particular, one has

$$S_2^{(j)} = \mathrm{ad}_X S_2^{(j-1)} + \mathrm{ad}_{Z_1^Y} S_1^{(j-1)} = \sum_{p=0}^{j-1} \mathrm{ad}_X^p \, \mathrm{ad}_{Z_1^Y} \, \mathrm{ad}_X^{j-p-1} Y, \tag{30}$$

and the operator $\mathrm{ad}_{Z_1^Y}$ in (30) can be evaluated as follows.

First, the Jacobi identity (8) for any three operators A, B, C can be restated in term of the adjoint operator as

$$\mathrm{ad}_{[A,B]} C = [\mathrm{ad}_A, \mathrm{ad}_B] C$$

or $\mathrm{ad}_{[A,B]} = [\mathrm{ad}_A, \mathrm{ad}_B]$. In general, it can be shown by induction that

$$\mathrm{ad}_{[A,[A,\ldots[A,B]]]} \equiv \mathrm{ad}_{\mathrm{ad}_A^n B} = [\mathrm{ad}_A, [\mathrm{ad}_A, \ldots [\mathrm{ad}_A, \mathrm{ad}_B]]].$$

On the other hand, a simple calculation leads to

$$\mathrm{ad}_A^n B = \sum_{p=0}^{n} (-1)^p \binom{n}{p} A^{n-p} B A^p,$$

so that

$$\mathrm{ad}_{\mathrm{ad}_A^n B} = [\mathrm{ad}_A, [\mathrm{ad}_A, \ldots [\mathrm{ad}_A, \mathrm{ad}_B]]] = \sum_{p=0}^{n} (-1)^p \binom{n}{p} \mathrm{ad}_A^{n-p} \, \mathrm{ad}_B \, \mathrm{ad}_A^p.$$

Therefore, from (26),

$$\mathrm{ad}_{Z_1^Y} = \sum_{k=0}^{\infty} (-1)^k \frac{B_k}{k!} \mathrm{ad}_{\mathrm{ad}_X^k Y} = \sum_{k=0}^{\infty} (-1)^k \frac{B_k}{k!} \sum_{j=0}^{k} (-1)^j \binom{k}{j} \mathrm{ad}_X^{k-j} \mathrm{ad}_Y \mathrm{ad}_X^j$$

and this expression, once inserted into $S_2^{(j)}$, Eq. (30), gives us Z_2^Y explicitly. This procedure was first applied with identical goal in [41].

Of course, these results have a dual version, i.e, it is possible to get analogous expressions for the homogeneous terms Z_n^X of degree n with respect to X [14, 57].

2.4 The Symmetric BCH Formula

In some applications it is required to compute the operator W defined by

$$\exp(\frac{1}{2}X)\exp(Y)\exp(\frac{1}{2}X) = \exp(W). \tag{31}$$

This is the so-called symmetric BCH formula. Two applications of the usual BCH formula lead to the expression of W in a given basis of $\mathscr{L}(X,Y)$. More efficient procedures exist, however, that allow one to construct explicitly the series $\sum_{n\geq 1} W_n$ defining W in terms of independent commutators involving X and Y up to an arbitrarily high degree. These are based on deriving a recurrence for the successive terms in the Lie series W through a differential equation and expressing it as

$$W = \sum_{i\geq 1} w_i E_i \tag{32}$$

as in the previous case.

Introducing a parameter t in (31),

$$W(t) = \log(\mathrm{e}^{tX/2}\,\mathrm{e}^{Y}\,\mathrm{e}^{tX/2}), \tag{33}$$

it can be shown that $W(t)$ satisfies the initial value problem

$$\frac{dW}{dt} = X + \sum_{n=2}^{\infty} \frac{B_n}{n!}\mathrm{ad}_W^n X, \qquad W(0) = Y. \tag{34}$$

Inserting here the series $W(t) = \sum_{k=0}^{\infty} W_k(t)$ we arrive at

$$\begin{aligned} W_1(t) &= Xt + Y \\ W_2(t) &= 0 \\ W_\ell(t) &= \sum_{j=2}^{\ell-1} \frac{B_j}{j!} \int_0^\tau (\mathrm{ad}_W^j X)_\ell \, ds, \qquad \ell \geq 3. \end{aligned} \tag{35}$$

The Lie series W is recovered by taking $t = 1$. In general, $W_{2m} = 0$ for $m \geq 1$, whereas terms W_{2m+1} up to W_{19} in both Hall and Lyndon bases have been constructed in [20]. Specifically, for the first terms in the Lyndon basis one has

$$
\begin{aligned}
W_1 &= X + Y \\
W_3 &= \frac{1}{12}[[X, Y], Y] - \frac{1}{24}[X, [X, Y]] \\
W_5 &= -\frac{1}{720}[[[[X, Y], Y], Y], Y] + \frac{1}{360}[[X, [X, Y]], [X, Y]] + \frac{1}{120}[[X, Y], [[X, Y], Y]] \\
&\quad + \frac{1}{180}[X, [[[X, Y], Y], Y]] - \frac{7}{1440}[X, [X, [[X, Y], Y]]] + \frac{7}{5760}[X, [X, [X, [X, Y]]]] \\
W_7 &= \frac{1}{30240}[[[[[[X, Y], Y], Y], Y], Y], Y] - \frac{1}{5040}[[[X, [X, Y]], [X, Y]], [X, Y]] \\
&\quad - \frac{1}{1512}[[X, [[[X, Y], Y], Y]], [X, Y]] + \frac{1}{10080}[[X, [[X, Y], Y]], [[X, Y], Y]] \\
&\quad + \frac{1}{10080}[[X, [X, [X, Y]]], [X, [X, Y]]] - \frac{1}{5040}[[[X, Y], Y], [[[X, Y], Y], Y]] \\
&\quad + \frac{1}{2016}[[X, [X, Y]], [X, [[X, Y], Y]]] - \frac{1}{2016}[[X, Y], [[[[X, Y], Y], Y], Y]] \\
&\quad + \frac{1}{1260}[[X, Y], [[X, Y], [[X, Y], Y]]] - \frac{1}{5040}[X, [[[[X, Y], Y], Y], Y], Y]] \\
&\quad + \frac{1}{1260}[X, [[X, [[X, Y], Y]], [X, Y]]] + \frac{13}{15120}[X, [[X, Y], [[[X, Y], Y], Y]]] \\
&\quad + \frac{53}{120960}[X, [X, [[[[X, Y], Y], Y], Y]]] - \frac{1}{4032}[X, [X, [[X, [X, Y]], [X, Y]]]] \\
&\quad - \frac{1}{2240}[X, [X, [[X, Y], [[X, Y], Y]]]] - \frac{13}{30240}[X, [X, [X, [[[X, Y], Y], Y]]]] \\
&\quad + \frac{31}{161280}[X, [X, [X, [X, [[X, Y], Y]]]]] - \frac{31}{967680}[X, [X, [X, [X, [X, [X, Y]]]]]].
\end{aligned}
$$

2.5 About the Convergence

The previous results are globally valid in the free Lie algebra $\mathscr{L}(X, Y)$, whereas if X and Y are elements in a normed Lie algebra then the resulting series are not guaranteed to converge except in a neighborhood of zero. We next review some results on the convergence domain of the different presentations and refer the reader to [8, 14, 20] and references therein for a more detailed treatment.

Assume $X, Y \in \mathfrak{g}$, where $\mathfrak{g}$ is a complete normed Lie algebra with a norm such that $\|XY\| \le \|X\| \, \|Y\|$ for all X, Y, so that

$$
\|[X, Y]\| \le 2\|X\| \, \|Y\|. \tag{36}
$$

Then, it is shown in [63] that the series (9) is absolutely convergent if $\|X\| < 1$ and $\|Y\| < 1$, whereas the domain of absolute convergence of the Dynkin presentation (6) contains the open set (X, Y) such that $\|X\| + \|Y\| < \frac{1}{2}\log 2$ [15, 24, 25].

Much enlarged domains of convergence can be ensured by analyzing the differential equations (11) and (13). Thus, in [52] it is shown that the series (7) with the terms computed by the recursion (12) converges absolutely if $\|X\| < 0.54343435$, $\|Y\| < 0.54343435$, whereas in [8], an analysis of the recurrence (14) leads to the convergence domain $D_1 \cup D_2$ of $\mathfrak{g} \times \mathfrak{g}$, where

$$D_1 = \left\{ (X, Y) \ : \ \|X\| < \frac{1}{2} \int_{2\|Y\|}^{2\pi} \frac{1}{g(x)} dx \right\}$$

$$D_2 = \left\{ (X, Y) \ : \ \|Y\| < \frac{1}{2} \int_{2\|X\|}^{2\pi} \frac{1}{g(x)} dx \right\} \tag{37}$$

and $g(x) = 2 + \frac{x}{2}(1 - \cot \frac{x}{2})$. It is stated in [46] that an analogous result was obtained by Mérigot.

Finally, if $Z = \log(\mathrm{e}^X \mathrm{e}^Y)$ is expressed as in (24) with $\varepsilon = 1$, i.e., as the sum of homogeneous components of degree n in Y, then, by analyzing (26) and (27), it is possible to show that the corresponding series converges absolutely for $\|X\| < 0.6178$, $\|Y\| < 0.6178$ [23].

3 The Zassenhaus Formula

3.1 General Considerations

In reference [43], Magnus cites an unpublished reference by Zassenhaus, reporting that there exists a formula which may be called the dual of the BCH formula. The result can be stated as follows, and constitutes in fact the **Problem 2** posed in the Introduction.

Theorem 1 (Zassenhaus formula) *The exponential* e^{X+Y}, *when* $X, Y \in \mathscr{L}(X, Y)$, *can be uniquely decomposed as*

$$\mathrm{e}^{X+Y} = \mathrm{e}^X \, \mathrm{e}^Y \prod_{n=2}^{\infty} \mathrm{e}^{C_n(X,Y)} = \mathrm{e}^X \, \mathrm{e}^Y \, \mathrm{e}^{C_2(X,Y)} \, \mathrm{e}^{C_3(X,Y)} \cdots \mathrm{e}^{C_n(X,Y)} \cdots, \tag{38}$$

where $C_n(X, Y) \in \mathscr{L}(X, Y)$ *is a homogeneous Lie polynomial in* X *and* Y *of degree* n.

That such a result does exist can be seen as a consequence of the BCH formula. In fact, one finds that $\mathrm{e}^{-X}\mathrm{e}^{X+Y} = \mathrm{e}^{Y+D}$, where D involves Lie polynomials of degree >1. Then $\mathrm{e}^{-Y}\mathrm{e}^{Y+D} = \mathrm{e}^{C_2+\tilde{D}}$, where $\tilde{D}$ involves Lie polynomials of degree >2, etc. The general result is then achieved by induction.

It is possible to obtain the first terms of the formula (38) just by comparing with the BCH formula. Specifically,

$$C_2(X,Y) = -\frac{1}{2}[X,Y], \qquad C_3(X,Y) = \frac{1}{3}[Y,[X,Y]] + \frac{1}{6}[X,[X,Y]],$$

but this process is increasingly difficult for higher values of n.

The Zassenhaus formula has found application in several fields, ranging from q-analysis in quantum groups [55] to the numerical analysis of the Schrödinger equation in the semiclassical regime [5], the treatment of hypoelliptic differential equations [35] and splitting methods [29, 30]. For this reason, several systematic computations of the terms C_n for $n > 3$ have been tried, starting with the work of Wilcox [71], where a recursive procedure is presented that has been subsequently used to get explicit expressions up to C_6 in terms of nested commutators [55]. See [21] and references therein for some historical background.

As with the BCH formula, the Zassenhaus terms C_n can be expressed either as a linear combination of elements of the homogeneous subspace $\mathcal{L}(X,Y)_n$ or as a linear combination of words in X and Y,

$$C_n = \sum_{w,|w|=n} g_w\, w, \tag{39}$$

where g_w is a rational coefficient and the sum is taken over all words w with length $|w| = n$ in the symbols X and Y. In the later case expressions of C_n up to $n = 15$ have been obtained in [70]. As we know, by applying the Dynkin–Specht–Wever theorem [37], C_n can also be written in terms of Lie elements of degree n, but the resulting expression is far from optimal.

For this reason, in [21] a recursive algorithm is designed that allows one to express the Zassenhaus terms C_n directly as a linear combination of independent elements of the homogeneous subspace $\mathcal{L}(X,Y)_n$. In other words, the procedure gives C_n up to a prescribed degree directly in terms of the minimum number of independent commutators involving n operators X and Y. In this way, no rewriting process in a Hall–Viennot basis of $\mathcal{L}(X,Y)$ is necessary, thus saving considerable computing time and memory resources. The algorithm can be easily implemented in a symbolic algebra system without any special requirement, beyond the linearity property of the commutator.

The following observation is worth remarking. Sometimes one finds the "left-oriented" Zassenhaus formula

$$e^{X+Y} = \cdots\, e^{\hat{C}_4(X,Y)}\, e^{\hat{C}_3(X,Y)}\, e^{\hat{C}_2(X,Y)}\, e^{Y}\, e^{X} \tag{40}$$

instead of (38). Since the respective exponents $\hat{C}_i$ and C_i are related through

$$\hat{C}_i(X,Y) = (-1)^{i+1} C_i(X,Y), \qquad i \geq 2,$$

any algorithm to generate the terms C_i allows one to get also the corresponding $\hat{C}_i$.

3.2 An Efficient Algorithm to Generate the Terms C_n

As usual, a parameter $\lambda > 0$ is introduced in (38) multiplying each operator X and Y,

$$e^{\lambda(X+Y)} = e^{\lambda X}\, e^{\lambda Y}\, e^{\lambda^2 C_2}\, e^{\lambda^3 C_3}\, e^{\lambda^4 C_4} \cdots \tag{41}$$

so that the original Zassenhaus formula is recovered when $\lambda = 1$. One considers then the products

$$\begin{aligned} R_1(\lambda) &= e^{-\lambda Y}\, e^{-\lambda X}\, e^{\lambda(X+Y)}, \\ R_n(\lambda) &= e^{-\lambda^n C_n} \cdots e^{-\lambda^2 C_2}\, e^{-\lambda Y}\, e^{-\lambda X}\, e^{\lambda(X+Y)} = e^{-\lambda^n C_n}\, R_{n-1}(\lambda), \qquad n \geq 2. \end{aligned} \tag{42}$$

It is clear from (41) that

$$R_n(\lambda) = e^{\lambda^{n+1} C_{n+1}}\, e^{\lambda^{n+2} C_{n+2}} \cdots. \tag{43}$$

Finally, we introduce

$$F_n(\lambda) \equiv \left(\frac{d}{d\lambda} R_n(\lambda)\right) R_n(\lambda)^{-1}, \qquad n \geq 1. \tag{44}$$

When $n = 1$, and taking into account the expression of $R_1(\lambda)$ given in (42), we get

$$\begin{aligned} F_1(\lambda) &= -Y - e^{-\lambda Y} X e^{\lambda Y} + e^{-\lambda Y} e^{-\lambda X}(X+Y) e^{\lambda X} e^{\lambda Y} \\ &= -Y - e^{-\lambda \mathrm{ad}_Y} X + e^{-\lambda \mathrm{ad}_Y} e^{-\lambda \mathrm{ad}_X}(X+Y) \\ &= e^{-\lambda \mathrm{ad}_Y}(e^{-\lambda \mathrm{ad}_X} - I)Y, \end{aligned}$$

that is,

$$F_1(\lambda) = \sum_{i=0}^{\infty} \sum_{j=1}^{\infty} \frac{(-\lambda)^{i+j}}{i!\,j!} \mathrm{ad}_Y^i \mathrm{ad}_X^j Y \tag{45}$$

or equivalently

$$F_1(\lambda) = \sum_{k=1}^{\infty} f_{1,k}\, \lambda^k, \qquad \text{with} \qquad f_{1,k} = \sum_{j=1}^{k} \frac{(-1)^k}{j!(k-j)!} \mathrm{ad}_Y^{k-j} \mathrm{ad}_X^j Y. \tag{46}$$

Here we have used the well known formula

$$e^A B e^{-A} = e^{\mathrm{ad}_A} B = \sum_{n \geq 0} \frac{1}{n!} \mathrm{ad}_A^n B.$$

A similar expression can be obtained for $F_n(\lambda)$, $n \geq 2$, by considering the expression of $R_n(\lambda)$ given in (42) and the relation (43). On the one hand, from (42) we get

$$\begin{aligned} F_n(\lambda) &= -n\, C_n\, \lambda^{n-1} + \mathrm{e}^{-\lambda^n C_n} \left(\frac{d}{d\lambda} R_{n-1}(\lambda) \right) R_{n-1}(\lambda)^{-1}\, \mathrm{e}^{\lambda^n C_n} \\ &= -n\, C_n\, \lambda^{n-1} + \mathrm{e}^{-\lambda^n C_n}\, F_{n-1}(\lambda)\, \mathrm{e}^{\lambda^n C_n} = -n\, C_n\, \lambda^{n-1} + \mathrm{e}^{-\lambda^n \mathrm{ad}_{C_n}}\, F_{n-1}(\lambda) \\ &= \mathrm{e}^{-\lambda^n \mathrm{ad}_{C_n}} (F_{n-1}(\lambda) - n\, C_n\, \lambda^{n-1}). \end{aligned} \tag{47}$$

Working out this recursion it is possible to write ($n \geq 2$)

$$F_n(\lambda) = \sum_{k=n}^{\infty} f_{n,k}\, \lambda^k, \qquad \text{with} \qquad f_{n,k} = \sum_{j=0}^{[k/n]-1} \frac{(-1)^j}{j!} \mathrm{ad}_{C_n}^j\, f_{n-1,k-nj}, \qquad k \geq n, \tag{48}$$

where $[k/n]$ denotes the integer part of k/n.

On the other hand, differentiating (43) with respect to λ and taking into account (44) we get

$$\begin{aligned} F_n(\lambda) &= (n+1)\, C_{n+1}\, \lambda^n + \sum_{j=n+2}^{\infty} j\, \lambda^{j-1}\, \mathrm{e}^{\lambda^{n+1} \mathrm{ad}_{C_{n+1}}} \cdots \mathrm{e}^{\lambda^{j-1} \mathrm{ad}_{C_{j-1}}}\, C_j \\ &= (n+1) C_{n+1} \lambda^n + (n+2) C_{n+2} \lambda^{n+1} + \cdots \\ &\quad + (2n+2) C_{2n+2} \lambda^{2n+1} + \lambda^{2n+2} [C_{n+1}, C_{n+2}] + \cdots \\ &= \sum_{j=n+1}^{2n+2} j\, C_j\, \lambda^{j-1} + \lambda^{2n+2} H_n(\lambda), \qquad n \geq 1, \end{aligned} \tag{49}$$

where $H_n(\lambda)$ involves commutators of C_j, $j \geq n+1$.

Notice that the terms $C_2, C_3, \ldots$ of the Zassenhaus formula can be then directly obtained by comparing (49) with the series expansions (46) and (48) for $F_n(\lambda)$, $n \geq 1$. Specifically, for the first terms we have

$$\begin{aligned} F_1(\lambda) &= f_{1,1}\lambda + f_{1,2}\lambda^2 + f_{1,3}\lambda^3 + \cdots = 2C_2\lambda + 3C_3\lambda^2 + 4C_4\lambda^3 + H_4(\lambda)\lambda^4 \\ F_2(\lambda) &= f_{2,2}\lambda^2 + f_{2,3}\lambda^3 + f_{2,4}\lambda^4 + \cdots = 3C_3\lambda^2 + 4C_4\lambda^3 + 5C_4\lambda^4 + \cdots \\ F_3(\lambda) &= f_{3,3}\lambda^3 + f_{3,4}\lambda^4 + \cdots = 4C_4\lambda^3 + 5C_5\lambda^4 + \cdots, \end{aligned}$$

whence

$$\begin{aligned} 2C_2 &= f_{1,1}, \\ 3C_3 &= f_{2,2} = f_{1,2}, \\ 4C_4 &= f_{3,3} = f_{2,3} = f_{1,3}, \\ 5C_5 &= f_{4,4} = f_{3,4} = f_{2,4}, \end{aligned}$$

and so, proceeding by induction, we finally arrive at the following recursive algorithm:

$$\begin{aligned}&\text{Define } f_{1,k} \text{ by eq. (46)}\\ &C_2 = (1/2)\, f_{1,1},\\ &\text{Define } f_{n,k} \quad n \geq 2,\ k \geq n \ \text{ by eq. (48)}\\ &C_n = (1/n)\, f_{[(n-1)/2],n-1} \quad n \geq 3.\end{aligned} \tag{50}$$

One of the remarkable features of this procedure is that the generated exponents C_n are expressed only in terms of linearly independent elements in the subspace $\mathscr{L}(X, Y)_n$. This can be shown by repeatedly applying the Lazard elimination principle [21]. As a result, the implementation in a symbolic algebra package is particularly easy, since one does not need to use the Jacobi identity and/or the antisymmetry property of the commutator. Moreover, the computation times and especially the memory requirements are much smaller than other previous procedures (see [21] for more details). For the sake of illustration, we next collect a *Mathematica* code of the preceding algorithm.

```
Cmt[a_, a_]:= 0;
Cmt[a___, 0, b___]:= 0;
Cmt[a___, c_ + d_, b___]:= Cmt[a, c, b] + Cmt[a, d, b];
Cmt[a___, n_ c_Cmt, b___]:= n Cmt[a, c, b];
Cmt[a___, n_ X, b___]:= n Cmt[a, X, b];
Cmt[a___, n_ Y, b___]:= n Cmt[a, Y, b];
Cmt /: Format[Cmt[a_, b_]]:=
   SequenceForm["[", a, ",", b, "]"];

ad[a_, 0, b_]:= b;
ad[a_, j_Integer, b_]:= Cmt[a, ad[a, j-1, b]];
ff[1, k_]:= ff[1, k] =
   Sum[((-1)^k/(j! (k-j)!)) ad[Y, k-j, ad[X, j, Y]],
      {j, 1, k}];
cc[2] = (1/2) ff[1, 1];
ff[p_, k_]:= ff[p, k] =
   Sum[((-1)^j/j!) ad[cc[p], j, ff[p-1, k - p j]],
      {j, 0, IntegerPart[k/p] - 1}];
cc[p_Integer]:= cc[p] =
   Expand[(1/p) ff[IntegerPart[(p-1)/2], p-1]];
```

The first six lines of the code define the commutator. It has attached just the linearity property (there is no need to attach to it the antisymmetry property and the Jacobi identity). The seventh line gives the correct format for output. Next, the symbol `ad` represents the adjoint operator and its powers $\mathrm{ad}_a^j b$, whereas `ff[1,k]`, `ff[p,k]` correspond to expressions (46) and (48), respectively. Finally `cc[p]` provides the explicit expression of C_p. In particular, we get as output

$$C_4 = -\frac{1}{24}[X,[X,[X,Y]]] - \frac{1}{8}[Y,[X,[X,Y]]] - \frac{1}{8}[Y,[Y,[X,Y]]]$$
$$C_5 = \frac{1}{120}[X,[X,[X,[X,Y]]]] + \frac{1}{30}[Y,[X,[X,[X,Y]]]] + \frac{1}{20}[Y,[Y,[X,[X,Y]]]]$$
$$+ \frac{1}{30}[Y,[Y,[Y,[X,Y]]]] + \frac{1}{20}[[X,Y],[X,[X,Y]]] + \frac{1}{10}[[X,Y],[Y,[X,Y]]].$$

We stress again that, by construction, all the commutators appearing in C_n are independent.

3.3 About the Convergence

Whereas the Zassenhaus formula is well defined in the free Lie algebra $\mathscr{L}(X,Y)$, if X and Y are elements of a complete normed Lie algebra, it has only a finite radius of convergence. This issue has also been considered in the literature, typically obtaining sufficient conditions for convergence of the form $\|X\| + \|Y\| < r$ for a given $r > 0$. In other words, if X, Y are such that $\|X\| + \|Y\| < r$, then

$$\lim_{n\to\infty} \mathrm{e}^X\,\mathrm{e}^Y\,\mathrm{e}^{C_2}\cdots\mathrm{e}^{C_n} = \mathrm{e}^{X+Y}. \tag{51}$$

Thus, the value $r = \log 2 - \frac{1}{2} \approx 0.1931$ was given in [61] and $r = 0.59670569\ldots$ in [7].

It turns out that the recursion of the previous section also allows one to improve the domain of convergence, as shown in [21]. By bounding appropriately the terms $F_n(\lambda)$ and also the C_n, i.e, by showing that

$$\|F_n(\lambda)\| \le f_n(\lambda), \qquad \|C_n\| \le \delta_n$$

and analyzing (numerically) the convergence of the series $\sum_{n=2}^{\infty} \delta_n$, it is possible to get a new convergence domain including, in particular, the region $\|X\| + \|Y\| < 1.054$, and extending to the points $(\|X\|, 0)$ and $(0, \|Y\|)$ with arbitrarily large value of $\|X\|$ or $\|Y\|$.

3.4 A Generalization

In certain physical problems one has to deal with the exponential of the infinite series

$$S(\lambda) = \sum_{n=1}^{\infty} \lambda^n A_n = \lambda A_1 + \lambda^2 A_2 + \cdots, \tag{52}$$

where A_i are generic non commuting indeterminates and $\lambda > 0$. The problem consists then in factorizing this exponential as

$$\mathrm{e}^{S(\lambda)} = \mathrm{e}^{\lambda C_1}\,\mathrm{e}^{\lambda^2 C_2}\cdots\,\mathrm{e}^{\lambda^n C_n}\cdots, \tag{53}$$

i.e., in obtaining a Zassenhaus-like formula adapted to this setting. It turns out that the algorithm developed in Sect. 3.2 can also be applied here with only minor changes.

As usual, we start by differentiating both sides of Eq. (53) and multiplying the result by $\mathrm{e}^{-S(\lambda)}$. From the left hand side we have

$$\left(\frac{d}{d\lambda}\mathrm{e}^{S(\lambda)}\right)\mathrm{e}^{-S(\lambda)} = \sum_{k=0}^{\infty}\frac{1}{(k+1)!}\mathrm{ad}^k_{S(\lambda)}S'(\lambda), \tag{54}$$

where $S'(\lambda) = \sum_{i=1}^{\infty} i\,\lambda^{i-1}A_i$. From the right hand side,

$$\frac{d}{d\lambda}\left(\mathrm{e}^{\lambda C_1}\mathrm{e}^{\lambda^2 C_2}\cdots\mathrm{e}^{\lambda^n C_n}\cdots\right)\mathrm{e}^{-S(\lambda)} = C_1 + \sum_{j=2}^{\infty} j\lambda^{j-1}\mathrm{e}^{\mathrm{ad}_{\lambda C_1}}\cdots\mathrm{e}^{\mathrm{ad}_{\lambda^{j-1}C_{j-1}}}C_j. \tag{55}$$

Next we express (54) as a power series in λ so that comparison with (55) gives us recursively the expression of each term C_n. Specifically, if we denote $S_k = \mathrm{ad}^k_S S'$, then

$$S_k = \sum_{j=k+1}^{\infty}\lambda^j S_{k,j}, \qquad \text{with} \qquad S_{k,j} = \sum_{\ell=1}^{j-k}\mathrm{ad}_{A_\ell}S_{k-1,j-\ell}, \quad k \geq 2,$$

whereas

$$S_{1,j} = \sum_{k=1}^{[j/2]}(j-2k+1)\mathrm{ad}_{A_k}A_{j+k-1}.$$

In this way,

$$\sum_{k=0}^{\infty}\frac{1}{(k+1)!}\mathrm{ad}^k_S S' = A_1 + 2\lambda A_2 + \sum_{j=2}^{\infty}\lambda^j\left((j+1)A_{j+1} + \sum_{\ell=1}^{j-1}\frac{1}{(\ell+1)!}S_{\ell,j}\right). \tag{56}$$

Now, comparing (55) with (56) it is clear that $C_1 = A_1$. Introduce now the functions

$$h_1 = 2A_2$$

$$h_n = (n+1)A_{n+1} + \sum_{\ell=1}^{n-1}\frac{1}{(\ell+1)!}S_{\ell,n}, \quad n \geq 2,$$

so that

$$\sum_{k=0}^{\infty} \frac{1}{(k+1)!} \mathrm{ad}_{S(\lambda)}^{k} S'(\lambda) - A_1 = \sum_{n=1}^{\infty} \lambda^n h_n$$

and

$$F_1 \equiv \mathrm{e}^{-\mathrm{ad}_{\lambda C_1}} \sum_{n=1}^{\infty} \lambda^n h_n.$$

Then

$$F_1 = \sum_{\ell=1}^{\infty} f_{1,\ell} \lambda^{\ell} \qquad \text{where} \qquad f_{1,\ell} \equiv \sum_{j=1}^{\ell} \frac{(-1)^{\ell-j}}{(\ell-j)!} \mathrm{ad}_{A_1}^{\ell-j} h_j,$$

whence, finally

$$C_2 = \frac{1}{2} f_{1,1}.$$

Carrying out this process for higher values of n, we define recursively the functions

$$F_n(\lambda) = \sum_{k=1}^{\infty} f_{n,k} \lambda^k, \qquad f_{n,k} = \sum_{j=0}^{[k/n]-1} \frac{(-1)^j}{j!} \mathrm{ad}_{C_n}^{j} f_{n-1,k-nj},$$

so that, analogously,

$$C_n = \frac{1}{n} f_{[\frac{n-1}{2}],n-1}, \quad n \geq 3.$$

Again, the implementation of this algorithm in a symbolic algebra package is straightforward, but now more computation time and memory resources are required if one aims to get high order terms. This can be clearly seen when considering the number of terms involved. Whereas in the usual Zassenhaus expansion C_{16} has 3711 terms, now there are 22322. The first terms of the expansion read explicitly

$$\begin{aligned}
C_2 &= A_2 \\
C_3 &= A_3 - \frac{1}{2}[A_1, A_2] \\
C_4 &= A_4 - \frac{1}{2}[A_1, A_3] + \frac{1}{6}[A_1, [A_1, A_2]] \\
C_5 &= A_5 - \frac{1}{2}[A_1, A_4] + \frac{1}{6}[A_1, [A_1, A_3]] - \frac{1}{24}[A_1, [A_1, [A_1, A_2]]] - \frac{1}{2}[A_2, A_3] \\
&\quad + \frac{1}{3}[A_2, [A_1, A_2]].
\end{aligned}$$

A detailed study of this expansion is carried out in [51].

4 The Magnus Expansion

4.1 The Procedure

As stated in the Introduction, **Problem 3** concerns the feasibility of expressing the fundamental matrix of the linear differential matrix equation

$$\frac{dY}{dt} = A(t)Y, \qquad Y(0) = I \tag{57}$$

as an exponential representation. In other words, the problem consists in defining, in terms of A, an operator $\Omega(t)$ such that $Y(t) = \exp(\Omega(t))$.

Equation (57) can be solved of course by applying the Neumann (Dyson) iterative procedure, thus expressing the solution as an infinite series whose first terms are

$$Y(t) = I + \int_0^t A(s)ds + \int_0^t A(s_1) \int_0^{s_1} A(s_2)ds_2 ds_1 + \cdots .$$

In general,

$$Y(t) = I + \sum_{n=1}^{\infty} P_n(t), \qquad \text{where} \qquad P_n(t) = \int_0^t ds_1 \cdots \int_0^{s_{n-1}} ds_n A_1 A_2 \cdots A_n \tag{58}$$

and $A_i \equiv A(s_i)$. The series in (58) has the obvious drawback that, when truncated, the resulting approximation may loose some properties the exact solution has. Suppose, for instance, that $A(t)$ is skew-Hermitian, as is the case in quantum mechanical problems. Then the exact solution is unitary, whereas any truncation of the series (58) is no longer so. As a result, the computation of e.g. transition probabilities may be problematic.

Motivated by this issue, Magnus proposed in his seminal paper [43] to write the solution as the exponential

$$Y(t) = \exp(\Omega(t)), \tag{59}$$

where Ω is itself an infinite series,

$$\Omega(t) = \sum_{m=1}^{\infty} \Omega_m(t). \tag{60}$$

In this way, "the partial sums of this series become Hermitian after multiplication by i if iA is a Hermitian operator" [43].

Starting from (57) and taking into account the derivative of the exponential, one obtains the differential equation satisfied by Ω, namely

$$\frac{d\Omega}{dt} = \sum_{n=0}^{\infty} \frac{B_n}{n!} \operatorname{ad}_{\Omega}^n A, \qquad \Omega(0) = 0. \tag{61}$$

Notice that, in contrast with (57), this equation is nonlinear. In any event, by defining

$$\Omega^{[0]} = 0, \qquad \Omega^{[1]}(t) = \int_0^t A(s_1) ds_1,$$

and applying Picard fixed point iteration, one gets

$$\Omega^{[n]}(t) = \int_0^t \left(A - \frac{1}{2}[\Omega^{[n-1]}, A] + \frac{1}{12}[\Omega^{[n-1]}, [\Omega^{[n-1]}, A]] + \cdots \right) ds_1$$

so that $\lim_{n\to\infty} \Omega^{[n]}(t) = \Omega(t)$ in a (presumably small) neighborhood of $t = 0$.

Inserting the series (60) into (61) it is possible to get the first few terms in closed form. Specifically,

$$\begin{aligned}
\Omega_1(t) &= \int_0^t A(t_1)\, \mathrm{d}t_1, \\
\Omega_2(t) &= \frac{1}{2} \int_0^t \mathrm{d}t_1 \int_0^{t_1} \mathrm{d}t_2\ [A(t_1), A(t_2)] \\
\Omega_3(t) &= \frac{1}{6} \int_0^t \mathrm{d}t_1 \int_0^{t_1} \mathrm{d}t_2 \int_0^{t_2} \mathrm{d}t_3\ ([A(t_1), [A(t_2), A(t_3)]] + [A(t_3), [A(t_2), A(t_1)]]).
\end{aligned} \tag{62}$$

Other, more systematic approaches are required to obtain the terms in the series (60) for any m. Thus, for instance, by using graph theory it is possible to get explicit formulae for $\Omega_m(t)$ at all orders, whereas the recursive procedure proposed in [39] is well suited for computations up to high order. It is given by

$$\begin{aligned}
S_m^{(1)} &= [\Omega_{m-1}, A], \qquad S_m^{(j)} = \sum_{n=1}^{m-j} [\Omega_n, S_{m-n}^{(j-1)}], \quad 2 \le j \le m-1 \\
\Omega_1 &= \int_0^t A(t_1) dt_1, \qquad \Omega_m = \sum_{j=1}^{m-1} \frac{B_j}{j!} \int_0^t S_m^{(j)}(t_1) dt_1, \quad m \ge 2.
\end{aligned} \tag{63}$$

Working out this recurrence one arrives at the alternative expression

$$\Omega_m(t) = \sum_{j=1}^{m-1} \frac{B_j}{j!} \sum_{\substack{k_1+\cdots+k_j=m-1 \\ k_1\geq 1,\ldots,k_j\geq 1}} \int_0^t \mathrm{ad}_{\Omega_{k_1}(s)}\, \mathrm{ad}_{\Omega_{k_2}(s)} \cdots \mathrm{ad}_{\Omega_{k_j}(s)} A(s)\, ds \qquad m \geq 2. \tag{64}$$

Notice that each term $\Omega_m(t)$ in the Magnus series (60) is a multiple integral of combinations of $m-1$ nested commutators containing m operators $A(t)$. In consequence, as pointed out before, if A is skew-Hermitian, any approximation obtained by truncating the Magnus series is unitary (as long as the exponential is correctly evaluated). More generally, if $A(t)$ belongs to some Lie algebra $\mathfrak{g}$, then it is clear that $\Omega(t)$ (and in fact any truncation of the Magnus series) also stays in $\mathfrak{g}$ and therefore $\exp(\Omega) \in \mathcal{G}$, where $\mathcal{G}$ denotes the Lie group whose corresponding Lie algebra is $\mathfrak{g}$.

The Magnus expansion shares another appealing property with the exact flow of (57), namely its time symmetry. Consider with greater generality the problem

$$\frac{dY}{dt} = A(t)Y, \qquad Y(t_0) = Y_0. \tag{65}$$

The flow $\varphi_t : Y(t_0) \to Y(t)$ corresponding to (65) is time-symmetric, $\varphi_{-t} \circ \varphi_t = \mathrm{Id}$, since integrating (65) from t_0 to any $t_f \geq t_0$ and back to t_0 leads to the original initial value $Y(t_0) = Y_0$. On the other hand, the Magnus expansion can be written as $Y(t+h) = \exp(\Omega(t,h))Y(t)$, so that time-symmetry implies that

$$\Omega(t+h, -h) = -\Omega(t,h). \tag{66}$$

If $A(t)$ is an analytic function and its Taylor series around $t + h/2$ is considered, then $\Omega(t,h)$ does not contain even powers of h. More specifically, if

$$A\left(t + \frac{h}{2} + \tau\right) = a_0 + a_1\tau + a_2\tau^2 + \cdots \quad \text{with} \quad a_i = \frac{1}{i!}\frac{d^i A(s)}{ds^i}\bigg|_{s=t+h/2}, \tag{67}$$

then the terms Ω_m in (63) computed at $t+h$ read

$$\begin{aligned}
\Omega_1 &= h a_0 + h^3 \frac{1}{12} a_2 + h^5 \frac{1}{80} a_4 + \mathcal{O}(h^7) \\
\Omega_2 &= h^3 \frac{-1}{12}[a_0, a_1] + h^5 \Big(\frac{-1}{80}[a_0, a_3] + \frac{1}{240}[a_1, a_2]\Big) + \mathcal{O}(h^7) \\
\Omega_3 &= h^5 \Big(\frac{1}{360}[a_0, a_0, a_2] - \frac{1}{240}[a_1, a_0, a_1]\Big) + \mathcal{O}(h^7) \\
\Omega_4 &= h^5 \frac{1}{720}[a_0, a_0, a_0, a_1] + \mathcal{O}(h^7),
\end{aligned} \tag{68}$$

whereas $\Omega_5 = \mathcal{O}(h^7)$, $\Omega_6 = \mathcal{O}(h^7)$ and $\Omega_7 = \mathcal{O}(h^9)$. Here we write for clarity $[a_{i_1}, a_{i_2}, \ldots, a_{i_{l-1}}, a_{i_l}] \equiv [a_{i_1}, [a_{i_2}, [\ldots, [a_{i_{l-1}}, a_{i_l}]\ldots]]]$. Notice that, as anticipated, only odd powers of h appear in Ω_k and, in particular, $\Omega_{2i+1} = \mathcal{O}(h^{2i+3})$ for $i > 1$.

This feature has shown to be very useful when designing numerical integrators based on the Magnus expansion [11, 12, 36].

Although one might think of (59) and (60) only as a formal representation of the solution $Y(t)$ of (57), it has been shown that by imposing certain conditions on the operator $A(t)$, the exponent $\Omega(t)$ is a continuous differentiable function of $A(t)$ and t verifying (61). Moreover it can be determined by a convergent series (60) whose terms are computed by applying the recursion (64). Specifically, the following result is proved in [18].

Theorem 2 *Let the equation $Y' = A(t)Y$ be defined in a Hilbert space $\mathscr{H}$ with $Y(0) = I$. Let $A(t)$ be a bounded operator on $\mathscr{H}$. Then, the Magnus series $\Omega(t) = \sum_{m=1}^{\infty} \Omega_m(t)$, with Ω_m given by the recursion (63), converges in the interval $t \in [0, T)$ such that*

$$\int_0^T \|A(s)\| \, ds < \pi$$

and the sum $\Omega(t)$ satisfies $\exp \Omega(t) = Y(t)$. The statement also holds when $\mathscr{H}$ is infinite-dimensional if Y is a normal operator (in particular, if Y is unitary).

This theorem, in fact, provides the optimal convergence domain, in the sense that π is the largest constant for which the result holds without any further restrictions on the operator $A(t)$. Nevertheless, it is quite easy to construct examples for which the bound estimate $r_c = \pi$ is still conservative: the Magnus series converges indeed for a larger time interval than that given by the theorem [18, 47]. Consequently, condition $\int_0^T \|A(s)\| ds < \pi$ is *not* necessary for the convergence of the expansion.

A more precise characterization of the convergence can be obtained in the case of $n \times n$ complex matrices $A(t)$. Specifically, in [18] the connection between the convergence of the Magnus series and the existence of multiple eigenvalues of the fundamental solution $Y(t)$ is analyzed. Let us introduce a new parameter $\varepsilon \in \mathbb{C}$ and denote by $Y_t(\varepsilon)$ the fundamental matrix of $Y' = \varepsilon A(t) Y$. Then, if the analytic matrix function $Y_t(\varepsilon)$ has an eigenvalue $\rho_0(\varepsilon_0)$ of multiplicity $\ell > 1$ for a certain ε_0 such that: (a) there is a curve in the ε-plane joining $\varepsilon = 0$ with $\varepsilon = \varepsilon_0$, and (b) the number of equal terms in $\log \rho_1(\varepsilon_0), \log \rho_2(\varepsilon_0), \ldots, \log \rho_\ell(\varepsilon_0)$ such that $\rho_k(\varepsilon_0) = \rho_0$, $k = 1, \ldots, \ell$ is less than the maximum dimension of the elementary Jordan block corresponding to ρ_0, then the radius of convergence of the series $\Omega_t(\varepsilon) \equiv \sum_{k \geq 1} \varepsilon^k \Omega_{t,k}$ verifying $\exp \Omega_t(\varepsilon) = Y_t(\varepsilon)$ is precisely $r = |\varepsilon_0|$. Notice that this obstacle to convergence is due just to the logarithmic function. If $A(t)$ itself has singularities in the complex plane, then they also restrict the convergence of the procedure.

Since the 1960s, the Magnus expansion has been extensively applied in mathematical physics, quantum physics and chemistry, control theory, nuclear, atomic and molecular physics, optics, etc., essentially as a tool to construct explicit analytical approximations for the corresponding solution. More recently, it has also been used as the starting point to design new and very efficient numerical integrators for the initial value problem defined by (65). The idea consists in dividing the time interval

$[t_0, t_f]$ into N subintervals steps and construct an approximation in each subinterval $[t_{n-1}, t_n]$, $n = 1, \ldots, N$, by truncating appropriately the exponent $\Omega(t_n, h)$, with $h = t_n - t_{n-1}$. This is done by analyzing the time-dependency in each term Ω_m of the Magnus series (60) and approximating the successive integrals appearing in Ω_m by a single quadrature up to the desired order. The resulting schemes, by construction, provide numerical approximations lying in the same Lie group $\mathscr{G}$ where the differential equation is defined: in the case of quantum mechanics, if (65) corresponds to the time-dependent Schrödinger equation, then the numerical solution is unitary and thus provides transition probabilities in the correct range of values for all times. Integration methods of this class are particular examples of geometric integrators: numerical schemes that preserve geometric properties of the continuous system, thus granting them with an improved qualitative behavior in comparison with general-purpose algorithms [9, 11, 32, 36].

The Magnus expansion can also be generalized to get useful approximations to the nonlinear time-dependent differential equation

$$\frac{dY}{dt} = A(t, Y)Y, \qquad Y(t_0) = Y_0 \tag{69}$$

defined in a Lie group $\mathscr{G}$ [19]. As in the linear case, the solution is represented by

$$Y(t) = \exp(\Omega(t, Y_0)Y_0, \tag{70}$$

where Ω satisfies the differential equation

$$\frac{d\Omega}{dt} = \sum_{k=0}^{\infty} \frac{B_k}{k!} \mathrm{ad}^k_{\Omega(s)} A(s, \mathrm{e}^{\Omega(s)} Y_0) ds, \qquad \Omega(0) = 0. \tag{71}$$

We can solve this equation by iteration ($\Omega^{[0]} = 0$), thus giving

$$\Omega^{[m]}(t) = \int_0^t \sum_{k=0}^{\infty} \frac{B_k}{k!} \mathrm{ad}^k_{\Omega^{[m-1]}(s)} A(s, \mathrm{e}^{\Omega^{[m-1]}(s)} Y_0) ds, \qquad m \geq 1.$$

It is then clear that

$$\Omega^{[1]}(t) = \int_0^t A(s, Y_0) ds = \Omega(t, Y_0) + \mathscr{O}(t^2), \tag{72}$$

whereas the truncation

$$\Omega^{[m]}(t) = \sum_{k=0}^{m-2} \frac{B_k}{k!} \int_0^t \mathrm{ad}^k_{\Omega^{[m-1]}(s)} A(s, \mathrm{e}^{\Omega^{[m-1]}(s)} Y_0) ds, \qquad m \geq 2, \tag{73}$$

once inserted in (70), provides an explicit approximation $Y^{[m]}(t)$ for the solution of (69) that is correct up to terms $\mathcal{O}(t^{m+1})$ [19]. In addition, $\Omega^{[m]}(t)$ reproduces exactly the sum of the first m terms in the Ω series of the usual Magnus expansion for the linear equation $Y' = A(t)Y$ [9].

4.2 The Magnus Expansion and pre-Lie Algebras

A careful analysis of the recursion (63) and (64) for the Magnus expansion shows that the object

$$A \rhd A(s) := \left[\int_0^t A(u)du, A(s) \right], \tag{74}$$

involving integration and the commutator operations, allows one to get more compact expressions for the successive terms in the series of Ω [26]. Specifically, we may write

$$S_2^{(1)} = [\Omega_1, A] = A \rhd A, \quad \text{so that} \quad \Omega_2' = -\frac{1}{2}\,(A \rhd A).$$

Analogously,

$$S_3^{(1)} = -\frac{1}{2}\,(A \rhd A) \rhd A, \qquad \left[\int_0^t S_3^{(1)}, A \right] = -\frac{1}{2}\,((A \rhd A) \rhd A) \rhd A,$$

$$[\Omega_1, S_3^{(1)}] = -\frac{1}{2}\, A \rhd ((A \rhd A) \rhd A)$$

and thus

$$\Omega_3' = -\frac{1}{2} S_3^{(1)} + \frac{1}{12} [\Omega_1, S_2^{(1)}]$$

$$\Omega_4' = \frac{1}{3} \left[\int_0^t S_3^{(1)}, A \right] + \frac{1}{6} [\Omega_1, S_3^{(1)}]$$

Alternatively, we have

$$\Omega_2 = -\frac{1}{2} \int_0^t A \rhd A(s) ds$$

$$\Omega_3 = \frac{1}{12} \int_0^t (A \rhd (A \rhd A))(s) ds + \frac{1}{4} \int_0^t ((A \rhd A) \rhd A)(s) ds$$

$$\Omega_4 = -\frac{1}{6}\int_0^t ((A \rhd A) \rhd A) \rhd A(s)ds + \frac{1}{12}\int_0^t A \rhd ((A \rhd A) \rhd A)(s)ds.$$

The bilinear operation (74) is a particular example of a *pre-Lie product* based on the *dendriform products*

$$(A \succ B)(s) \equiv \left(\int_0^s A(u)du\right) B(s), \qquad (A \prec B)(s) \equiv A(s) \left(\int_0^s B(u)du\right), \tag{75}$$

where A and B are two given matrices. More generally, a left pre-Lie algebra $(\mathscr{A}, \rhd)$ is a vector space $\mathscr{A}$ equipped with an operation $\rhd$ subject to the following relation [44]:

$$(a \rhd b) \rhd c - a \rhd (b \rhd c) = (b \rhd a) \rhd c - b \rhd (a \rhd c),$$

whereas a dendriform algebra is a vector space endowed with two bilinear operations $\succ$ and $\prec$ satisfying the following three axioms:

$$\begin{aligned}
(a \prec b) \prec c &= a \prec (b \prec c + b \succ c), \\
(a \succ b) \prec c &= a \succ (b \prec c), \\
a \succ (b \succ c) &= (a \prec b + a \succ b) \succ c.
\end{aligned}$$

Clearly, a dendriform algebra is at the same time a pre-Lie algebra, since

$$a \rhd b \equiv a \succ b - b \prec a$$

is a left pre-Lie product. Of course, a right pre-Lie algebra can be defined analogously [28]. Defining the operator L_a as $L_a b = a \rhd b$, we can formally express Eq. (61) for Ω as [28]

$$\frac{d\Omega}{dt} = \frac{L_\Omega}{e^{L_\Omega} - I}(A) = \sum_{n \geq 0} \frac{B_n}{n!} L_\Omega^n(A).$$

This allows one to generalize the Magnus expansion to pre-Lie and dendriform algebras, and analyze their purely algebraic and combinatorial features in a more abstract setting, with applications in other areas, such as Jackson's q-integral and linear q-difference equations [27].

4.3 The Magnus Expansion and the BCH Series

The Magnus expansion can also be used to get explicitly the terms of the series Z in

$$Z = \log(e^{X_1} e^{X_2})$$

when X_1 and X_2 are two non commuting indeterminate variables, i.e., we can use it to construct term by term the Baker–Campbell–Hausdorff series. This can be done simply by considering the initial value problem (57) with the piecewise constant matrix-valued function

$$A(t) = \begin{cases} X_2 & 0 \le t \le 1 \\ X_1 & 1 < t \le 2. \end{cases} \tag{76}$$

The exact solution at $t = 2$ is $Y(2) = \mathrm{e}^{X_1}\,\mathrm{e}^{X_2}$. Now we can use the recursion (63) to compute the exponent $\Omega(t)$ at $t = 2$ so that $Y(2) = \mathrm{e}^{\Omega(2)}$. In this way we generate the BCH series in the form (7). Although this procedure does not constitute a better alternative in practice with respect to the algorithm presented in Sect. 2.2, it does allow one to get a sharper bound on the convergence domain of the series: by applying Theorem 2 to this case we obtain the following result [20].

Theorem 3 *The Baker–Campbell–Hausdorff series in the form (7) converges absolutely when* $\|X_1\| + \|X_2\| < \pi$.

This result can be generalized, of course, to any number of non commuting operators $X_1, X_2, \ldots, X_q$. Specifically, the series

$$Z = \log(\mathrm{e}^{X_1}\,\mathrm{e}^{X_2}\,\cdots\,\mathrm{e}^{X_q}),$$

converges absolutely if $\|X_1\| + \|X_2\| + \cdots + \|X_q\| < \pi$. This connection allows one to relate in a natural way the underlying pre-Lie structure of the Magnus expansion with the BCH series and the set of rooted trees used in its derivation.

5 Some Applications

The previous exponential identities have found applications in many different fields ranging from pure and applied mathematics to physics and physical chemistry. It is our purpose in this section to review three of them, perhaps not sufficiently well known: the role of the BCH formula for obtaining splitting and composition methods for differential equations, a particular form of the so-called Kashiwara–Vergne conjecture (now a theorem) and the existence of non-trivial identities involving commutators in a free Lie algebra. For a comprehensive list of applications we refer the reader to e.g. [11, 14, 32] and references therein.

5.1 Splitting Methods

The BCH formula is widely used in the design and analysis of numerical integration methods for differential equations, specifically to obtain the order conditions in splitting and composition methods [32, 45]. Let us consider an initial value problem of the form

$$\frac{dx}{dt} = f(x), \qquad x(0) = x_0 \in \mathbb{R}^d \tag{77}$$

whose vector field can be decomposed as a sum of two contributions, $f(x) = f^{[1]}(x) + f^{[2]}(x)$, in such a way that each sub-problem

$$\frac{dx}{dt} = f^{[1]}(x), \qquad \frac{dx}{dt} = f^{[2]}(x), \qquad x(0) = x_0 \in \mathbb{R}^d$$

can be integrated exactly, with solutions $x(h) = \varphi_h^{[1]}(x_0)$, $x(h) = \varphi_h^{[2]}(x_0)$ at $t = h$. Then, by composing these solutions as

$$\chi_h = \varphi_h^{[2]} \circ \varphi_h^{[1]} \tag{78}$$

we get a first-order approximation to the exact solution. By introducing suitable (real) parameters α_i it is possible to construct higher-order approximations by means of the *composition method*

$$\psi_h = \chi_{\alpha_s h} \circ \chi_{\alpha_{s-1} h} \circ \cdots \circ \chi_{\alpha_1 h}. \tag{79}$$

Alternatively, we may consider more maps in (78) with additional parameters. In this case, one has a *splitting method* of the form

$$\psi_h = \varphi_{b_{s+1}h}^{[2]} \circ \varphi_{a_s h}^{[1]} \circ \varphi_{b_s h}^{[2]} \circ \cdots \circ \varphi_{b_2 h}^{[2]} \circ \varphi_{a_1 h}^{[1]} \circ \varphi_{b_1 h}^{[2]}. \tag{80}$$

In both cases, the coefficients α_i, a_i, b_i have to satisfy a set of conditions guaranteeing that the resulting schemes are of a prescribed order r in h, i.e.,

$$\psi_h(x_0) = \varphi_h(x_0) + \mathcal{O}(h^{r+1}),$$

where $\varphi_h(x_0)$ denotes the exact solution of (77) for a time step h. These *order conditions* are formulated as polynomial equations in the coefficients whose degree and complexity increase with the order of the method. Constructing particular integrators requires first obtaining and then solving these order conditions, and is here where the BCH formula has shown to be an extremely helpful tool [10, 32, 45].

In the following we illustrate how the BCH formula is used to get the order conditions for splitting methods of the form (80). The important point here is that we can introduce differential operators and series of differential operators associated with the vector fields f, $f^{[1]}$, $f^{[2]}$ and the numerical integrator ψ_h. Specifically, we can associate with the vector field f in (77) the first-order differential operator (or Lie derivative)

$$L_f = \sum_{i=1}^{d} f_i \frac{\partial}{\partial x_i}$$

and the Lie transformation $\exp(tL_f)$ in such a way that the exact solution of (77) can be formally written as

$$\varphi_h(x_0) = \sum_{k\geq 0} \frac{h^k}{k!}(L_f^k \,\mathrm{Id})(x_0) \equiv \exp(hL_f)[\mathrm{Id}](x_0), \tag{81}$$

where $\mathrm{Id}(x) = x$ denotes the identity map [32].

Analogously, the Lie derivatives corresponding to $f^{[1]}$ and $f^{[2]}$ read, respectively,

$$A \equiv L_{f^{[1]}} = \sum_{i=1}^{d} f_i^{[1]}(x)\frac{\partial}{\partial x_i}, \qquad B \equiv L_{f^{[2]}} = \sum_{i=1}^{d} f_i^{[2]}(x)\frac{\partial}{\partial x_i},$$

so that

$$\begin{aligned}&\left(\varphi_{b_{s+1}h}^{[2]} \circ \varphi_{a_s h}^{[1]} \circ \varphi_{b_s h}^{[2]} \circ \cdots \circ \varphi_{b_2 h}^{[2]} \circ \varphi_{a_1 h}^{[1]} \circ \varphi_{b_1 h}^{[2]}\right)(x_0) = \\ &\exp(b_1 hB)\,\exp(a_1 hA)\cdots\,\exp(a_s hA)\,\exp(b_{s+1}hB)[\mathrm{Id}](x_0).\end{aligned} \tag{82}$$

Notice the opposite order of the operators with respect to the maps in (82). Now, by formally applying the BCH formula in sequence to the series of differential operators

$$\Psi(h) = \exp(b_1 hB)\,\exp(a_1 hA)\exp(b_2 hB)\cdots\,\exp(b_s hB)\,\exp(a_s hA)\,\exp(b_{s+1}hB)$$

we end up with

$$\Psi(h) = \exp(F(h)), \qquad \text{where} \qquad F = \sum_{n\geq 1} h^n F_n,$$

so that the integrator (80) is of order r if

$$F_1 = L_f = L_{f^{[1]}} + L_{f^{[2]}} = A + B, \qquad \text{and} \qquad F_k = 0 \quad \text{for} \quad 2 \leq k \leq r.$$

In more detail [9],

$$\begin{aligned}F(h) = {} & h(v_a A + v_b B) + h^2 v_{ab}[A,B] + h^3(v_{aab}[A,[A,B]] + v_{bab}[B,[A,B]]) \\ & + h^4(v_{aaab}[A,[A,[A,B]]] + v_{baab}[B,[A,[A,B]]] + v_{bbab}[B,[B,[A,B]]]) \\ & + \mathcal{O}(h^5),\end{aligned} \tag{83}$$

where $v_a, v_b, v_{ab}, v_{aab}, v_{bab}, v_{aaab}, \ldots$ are polynomials in the parameters a_i, b_i of the scheme. In particular [10],

$$v_a = \sum_{i=1}^{s} a_i, \qquad v_b = \sum_{i=1}^{s+1} b_i, \qquad v_{ab} = \frac{1}{2}v_a v_b - \sum_{1\leq i\leq j\leq s} b_i a_j, \tag{84}$$

$$2v_{aab} = \frac{1}{6}v_a^2 v_b - \sum_{1 \le i < j \le k \le s} a_i b_j a_k, \qquad 2v_{bab} = -\frac{1}{6}v_a v_b^2 + \sum_{1 \le i \le j < k \le s-1} b_i a_j b_k,$$

and the order conditions for the splitting method are obtained by requiring

$$v_a = v_b = 1, \qquad v_{ab} = v_{aab} = v_{bab} = \cdots = 0$$

up to the order considered. These are necessary and sufficient conditions to achieve the desired order as long as $F(h)$ is expressed in terms of a basis in the free Lie algebra $\mathscr{L}(A, B)$ generated by $\{A, B\}$ (see [10] for more details).

Splitting methods have a long history both in the numerical analysis of ordinary and partial differential equations (sometimes with different names) and in applications arising in many different fields: celestial mechanics, chemical physics, molecular dynamics, quantum statistical mechanics, etc, especially in the context of geometric numerical integration [9, 32, 45].

5.2 *The Kashiwara–Vergne Conjecture*

In the course of their research on the transport of the convolution product by the exponential application for invariant distributions, Kashiwara and Vergne [38] announced in 1978 the following combinatorial conjecture related with a particular way of expressing the Baker–Campbell–Hausdorff formula.

Conjecture 1 (Kashiwara–Vergne) *Let us denote by $Z(X, Y) = \log(\mathrm{e}^X \mathrm{e}^Y)$ the BCH series. For any Lie algebra $\mathfrak{g}$ of finite dimension, there exist series $F(X, Y)$ and $G(X, Y)$ on $\mathfrak{g} \times \mathfrak{g}$ without constant term taking values in $\mathfrak{g}$ such that they satisfy*

$$X + Y - Z(Y, X) = (1 - \mathrm{e}^{-\mathrm{ad}_X}) F(X, Y) + (\mathrm{e}^{\mathrm{ad}_Y} - 1) G(X, Y) \tag{85}$$

and the trace identity

$$\mathrm{tr}\,(\mathrm{ad}_X \circ \partial_X F + \mathrm{ad}_Y \circ \partial_Y G) = \frac{1}{2}\mathrm{tr}\left(\frac{\mathrm{ad}_X}{\mathrm{e}^{\mathrm{ad}_X} - 1} + \frac{\mathrm{ad}_Y}{\mathrm{e}^{\mathrm{ad}_Y} - 1} - \frac{\mathrm{ad}_{Z(X,Y)}}{\mathrm{e}^{\mathrm{ad}_{Z(X,Y)}} - 1} - 1\right). \tag{86}$$

Here $\partial_X F, \partial_Y G \in \mathrm{End}(\mathfrak{g})$ are defined by

$$\partial_X F(X, Y) : U \longmapsto \frac{d}{dt} F(X + tU, Y)|_{t=0}, \quad \partial_Y G(X, Y) : U \longmapsto \frac{d}{dt} G(X, Y + tU)|_{t=0}$$

and tr *denotes the trace of an endomorphism of $\mathfrak{g}$.*

Several remarks are in order with respect to this statement. First, Eq. (85) is essentially equivalent to grouping together in the BCH formula all terms that are of the form $[X, \ldots]$, resp. $[Y, \ldots]$. Of course, F and G are not uniquely determined by this property [16]. The major difficulty of this conjecture lies then in the trace Eq. (86) [4]. Second, the conjecture establishes the existence of a pair of functions (F, G) satisfying (85) and (86). It turns out, however, that the pair

$$(G(-Y, -X), F(-Y, -X)) \tag{87}$$

also constitutes a solution. In consequence, it is possible to look only for symmetric solutions, i.e., functions verifying $G(X, Y) = F(-Y, -X)$.

Kashiwara and Vergne proposed a symmetric pair of universal Lie series and show that, for solvable Lie algebras, they verify the trace Eq. (86). These functions can be expressed as follows [60].

Let ψ be the function defined by

$$\psi(z) = \frac{e^z - 1 - z}{(e^z - 1)(1 - e^{-z})},$$

and denote $Z(t) = Z(tX, tY), 0 \le t \le 1$. Then the functions

$$F^1(X, Y) = \left(\int_0^1 \frac{1 - e^{-t\,\mathrm{ad}_X}}{1 - e^{-\mathrm{ad}_X}} \circ \psi(\mathrm{ad}_{Z(t)}) dt \right) (X + Y) \tag{88}$$

and $G^1(X, Y) = F^1(-Y, -X)$ verify Eq. (85) by construction. This is also true for the functions

$$\begin{aligned} F^0(X, Y) &= \frac{1}{2} \left(F^1(X, Y) + e^{\mathrm{ad}_X} F^1(-X, -Y) \right) + \frac{1}{4}(Z(X, Y) - X) \\ G^0(X, Y) &= F^0(-Y, -X) \end{aligned} \tag{89}$$

which, in addition, satisfy Eq. (86) when $\mathfrak{g}$ is solvable [38] and also when $\mathfrak{g} = \mathrm{sl}(2, \mathbb{R})$ [59].

In 2005 Alekseev and Meinrenken [3] proved the Kashiwara–Vergne combinatorial conjecture with complete generality by using a deformation of the Baker–Campbell–Hausdorff series proposed by Torossian [64]. Due, in particular, to property (87), there are many solutions to the Kashiwara–Vergne problem. Nevertheless, it has been shown by means of a computer code that the functions (89) do *not* satisfy the Kashiwara–Vergne conjecture in the case of a general Lie algebra [2]. In consequence, this solution is not universal, since it is not valid for all finite dimensional Lie algebras.

It is clear that the algorithm proposed in Sect. 2.2 can be applied here to construct explicitly the Lie series (F^0, G^0) in terms of X, Y up to high degree. The resulting expressions may then provide additional information on their structure and validity.

We have constructed a code based on the algorithm developed in Sect. 2.2 for the BCH series which allows us to generate the series (F^0, G^0) up to an arbitrary order in an efficient way. In particular, we reproduce the same results obtained in [2] up to order 8, verifying in this way that the expressions (89) do not satisfy the Kashiwara–Vergne conjecture at this order for general Lie algebras. As an illustration, G^0 up to order 6 reads in the Lyndon basis

$$
\begin{aligned}
G^0(X,Y) = {} & -\frac{X}{4} - \frac{1}{24}[X,Y] + \frac{1}{48}[[X,Y],Y] - \frac{1}{48}[X,[X,Y]] \\
& + \frac{1}{180}[[[X,Y],Y],Y] - \frac{1}{480}[X,[[X,Y],Y]] - \frac{1}{360}[X,[X,[X,Y]]] \\
& - \frac{1}{2880}[[[[X,Y],Y],Y],Y] + \frac{1}{1440}[[X,[X,Y]],[X,Y]] + \frac{1}{480}[[X,Y],[[X,Y],Y]] \\
& + \frac{1}{360}[X,[[[X,Y],Y],Y]] - \frac{1}{360}[X,[X,[[X,Y],Y]]] + \frac{1}{2880}[X,[X,[X,[X,Y]]]] \\
& - \frac{1}{5040}[[[[[X,Y],Y],Y],Y],Y] + \frac{1}{1260}[[X,[[X,Y],Y]],[X,Y]] \\
& + \frac{1}{840}[[X,Y],[[[X,Y],Y],Y]] + \frac{23}{40320}[X,[[[[X,Y],Y],Y],Y]] \\
& - \frac{1}{6720}[X,[[X,[X,Y]],[X,Y]]] - \frac{1}{6720}[[X,[X,Y]],[[X,Y],Y]] \\
& - \frac{1}{6048}[X,[X,[[[X,Y],Y],Y]]] - \frac{13}{40320}[X,[X,[X,[[X,Y],Y]]]] \\
& + \frac{1}{10080}[X,[X,[X,[X,[X,Y]]]]].
\end{aligned}
$$

These series are absolutely convergent in a neighborhood of the origin when a norm is introduced in $\mathfrak{g}$, as shown by Rouvière [60]. As a matter of fact, the domain of convergence can be enlarged by using the results obtained for the BCH series. For completeness, we reproduce here Rouvière's argument.

Since the function ψ is meromorphic in $\mathbb{C}$ with poles at $\pm 2ki\pi$, $k = 1, 2, \ldots$, it can be expanded in a power series in the disk $|z| < 2\pi$. With a norm on $\mathfrak{g}$ satisfying Eq. (36), it is clear that with the corresponding norm on $\mathrm{End}(\mathfrak{g})$ one has $\|\mathrm{ad}_X\| \leq 2\|X\|$. Then the integrand of the function F^1 in (88) can be expanded into a power series of t and the endomorphisms ad_X, $\mathrm{ad}_{Z(t)}$ if $\|\mathrm{ad}_X\| < 2\pi$ and $\|\mathrm{ad}_{Z(t)}\| < 2\pi$. These constraints are clearly satisfied when $(X, Y) \in D_1 \cup D_2$ as given by (37), and thus $F^1(X, Y)$ (and obviously $F^0(X, Y)$) is absolutely convergent in $D_1 \cup D_2$.

5.3 High Order Identities Involving Commutators

As we mentioned in Sect. 2.2, the BCH series expressed in different bases of the free Lie algebra $\mathscr{L}(X, Y)$ has a different number of non-vanishing coefficients. In [20] is shown that in the Lyndon basis this number is sensibly reduced with respect to the Hall basis (up to degree 20 by about a 30%). An interesting question could be to identify the particular basis where the number of non-vanishing terms is minimum, i.e., to get the most compact expression of the form (15). Partial results in this setting are given in [40, 53], where shortened versions of the BCH series are obtained up to degree 8 and 10, respectively. This is done by considering a right-normed basis in $\mathscr{L}(X, Y)$, i.e., one that consists of elements of the form $[a_{i_1}, [a_{i_2}, [\ldots, [a_{i_{t-1}}, a_{i_t}]\ldots]]]$, where a_{i_p} are the generators X, Y. Such a basis exists and can be constructed algorithmically, as shown in e.g. [13], although the process is by no means straightforward.

In [53], in particular, by comparing different procedures to obtain the BCH formula and some existing symmetries, several remarkable identities satisfied by right-nested commutators at high degree were unveiled which, in turn, allowed its author to identify independent commutators and eventually simplify the series up to order eight. Recognizing these generalized identities could thus be an essential ingredient to get more compact expressions for the BCH series and other exponential expansions.

It turns out that the Magnus expansion can also be used for this purpose, as we next show. If in (67) $a_1 = 0$, $a_i = 0$ for $i > 2$ and denote $a_0 = X$, $a_2 = Y$, i.e., we compute the Magnus expansion at $t + h$ with

$$A(t + h/2 + \tau) = X + Y\tau^2,$$

then clearly Ω_{2k} vanishes due to time-symmetry. Thus, when the recursion (63) is applied, all terms with even powers of h must be identically zero. These terms are linear combinations of right-nested commutators of the form $[a_{i_1}, [a_{i_2}, [\ldots, [a_{i_{l-1}}, a_{i_l}]\ldots]]]$ and give rise to non-trivial identities involving X and Y.

Proceeding in this way we obtain the following three identities arising in Ω_6:

$$\begin{aligned}
&(6.1): \quad 3[YXXYXY] + [XXYYXY] - 3[XYXYXY] - [YYXXXY] = 0;\\
&(6.2): \quad [YYXYXY] + [XYYYXY] - 2[YXYYXY] = 0;\\
&(6.3): \quad [XXXYXY] + [YXXXXY] - 2[XYXXXY] = 0,
\end{aligned}$$

whereas from Ω_8 we obtain four more identities:

$$
\begin{aligned}
(8.1): \quad & [XXXYXXXXY] - 3[XXYXXXXXY] + 3[XYXXXXXXY] \\
& - [YXXXXXXXY] = 0; \\
(8.2): \quad & [XXXXYYXXY] - 2[XXXYXYXXY] - 2[XXYYXXXXY] \\
& + 8[XYXYXXXXY] - 3[XYYXXXXXY] - 2[YXXXYXXXY] \\
& - [YXYXXXXXY] + [YYXXXXXXY] = 0; \\
(8.3): \quad & -29838[XXXYYYXXY] + 61125[XXYXYYXXY] \\
& - 4347[XYXXYYXXY] - 56778[XYXYXYXXY] \\
& - 1449[XYYYXXXXY] - 17477[YXXXYYXXY] \\
& + 56778[YXXYXYXXY] + 23273[YXYYXXXXY] \\
& - 61125[YYXYXXXXY] + 29838[YYYXXXXXY] = 0 \\
(8.4): \quad & 3[XXYYYYXXY] - 3[XYXYYYXXY] - 6[XYYXYYXXY] \\
& - 9[YXXYYYXXY] + 24[YXYXYYXXY] - 4[YYXXYYXXY] \\
& - 6[YYXYXYXXY] + [YYYYXXXXY] = 0.
\end{aligned}
$$

Here $[XXXYXXXXY]$ denotes the right-nested commutator $[X, [X, [X, [Y, [X, [X, [X, Y]]]]]]]$, etc. Identities involving three operators can be obtained in a similar way if instead we consider

$$A(t + h/2 + \tau) = X_1 + X_2\tau^2 + X_3\tau^4$$

and repeat the procedure.

Acknowledgements The author would like to thank the three referees for their insightful remarks that have helped him to improve the paper. This work has been partially supported by Ministerio de Economía y Competitividad (Spain) through the coordinated project MTM2013-46553-C3 (co-funded by FEDER).

Appendix

5.4 Lie Algebras

A *Lie algebra* is a vector space $\mathfrak{g}$ together with a map $[\cdot, \cdot]$ from $\mathfrak{g} \times \mathfrak{g}$ into $\mathfrak{g}$ called Lie bracket, with the following properties:

1. $[\cdot, \cdot]$ is bilinear.
2. $[X, Y] = -[Y, X]$ for all $X, Y \in \mathfrak{g}$.
3. $[X, [Y, Z]] + [Y, [Z, X]] + [Z, [X, Y]] = 0$ for all $X, Y, Z \in \mathfrak{g}$.

Condition 2 is called *skew symmetry* and Condition 3 is the *Jacobi identity*. One should remark that $\mathfrak{g}$ can be any vector space and that the Lie bracket operation $[\cdot, \cdot]$ can be any bilinear, skew-symmetric map that satisfies the Jacobi identity. Thus, in particular, the space of all $n \times n$ (real or complex) matrices is a Lie algebra with the Lie bracket defined as the commutator $[A, B] = AB - BA$.

Associated with any $X \in \mathfrak{g}$ we can define a linear map $\mathrm{ad}_X : \mathfrak{g} \longrightarrow \mathfrak{g}$ which acts according to

$$\mathrm{ad}_X Y = [X, Y], \qquad \mathrm{ad}_X^j Y = [X, \mathrm{ad}_X^{j-1} Y], \qquad \mathrm{ad}_X^0 Y = Y, \qquad j \in \mathbb{N} \tag{90}$$

for all $Y \in \mathfrak{g}$. The "ad" operator allows one to express nested Lie brackets in an easy way. Thus, for instance, $[X, [X, [X, Y]]]$ can be written as $\mathrm{ad}_X^3 Y$. Moreover, as a consequence of the Jacobi identity, one has the following properties:

1. $\mathrm{ad}_{[X,Y]} = \mathrm{ad}_X \mathrm{ad}_Y - \mathrm{ad}_Y \mathrm{ad}_X = [\mathrm{ad}_X, \mathrm{ad}_Y]$
2. $\mathrm{ad}_Z [X, Y] = [X, \mathrm{ad}_Z Y] + [\mathrm{ad}_Z X, Y]$.

For matrix Lie algebras one has the important relation (see e.g. [36])

$$\mathrm{e}^X \, Y \, \mathrm{e}^{-X} = \mathrm{e}^{\mathrm{ad}_X} Y = \sum_{k=0}^{\infty} \frac{1}{k!} \mathrm{ad}_X^k Y,$$

so that

$$\mathrm{e}^X \, \mathrm{e}^Y \, \mathrm{e}^{-X} = \mathrm{e}^Z, \qquad \text{with} \qquad Z = \mathrm{e}^{\mathrm{ad}_X} Y.$$

The derivative of the matrix exponential map also plays an important role in our treatment. Given a matrix $\Omega(t)$, then [36]

$$\frac{d}{dt} \exp(\Omega(t)) = d\exp_{\Omega(t)}(\Omega'(t)) \, \exp(\Omega(t)),$$
$$\frac{d}{dt} \exp(\Omega(t)) = \exp(\Omega(t)) \, d\exp_{-\Omega(t)}(\Omega'(t)),$$

where $d\exp_\Omega(C)$ is defined by the (everywhere convergent) power series

$$d\exp_\Omega(C) = \sum_{k=0}^{\infty} \frac{1}{(k+1)!} \mathrm{ad}_\Omega^k(C) \equiv \frac{\mathrm{e}^{\mathrm{ad}_\Omega} - I}{\mathrm{ad}_\Omega}(C).$$

If the eigenvalues of the linear operator ad_Ω are different from $2m\pi i$ with $m \in \{\pm 1, \pm 2, \ldots\}$ then the operator $d\exp_\Omega$ is invertible [11, 36] and

$$d\exp_\Omega^{-1}(C) = \frac{\mathrm{ad}_\Omega}{\mathrm{e}^{\mathrm{ad}_\Omega} - I}(C) = \sum_{k=0}^{\infty} \frac{B_k}{k!} \mathrm{ad}_\Omega^k(C),$$

where B_k are the Bernoulli numbers.

5.5 *Free Lie Algebras and Hall–Viennot Bases*

Very often it is necessary to carry out computations in a Lie algebra when no particular algebraic structure is assumed beyond what is common to all Lie algebras. It is in this context, in particular, where the notion of *free Lie algebra* plays a fundamental role. Given an arbitrary index set I (either finite or countably infinite), we can say that a Lie algebra $\mathfrak{g}$ is *free* over the set I if [49]

1. for every $i \in I$ there corresponds an element $X_i \in \mathfrak{g}$;
2. for any Lie algebra $\mathfrak{h}$ and any function $i \mapsto Y_i \in \mathfrak{h}$, there exists a unique Lie algebra homomorphism $\pi : \mathfrak{g} \to \mathfrak{h}$ satisfying $\pi(X_i) = Y_i$ for all $i \in I$.

If $\mathcal{T} = \{X_i : i \in I\} \subset \mathfrak{g}$, then the algebra $\mathfrak{g}$ can be viewed as the set of all Lie brackets of X_i. In this sense, we can say that $\mathfrak{g}$ is the free Lie algebra generated by $\mathcal{T}$ and we denote $\mathfrak{g} = \mathcal{L}(X_1, X_2, \ldots)$. Elements of $\mathcal{L}(X_1, X_2, \ldots)$ are called Lie polynomials.

It is important to remark that $\mathfrak{g}$ is a universal object, and that computations in $\mathfrak{g}$ can be applied in any particular Lie algebra $\mathfrak{h}$ via the homomorphism π [49], just by replacing each abstract element X_i with the corresponding Y_i.

In practical calculations, it is useful to represent a free Lie algebra by means of a basis (in the vector space sense). There are several systematic procedures to construct such a basis. Here, for simplicity, we will consider the free Lie algebra generated by just two elements $\mathcal{T} = \{X, Y\}$, and the so-called Hall–Viennot bases. A set $\{E_i : i = 1, 2, 3, \ldots\} \subset \mathcal{L}(X, Y)$ whose elements are of the form

$$E_1 = X, \quad E_2 = Y, \quad \text{and} \quad E_i = [E_{i'}, E_{i''}] \quad i \geq 3, \tag{91}$$

with some positive integers $i', i'' < i$ $(i = 3, 4, \ldots)$ is a Hall–Viennot basis if there exists a total order relation $\succ$ in the set of indices $\{1, 2, 3, \ldots\}$ such that $i \succ i''$ for all $i \geq 3$, and the map

$$d : \{3, 4, \ldots\} \longrightarrow \{(j, k) \in \mathbb{Z} \times \mathbb{Z} \; : \; j \succ k \succeq j''\}, \tag{92}$$

$$d(i) = (i', i'') \tag{93}$$

(with the convention $1'' = 2'' = 0$) is bijective.

In [57, 67], Hall–Viennot bases are indexed by a subset of words (a Hall set of words) on the alphabet $\{x, y\}$. Such Hall set of words $\{w_i : i \geq 1\}$ can be obtained by defining recursively w_i as the concatenation $w_{i'}w_{i''}$ of the words $w_{i'}$ and $w_{i''}$, with $w_1 = x$ and $w_2 = y$. In particular, if the map (92) is constructed in such a way that the total order relation $\succ$ is the natural order relation in $\mathbb{Z}$, i.e., $>$, then the first elements of the Hall set of words w_i associated to the indices $i = 1, 2, \ldots, 14$ are x, y, yx, yxx, yxy, $yxxx$, $yxxy$, $yxyy$, $yxxxx$, $yxxxy$, $yxxyy$, $yxyyy$, $yxxyx$, $yxyyx$. In consequence, the corresponding elements of the basis in $\mathcal{L}(X, Y)$ are

$$X,\ Y,\ [Y,X],\ [[Y,X],X],\ [[Y,X],Y],\ [[[Y,X],X],X],\ [[[Y,X],X],Y],\ [[[Y,X],Y],Y],$$
$$[[[[Y,X],X],X],X],\ [[[[Y,X],X],X],Y],\ [[[[Y,X],X],Y],Y],\ [[[[Y,X],Y],Y],Y],$$
$$[[[Y,X],X],[Y,X]],\ [[[Y,X],Y],[Y,X]]. \tag{94}$$

Notice that if the total order is chosen as $<$ instead, it results in the classical Hall basis as presented in [15].

On the other hand, the Lyndon basis can be constructed as a Hall–Viennot basis by considering the order relation $\succ$ as follows: $i \succ j$ if, in lexicographical order (i.e., the order used when ordering words in the dictionary), the Hall word w_i associated to i comes before than the Hall word w_j associated to j. The Hall set of words $\{w_i : i \geq 1\}$ corresponding to the Lyndon basis is the set of Lyndon words, which can be defined as the set of words w on the alphabet $\{x, y\}$ satisfying that, for arbitrary decompositions of w as the concatenation $w = uv$ of two non-empty words u and v, the word w is smaller than v in lexicographical order [42, 67]. The Lyndon words for $i = 1, 2, \ldots, 14$ are $x, y, xy, xyy, xxy, xyyy, xxyy, xxxy, xyyyy, xxyxy, xyxyy, xxyyy, xxxyy, xxxxy$ and the corresponding (Lyndon) basis in $\mathscr{L}(X, Y)$ is formed by

$$X,\ Y,\ [X,Y],\ [[X,Y],Y],\ [X,[X,Y]],\ [[[X,Y],Y],Y],\ [X,[[X,Y],Y]],\ [X,[X,[X,Y]]],$$
$$[[[[X,Y],Y],Y],Y],\ [[X,[X,Y]],[X,Y]],\ [[X,Y],[[X,Y],Y]],\ [X,[[[X,Y],Y],Y]],$$
$$[X,[X,[[X,Y],Y]]],\ [X,[X,[X,[X,Y]]]]. \tag{95}$$

It is possible to compute the dimension c_n of the linear subspace in the free Lie algebra generated by all the independent Lie brackets of order n, denoted by $\mathscr{L}_n(X, Y)$. This number is provided by the so-called Witt's formula [15, 45]:

$$c_n = \frac{1}{n} \sum_{d|n} \mu(d) 2^{n/d}, \tag{96}$$

where the sum is over all (positive) divisors d of the degree n and $\mu(d)$ is the Möbius function, defined by the rule $\mu(1) = 1$, $\mu(d) = (-1)^k$ if d is the product of k distinct prime factors and $\mu(d) = 0$ otherwise [45]. For $n \leq 12$ one has explicitly

n	1	2	3	4	5	6	7	8	9	10	11	12
c_n	1	1	2	3	6	9	18	30	56	99	186	335

References

1. Abramowitz, M., Stegun, I.A.: Handbook of Mathematical Functions. Dover (1965)
2. Albert, L., Harinck, P., Torossian, C.: Solution non universelle pour le probleme KV-78. J. Lie Theory 617–626 (2008)

3. Alekseev, A., Meinrenken, E.: On the Kashiwara-Vergne conjecture. Invent. Math. **164**, 615–634 (2006)
4. Alekseev, A., Torossian, C.: The Kashiwara-Vergne conjecture and Drinfeld's associators. Ann. Math. **175**, 415–463 (2012)
5. Bader, P., Iserles, A., Kropielnicka, K., Singh, P.: Effective approximation for the semiclassical Schrödinger equation. Found. Comput. Math. **14**, 689–720 (2014)
6. Baker, H.F.: Alternant and continuous groups. Proc. Lond. Math. Soc. (Second series) **3**, 24–47 (1905)
7. Bayen, F.: On the convergence of the Zassenhaus formula. Lett. Math. Phys. **3**, 161–167 (1979)
8. Blanes, S., Casas, F.: On the convergence and optimization of the Baker-Campbell-Hausdorff formula. Linear Algebra Appl. **378**, 135–158 (2004)
9. Blanes, S., Casas, F.: A Concise Introduction to Geometric Numerical Integration. CRC Press (2016)
10. Blanes, S., Casas, F., Murua, A.: Splitting and composition methods in the numerical integration of differential equations. Bol. Soc. Esp. Mat. Apl. **45**, 89–145 (2008)
11. Blanes, S., Casas, F., Oteo, J.A., Ros, J.: The Magnus expansion and some of its applications. Phys. Rep. **470**, 151–238 (2009)
12. Blanes, S., Casas, F., Ros, J.: High order optimized geometric integrators for linear differential equations. BIT **42**, 262–284 (2002)
13. Bokut, L., Chibrikov, E.S.: Lyndon–Shirshov words, Gröbner–Shirshov bases, and free Lie algebras. In: Sabinin, L., Sbitneva, L., Shestakov, I. (eds.) Non-Associative Algebra and its Applications, pp. 17–39. Chapman & Hall/CRC Press (2006)
14. Bonfiglioli, A., Fulci, R.: Topics in Noncommutative Algebra. The Theorem of Campbell, Baker, Hausdorff and Dynkin. Lecture Notes in Mathematics. Springer (2012)
15. Bourbaki, N.: Lie Groups and Lie Algebras, Chapters 1–3. Springer (1989)
16. Burgunder, E.: Eulerian idempotent and Kashiwara-Vergne conjecture. Ann. Inst. Fourier **58**, 1153–1184 (2008)
17. Campbell, J.E.: On a law of combination of operators. Proc. Lond. Math. Soc. **29**, 14–32 (1898)
18. Casas, F.: Sufficient conditions for the convergence of the Magnus expansion. J. Phys. A: Math. Theor. **40**, 15001–15017 (2007)
19. Casas, F., Iserles, A.: Explicit Magnus expansions for nonlinear equations. J. Phys. A: Math. Gen. **39**, 5445–5461 (2006)
20. Casas, F., Murua, A.: An efficient algorithm for computing the Baker-Campbell-Hausdorff series and some of its applications. J. Math. Phys. **50**, 033513 (2009)
21. Casas, F., Murua, A., Nadinic, M.: Efficient computation of the Zassenhaus formula. Comput. Phys. Commun. **183**, 2386–2391 (2012)
22. Coddington, E.A., Levinson, N.: Theory of Ordinary Differential Equations. McGraw Hill (1955)
23. Day, J., So, W., Thompson, R.C.: Some properties of the Campbell Baker Hausdorff series. Linear Multilinear Algebra **29**, 207–224 (1991)
24. Dynkin, E.B.: Evaluation of the coefficients of the Campbell-Hausdorff formula. Dokl. Akad. Nauk. SSSR **57**, 323–326 (1947)
25. Dynkin, E.B.: Calculation of the coefficients in the Campbell–Hausdorff series. In: Dynkin, E.B., Yushkevich, A.A., Seitz, G.M., Onishchik, A.L. (eds.) Selected Papers of E.B. Dynkin with Commentary, pp. 31–35. American Mathematical Society (2000)
26. Ebrahimi-Fard, K., Manchon, D.: A Magnus- and Fer-type formula in dendriform algebras. Found. Comput. Math. **9**, 295–316 (2009)
27. Ebrahimi-Fard, K., Manchon, D.: Twisted dendriform algebras and the pre-Lie Magnus expansion. J. Pure Appl. Algebra **215**, 2615–2627 (2011)
28. Ebrahimi-Fard, K., Manchon, D.: The Manus expansion, trees and Knuth's rotation correspondence. Found. Comput. Math. **14**, 1–25 (2014)
29. Geiser, J., Tanoglu, G.: Operator-splitting methods via the Zassenhaus product formula. Appl. Math. Comput. **217**, 4557–4575 (2011)

30. Geiser, J., Tanoglu, G., Guecueyenen, N.: Higher order operator-splitting methods via Zassenhaus product formula: theory and applications. Comput. Math. Appl. **62**, 1994–2015 (2011)
31. Goldberg, K.: The formal power series for $\log(e^x e^y)$. Duke Math. J. **23**, 13–21 (1956)
32. Hairer, E., Lubich, Ch., Wanner, G.: Geometric Numerical Integration. Structure-Preserving Algorithms for Ordinary Differential Equations, 2nd edn. Springer (2006)
33. Hall, B.C.: Lie Groups, Lie Algebras, and Representations, An Elementary Introduction. Springer (2003)
34. Hausdorff, F.: Die symbolische exponential formel in der Gruppen theorie. Leipziger Ber. **58**, 19–48 (1906)
35. Hörmander, L.: Hypoelliptic second order differential equations. Acta Math. **119**, 147–171 (1967)
36. Iserles, A., Munthe-Kaas, H.Z., Nørsett, S.P., Zanna, A.: Lie-group methods. Acta Numer. **9**, 215–365 (2000)
37. Jacobson, N.: Lie Algebras. Dover Publications Inc, New York (1979)
38. Kashiwara, M., Vergne, M.: The Campbell-Hausdorff formula and invariant hyperfunctions. Invent. Math. **47**, 249–272 (1978)
39. Klarsfeld, S., Oteo, J.A.: Recursive generation of higher-order terms in the Magnus expansion. Phys. Rev. A **39**, 3270–3273 (1989)
40. Kolsrud, M.: Maximal reductions in the Baker-Hausdorff formula. J. Math. Phys. **34**(1), 270–285 (1993)
41. Lévy, A.A.: Une généralisation de l'identité de Jacobi et son application á l'explicitation de la récurrence de Magnus pour la formule Baker-Hausdorff discrète. C. R. Acad. Sci. Paris A **275**, 877–880 (1972)
42. Lothaire, M.: Combinatorics on Words. Cambridge University Press (1983)
43. Magnus, W.: On the exponential solution of differential equations for a linear operator. Commun. Pure and Appl. Math. **VII**, 649–673 (1954)
44. Manchon, D.: A short survey on pre-Lie algebras. In: Carey, A. (ed.) Noncommutative Geometry and Physics: Renormalisation, Motives, Index Theory. European Mathematical Society (2011)
45. McLachlan, R.I., Quispel, R.: Splitting methods. Acta Numer. **11**, 341–434 (2002)
46. Michel, J.: Bases des algèbres de Lie libre et série de Hausdorff. Séminaire Dubreil. Algèbre **27**, 1–9 (1974)
47. Moan, P.C., Niesen, J.: Convergence of the Magnus series. Found. Comput. Math. **8**, 291–301 (2008)
48. Moler, C.B., Van Loan, C.F.: Nineteen dubious ways to compute the exponential of a matrix, twenty-five years later. SIAM Rev. **45**, 3–49 (2003)
49. Munthe-Kaas, H., Owren, B.: Computations in a free Lie algebra. Philos. Trans. R. Soc. A **357**, 957–981 (1999)
50. Murua, A.: The Hopf algebra of rooted trees, free Lie algebras, and Lie series. Found. Comput. Math. **6**, 387–426 (2006)
51. Nadinic, M.: Métodos de Integración Geométrica. Análisis y Algoritmos Numéricos, Ph.D. thesis, Universitat Jaume I (2016)
52. Newman, M., So, W., Thompson, R.C.: Convergence domains for the Campbell-Baker-Hausdorff formula. Linear Multilinear Algebra **24**, 301–310 (1989)
53. Oteo, J.A.: The Baker-Campbell-Hausdorff formula and nested commutator identities. J. Math. Phys. **32**(2), 419–424 (1991)
54. Perko, L.: Differential Equations and Dynamical Systems, 2nd edn. Springer (1996)
55. Quesne, C.: Disentangling q-exponentials: a general approach. Int. J. Theor. Phys. **43**, 545–559 (2004)
56. Reed, M., Simon, B.: Methods of Modern Mathematical Physics I: Functional Analysis. Academic Press (1980)
57. Reutenauer, C.: Free Lie Algebras, vol. 7. Oxford University Press (1993)
58. Rossmann, W.: Lie Groups. An Introduction Through Linear Groups. Oxford University Press (2002)

59. Rouvière, F.: Démonstration de la conjecture de Kashiwara-Vergne pour $SL_2(\mathbb{R})$. C. R. Acad. Sci. Paris **292**, 657–660 (1981)
60. Rouvière, F.: Espaces symétriques et méthode de Kashiwara-Vergne. Ann. Sci. École Norm. Sup. **19**, 553–581 (1986)
61. Suzuki, M.: On the convergence of exponential operators–the Zassenhaus formula, BCH formula and systematic approximants. Commun. Math. Phys. **57**, 193–200 (1977)
62. Thompson, R.C.: Cyclic relations and the Goldberg coefficients in the Campbell-Baker-Hausdorff formula. Proc. Am. Math. Soc. **86**, 12–14 (1982)
63. Thompson, R.C.: Convergence proof for Goldberg's exponential series. Linear Algebra Appl. **121**, 3–7 (1989)
64. Torossian, C.: Sur la conjecture combinatoire de Kashiwara-Vergne. J. Lie Theory **12**, 597–616 (2002)
65. Trotter, H.F.: On the product of semi-groups of operators. Proc. Am. Math. Soc. **10**, 545–551 (1959)
66. Varadarajan, V.S.: Lie Groups, Lie Algebras, and Their Representations. Springer (1984)
67. Viennot, X.G.: Algèbres de Lie Libres et monoïdes libres. Lecture Notes in Mathematics, vol. 691. Springer (1978)
68. Wermuth, E.M.E.: Commuting matrix exponentials: problem 88–1. SIAM Rev. **31**, 125–126 (1989)
69. Wermuth, E.M.E.: Two remarks on matrix exponentials. Linear Algebra Appl. **117**, 127–132 (1989)
70. Weyrauch, M., Scholz, D.: Computing the Baker-Campbell-Hausdorff series and the Zassenhaus product. Comput. Phys. Commun. **180**, 1558–1565 (2009)
71. Wilcox, R.M.: Exponential operators and parameter differentiation in quantum physics. J. Math. Phys. **8**, 962–982 (1967)

Post-Lie Algebras, Factorization Theorems and Isospectral Flows

Kurusch Ebrahimi-Fard and Igor Mencattini

Abstract In these notes we review and further explore the Lie enveloping algebra of a post-Lie algebra. From a Hopf algebra point of view, one of the central results, which will be recalled in detail, is the existence of second Hopf algebra structure. By comparing group-like elements in suitable completions of these two Hopf algebras, we derive a particular map which we dub post-Lie Magnus expansion. These results are then considered in the case of Semenov-Tian-Shansky's double Lie algebra, where a post-Lie algebra is defined in terms of solutions of modified classical Yang–Baxter equation. In this context, we prove a factorization theorem for group-like elements. An explicit exponential solution of the corresponding Lie bracket flow is presented, which is based on the aforementioned post-Lie Magnus expansion.

Keywords Post-Lie algebra · Universal enveloping algebra · Hopf algebra
Magnus expansion · Classical r-matrices · Classical Yang–Baxter equation
Factorization theorems · Isospectral flow

MSC Classification 16T05 · 16T10 · 16T25 · 16T30 · 17D25

1 Introduction

These notes are based on recent joint work [17–19] by the authors together with A. Lundervold and H. Z. Munthe-Kaas. They present an extended summary of a talk given by the first author at the Instituto de Ciencias Matemáticas (ICMAT) in

K. Ebrahimi-Fard
Department of Mathematical Sciences, Norwegian University of Science and Technology – NTNU, 7491 Trondheim, Norway
e-mail: kurusch.ebrahimi-fard@ntnu.no

I. Mencattini (✉)
Instituto de Ciências Matemáticas e de Computação, Universidade de São Paulo (USP), São Carlos, SP, Brazil
e-mail: igorre@icmc.usp.br

K. Ebrahimi-Fard and M. Barbero Liñán (eds.), *Discrete Mechanics, Geometric Integration and Lie–Butcher Series*, Springer Proceedings in Mathematics & Statistics 267, https://doi.org/10.1007/978-3-030-01397-4_7

Madrid.[1] The main aim is to explore a certain factorization problem for Lie groups in the framework of universal enveloping algebra from the perspective offered by the relatively new theory of post-Lie algebras. The latter provides another viewpoint on the notion of (finite dimensional) double Lie algebra, which is a Lie algebra $\mathfrak{g}$ over the ground field $\mathbb{F}$ endowed with a solution $R_+ \in \mathrm{End}_{\mathbb{F}}(\mathfrak{g})$ of the modified classical Yang–Baxter equation:

$$[R_+x, R_+y] = R_+([R_+x, y] + [x, R_+y] - [x, y]). \tag{1}$$

The identity implies that the bracket

$$[x, y]_{R_+} := [R_+x, y] - [R_+y, x] - [x, y]$$

satisfies the Jacobi identity and therefore yields another Lie algebra, denoted $\mathfrak{g}_R$, on the vector space underlying $\mathfrak{g}$. Thanks to the seminal work of Semenov-Tian-Shansky [33], solutions of (1), known as classical r-matrices, play an important role in studying solutions of Lax equations, which in turn are intimately related to a factorization problem in the Lie group corresponding to $\mathfrak{g}$. In the framework of the universal enveloping algebra of the Lie algebra $\mathfrak{g}$, this factorization problem has been studied in [31, 35]. In these works it is shown, among other things, that every solution of the modified classical Yang–Baxter equation gives rise to a factorization of group-like elements in (a suitable completion of) the universal enveloping algebra of $\mathfrak{g}$. On the other hand, in [2] it was shown that in a Lie algebra $\mathfrak{g}$ every solution of (1) gives rise to a post-Lie algebra.

A post-Lie algebra [23, 24, 39], which we denote by the triple $(V, \triangleright, [\cdot, \cdot])$, consists of a vector space V which is endowed with two bilinear operations, the Lie bracket $[\cdot, \cdot] : V \otimes V \to V$ and the magmatic post-Lie product $\triangleright : V \otimes V \to V$. The particular relations that the latter is supposed to satisfy with respect to the Lie bracket are such that

$$[\![x, y]\!] := x \triangleright y - y \triangleright x - [x, y] \tag{2}$$

yields another Lie bracket on V. The complete definition will given further below. However, the following geometric example [23, 24] may provide some insight into the interplay between the post-Lie product and the Lie bracket in post-Lie algebra. Recall that a *linear connection* is a $\mathbb{F}$-bilinear application $\nabla : \mathfrak{X}_M \times \mathfrak{X}_M \to \mathfrak{X}_M$ on $\mathfrak{X}_M$, the vector space of smooth vector fields on the manifold M, satisfying the Leibniz rule $\nabla_X(fY) = X(f)Y + f\nabla_X Y$, for all $f \in C^\infty(M)$ and all $X, Y \in \mathfrak{X}_M$. Clearly, a linear connection endows $\mathfrak{X}_M$ with a product, defined simply as $(X, Y) \mapsto X \curvearrowright Y := \nabla_X Y$. The *torsion* of ∇ is a skew-symmetric tensor $\mathrm{T}\colon TM \wedge TM \to TM$

[1]Brainstorming Workshop on "New Developments in Discrete Mechanics, Geometric Integration and Lie-Butcher Series", May 25-28, 2015, ICMAT, Madrid, Spain. Supported by a grant from Iceland, Liechtenstein and Norway through the EEA Financial Mechanism as well as the project "Mathematical Methods for Ecology and Industrial Management" funded by Ayudas Fundación BBVA a Investigadores, Innovadores y Creadores Culturales.

$$\mathrm{T}(X, Y) := X \curvearrowright Y - Y \curvearrowright X - [X, Y], \tag{3}$$

where $[\cdot, \cdot]$ denotes the *Jacobi-Lie bracket* of vector fields, defined by $[X, Y](f) = X(Y(f)) - Y(X(f))$, for every $X, Y \in \mathfrak{X}_M$ and every $f \in C^\infty(M)$. The *curvature* tensor $\mathrm{R}\colon TM \wedge TM \to \mathrm{End}(TM)$ satisfies the identity

$$\mathrm{R}(X, Y)Z = a_{\curvearrowright}(X, Y, Z) - a_{\curvearrowright}(Y, X, Z) + \mathrm{T}(X, Y) \curvearrowright Z, \tag{4}$$

where $a_{\curvearrowright}(X, Y, Z) := X \curvearrowright (Y \curvearrowright Z) - (X \curvearrowright Y) \curvearrowright Z$ is the usual associator with respect to the product $\curvearrowright$. Torsion and curvature are related by the *Bianchi identities*

$$\sum_{\circlearrowleft}(\mathrm{T}(\mathrm{T}(X, Y), Z) + (\nabla_X \mathrm{T})(Y, Z)) = \sum_{\circlearrowleft} \mathrm{R}(X, Y)Z \tag{5}$$

$$\sum_{\circlearrowleft}((\nabla_X \mathrm{R})(Y, Z) + \mathrm{R}(\mathrm{T}(X, Y), Z)) = 0, \tag{6}$$

where $\sum_{\circlearrowleft}$ denotes the sum over the three cyclic permutations of (X, Y, Z). If a connection is *flat*, $\mathrm{R} = 0$, and has *constant torsion*, $\nabla_X \mathrm{T} = 0$, then (5) reduces to the Jacobi identity, such that the torsion defines a Lie bracket $[X, Y]_{\mathrm{T}} := \mathrm{T}(X, Y)$, which is related to the Jacobi–Lie bracket by (3). The covariant derivation formula $\nabla_X(\mathrm{T}(Y, Z)) = (\nabla_X \mathrm{T})(Y, Z) + \mathrm{T}(\nabla_X Y, Z) + \mathrm{T}(Y, \nabla_X Z)$ together with $\nabla_X \mathrm{T} = 0$ imply

$$X \curvearrowright [Y, Z]_{\mathrm{T}} = [X \curvearrowright Y, Z]_{\mathrm{T}} + [Y, X \curvearrowright Z]_{\mathrm{T}}. \tag{7}$$

On the other hand, (4) together with $\mathrm{R} = 0$ yield

$$[X, Y]_{\mathrm{T}} \curvearrowright Z = a_{\curvearrowright}(X, Y, Z) - a_{\curvearrowright}(Y, X, Z). \tag{8}$$

Relations (7) and (8) define the post-Lie algebra $(\mathfrak{X}_M, \curvearrowright, [\cdot, \cdot]_{\mathrm{T}})$, see Proposition 7.

Note that for a connection which is both flat and torsion free ($\mathrm{T} = 0 = \mathrm{R}$), equation (4) implies $a_{\curvearrowright}(X, Y, Z) = a_{\curvearrowright}(Y, X, Z)$. This is the characterizing identity of a (left) pre-Lie algebra, which is Lie admissible, i.e., by skew-symmetrization one obtains a Lie algebra. We refer the reader to [4, 7, 10, 26] for details.

Returning to the abstract definition of post-Lie algebra, $(V, \rhd, [\cdot, \cdot])$, we consider the lifting of the post-Lie product to the universal enveloping algebra, $\mathcal{U}(\mathfrak{g})$, of the Lie algebra $\mathfrak{g} := (V, [\cdot, \cdot])$. It turns out that it allows to define another Hopf algebra, $\mathcal{U}_*(\mathfrak{g})$, on the underlying vector space of $\mathcal{U}(\mathfrak{g})$, which is isomorphic, as a Hopf algebra, to the universal enveloping algebra corresponding to the Lie algebra $\bar{\mathfrak{g}} := (V, [\![\cdot, \cdot]\!])$ defined in terms of the Lie bracket (2). The Hopf algebra isomorphism between $\mathcal{U}(\bar{\mathfrak{g}})$ and $\mathcal{U}_*(\mathfrak{g})$ is an extension of the identity between the Lie algebras $\mathfrak{g}$ and $\bar{\mathfrak{g}}$. Moreover, for every $x \in \mathfrak{g}$ there exists a unique element $\chi(x) \in \mathfrak{g}$, such that $\exp(x) = \exp^*(\chi(x))$ with respect to (suitable completions of) $\mathcal{U}(\mathfrak{g})$ respectively $\mathcal{U}_*(\mathfrak{g})$. The map $\chi : \mathfrak{g} \to \mathfrak{g}$ is called post-Lie Magnus expansion and is defined as the solution of a particular differential equation.

From [2] we know that every solution of (1) turns a double Lie algebra [33] into a post-Lie algebra. The Lie bracket $[\cdot, \cdot]_{R_+}$ on $\mathfrak{g}_R$ is a manifestation of (2), and the aforementioned Hopf algebra isomorphism between $\mathscr{U}_*(\mathfrak{g})$ and $\mathscr{U}(\bar{\mathfrak{g}}) = \mathscr{U}(\mathfrak{g}_R)$ can be realized in terms of the solution R_+ and the Hopf algebra structures of these two universal enveloping algebras. The role of post-Lie algebra in the context of the factorization problem on $\mathscr{U}(\mathfrak{g})$, mentioned above, becomes clear from the fact that any group-like element $\exp(x)$ in (a suitable completion of) $\mathscr{U}(\mathfrak{g})$ factorizes into the product of two group-like elements, $\exp(\chi_+(x))$ and $\exp(\chi_-(x))$, with $\chi_\pm(x) := \pm R_\pm \chi(x)$, where $R_- := R_+ - \mathrm{id}$.

In what follows $\mathbb{F}$ denotes the ground field of characteristic zero over which all algebraic structures are considered. Unless stated otherwise, $\mathbb{F}$ will be either the complex numbers $\mathbb{C}$ or the real numbers $\mathbb{R}$.

2 Basic Lie Theory

In this section we present some background on Lie theory. The aim is to recall notions and to fix notations, which will be used in later sections. We will discuss the construction of the so-called *I-adic* completion of an augmented algebra since it plays a central role in these notes. For details the reader is referred to [14, 28, 30, 40].

2.1 Lie Groups and Lie Algebras

A Lie group G is a smooth manifold endowed with the structure of an abstract group, which is compatible with the underlying differentiable structure of G. This means that both maps the multiplication $m : G \times G \to G$ and the inversion $i : G \to G$ are smooth applications. A map $\psi : G_1 \to G_2$ is a morphism of Lie groups if it is a smooth homomorphism. An important class of examples of Lie groups is obtained as follows. Let V be a finite dimensional $\mathbb{F}$-vector space, $n := dim\ V < \infty$. Then the linear isomorphisms of V form a Lie group, which will be denoted $\mathrm{GL}(V)$. The smooth structure on $\mathrm{GL}(V)$ is the one induced by the Euclidean structure defined on V following the choice of a basis. In this way $\mathrm{GL}(V)$ can be identified with an open subset of $\mathrm{End}_{\mathbb{F}}(V)$, the vector space of all linear endomorphisms of V. Finally, note that the choice of a basis of V induces a diffeomorphism between $\mathrm{GL}(V)$ and $\mathrm{GL}_n(\mathbb{F})$, respectively, between $\mathrm{End}_{\mathbb{F}}(V)$ and $\mathrm{Mat}_n(\mathbb{F})$. Here $\mathrm{Mat}_n(\mathbb{F})$ is the set of $n \times n$ matrices with entries in $\mathbb{F}$, and $\mathrm{GL}_n(\mathbb{F})$ consists of invertible $n \times n$ matrices. If G is a Lie group, then the connected component G^0 of the identity $e \in G$ is a normal subgroup of G, whose index is equal to the number of connected components of G, and it is generated by a suitable open neighborhood U of the identity element of G. More precisely, there exists $e \in U \subset G$ such that $G^0 = \bigcup_{i=1}^{\infty} U^i$, i.e., every element of G^0 can be written as a product of finitely many elements of U. For each element

$g \in G$, one can define the diffeomorphisms $L_g : G \to G$ and $R_g : G \to G$, called the *left*-, respectively, *right-translation* by g. For each element x in the tangent space at the identity of G, $x \in T_eG$, let $X_x : G \to TG$ be the map defined for all $g \in G$ by $X_x : g \rightsquigarrow (L_g)_{*,e}x$.[2] Then X_x is smooth and it satisfies $\pi \circ X_x = \mathrm{id}_G$, where $\pi : TG \to G$ is the canonical projection. In other words, X_x is a *left-invariant, smooth vector field* on G. Recall that a vector field X on G is called *left-invariant*, if for each $g \in G$,

$$(L_gX)(h) = X(h), \quad \forall h \in G, \tag{9}$$

where

$$(L_gX)(h) := (L_g)_{*,g^{-1}h}\big(X(g^{-1}h)\big) \tag{10}$$

for all $g, h \in G$. Formula (10) defines a *left-action* of G on $\mathfrak{X}(G)$, the Lie algebra of smooth vector fields on G, and together with (9), it implies that X is left-invariant if and only if, for all $g \in G$, $(L_g)_{*,h}X(h) = X(gh)$, for all $h \in G$. In fact, given a left-invariant vector field X, one has that for all $g \in G$:

$$\begin{aligned}\big(L_{g^{-1}}\big)_{*,h}X(h) &= \big(L_{g^{-1}}\big)_{*,h}\big(L_gX\big)(h)\\ &= \big(L_{g^{-1}}\big)_{*,h}\big(L_g\big)_{*,g^{-1}h}\big(X(g^{-1}h)\big) = X(g^{-1}h), \quad \forall h \in G.\end{aligned}$$

The set $\mathfrak{X}(G)^G \subset \mathfrak{X}(G)$ of all left-invariant vector fields is a Lie subalgebra of $\mathfrak{X}(G)$ of dimension equal to the dimension of G and, in particular, $X \in \mathfrak{X}(G)$ is left-invariant if and only if $X = X_x$ for some $x \in T_eG$. This observation let to introduce the structure of a Lie algebra on T_eG, i.e., a bilinear, skew-symmetric bracket $[\cdot,\cdot] : T_eG \times T_eG \to T_eG$ satisfying the *Jacobi identity*

$$[[x, y], z] + [[z, x], y] + [[y, z], x] = 0, \quad \forall x, y, z \in T_eG.$$

By definition, $[x, y] := [X_x, X_y](e)$, for all $x, y \in T_eG$. The pair $(T_eG, [\cdot,\cdot])$ will be denoted by $\mathfrak{g}$. To every homomorphism of Lie groups corresponds a homomorphism between the corresponding Lie algebras, i.e., if $\psi : G_1 \to G_2$ is a homomorphism between groups G_i, $i = 1, 2$, with corresponding Lie algebras $\mathfrak{g}_i = (T_eG_i, [\cdot,\cdot]_i)$, $i = 1, 2$, then its differential evaluated at e satisfies $\psi_{*,e}[x, y]_1 = [\psi_{*,e}(x), \psi_{*,e}(y)]_2$, for all $x, y \in \mathfrak{g}_1$. If V is a finite dimensional vector space, it is easy to prove that the Lie algebra of $\mathrm{GL}(V)$ is $\mathfrak{gl}(V) = (\mathrm{End}_{\mathbb{F}}(V), [\cdot,\cdot])$, where the bracket $[\cdot,\cdot]$ is obtained from skew-symmetrizing the associative product of $\mathrm{End}_{\mathbb{F}}(V)$. Since every left (right) invariant vector field is complete, for every $x \in \mathfrak{g}$ the integral curve of X_x, going through the identity $e \in G$ at $t = 0$, defines a smooth map $\gamma_e^x : \mathbb{R} \to G$. It can be shown that γ_e^x is a Lie group homomorphism from $(\mathbb{R}, +)$ to G, and every

---

[2] In the following we will use $*$-notation to denote the differential of a smooth application $\phi : M_1 \to M_2$, if $M_2 \neq \mathbb{F}$. More precisely the differential of ϕ at $m \in M_1$ will be denoted as $\phi_{*,m}$. Recall that this is a linear map between T_mM_1 and $T_{\phi(m)}M_2$ such that $(\phi_{*,m}v)f = v(f \circ \phi)$, for all $v \in T_mM_1$ and all $f \in C^\infty(M_2)$. On the other hand, if $M_2 = \mathbb{F}$, i.e., if $\phi = H : M \to \mathbb{F}$ is a smooth function, we will write its differential at the point $m \in M$ as dH_m. Note that $dH_m \in T_m^*M$.

continuous group homomorphism between $(\mathbb{R}, +)$ and G is of this form. For all $x \in \mathfrak{g}$ the curve γ_e^x is called a 1-*parameter group homomorphism* of G. Given such a 1-parameter group homomorphism, one can define the *exponential map* of G, $\exp : \mathfrak{g} \to G$

$$x \in \mathfrak{g} \rightsquigarrow \exp x = \gamma_e^x(1). \tag{11}$$

For example if $G = \mathrm{GL}_n(\mathbb{F})$ and $\mathfrak{g} = \mathfrak{gl}_n(\mathbb{F})$, then the exponential map defined in (11) is the *usual exponential map*, i.e., $\exp x = \mathrm{e}^x := \sum_{k\geq 0} \frac{x^k}{k!}$, for all $x \in \mathfrak{g}$. The exponential map (11) has the following properties:

1. $\exp x$ is smooth;
2. $(\exp)_{*,0} = \mathrm{id}_{\mathfrak{g}}$. In particular, there exist $U \subset \mathfrak{g}$ and $V \subset G$, open neighborhoods of $0 \in \mathfrak{g}$ respectively $e \in G$, such that $\exp|_U : U \to V$ is a diffeomorphism.
3. Let G_1 and G_2 be two Lie groups and $\mathfrak{g}_1$ respectively $\mathfrak{g}_2$ the corresponding Lie algebras. If $\phi : G_1 \to G_2$ is a Lie group morphism, then:

$$\exp \circ \phi_{*,e} = \phi \circ \exp . \tag{12}$$

For each $g \in G$, let $c_g : G \to G$ be defined by $c_g = R_{g^{-1}} L_g = L_g R_{g^{-1}}$. Then c_g is a diffeomorphism of G such that $c_g(e) = e$ and $c_{g_1 g_2} = c_{g_1} c_{g_2}$ for all $g_1, g_2 \in G$. From these observations, it follows that, for each $g \in G$, $\mathrm{Ad}_g = (c_g)_{*,e} \in \mathrm{GL}(\mathfrak{g})$ is an automorphism of $\mathfrak{g}$. In this way it is defined a homomorphism of Lie groups, $\mathrm{Ad} : G \to \mathrm{Aut}(\mathfrak{g})$, taking $g \in G$ to $\mathrm{Ad}_g \in \mathrm{Aut}(\mathfrak{g})$, called the *adjoint representation* of G. The differential of Ad at the identity $e \in G$ is a Lie algebra homomorphism $\mathrm{ad} := \mathrm{Ad}_{*,e} : \mathfrak{g} \to \mathrm{End}(\mathfrak{g})$, called the *adjoint representation* of $\mathfrak{g}$. Note that $\mathrm{ad}_x\, y = [x, y]$, for all $x, y \in \mathfrak{g}$. Recall that $\mathfrak{g}^*$ is the dual of $\mathfrak{g}$. Together with the adjoint representation one can introduce the *co-adjoint* representations of G and $\mathfrak{g}$. More precisely, one can define the morphism of Lie groups $\mathrm{Ad}^\sharp : G \to \mathrm{GL}(\mathfrak{g}^*)$, via

$$\langle \mathrm{Ad}_g^\sharp\, \alpha, x\rangle = \langle \alpha, \mathrm{Ad}_{g^{-1}}\, x\rangle, \quad \forall g \in G, \alpha \in \mathfrak{g}^*,\ x \in \mathfrak{g},$$

whose differential at the identity $e \in G$ is the morphism of Lie algebras $\mathrm{ad}^\sharp : \mathfrak{g} \to \mathrm{End}_{\mathbb{F}}(\mathfrak{g})$, defined by $\langle \mathrm{ad}_x^\sharp\, \alpha, y\rangle = -\langle \alpha, \mathrm{ad}_x\, y\rangle$, for all $x, y \in \mathfrak{g}$, $\alpha \in \mathfrak{g}^*$. Finally, for each $g \in G$ and $x \in \mathfrak{g}$, one has that

$$\begin{aligned} c_g(\exp x) &= \exp\big(\mathrm{Ad}_g\, x\big), \\ \mathrm{Ad}_{\exp x} &= \mathrm{e}^{\mathrm{ad}_x}, \end{aligned}$$

where $\mathrm{e}^A = \sum_{k\geq 0} \frac{A^k}{k!}$, for all $A \in \mathrm{End}_{\mathbb{F}}(\mathfrak{g})$.

As mentioned before, $\exp_{*,0} = \mathrm{id}_{\mathfrak{g}}$. A closed formula for the differential of the exponential map at general $x \in \mathfrak{g}$ is:

$$\exp_{*,x} = \big(L_{\exp x}\big)_{*,e} \circ \frac{\mathrm{id}_{\mathfrak{g}} - \mathrm{e}^{-\mathrm{ad}_x}}{\mathrm{ad}_x}, \tag{13}$$

where the formal expression $\frac{\mathrm{id}_{\mathfrak{g}} - e^{-\mathrm{ad}_x}}{\mathrm{ad}_x}$ represents the element of $\mathrm{End}_{\mathbb{F}}(\mathfrak{g})$ defined by

$$\frac{\mathrm{id}_{\mathfrak{g}} - e^{-\mathrm{ad}_x}}{\mathrm{ad}_x} = \int_0^1 e^{-s\,\mathrm{ad}_x}\, ds.$$

In particular one can prove that the exponential map is a local diffeomorphism in a neighborhood of $x \in \mathfrak{g}$ if and only if the linear operator ad_x has no eigenvalues in the set $2\pi\,\imath\,\mathbb{Z}\backslash\{0\}$. Choosing $x, y \in \mathfrak{g}$ belonging to a sufficiently small open neighborhood U of $0 \in \mathfrak{g}$, such that $\exp x \exp y \in V \subset G$, where V is a small neighborhood of $e \in G$, one is able to find an element $\mathrm{BCH}(x, y) \in \mathfrak{g}$ such that

$$\exp \mathrm{BCH}(x, y) = \exp x\, \exp y,$$

or, what is equivalent, such that

$$\mathrm{BCH}(x, y) = \log\big(\exp x\, \exp y\big),$$

where $\log : V \to U$ denotes the inverse of the restriction of the exponential map. An explicit formula for $\mathrm{BCH}(x, y)$ is given by the so-called *Baker–Campbell–Hausdorff* series (BCH-series), as stated in the following theorem.

Theorem 1 *Let G be a Lie group and $\mathfrak{g}$ its Lie algebra. For $x, y \in \mathfrak{g}$ sufficiently close to $0 \in \mathfrak{g}$, the following formula holds:*

$$\mathrm{BCH}(x, y) = x + y +$$
$$\sum_{k=1}^{\infty} \frac{(-1)^k}{k+1} \sum_{\substack{l_1,\ldots,l_k \geq 0 \\ m_1,\ldots,m_k \geq 0 \\ l_j+m_j>0}} \frac{1}{l_1 + \cdots + l_k + 1}\Big(\frac{(\mathrm{ad}_x)^{l_1}}{l_1!} \circ \frac{(\mathrm{ad}_y)^{m_1}}{m_1!} \circ \cdots \circ \frac{(\mathrm{ad}_x)^{l_k}}{l_k!} \circ \frac{(\mathrm{ad}_y)^{m_k}}{m_k!}\Big)(x).$$

The first few terms are:

$$\mathrm{BCH}(x, y) = x + y + \frac{1}{2}[x, y] + \frac{1}{12}([x, [x, y]] + [y, [y, x]]) + \cdots.$$

The reduced BCH-series is defined by $\overline{\mathrm{BCH}}(x, y) := \mathrm{BCH}(x, y) - x - y$. From Theorem 1 it is clear that $\mathrm{BCH}(x, y)$ is a *Lie series* in the non-commutative variables x, y, i.e., it is a series of iterated Lie brackets of the elements $x, y \in \mathfrak{g}$.

Remark 1 Note that Theorem 1 is an analytical result addressing convergence of the BCH-series. One may consider $\mathrm{BCH}(x, y)$ for arbitrary elements of $\mathfrak{g}$ and ignore convergence issues. However, in this case one must consider $\mathrm{BCH}(x, y)$ as a formal object, see also Example 5 further below.

2.2 Universal Enveloping Algebras and Hopf Algebras

We follow references [28, 32]. Let $\mathfrak{g}$ be a finite dimensional Lie algebra and let $T(\mathfrak{g}) = T_{\mathfrak{g}} = (T^{\bullet}\mathfrak{g}, \otimes)$ be its *tensor algebra*, which is a *graded*, *associative* and *non–commutative* algebra, whose homogeneous sub-space of degree n, $\mathfrak{g}^n := \mathfrak{g} \otimes \mathfrak{g}^{n-1}$, $\mathfrak{g}^0 := 1$, is generated, as vector space, by monomials of the form $x_{i_1} \otimes \cdots \otimes x_{i_n}$. Note that we will identify $x_{i_1} \otimes \cdots \otimes x_{i_n}$ with words $x_{i_1} \cdots x_{i_n}$. Consider the 2-*sided ideal*

$$J = \langle x \otimes y - y \otimes x - [x, y] \rangle := T_{\mathfrak{g}}\big(x \otimes y - y \otimes x - [x, y]\big)T_{\mathfrak{g}}.$$

Definition 1 *(Universal enveloping algebra)* The *universal enveloping algebra* $\mathscr{U}(\mathfrak{g})$ of $\mathfrak{g}$ is the algebra $T_{\mathfrak{g}}/J$ whose product is induced on the quotient vector space by $\otimes$, i.e., if $\overline{X}, \overline{Y} \in \mathscr{U}(\mathfrak{g})$ are the classes of the monomials $X \in \mathfrak{g}^k$ and $Y \in \mathfrak{g}^l$, then $\overline{X} \cdot \overline{Y}$ is the class of the monomial $X \otimes Y \in \mathfrak{g}^{k+l}$.

Note that $\mathscr{U}(\mathfrak{g})$ is a *unital*, *associative* algebra. In general it is not graded, since the ideal J is non–homogeneous. However, $\mathscr{U}(\mathfrak{g})$ is a *filtered* algebra, that is, it is endowed with the filtration $\mathbb{F} = \mathscr{U}_0(\mathfrak{g}) \subset \mathscr{U}_1(\mathfrak{g}) \subset \cdots \subset \mathscr{U}_n(\mathfrak{g}) \subset \cdots$, where $\mathscr{U}_n(\mathfrak{g})$ is the subspace of $\mathscr{U}(\mathfrak{g})$ generated by *monomials of length at most* n, i.e., monomials like $x_{i_1} \cdots x_{i_n}$ with $x_{i_1}, \ldots, x_{i_n} \in \mathfrak{g}$. Note that $\mathscr{U}_i(\mathfrak{g}) \cdot \mathscr{U}_j(\mathfrak{g}) \subset \mathscr{U}_{i+j}(\mathfrak{g})$, for all $i, j \geq 0$, and that $\mathscr{U}(\mathfrak{g}) = \cup_{k \geq 0} \mathscr{U}_k(\mathfrak{g})$. Observe that $\mathscr{U}_1(\mathfrak{g}) \simeq \mathfrak{g}$, so that there is a *natural* homomorphism of Lie algebras, $i : \mathfrak{g} \to \mathscr{U}(\mathfrak{g})_{\text{Lie}}$. The adjective universal emphasizes the fact that $\mathscr{U}(\mathfrak{g})$ has the following property: suppose that $\mathscr{A}$ is an associative algebra and that $j : \mathfrak{g} \to \mathscr{A}_{\text{Lie}}$ is a morphism of Lie algebras. Then there exists a *unique* morphism of unital associative algebras, $\phi : \mathscr{U}(\mathfrak{g}) \to \mathscr{A}$, which makes the following diagram of Lie algebras commutative:

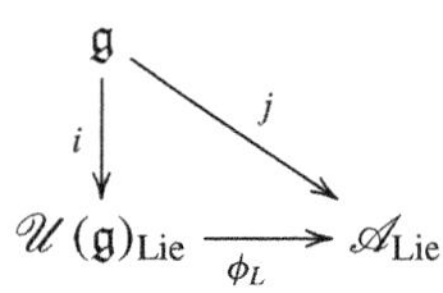

The *graded algebra* associated to $\mathscr{U}(\mathfrak{g})$ is:

$$\text{gr}(\mathscr{U}(\mathfrak{g})) = \bigoplus_{k \geq 0} \frac{\mathscr{U}_k(\mathfrak{g})}{\mathscr{U}_{k-1}(\mathfrak{g})}, \quad \mathscr{U}_{-1}(\mathfrak{g}) = \{0\}.$$

Furthermore, note that $\mathfrak{g} \simeq \dfrac{\mathscr{U}_1(\mathfrak{g})}{\mathscr{U}_0(\mathfrak{g})}$ so that there exists a linear map $i : \mathfrak{g} \to \text{gr}(\mathscr{U}(\mathfrak{g}))$ and $\text{gr}(\mathscr{U}(\mathfrak{g}))$, endowed with the obvious multiplication, is a *commutative algebra*. In fact, for every $k \geq 0$, $x_{i_1} \cdots x_{i_k} - x_{\sigma(i_1)} \cdots x_{\sigma(i_k)} \in \mathscr{U}_{k-1}(\mathfrak{g})$, for all $\sigma \in \Sigma_k$, the permutation group of k elements. This is clear when σ is a transposition. For a general σ, the statement follows from the fact that every permutation is the product of transpositions. Observe that since a general $x_{i_1}\cdots x_{i_k}\in\mathscr{U}_k(\mathfrak{g})$ can be written as

$$x_{i_1}\cdots x_{i_k}=\frac{1}{k!}\sum_{\sigma\in\Sigma_k}x_{\sigma(i_1)}\cdots x_{\sigma(i_k)}+\frac{1}{k!}\sum_{\sigma\in\Sigma_k}\big(x_{i_1}\cdots x_{i_k}-x_{\sigma(i_1)}\cdots x_{\sigma(i_k)}\big),\qquad(14)$$

where each summand of the second sum is an element of $\mathscr{U}_{k-1}(\mathfrak{g})$, for each $k\geq 1$ one has the following exact sequence of vector spaces

$$0\longrightarrow\mathscr{U}_{k-1}(\mathfrak{g})\xrightarrow{\ i_{k-1}\ }\mathscr{U}_k(\mathfrak{g})\xrightarrow{\ \sigma_k\ }\frac{\mathscr{U}_k(\mathfrak{g})}{\mathscr{U}_{k-1}(\mathfrak{g})}\longrightarrow 0\qquad(15)$$

where $\sigma_k(x_{i_1}\cdots x_{i_k})$ is the class in $\frac{\mathscr{U}_k(\mathfrak{g})}{\mathscr{U}_{k-1}(\mathfrak{g})}$ of the sum $\frac{1}{k!}\sum_{\sigma\in\Sigma_k}x_{\sigma(i_1)}\cdots x_{\sigma(i_k)}$, see Formula (14).

Together with the universal enveloping algebra, one can introduce the *symmetric algebra* of $\mathfrak{g}$, $S(\mathfrak{g})=S_{\mathfrak{g}}:=T_{\mathfrak{g}}/J'$ where J' is the 2-sided ideal $T_{\mathfrak{g}}\big(x\otimes y-y\otimes x\big)T_{\mathfrak{g}}$. $S_{\mathfrak{g}}$ is a *graded commutative algebra* endowed with a natural *injective* linear map $j:\mathfrak{g}\to S_{\mathfrak{g}}$, having the following *universal property*: Given a commutative algebra $\mathscr{C}$ and a linear map $f:\mathfrak{g}\to\mathscr{C}$ there exists a *unique* map of commutative algebras $\phi:S_{\mathfrak{g}}\to\mathscr{C}$ which closes the following to a commutative diagram:

$$\begin{array}{ccc}\mathfrak{g} & & \\ {\scriptstyle j}\downarrow & \searrow^{f} & \\ S_{\mathfrak{g}} & \xrightarrow[\phi]{} & \mathscr{C}\end{array}$$

For each $k\geq 0$, $S_k(\mathfrak{g})$ denotes the homogeneous component of degree k of $S_{\mathfrak{g}}$. More precisely,

$$S_{\mathfrak{g}}=\oplus_{k\geq 0}S_k(\mathfrak{g}),$$

and $S_0(\mathfrak{g}):=\mathbb{F}$ and $S_{-1}(\mathfrak{g})=\{0\}$. Letting $\mathscr{C}=\mathrm{gr}(\mathscr{U}(\mathfrak{g}))$ and $f=i:\mathfrak{g}\to\mathrm{gr}(\mathscr{U}(\mathfrak{g}))$, one can state the following important result.

Theorem 2 (Poincaré–Birkhoff–Witt) *The corresponding map $\phi:S_{\mathfrak{g}}\to\mathrm{gr}(\mathscr{U}(\mathfrak{g}))$ in the above diagram is an isomorphism of graded commutative algebras. In particular, for each $k\geq 0$ one has that*

$$\phi_k:=\phi|_{S_k(\mathfrak{g})}:S_k(\mathfrak{g})\to\frac{\mathscr{U}_k(\mathfrak{g})}{\mathscr{U}_{k-1}(\mathfrak{g})}\qquad(16)$$

is an isomorphism of vector spaces.

Note that ϕ_k in (16) maps every monomial $x_{i_1}\cdots x_{i_k}\in S_k(\mathfrak{g})$ to the class of $x_{i_1}\cdots x_{i_k}$ in $\frac{\mathscr{U}_k(\mathfrak{g})}{\mathscr{U}_{k-1}(\mathfrak{g})}$, i.e., $\phi(x_{i_1}\cdots x_{i_k})=x_{i_1}\cdots x_{i_k}\bmod\mathscr{U}_{k-1}(\mathfrak{g})$, where the product on the l.h.s is the one in the symmetric algebra while the product on the r.h.s is the one in the universal enveloping algebra. Since for each $x_{i_1}\cdots x_{i_k}\in\mathscr{U}_k(\mathfrak{g})$

$$x_{i_1} \cdots x_{i_k} = \frac{1}{k!} \sum_{\sigma \in \Sigma_k} x_{\sigma(i_1)} \cdots x_{\sigma(i_k)} \bmod \mathscr{U}_{k-1}(\mathfrak{g}),$$

see (14), for each k, one can use the (inverse) of the map ϕ_k together with (15), to define the following exact sequence

$$0 \longrightarrow \mathscr{U}_{k-1}(\mathfrak{g}) \xrightarrow{i_{k-1}} \mathscr{U}_k(\mathfrak{g}) \xrightarrow{s_k} S_k(\mathfrak{g}) \longrightarrow 0 \tag{17}$$

where $s_k = \phi_k^{-1} \circ \sigma_k$ is defined by

$$s_k(x_{i_1} \cdots x_{i_k}) = x_{i_1} \cdots x_{i_k}. \tag{18}$$

The map s_k defined above is called the (degree k) *symbol map*. Since (17) is an exact sequence of vector spaces it splits. The linear map $\mathrm{symm}_k : S_k(\mathfrak{g}) \to \mathscr{U}_k(\mathfrak{g})$ defined by

$$\mathrm{symm}_k(x_{i_1} \cdots x_{i_k}) = \frac{1}{k!} \sum_{\sigma \in \Sigma_k} x_{\sigma(i_1)} \cdots x_{\sigma(i_k)} \tag{19}$$

and called the (degree k) *symmetrization map*, is a *section* of (17), i.e., for each k, $s_k \circ \mathrm{symm}_k = \mathrm{id}_{S_k(\mathfrak{g})}$.

Note that both products in (18) and (19) on the right and left side should be interpreted accordingly to the meaning of the monomials.

Observe that when $\mathfrak{g}$ is abelian we have $\mathscr{U}(\mathfrak{g}) \simeq S_\mathfrak{g}$, while for general $\mathfrak{g}$ the Poincaré–Birkhoff–Witt theorem tells us that we still have an isomorphism $\mathscr{U}(\mathfrak{g}) \simeq S_\mathfrak{g}$ but *only* at the level of vector spaces.

The universal enveloping algebra is an example of a *quasi-commutative algebra*, i.e., an associative, unital and filtered algebra $\mathscr{A}$, whose associated graded algebra $\mathrm{gr}(\mathscr{A})$ is commutative. One can prove that if $\mathscr{A}$ is a quasi-commutative algebra, then $\mathrm{gr}(\mathscr{A})$ is a *Poisson algebra*, see Sect. 4. To define the Poisson bracket on $\mathrm{gr}(\mathscr{A})$ it suffices to define it on the homogeneous components of the associated algebra. To this end, let:

$$\{\cdot,\cdot\} : \frac{\mathscr{A}_i}{\mathscr{A}_{i-1}} \times \frac{\mathscr{A}_j}{\mathscr{A}_{j-1}} \to \frac{\mathscr{A}_{i+j-1}}{\mathscr{A}_{i+j-2}}, \quad (\overline{x}, \overline{y}) \rightsquigarrow (xy - yx) \bmod \mathscr{A}_{i+j-2}, \tag{20}$$

where $x \in \mathscr{A}_i$ and $y \in \mathscr{A}_j$ are two lifts of $\overline{x}$ respectively $\overline{y}$. The proof follows at once after showing that such a bracket is well defined, in particular, that given x, y as above $xy - yx \in \mathscr{A}_{i+j-1}$, and that the result does not depend on the choice of the two lifts. Given that, the proof that the above bracket is Poisson follows from the fact that $\mathscr{A}$ is an associative algebra. Then, in particular, given a Lie algebra $\mathfrak{g}$, the graded algebra associated to $\mathscr{U}(\mathfrak{g})$ is a Poisson algebra. Using the Poincaré–Birkhoff–Witt theorem, such a Poisson structure can be transferred to the symmetric algebra $S_\mathfrak{g}$. In this framework it is worth to note that the Poisson bracket induced on $S_\mathfrak{g}$ by the one defined on $\mathrm{gr}(\mathscr{U}(\mathfrak{g}))$ coincides with the linear Poisson structure of $\mathfrak{g}$, see (37) further below. To prove this statement, it suffices to check it on the restriction of (20) to the components of degree 1:

$$\{\cdot,\cdot\}: \frac{\mathscr{U}_1(\mathfrak{g})}{\mathscr{U}_0(\mathfrak{g})} \times \frac{\mathscr{U}_1(\mathfrak{g})}{\mathscr{U}_0(\mathfrak{g})} \to \frac{\mathscr{U}_1(\mathfrak{g})}{\mathscr{U}_0(\mathfrak{g})}, \quad (\overline{x},\overline{y}) \rightsquigarrow (xy-yx) \bmod \mathscr{U}_0(\mathfrak{g}),$$

which shows that $\{\overline{x},\overline{y}\} = \overline{[x,y]}$ (remember that $\frac{\mathscr{U}_1(\mathfrak{g})}{\mathscr{U}_0(\mathfrak{g})} \simeq \mathfrak{g}$ and that $\mathscr{U}_0(\mathfrak{g}) \simeq \mathbb{F}$). Furthermore, one can prove that if A is a positively filtered algebra, such that

(i) $A_0 = \mathbb{F}$,
(ii) A is generated as a ring by A_1 and
(iii) A is almost-commutative, then

there exists a Lie algebra $\mathfrak{g}$ and an ideal I of $\mathscr{U}(\mathfrak{g})$, such that $A \simeq \mathscr{U}(\mathfrak{g})/I$.
Universal enveloping algebra as a Hopf algebra. The universal enveloping algebra $\mathscr{U}(\mathfrak{g})$ of a Lie algebra $\mathfrak{g}$ carries the structure of a *Hopf algebra*. See [6, 38].

Recall that an associative $\mathbb{F}$-algebra with unit is a triple (A, m, i), consisting of the vector space A, together with the map $m : A \otimes A \to A$ (multiplication) and map $i : \mathbb{F} \to A$ (unit), such that

$$\begin{aligned} m \circ (m \otimes \mathrm{id}) = m \circ (\mathrm{id} \otimes m) &: A \otimes A \otimes A \to A \qquad \textit{associativity} \\ m \circ (i \otimes \mathrm{id}) = \mathrm{id} = m \circ (\mathrm{id} \otimes i) &: A \otimes \mathbb{F} \simeq \mathbb{F} \otimes A \to A \qquad \textit{unitproperty}. \end{aligned}$$

If $\tau : A \otimes A \to A \otimes A$ is defined by $\tau(a \otimes b) = b \otimes a$, then (A, m, i) is called *commutative* if $m \circ \tau = m$.

A *co-algebra* is defined as a triple (C, Δ, ε), where C is a vector space, and $\Delta : C \to C \otimes C$, $\varepsilon : C \to \mathbb{F}$ are two linear maps, the first is called *co-product* and the second is called *co-unit*. Co-product and co-unit satisfy the following properties:

$$\begin{aligned} (\Delta \otimes \mathrm{id}) \circ \Delta = (\mathrm{id} \otimes \Delta) \circ \Delta &: C \to C \otimes C \otimes C \qquad \textit{co-associativity} \\ (\varepsilon \otimes \mathrm{id}) \circ \Delta = \mathrm{id} = (\mathrm{id} \otimes \varepsilon) \circ \Delta &: C \to C \qquad \textit{co-unitproperty}. \end{aligned}$$

A co-algebra is *co-commutative* if $\tau \circ \Delta = \Delta$. Note that the notions of algebra and co-algebra are *almost* dual to each other. More precisely, the dual of a co-algebra is an algebra whose multiplication and unit maps are obtained by reversing arrows of co-multiplication and co-unit. On the other hand, taking the dual of an algebra and reversing the arrows of multiplication and unit map, one obtains a co-algebra (A^*, m^*, i^*) if $\dim A < \infty$. Otherwise, the co-multiplication obtained by reversing the arrows of multiplication, is a map $m^* : A^* \to (A \otimes A)^*$, which, in general, is a proper vector sub-space of $A^* \otimes A^*$. Finally, a *Hopf algebra* is a vector space endowed with compatible algebra and co-algebra structures as well as an *antipode* $S : H \to H$. That is, a Hopf algebra is a quintuplet $(H, m, i, \Delta, \varepsilon, S)$ consisting of a vector space H together with maps:

$$m : H \otimes H \to H \qquad \textit{multiplication}$$

$$
\begin{aligned}
\Delta : H \to H \otimes H &\quad co-multiplication \\
i : \mathbb{F} \to H &\quad unit \\
\varepsilon : H \to \mathbb{F} &\quad co-unit \\
S : H \to H &\quad antipode
\end{aligned}
$$

such that (H, m, i) is an algebra and (H, Δ, ε) is a co-algebra, which are compatible

$$\Delta \circ m = (m \otimes m) \circ (\mathrm{id} \otimes \tau \otimes \mathrm{id}) \circ \Delta \otimes \Delta \tag{21}$$

$$\varepsilon \otimes \varepsilon = \varepsilon \circ m. \tag{22}$$

Note that these conditions are equivalent to saying that (Δ, ε) are algebra morphisms—equivalently, (m, i) are co-algebra morphisms. The antipode S is a linear map satisfying:

$$m \circ (\mathrm{id} \otimes S) \circ \Delta = i \circ \varepsilon = m \circ (S \otimes \mathrm{id}) \circ \Delta,$$

and it is easy to show that it is a co-algebra and algebra anti-homorphism, such that $S \circ i = i$ and $\varepsilon \circ S = \varepsilon$. Without antipode, a Hopf algebra reduces to a *bialgebra*.

An element $x \in H$ will be called *primitive* if $\Delta x = x \otimes 1 + 1 \otimes x$, while $g \in H$ will be called *group-like* if $\Delta g = g \otimes g$. Let $\mathscr{P}(H)$ and $\mathscr{G}(H)$ be the sets of primitive respectively group-like elements in the Hopf algebra $(H, m, i, \Delta, \varepsilon, S)$. Note that if $g_1, g_2 \in \mathscr{G}(H)$, then $g_1 \cdot g_2 := m(g_1, g_2) \in \mathscr{G}(H)$, and if $x_1, x_2 \in \mathscr{P}(H)$, then $[x_1, x_2] := x_1 \cdot x_2 - x_2 \cdot x_1 \in \mathscr{P}(H)$. In particular, $(\mathscr{P}(H), [\cdot, \cdot])$ is a Lie algebra. Furthermore, defining $e := i(1)$ and $g^{-1} := S(g)$ for all $g \in \mathscr{G}(H)$, one can show that $g \cdot e = g = e \cdot g$, and $g^{-1} \cdot g = e = g \cdot g^{-1}$, for all $g \in \mathscr{G}(H)$. In other words, $(\mathscr{G}(H), \cdot)$ is a group whose identity element is e, such that for each $g \in \mathscr{G}(H)$, $g^{-1} = S(g)$.

Proposition 1 *Let $\mathfrak{g}$ be a Lie algebra. Its universal enveloping algebra $\mathscr{U}(\mathfrak{g})$ is a co-commutative Hopf algebra.*

Proof To prove the first part of the statement it suffices to define the antipode and a co-algebra structure compatible with the algebra structure of $\mathscr{U}(\mathfrak{g})$. Let $\mathfrak{G} = \mathfrak{g} \oplus \mathfrak{g}$ be endowed with the structure of direct product Lie algebra and let $\Delta : \mathfrak{g} \to \mathfrak{G}$ be the diagonal embedding, i.e., $\Delta(x) = (x, x)$, for all $x \in \mathfrak{g}$. Then, by the universal property, Δ extends uniquely to an associative algebra morphism $\Delta : \mathscr{U}(\mathfrak{g}) \to \mathscr{U}(\mathfrak{G})$, which, composed with the canonical isomorphism $\mathscr{U}(\mathfrak{G}) \simeq \mathscr{U}(\mathfrak{g}) \otimes \mathscr{U}(\mathfrak{g})$, defines a linear map $\Delta : \mathscr{U}(\mathfrak{g}) \to \mathscr{U}(\mathfrak{g}) \otimes \mathscr{U}(\mathfrak{g})$, defined by

$$
\begin{aligned}
\Delta(x_1 \cdots x_n) &= x_1 \cdots x_n \otimes 1 + 1 \otimes x_1 \cdots x_n \\
&+ \sum_{k=1}^{n-1} \sum_{\sigma \in \Sigma_{k,n-k}} x_{\sigma(1)} \cdots x_{\sigma(k)} \otimes x_{\sigma(k+1)} \cdots x_{\sigma(n)}
\end{aligned} \tag{23}
$$

where for each $k = 1, \dots, n-1$, $\Sigma_{k,n-k}$ is the subgroup of Σ_n of the $(k, n-k)$ shuffles. Starting now from the trivial map $\mathfrak{g} \to 0$ and using again the universal property of $\mathscr{U}(\mathfrak{g})$, one can define the co-unit map $\varepsilon : \mathscr{U}(\mathfrak{g}) \to \mathbb{F}$. It is again the universal property of the enveloping algebra, that permits to show that $(\mathscr{U}(\mathfrak{g}), \Delta, \varepsilon)$ is a co-algebra.

On the other hand, the map $S : \mathfrak{g} \to \mathfrak{g}$, defined by $S(x) = -x$ for all $x \in \mathfrak{g}$, is a Lie algebra anti-homomorphism, which extends in a unique way to an associative algebra homomorphism $S : \mathscr{U}(\mathfrak{g}) \to \mathscr{U}(\mathfrak{g})$, such that $S(x_{i_1} \cdots x_{i_n}) = (-1)^n x_{i_n} \cdots x_{i_1}$ for each monomial $x_{i_1} \cdots x_{i_n}$, and it satisfies the antipode property. Finally, the proof of co-commutativity follows at once from the universal property of $\mathscr{U}(\mathfrak{g})$ and noticing that the maps $(\Delta \otimes \mathrm{id}) \circ \Delta$ and $(\mathrm{id} \otimes \Delta) \circ \Delta : \mathscr{U}(\mathfrak{g}) \to \mathscr{U}(\mathfrak{g}) \otimes \mathscr{U}(\mathfrak{g}) \otimes \mathscr{U}(\mathfrak{g})$ are both obtained from the embeddings of $\mathfrak{g}$ into $\mathfrak{g} \oplus \mathfrak{G} \simeq \mathfrak{g} \oplus \mathfrak{g} \oplus \mathfrak{g}$ respectively $\mathfrak{g} \to \mathfrak{G} \oplus \mathfrak{g} \simeq \mathfrak{g} \oplus \mathfrak{g} \oplus \mathfrak{g}$.

Note that every $x \in \mathfrak{g} = \mathscr{U}_1(\mathfrak{g})$ is a primitive element. Furthermore, it can be shown that if $\xi \in \mathscr{U}(\mathfrak{g})$ is primitive then $\xi \in \mathfrak{g}$. In other words, one can prove that $\mathscr{P}(\mathscr{U}(\mathfrak{g})) = \mathfrak{g}$. On the other hand, it is simple to see that in $\mathscr{U}(\mathfrak{g})$ there are no group-like elements of degree greater than zero, i.e., $\mathscr{G}(\mathscr{U}(\mathfrak{g})) = \mathbb{F}$. To associate non-trivial group-like elements to $\mathscr{U}(\mathfrak{g})$ one needs to consider instead of $\mathscr{U}(\mathfrak{g})$ a suitable *completion* of it.

Remark 2 Since $\mathfrak{g} = \mathscr{P}(\mathscr{U}(\mathfrak{g}))$, every Lie polynomial is still a primitive element of $\mathscr{U}(\mathfrak{g})$.

Complete Hopf algebras. We follow reference [30]. In what follows all algebras are unital and defined over the field $\mathbb{F}$. A will be called an *augmented* algebra, if it comes endowed with an algebra morphism $\varepsilon : A \to \mathbb{F}$ called the *augmentation map*. In this case its kernel $\ker \varepsilon$ will be called the *augmentation ideal* and it will be denoted by I.

Example 1 $A = \mathscr{U}(\mathfrak{g})$ is an example of augmented algebra. In fact the co-unit $\varepsilon : \mathscr{U}(\mathfrak{g}) \to \mathbb{F}$ is a augmentation map and its kernel, $I = \cup_{k>0} \mathscr{U}_k(\mathfrak{g})$, is the corresponding augmentation ideal.

A *decreasing filtration* of A is a decreasing sequence $A = F_0 A \supset F_1 A \supset \cdots$ of sub-vector spaces, such that $1 \in F_0 A$ and $F_p A \cdot F_q A \subset F_{p+q} A$, and $\mathrm{gr}\, A = \oplus_{n=0}^{\infty} F_n A / F_{n+1} A$ has a natural structure of a graded algebra. Note that for each k, $F_k A$ is a two-side ideal of A. We can now define the notion of a *complete augmented algebra*.

Definition 2 A *complete augmented algebra* is an augmented algebra A endowed with a decreasing filtration $\{F_k A\}_{k \in \mathbb{N}}$ such that:

(1) $F_1 A = I$,
(2) $\mathrm{gr}\, A$ is generated as an algebra by $\mathrm{gr}_1\, A$,

(3) As an algebra, A is the *inverse limit* $A = \varprojlim A/F_nA$.[3]

Example 2 Let A be an augmented algebra. Then $\hat{A} = \varprojlim A/I^n$ is a complete augmented algebra where, for each $n \geq 0$, $F_n\hat{A} = \widehat{I^n} = \varprojlim I^n/I^k$, $k \geq n$. It is worth to recall that $\hat{A}$ is also called the *I-adic completion* of A. Note that in this case $F_nA := I^n$ if $n \geq 1$ and $F_0A = A$ and the inverse system defining the completion is given by the data $(\{A_i\}_{i\in I}, \{f_{ij}\}_{i,j\in\mathscr{I}})$ where $\mathscr{I} = \mathbb{N}$, $A_n = A/I^n$ and $f_{ij} : A_j \to A_i$ is the application that, for all $a \in A$, maps $a \bmod A_j$ to $a \bmod A_i$, for all $j \leq i$.

Since $A/I^n \simeq \hat{A}/\widehat{I^n}$, one has that $\operatorname{gr}\hat{A} \simeq \operatorname{gr} A = \oplus_{n\geq 0} I^n/I^{n+1}$, which implies that $\hat{A}$ satisfies property (2) in the definition above. Property (1) is clear from the definition of the filtration of $\hat{A}$, while Property (3) follows from the isomorphism $\hat{\hat{A}} \simeq \hat{A}$, for each $\hat{A} = \varprojlim A_n$, where $\hat{\hat{A}} = \varprojlim \widehat{A_n}$ and $\widehat{A_n} = \varprojlim A_k$, $k \geq n$. In particular, taking $A = \mathscr{U}(\mathfrak{g})$ and $I = \cup_{k>0}\mathscr{U}_k(\mathfrak{g})$, see Example 1, one can define the complete augmented algebra

$$\hat{\mathscr{U}}(\mathfrak{g}) = \varprojlim \mathscr{U}(\mathfrak{g})/I^n, \tag{24}$$

which will be simply called in the following the *completion* of $\mathscr{U}(\mathfrak{g})$.

Let $V = F_0V \supset F_1V \supset F_2V \supset \cdots$ be a filtered vector space and $\pi_n : F_nV \to \operatorname{gr}_n V$ be the canonical surjection. If W is another filtered vector space, then one can define a filtration on $V \otimes W$ declaring that $F_n(V \otimes W) = \sum_{i+j=n} F_iV \otimes F_jV \subset V \otimes W$, for all $n \geq 0$, where one identifies $F_iV \otimes F_jV$ with its image in $V \otimes W$ via the canonical injection. If V and W are complete, i.e. if $V = \varprojlim V/F_nV$ and $W = \varprojlim W/F_nW$, then we denote by $V\hat{\otimes}W$ the completion of $V \otimes W$ with respect to the filtrations defined above, and we denote with $x\hat{\otimes}y$ the image of $x \otimes y$ via the canonical morphism between $V \otimes W$ and $V\hat{\otimes}W$. Note that, since $F_{2n}(V \otimes W) \subset F_nV \otimes W + V \otimes F_nW \subset F_n(V \otimes W)$, one has that

$$V\hat{\otimes}W = \varprojlim(V_n \otimes W_n),$$

where, given the filtered vector space $V = F_0V \supset F_1V \supset F_2V \supset \cdots$, $V_n = V/F_nV$.

Definition 3 The vector space $V\hat{\otimes}W$ so defined is called the *complete tensor product* of the complete vector spaces V and W.

---

[3]Let $(\mathscr{I}, \leq)$ be a *directed poset*. Recall that a pair $(\{A_i\}_{i\in\mathscr{I}}, \{f_{ij}\}_{i,j\in\mathscr{I}})$ is called an *inverse* or *projective system* of sets over $\mathscr{I}$, if A_i is a set for each $i \in \mathscr{I}$, $f_{ij} : A_i \to A_j$ is a map defined for all $j \leq i$ such that $f_{ij} \circ f_{jk} = f_{ik} : A_i \to A_k$, every time the corresponding maps are defined and $f_{ii} = \mathrm{id}_{A_i}$. Then the *inverse* or the *projective limit* of the inverse system $(\{A_i\}_{i\in\mathscr{I}}, \{f_{ij}\}_{i,j\in\mathscr{I}})$ is

$$\varprojlim A_i = \{\xi \in \prod_{i\in\mathscr{I}} A_i \mid f_{ij}(p_i(\xi)) = p_j(\xi),\ \forall j \leq i\},$$

where, for each $i \in \mathscr{I}$, $p_i : \prod_{i\in\mathscr{I}} A_i \to A_i$ is the canonical projection. This definition is easily specialized to define the inverse limit in the category of algebras, co-algebras and Hopf algebras.

Remark 3 A couple of remarks are in order.

1. Let V and W be two filtered vector spaces. Then the map $\rho : \mathrm{gr}\, V \otimes \mathrm{gr}\, W \to \mathrm{gr}\,(V \otimes W)$, defined by $\rho(\pi_p x \otimes \pi_q y) = \pi_{p+q}(x \otimes y)$, is an isomorphism, which, if V and W are complete, induces an isomorphism, still denoted by ρ, between $\mathrm{gr}\, V \otimes \mathrm{gr}\, W$ and $\mathrm{gr}\, V \hat{\otimes}\, \mathrm{gr}\, W$, and which takes $\pi_p x \otimes \pi_q y$ to π_{p+q} $(x \hat{\otimes} y)$, for all $p, q \in \mathbb{N}$ and for all $x \in V$ and $y \in W$.
2. If A and A' are two complete augmented algebras, then $F_n(A \otimes A')$ is a filtration of $A \otimes A'$ and the corresponding completed tensor product $A \hat{\otimes} A'$ becomes a complete augmented algebra. The complete tensor product of complete algebras has the following property. If A and B are augmented algebras then, $\hat{A} \hat{\otimes} \hat{B} = \widehat{A \otimes B}$.

Finally we can introduce the following concept.

Definition 4 A *complete Hopf algebra* $(H, m, i, \Delta, \varepsilon, S)$ is a complete augmented algebra (H, m, i), where $\Delta : H \to H \hat{\otimes} H$ and $S : H \to H$ are morphisms of complete augmented algebras, and $\varepsilon : H \to \mathbb{F}$ is the augmentation map. These morphisms satisfy the same properties as in the usual definition of Hopf algebra, with the usual tensor product replaced by the complete tensor product.

Note that $(H, m, i, \Delta, \varepsilon, S)$ is co-commutative if $\tau \circ \Delta = \Delta$. Furthermore, to every Hopf algebra $(H, \hat{m}, \hat{i}, \hat{\Delta}, \hat{\varepsilon}, \hat{S})$ one can associate a complete Hopf algebra by considering $\hat{H}$ and $\hat{\Delta} : \hat{H} \to \widehat{H \otimes H} \simeq \hat{H} \hat{\otimes} \hat{H}$, see Remark 3 above.

Example 3 Let $\mathfrak{g}$ be a finite dimensional Lie algebra. Then $\hat{\mathscr{U}}(\mathfrak{g})$ carries a structure of complete Hopf algebra, see Example 2.

Given a complete Hopf algebra $(H, \hat{m}, \hat{i}, \hat{\Delta}, \hat{\varepsilon}, \hat{S})$, one can define:

$$\mathscr{P}(H) := \{x \in I_H \mid \Delta x = x \hat{\otimes} 1 + 1 \hat{\otimes} x\}$$
$$\mathscr{G}(H) := \{x \in 1 + I_H \mid \Delta x = x \hat{\otimes} x\},$$

i.e. the set of *primitive* and, respectively, of *group-like* elements of H.

Note that if A is a complete augmented algebra and if $x \in A$, $e^x = \sum_{n\geq 0} \frac{x^n}{n!}$ belongs to A. This follows from Property 3) in Definition 2, noticing that, if $S_n = \sum_{k=0}^{n} \frac{x^k}{k!}$ for all $n \geq 0$, then the sequence $\{S_n\}_{n\in\mathbb{N}}$ is *convergent*, since it is *Cauchy*.[4]

Let $(H, m, i, \Delta, \varepsilon, S)$ be a complete Hopf algebra. Then

Proposition 2 $x \in \mathscr{P}(H) \iff e^x \in \mathscr{G}(H)$.

---

[4]Recall that if M is a $\mathbb{Z}$-module endowed with a decreasing filtration, $M = M_0 \supset M_1 \supset M_2 \supset \cdots$, then a sequence $(x_k)_{k\in\mathbb{N}}$ is called a *Cauchy sequence* if for each r there exists N_r, such that, if $n, m > N_r$, then $x_n - x_m \in M_r$. This amounts to saying, that if n, m are sufficiently large, then $x_n + M_r = x_m + M_r$. This implies that $(x_k)_{k\in\mathbb{N}}$ is a coherent sequence, i.e., it belongs to $\hat{M} = \varprojlim M/M_k$. In other words, every Cauchy sequence is convergent in $\hat{M}$. These considerations can be extended verbatim to the case of complete augmented algebras.

Proof In fact $x \in \mathscr{P}(H) \iff \Delta x = x\hat{\otimes}1 + 1\hat{\otimes}x \iff \mathrm{e}^{\Delta x} = \mathrm{e}^{x\hat{\otimes}1+1\hat{\otimes}x}$ and, since $(x\hat{\otimes}1)(1\hat{\otimes}x) - (1\hat{\otimes}x)(x\hat{\otimes}1) = 0$, one has that

$$\mathrm{e}^{x\hat{\otimes}1+1\hat{\otimes}x} = \mathrm{e}^{x\hat{\otimes}1} \cdot \mathrm{e}^{1\hat{\otimes}x} = (\mathrm{e}^x \hat{\otimes} 1)(1 \hat{\otimes} \mathrm{e}^x) = \mathrm{e}^x \hat{\otimes} \mathrm{e}^x,$$

which implies the statement since $\Delta\, \mathrm{e}^x = \mathrm{e}^{\Delta x}$.

Corollary 1 *The exponential map* $\exp : \mathscr{P}(H) \to \mathscr{G}(H)$, $\exp : x \rightsquigarrow \mathrm{e}^x$, *defines an isomorphism of sets, whose inverse is the* logarithmic series, *defined by* $\log(1+x) = \sum_{n\geq 1}(-1)^{n-1}\frac{x^n}{n}$, $\forall x \in I_H$.

Example 4 Let $\mathfrak{g}$ be a finite dimensional Lie algebra and let $\hat{\mathscr{U}}(\mathfrak{g})$ be the corresponding complete universal enveloping algebra, see Example 2. Then, given $\xi \in \hat{\mathscr{U}}(\mathfrak{g})$, e^ξ is a group-like element *if and only if* $\xi \in \mathfrak{g}$. Moreover, from the previous corollary, one knows that if $x \in \cup_{k\geq 1}\hat{\mathscr{U}}_k(\mathfrak{g})$ and $y = 1 + x$ such that $\Delta y = y\hat{\otimes}y$, then there exists $z \in \mathscr{P}(\hat{\mathscr{U}}(\mathfrak{g}))$ such that $y = \mathrm{e}^z$, see Example 2.

We conclude this part by noticing that on every complete Hopf algebra, both the Lie algebra of primitive elements and the group of group-like elements inherit a filtration. More precisely one has the

Proposition 3 *If for all* $k \geq 0$

$$\begin{aligned} F_k\mathscr{G}(H) &= \{x \in \mathscr{G}(H) \mid x - 1 \in F_kH\} \\ F_k\mathscr{P}(H) &= \mathscr{P}(H) \cap F_kH \end{aligned}$$

then $\{F_k\mathscr{G}(H)\}$ *and* $\{F_k\mathscr{P}(H)\}$ *are filtrations of* $\mathscr{G}(H)$ *respectively* $\mathscr{P}(H)$. *Moreover:*

1. *The exponential map induces an isomorphism of graded algebra* $gr\,\mathscr{P}(H) \to gr\,\mathscr{G}(H)$.
2. $$\begin{aligned} \mathscr{P}(H) &\simeq \varprojlim \mathscr{P}(H)/F_k\mathscr{P}(H) \\ \mathscr{G}(H) &\simeq \varprojlim \mathscr{G}(H)/F_k\mathscr{G}(H). \end{aligned}$$

Example 5 If $H = \hat{\mathscr{U}}(\mathfrak{g})$, the previous proposition implies that, for all $x, y \in \mathfrak{g}$, $\mathrm{BCH}(x, y) \in \mathscr{P}\big(\hat{\mathscr{U}}(\mathfrak{g})\big)$, i.e., $\mathrm{BCH}(x, y)$ is *convergent* for all $x, y \in \mathfrak{g}$. The proof of this statement is based on two observations. First, $\mathrm{BCH}(x, y)$ is a Lie series in x, y that, seen as an element of $\hat{\mathscr{U}}(\mathfrak{g})$ can be written as

$$\mathrm{BCH}(x, y) = \sum_{n=0}^{\infty} z_n(x, y), \tag{25}$$

where, for each $n \geq 0$, $z_n(x, y)$ is the *non-commutative homogeneous polynomial* of degree n in x, y, obtained from the corresponding Lie polynomial in

$\mathrm{BCH}(x, y)$ using the relation $[x, y] = xy - yx$. Second, the sequence $\{S_n\}_{n\geq 0}$, where $S_n = \sum_{k=0}^{n} z_k(x, y)$ is *Cauchy*.

3 Pre- and Post–Lie Algebras

In this section we will introduce the definitions and the main properties of a pre- and post-Lie algebra, stressing the relevance of these notions in the theory of smooth manifolds and Lie groups.

An algebra $(A, \cdot)$ is called *Lie admissible* if the bracket $[\cdot, \cdot] : A \otimes A \to A$ defined by anti-symmetrization, $[x, y] := x \cdot y - y \cdot x$, for all $x, y \in A$, is a Lie bracket, i.e., if it satisfies the Jacobi identity. For example every associative algebra is Lie admissible. Given $(A, \cdot)$, let

$$a_{\cdot}(x, y, z) := (x \cdot y) \cdot z - x \cdot (y \cdot z), \quad \forall x, y, z \in A \tag{26}$$

be the *associator* defined for the product $\cdot$. Note that $(A, \cdot)$ is associative if and only if $a_{\cdot}(x, y, z) = 0$, for all $x, y, z \in A$. In the next two subsections the notions of pre- and post-Lie algebras are introduced. Such algebras are rather natural from the viewpoint of geometry. Moreover, later we will see that they are closely related to solutions of classical Yang–Baxter equations.

3.1 Pre-Lie Algebra

Weakening the condition $a_{\cdot}(x, y, z) = 0$, one arrives at a class of Lie admissible algebras, which is more general than that of associative algebras.

Definition 5 $(A, \cdot)$ is a *left pre-Lie algebra* if, for all $x, y, z \in A$

$$a_{\cdot}(x, y, z) = a_{\cdot}(y, x, z). \tag{27}$$

Note that together with the notion of left pre-Lie algebra one can introduce that of a *right pre-Lie algebra* where condition (27) is traded for $a_{\cdot}(x, y, z) = a_{\cdot}(x, z, y)$, for all $x, y, z \in A$. The notions of right and left pre-Lie algebras are equivalent. Indeed, if $(A, \cdot)$ is a left (right) pre-Lie algebra, then $(A, \cdot^{\mathrm{op}})$ is a right (left) pre-Lie algebra, where $x \cdot^{\mathrm{op}} y = y \cdot x$. For this reason, from now on, we will focus on the case of left pre-Lie algebras, which will be called simply pre-Lie algebras.

Let $(A, \cdot)$ be a pre-Lie algebra and let $\nabla : A \to \mathrm{End}(A)$ be the morphism defined by $\nabla(x) := \nabla_x : A \to A$, $\nabla_x y = x \cdot y$. Then, the pre-Lie condition implies that

$$[\nabla_x, \nabla_y] = \nabla_{[x,y]}, \quad \forall x, y \in A,$$

that is, $\nabla : A \to \mathrm{End}(A)$ is a morphism of Lie algebras, where the Lie brackets of A and of $\mathrm{End}(A)$ are defined by skew-symmetrizing the pre-Lie product of A, respectively, the associative product of $\mathrm{End}(A)$. It is worth to recall that the Lie algebra structure on A defined by skew-symmetrizing the pre-Lie product is called *subordinate* to it. Furthermore, defining for $x, y \in A$ the expression $\mathrm{T}(x, y) = \nabla_x y - \nabla_y x - [x, y]$, it is obvious from the definition that $\mathrm{T}(x, y) = 0$. From these observations, as it was already remarked in the in the Introduction, a source of examples of pre-Lie algebras can be found looking at locally flat manifolds, i.e. manifolds endowed with a linear flat and torsion free connection. It is worth to note that a n-dimensional manifold M admits a (linear) torsion-free and flat connection *if and only if* it admits an *affine structure*, i.e., a (maximal) atlas whose transition functions are constant and take values in $\mathrm{GL}_n(\mathbb{F}) \ltimes \mathbb{F}^n$. In fact, given such a ∇, for all $m \in M$ one can find an open neighborhood $m \in U$ and $X_1, \ldots, X_n \in \mathfrak{X}_M(U)$ a local frame for TM such that $\nabla_{X_i} X_j = 0$ for all $i, j = 1, \ldots, n$. Then, if $\alpha_1, \ldots, \alpha_n$ is the *dual* local frame, one has that $d\alpha_i = 0$ for all i. Indeed, one verifies that

$$d\alpha_i(X_j, X_k) = X_j\alpha_i(X_k) - X_k\alpha_i(X_j) - \alpha_i([X_j, X_k]) = -\alpha_i([X_j, X_k]) = 0,$$

since $\alpha_i(X_j) = \delta_{ij}$, and $\alpha_i([X_j, X_k]) = 0$ due to the fact that $[X_j, X_k] = \nabla_{X_j} X_k - \nabla_{X_k} X_j - \mathrm{T}_\nabla(X_j, X_k) = 0$. Then, on a neighborhood V of $m \in M$, eventually contained in U, one can find $x_1, \ldots, x_n \in C^\infty_M(V)$, such that $dx_i = \alpha_i$, for all $i = 1, \ldots, n$. The local functions $x_1, \ldots, x_n$ so defined form a system of local coordinates on (a neighborhood of M eventually smaller than) V. In this way one defines a system of local coordinates on M such that, if $(V, x_1, \ldots, x_n)$ and $(W, y_1, \ldots, y_n)$ are two overlapping local charts, $dy_i = \sum_{k=1}^n T_i^k dx_k$, where T_i^k, $k, i = 1, \ldots, n$, are the trasition functions between the two local charts. Then $0 = \nabla dy_i = \sum_{k=1}^n dT_i^k \wedge dx_k$, which implies that $dT_i^k = 0$, for all $i, k = 1, \ldots, n$. From this it follows that the functions T_i^k are (locally) constant, i.e., $T_i^k = \frac{\partial y_i}{\partial x_k} \in \mathbb{F}$ for all $i, k = 1, \ldots, n$, which implies that $y_i = \sum_{k=1}^n T_i^k x_k + C_i$, $C_i \in \mathbb{F}$, proving the statement. To prove that to every affine structure corresponds a flat and torsion-free linear connection one should follow backward all the steps of the argument just presented. A class of examples of manifolds endowed with an affine structure is presented in the following example.

Example 6 (Invariant affine structures on Lie groups) First, recall that given a vector field X on a smooth manifold M, one can define the *Lie derivative* $\mathscr{L}_X$ and the *interior product* i_X, which are derivations of the full tensor algebra of M. Once restricted to the exterior algebra defined by T^*M, they become derivations of degree 0 and degree -1, respectively. They are related by the formula $\mathscr{L}_X = i_X \circ d + d \circ i_X$, where d is Cartan's differential. In particular, given a differential k-form $\eta \in \Omega^k(M)$, then $\mathscr{L}_X\eta \in \Omega^k(M)$ and for all $m \in M$

$$(\mathscr{L}_X\eta)_m = \frac{d}{dt}\bigg|_{t=0} (\varphi^*_{X,t}\eta)_m,$$

where $\{\varphi_{X,t}\}_{t\in\mathbb{R}}$ is the *local* 1*-parameter group of diffeomorphisms* defined by X.

A *symplectic form* on a manifold M is a 2-form which is *closed* and *non-degenerate*. The pair (M,ω) is called a *symplectic manifold*. Given a symplectic manifold (M,ω) and Lie group G acting on M via $\varphi : G\times M \to M$, ω will be called G-invariant if $\varphi_g^*\omega=\omega$, for all $g\in G$. In particular, a *symplectic Lie group* is a pair (G,ω) consisting of a Lie group and a *left-invariant symplectic form*, i.e., a symplectic form invariant with respect to *left-translations*. Let (G,ω) be a symplectic Lie group and let x, y be elements in the Lie algebra $\mathfrak{g}$ of G. Then, $\mathscr{L}_{X_x}(i_{X_y}\omega)$ is a left-invariant 1-form on G, to which corresponds the unique left-invariant vector field X_z, such that $-i_{X_z}\omega=\mathscr{L}_{X_x}i_{X_y}\omega$. Note that, since $\mathscr{L}_{X_x}i_{X_y}\omega=i_{X_x}di_{X_y}\omega$, for each $f\in C^\infty(G)$ and for all $x,y\in\mathfrak{g}$,

$$\mathscr{L}_{fX_x}i_{X_y}\omega=f\mathscr{L}_{X_x}i_{X_y}\omega \quad\text{and}\quad \mathscr{L}_{X_x}i_{fX_y}\omega=\langle df, X_x\rangle i_{X_y}\omega+f\mathscr{L}_{X_x}i_{X_y}\omega.$$

In other words, defining $\nabla_{X_x}X_y$ as the unique left-invariant vector field such that

$$-i_{\nabla_{X_x}X_y}\omega=\mathscr{L}_{X_x}i_{X_y}\omega, \tag{28}$$

for all X_x, X_y left-invariant vector fields, one sees that ∇ admits a unique extension to a G-invariant linear connection on G. If one denotes still with ∇ this connection, then ∇ is *flat* and *torsion-free*. To prove this statement it suffices to show that $T_\nabla(X_x,X_y)=0$ and $R_\nabla(X_x,X_y)=0$ for all $x,y\in\mathfrak{g}$. Let us compute

$$\begin{aligned}
&\mathscr{L}_{X_x}i_{X_y}\omega-\mathscr{L}_{X_y}i_{X_x}\omega-i_{[X_x,X_y]}\omega\\
&=\mathscr{L}_{X_x}i_{X_y}\omega-\mathscr{L}_{X_y}i_{X_x}\omega-\mathscr{L}_{X_x}i_{X_y}\omega+i_{X_y}\mathscr{L}_{X_x}\omega\\
&=-i_{X_y}di_{X_x}\omega+i_{X_y}di_{X_x}\omega\\
&=0,
\end{aligned}$$

where we used that $d\omega=0$ and that $\mathscr{L}_X\alpha=di_X\alpha+i_Xd\alpha$, for all forms α and all vector fields X. Since $T_\nabla(X_x,X_y)$ is the unique left-invariant vector field such that $-i_{T_\nabla(X_x,X_y)}\omega=\mathscr{L}_{X_x}i_{X_y}\omega-\mathscr{L}_{X_y}i_{X_x}\omega-i_{[X_x,X_y]}\omega$, the non-degeneracy of ω forces $T_\nabla(X_x,X_y)=0$. Let us now observe that if $x,y,z\in\mathfrak{g}$ then $\nabla_{X_x}\nabla_{X_y}X_z$ and $\nabla_{[X_x,X_y]}X_z$ are the unique left-invariant vector fields such that

$$-i_{\nabla_{X_x}\nabla_{X_y}X_z}\omega=i_{X_x}d\big(i_{X_y}d(i_{X_z}\omega)\big)$$

and, respectively,

$$-i_{\nabla_{[X_x,X_y]}X_z}\omega=i_{[X_x,X_y]}di_{X_z}\omega.$$

One sees that

$$i_{X_x}d\big(i_{X_y}d(i_{X_z}\omega)\big)-i_{X_y}d\big(i_{X_x}d(i_{X_z}\omega)\big)=i_{[X_x,X_y]}di_{X_z}\omega, \quad \forall x,y,z\,\mathfrak{g},$$

which again, by the non-degeneracy of ω, is equivalent to

$$\nabla_{X_x}\nabla_{X_y}X_z - \nabla_{X_x}\nabla_{X_y}X_z = \nabla_{[X_x,X_y]}X_z, \quad \forall x, y, z \in \mathfrak{g},$$

proving the flatness of ∇. In other words we have shown that

Theorem 3 *Every symplectic Lie group (G, ω) admits an affine structure.*

In particular, since for all $x, y \in \mathfrak{g}$ there exists a (unique) $z \in \mathfrak{g}$ such that $\nabla_{X_x}X_y = X_z$, the underlying vector space of the Lie algebra $\mathfrak{g}$ results being endowed with a product $\cdot : \mathfrak{g} \otimes \mathfrak{g} \to \mathfrak{g}$ defined by

$$x \cdot y = z, \quad \forall x, y, z \ \text{ s.t. } \ X_z = \nabla_{X_x}X_y. \tag{29}$$

Since ∇ is flat and torsion-free, it is easy to show that $\cdot$ is a *pre-Lie* product on the vector space underlying $\mathfrak{g}$ and that, for all $x, y \in \mathfrak{g}$, $x \cdot y - y \cdot x = [x, y]$. In other words.

Corollary 2 *The Lie algebra of a symplectic Lie group is subordinate to the pre-Lie product defined in* (29).

Finally, since $d\omega = 0$, $\omega_e \in \mathscr{Z}^2(\mathfrak{g}, \mathbb{F})$, where $\mathscr{Z}^2(\mathfrak{g}, \mathbb{F})$ is the group of 2-cocycles of $\mathfrak{g}$ with values in the trivial $\mathfrak{g}$-module $\mathbb{F}$, with respect to the cohomology of Cartan–Eilenberg of $\mathfrak{g}$ with coefficients in the trivial $\mathfrak{g}$-module $\mathbb{F}$. See for example [21]. Hence, $\omega_e \in \mathrm{Hom}_{\mathbb{F}}(\Lambda^2\mathfrak{g}, \mathbb{F})$ such that

$$\omega_e(x, [y, z]) + \omega_e(z, [x, y]) + \omega_e(y, [z, x]) = 0, \quad \forall x, y, z \in \mathfrak{g},$$

and since ω is non-degenerate, ω_e is also non-degenerate. On the other hand, if $\eta \in \mathscr{Z}^2(\mathfrak{g}, \mathbb{F})$ is non-degenerate, it defines a unique left-invariant symplectic form ω_η on G via the formula:

$$\omega_{\eta_g} = (L_g)_e^*\eta, \quad \forall g \in G.$$

In other words, the left-invariant symplectic forms on G are in one-to-one correspondence with the non-degenerate elements of $\mathscr{Z}^2(\mathfrak{g}, \mathbb{F})$. See also Subsection 4.4 for a more general approach to this kind of structures.

3.2 Post-Lie Algebra

The second class of algebras playing an central role in the present work is introduced in the following definition.

Definition 6 Let $(\mathfrak{g}, [\cdot, \cdot])$ be a Lie algebra, and let $\triangleright : \mathfrak{g} \otimes \mathfrak{g} \to \mathfrak{g}$ be a binary product such that for all $x, y, z \in \mathfrak{g}$

$$x \triangleright [y, z] = [x \triangleright y, z] + [y, x \triangleright z], \tag{30}$$

and

$$[x, y] \triangleright z = \mathrm{a}_{\triangleright}(x, y, z) - \mathrm{a}_{\triangleright}(y, x, z). \tag{31}$$

Then $(\mathfrak{g}, [\cdot, \cdot], \triangleright)$ is called a *left post-Lie algebra.*

Relation (30) implies that for every left post-Lie algebra the natural linear map $d_{\triangleright} : \mathfrak{g} \to \mathrm{End}_{\mathbb{F}}(\mathfrak{g})$ defined by $d_{\triangleright}(x)(y) \to x \triangleright y$ takes values in the derivations of the Lie algebra $(\mathfrak{g}, [\cdot, \cdot])$.

Together with the notion of left post-Lie algebra one can introduce that of *right post-Lie algebra* $(\mathfrak{g}, [\cdot, \cdot], \triangleleft)$. Also in this case $(\mathfrak{g}, [\cdot, \cdot])$ is a Lie algebra and $\triangleleft : \mathfrak{g} \otimes \mathfrak{g} \to \mathfrak{g}$ is a binary product such that for each $x \in \mathfrak{g}$, $d_{\triangleleft}(x)(y) = x \triangleleft y$ is a derivation of $(\mathfrak{g}, [\cdot, \cdot])$ and the analogue of (31) is

$$[x, y] \triangleleft z = \mathrm{a}_{\triangleleft}(y, x, z) - \mathrm{a}_{\triangleleft}(x, y, z), \quad \forall x, y, z \in \mathfrak{g}.$$

Proposition 4 *If* $(\mathfrak{g}, [\cdot, \cdot], \triangleright)$ *is a left post-Lie algebra, then* $(\mathfrak{g}, [\cdot, \cdot], \triangleleft)$*, where*

$$x \triangleleft y := x \triangleright y - [x, y]$$

is a right post-Lie algebra.

Proof First, we show that

$$\begin{aligned} x \triangleleft [y, z] &= x \triangleright [y, z] - [x, [y, z]] \\ &= [x \triangleright y, z] + [y, x \triangleright z] - [[x, y], z] - [y, [x, z]] \\ &= [x \triangleright y - [x, y], z] + [y, x \triangleright z - [x, z]] \\ &= [x \triangleleft y, z] + [y, x \triangleleft z]. \end{aligned}$$

From

$$[x, y] \triangleleft z = [x, y] \triangleright z - [[x, y], z], \tag{32}$$

and

$$(y \triangleleft x) \triangleleft z = (y \triangleright x) \triangleright z - [y \triangleright x, z] - [y, x] \triangleright z + [[y, x], z] \tag{33}$$

$$y \triangleleft (x \triangleleft z) = y \triangleright (x \triangleright z) - [y, x \triangleright z] - y \triangleright [x, z] + [y, [x, z]]. \tag{34}$$

one deduces that

$$a_{\triangleleft}(y, x, z) - a_{\triangleleft}(x, y, z) = [x, y] \triangleright z - [[x, y], z],$$

which is what we needed to show, see Formula (32).

Moreover, though post-Lie algebras are not Lie-admissible, one can prove the following proposition.

Proposition 5 *Let* $(\mathfrak{g}, [\cdot,\cdot], \rhd)$ *be a left post-Lie algebra. The bracket*

$$[\![x, y]\!] := x \rhd y - y \rhd x - [x, y] \tag{35}$$

satisfies the Jacobi identity for all $x, y \in \mathfrak{g}$, *and it defines on* $\mathfrak{g}$ *the structure of a Lie algebra.*

Proof It follows from a direct computation using the identities (30) and (31).

In particular, as consequence of the previous result one has

Corollary 3 *Given a left post-Lie algebra* $(\mathfrak{g}, [\cdot,\cdot], \rhd)$, *the product* $\succ: \mathfrak{g} \otimes \mathfrak{g} \to \mathfrak{g}$, *defined by*

$$x \succ y := x \rhd y + \frac{1}{2}[x, y], \quad \forall x, y \in \mathfrak{g}$$

defines on $\mathfrak{g}$ *the structure of Lie admissible algebra.*

Clearly, both the proposition and the corollary can be easily adapted to the case of right post-Lie algebra.

Remark 4 A few remarks are in order.

1. From now on, given a post-Lie algebra $(\mathfrak{g}, \rhd, [\cdot,\cdot])$, we will denote by $\mathfrak{g}$ the Lie algebra with barcket $[\cdot,\cdot]$ and by $\overline{\mathfrak{g}}$ the Lie algebra with bracket $[\![\cdot,\cdot]\!]$.
2. Pre- and post-Lie algebras are important in the theory of numerical methods for differential equations. We refer the reader to [7, 10, 17, 23, 26] for background and details.
3. It is worth noting that if $(\mathfrak{g}, \rhd, [\cdot,\cdot])$ is an *abelian* left post-Lie algebra, i.e., $[\cdot,\cdot] \equiv 0$, then it reduces to the pre-Lie algebra $(\mathfrak{g}, \rhd)$, whose underlying Lie algebra is $(\mathfrak{g}, [\![\cdot,\cdot]\!])$, see (31) and Definition 5.

As for the case of pre-Lie algebras, differential geometry is a natural place to look for examples of post-Lie algebras. This is based on the well known result that states that if ∇ is a linear connection on M then, for all $X, Y, Z \in \mathfrak{X}_M$.

Proposition 6[5]

$$\sum_{\circlearrowleft}(\mathrm{R}(X, Y)Z - \mathrm{T}(\mathrm{T}(X, Y), Z) - (\nabla_X \mathrm{T})(Y, Z)) = 0. \tag{36}$$

[5]Formula (36) is known as the Bianchi's 1st identity. Among many other identities fulfilled by the covariant derivatives of the torsion and curvature of a linear connection, the so called Bianchi's 2nd identity is worth to recall:

$$\sum_{\circlearrowleft}\big((\nabla_X \mathrm{R})(Y, Z) + \mathrm{R}(\mathrm{T}(X, Y), Z)\big) = 0, \quad \forall X, Y, Z \in \mathfrak{X}_M.$$

Proof Since all the terms in (36) are tensors, it suffices to prove it for $X = \partial_i, Y = \partial_j$ and $Z = \partial_k$ where ∂_i, ∂_j and ∂_k are elements of a local frame. The formula follows now by a direct computation, noticing that $[\partial_i, \partial_j] = [\partial_i, \partial_k] = [\partial_j, \partial_k] = 0$ and that $(\nabla_X \mathrm{T})(Y, Z) = \nabla_X \mathrm{T}(Y, Z) - \mathrm{T}(\nabla_X Y, Z) - \mathrm{T}(Y, \nabla_X Z)$, for all $X, Y, Z \in \mathfrak{X}_M$.

Then, if ∇ is flat and has constant torsion, this formula implies that $[\cdot,\cdot]_\mathrm{T} : \mathfrak{X}_M \times \mathfrak{X}_M \to \mathfrak{X}_M$, defined by $[X, Y]_\mathrm{T} = \mathrm{T}(X, Y)$, for all $X, Y \in \mathfrak{X}_M$ is a Lie bracket on $\mathfrak{X}_M$. In particular, defining $X \triangleright Y := \nabla_X Y$ for all $X, Y \in \mathfrak{X}_M$, then

$$X \triangleright [Y, Z]_\mathrm{T} = \nabla_X \mathrm{T}(Y, Z) = \mathrm{T}(\nabla_X Y, Z) + \mathrm{T}(Y, \nabla_X Z) = [X \triangleright Y, Z]_\mathrm{T} + [Y, X \triangleright Z]_\mathrm{T}$$

and

$$\begin{aligned}
[X, Y]_\mathrm{T} \triangleright Z &= \nabla_{\mathrm{T}(X,Y)} Z \\
&= \nabla_{\nabla_X Y} Z - \nabla_{\nabla_Y X} Z - \nabla_{[X,Y]} Z \\
&= \nabla_{\nabla_X Y} Z - \nabla_{\nabla_Y X} Z - \nabla_X \nabla_Y Z + \nabla_Y \nabla_X Z \\
&= (X \triangleright Y) \triangleright Z - (Y \triangleright X) \triangleright Z - X \triangleright (Y \triangleright X) + Y \triangleright (X \triangleright Z) \\
&= a_\triangleright(X, Y, Z) - a_\triangleright(Y, X, Z),
\end{aligned}$$

for all $X, Y, Z \in \mathfrak{X}_M$. In the third equality we used $\mathrm{R}_\nabla = 0$. Moreover

$$[X, Y]_\mathrm{T} = \mathrm{T}(X, Y) = \nabla_X Y - \nabla_Y X - [X, Y] = X \triangleright Y - Y \triangleright X - [X, Y].$$

Summarizing, under the assumptions on the linear connection ∇, one has that:

$$\begin{aligned}
X \triangleright [Y, Z]_\mathrm{T} &= [X \triangleright Y, Z]_\mathrm{T} + [Y, X \triangleright Z]_\mathrm{T} \\
[X, Y]_\mathrm{T} \triangleright Z &= a_\triangleright(X, Y, Z) - a_\triangleright(Y, X, Z) \\
[X, Y] &= X \triangleright Y - Y \triangleright X - [X, Y]_\mathrm{T},
\end{aligned}$$

for all $X, Y, Z \in \mathfrak{X}_M$. In other words

Proposition 7 [23] *If ∇ is a flat linear connection on the manifold M, with constant-torsion, then $(\mathfrak{X}_M, \triangleright, [\cdot,\cdot]_T)$ is a left post-Lie algebra.*

Remark 5 A few remarks are in order.

1. Note that in the previous proposition the Lie-Jacobi bracket between vector fields, plays the role of the Lie bracket $[\![\cdot,\cdot]\!]$ in the post-Lie structure, while the role of the bracket $[\cdot,\cdot]$ is taken by $[\cdot,\cdot]_\mathrm{T}$, i.e., the one defined by the torsion tensor.
2. If one had defined $[X, Y]_\mathrm{T} = -\mathrm{T}(X, Y)$ and $X \triangleleft Y = \nabla_X Y$, which is the same product one has in the previous proposition, then $(\mathfrak{X}_M, \triangleleft, [\cdot,\cdot]_\mathrm{T})$ is a right post-Lie algebra.

At this point, it is worth recalling a classical result from differential geometry due to Cartan and Schouten. See Refs. [5, 29]. Let G be a Lie group and $\mathfrak{g}$ its corresponding Lie algebra. A linear connection ∇ on G is called left-invariant if for all left-invariant vector fields, X, Y, $\nabla_X Y$ is a left-invariant vector field. Then

Proposition 8 *There is a one-to-one correspondence between the set of left-invariant connections on G and the set* $\mathrm{Hom}_{\mathbb{F}}(\mathfrak{g} \otimes \mathfrak{g}, \mathfrak{g})$.

Given a left-invariant connection, ∇, let $\alpha \in \mathrm{Hom}_{\mathbb{F}}(\mathfrak{g} \otimes \mathfrak{g}, \mathfrak{g})$ be the corresponding bilinear form, and let s and a be the symmetric, respectively skew-symmetric summands of α, i.e., $s = \frac{\alpha+\sigma\alpha}{2}$ and $a = \frac{\alpha-\sigma\alpha}{2}$, where $\sigma\alpha(x, y) := \alpha(y, x)$ for all $x, y \in \mathfrak{g}$.

Corollary 4 *The connection* ∇ *is torsion-free if and only if* $a(\cdot,\cdot) = \frac{1}{2}[\cdot,\cdot]$, *where* $[\cdot,\cdot]$ *is the Lie bracket on* $\mathfrak{g}$.

A left-invariant connection ∇ is called a *Cartan connection* if there exists a one-to-one correspondence between the set of the geodesics of ∇ going through the unit e and the 1-parameter subgroups of G.

Theorem 4 *A left-invariant connection* ∇ *on G is a Cartan connection if and only if the symmetric part of the bilinear form* α *corresponding to* ∇ *is zero. In other words, Cartan's connections on G are in one-to-one correspondence with* $\mathrm{Hom}(\Lambda^2\mathfrak{g}, \mathfrak{g})$.

Let $\lambda \in \mathbb{F}$ and define $\alpha_\lambda : \mathfrak{g} \otimes \mathfrak{g} \to \mathfrak{g}$ by $\alpha_\lambda(x, y) = \lambda[x, y]$ for all $x, y \in \mathfrak{g}$. Then the curvature and the torsion of the left-invariant connection defined by α_λ are

$$\begin{aligned} \mathrm{R}_\lambda(X, Y)Z &= (\lambda^2 - \lambda)[[X, Y], Z] \\ \mathrm{T}_\lambda(X, Y) &= (2\lambda - 1)[X, Y], \end{aligned}$$

for all $X, Y, Z \in \mathfrak{X}_G$. In particular, the Cartan connection defined by $\alpha_\lambda(\cdot,\cdot) = \lambda[\cdot,\cdot]$ is flat if and only if $\lambda = 1$ or $\lambda = 0$. Then, going back to our main topic, one finds

Corollary 5 *The Cartan connections defined by* $\nabla_X Y = [X, Y]$ *and* $\nabla_X Y = 0$, *for all* $X, Y \in \mathfrak{X}_G$ *define a (left) post-Lie algebra structure on* $\mathfrak{X}_G$.

Proof Note that the two cases correspond to $\lambda = 1$, $\lambda = 0$, respectively. For $\lambda = 1$ one has $\mathrm{T}(\cdot,\cdot) = [\cdot,\cdot]$, while for $\lambda = 0$ one has $\mathrm{T}(\cdot,\cdot) = -[\cdot,\cdot]$. On the other hand,

$$(\nabla_X \mathrm{T})(Y, Z) = \nabla_X(\mathrm{T}(Y, Z)) - \mathrm{T}(\nabla_X Y, Z) - \mathrm{T}(Y, \nabla_X Z).$$

Then, when $\lambda = 0$, one has $(\nabla_X \mathrm{T})(Y, Z) = 0$, since $\nabla_X Y = 0$ for all X, Y and when $\lambda = 1$, one has

$$\begin{aligned} (\nabla_X \mathrm{T})(Y, Z) &= \nabla_X(\mathrm{T}(Y, Z)) - \mathrm{T}(\nabla_X Y, Z) - \mathrm{T}(Y, \nabla_X Z) \\ &= \nabla_X([Y, Z]) - [[X, Y], Z] - [Y, [X, Z]] \\ &= [X, [Y, Z]] - [[X, Y], Z] - [Y, [X, Z]] = 0, \end{aligned}$$

thanks to the Jacobi identity. Then the statement follows from Proposition 7, observing that $[\cdot,\cdot]_{\mathrm{T}} = [\cdot,\cdot]$.

4 Poisson Structures and r-matrices

This section has two main goals. First to introduce the theory of classical r-matrices and, second, the one of isospectral flows. To this end, we will introduce the reader to the theory of the classical integrable systems, where both classical r-matrices and isospectral flows play a central role. Classical r-matrices will be used to produce examples of pre and post-Lie algebras, while isospectral flows will be studied in the last section from the point of view of post-Lie algebra. We will also discuss in some details the factorization of (suitable) elements of a Lie group whose Lie algebra is endowed with a classical r-matrix. The analogue of this construction, applied to the group-like elements of (the I-adic completion of) the universal enveloping algebra of a finite dimensional Lie algebra, will be discuss at the end of these notes.

4.1 Poisson Manifolds and Isospectral Flows

We follow references [25, 33, 34]. Let P be a smooth manifold. A *Poisson bracket* on P is a Lie bracket $\{\cdot,\cdot\}$ on $C^\infty(P)$, such that

$$\{f_1 f_2, f_3\} = f_1\{f_2, f_3\} + \{f_1, f_3\} f_2, \quad \forall f_1, f_2, f_3 \in C^\infty(P).$$

The pair $(P, \{\cdot,\cdot\})$ is called a *Poisson manifold* and the pair $(C^\infty(P), \{\cdot,\cdot\})$ a *Poisson algebra*.[6] Since every Poisson bracket is a skew-symmetric bi-derivation of $C^\infty(P)$, it can be obtained in terms of a bi-vector field Π, i.e., by a smooth section of $\Lambda^2 TP$, the second exterior power of the tangent bundle TP. More precisely, given such a section Π, one can define $\{f, g\} = \Pi(df, dg)$, for all $f, g \in C^\infty(P)$. Such a bracket is by definition a skew-symmetric bi-derivation of $C^\infty(P)$. Moreover, defining

$$\begin{aligned}[\Pi_1, \Pi_2]_{SN}(f, g, h) = {} & \{\{f, g\}_1, h\}_2 + \{\{h, f\}_1, g\}_2 + \{\{g, h\}_1, f\}_2 \\ & + \{\{f, g\}_2, h\}_1 + \{\{h, f\}_2, g\}_1 + \{\{g, h\}_2, f\}_1,\end{aligned}$$

for all $f, g, h \in C^\infty(P)$, one has that $[\Pi, \Pi]_{SN} = 0$ if and only if the corresponding bracket satisfies the Jacobi identity, i.e., $[\Pi, \Pi]_{SN} = 0$ if and only if the corresponding bracket is Poisson.[7]

Example 7 (Symplectic structures vs. Poisson structures) Let M be a smooth manifold and let $\omega \in \Omega^2(M)$ be a *non-degenerate* 2-form, i.e., a smooth section of

---

[6]One can define the notion of a Poisson algebra in a more algebraic setting, without any reference to some underlying manifold. More precisely one say that a associative and commutative algebra $(A, \cdot)$ is a Poisson algebra if there exists a $\mathbb{F}$-bilinear, skew-symmetric map $\{\cdot,\cdot\} : A \otimes_{\mathbb{F}} A \to A$, which is a bi-derivation of $(A, \cdot)$ and which fulfills the Jacobi identity. Such a bilinear map is called a Poisson bracket.

[7]The bracket $[\cdot,\cdot]_{SN}$ just introduced extends to $\Lambda^\bullet TP$, the full exterior algebra of TP, and is called the *Schouten–Nijenhuis bracket*.

$\Lambda^2 T^*M$, the second exterior power of T^*M, such that $\omega_m : T^*_mM \times T^*_mM \to \mathbb{F}$ is non-degenerate, for all $m \in M$. Then, ω defines an isomorphism between $\Omega^1(M)$ and $\mathfrak{X}(M)$ which associates to each $f \in C^\infty(M)$ its *Hamiltonian vector field* X_f, defined by the condition $df = -\omega(X_f, \cdot)$. In this way, for each $f, g \in C^\infty(M)$ one can define $\{f, g\}_\omega = \omega(X_f, X_g)$, which is a skew-symmetric bi-derivation of $C^\infty(M)$. Furthermore, $\{\cdot,\cdot\}_\omega$ is a Poisson bracket if and only if $d\omega = 0$. In this case ω is called a *symplectic form* and the Poisson tensor corresponding to $\{\cdot,\cdot\}_\omega$ is the *inverse* of ω.

Let $\mathfrak{g}$ be a finite dimensional Lie algebra and $\mathfrak{g}^*$ its dual vector space. Let $\mathbb{F}[\mathfrak{g}]$ be the algebra of polynomial functions on $\mathfrak{g}^*$. For $x, y \in \mathfrak{g}$, let

$$\{x, y\}(\alpha) := \langle \alpha, [x, y] \rangle, \quad \forall \alpha \in \mathfrak{g}^*. \tag{37}$$

Since $\mathbb{F}[\mathfrak{g}]$ is generated in degree 1 by $\mathfrak{g}$, the bracket in (37) admits a unique extension to a skew-symmetric bi-derivation of $\mathbb{F}[\mathfrak{g}]$. This bracket satisfies the Jacobi identity, and therefore yields a Poisson bracket on the algebra of polynomial functions on $\mathfrak{g}^*$. On a basis $x_1, \ldots, x_n$ of $\mathfrak{g}$, (37) reads $\{x_i, x_j\} = \sum_{k=1}^n C_{ij}^k x_k$, for $i, j = 1, \ldots, n$, where $\{C_{ij}^k\}_{i,j,k=1,\ldots,n}$, are the *structure constants* of $\mathfrak{g}$, defined by $[x_i, x_j] = C_{ij}^k x_k$, for $i, j = 1, \ldots, n$. This implies that $\mathfrak{g}$, seen as the vector sub-space of $\mathbb{F}[\mathfrak{g}]$ of linear functions on $\mathfrak{g}^*$, is closed with respect to (37), i.e., it implies that the Poisson bracket of two linear functions is still a linear function. For this reason, the Poisson bracket induced on $\mathbb{F}[\mathfrak{g}]$ by (37) is called a *linear Poisson bracket*. From the discussion above it follows that giving a Lie bracket on $\mathfrak{g}$ is equivalent to giving a linear Poisson structure on $\mathfrak{g}^*$. Let us recall that, given a Poisson manifold $(P, \{\cdot,\cdot\})$ and a smooth function $H : P \to \mathbb{F}$, one can define the Hamiltonian vector field $X_H \in \mathfrak{X}(P)$ whose Hamiltonian is H:

$$X_H(m) = \{H, \cdot\}(m), \quad \forall m \in P,$$

or, equivalently, $X_H(m) = \Pi(dH, \cdot)(m)$, for $m \in P$.

Now let $H \in C^\infty(\mathfrak{g}^*)$. Then the Hamiltonian vector field X_H with respect to the linear Poisson bracket defined on $\mathfrak{g}^*$ is given by:

$$X_H(\alpha) = -\operatorname{ad}^\sharp_{dH_\alpha}(\alpha), \quad \forall \alpha \in \mathfrak{g}^*, \tag{38}$$

where $\operatorname{ad}^\sharp$ is the co-adjoint representation of $\mathfrak{g}$. The Hamiltonian equations, corresponding to the integral curves of the vector field X_H in (38) can be written as

$$\dot{\alpha} = -\operatorname{ad}^\sharp_{dH_\alpha}(\alpha). \tag{39}$$

Recall now that $f \in C^\infty(P)$ is called a *Casimir* of $(P, \{\cdot,\cdot\})$ if $\{f, g\} = 0$, for all $g \in C^\infty(P)$, which forces X_f to be the zero-vector field, i.e., X_f is such that $X_f(m) = 0$ for all $m \in P$. Moreover, this condition implies that Casimir functions are constant along the leaves of the *symplectic foliation* associated to $(P, \{\cdot,\cdot\})$. [8] In

---

[8]The symplectic foliation of a Poisson manifold $(P, \{\cdot,\cdot\})$ is the generalized distribution in the sense of Sussmann defined on P by the Hamiltonian vector fields. Each leaf of this distribution

the particular case of a linear Poisson structure, if G is assumed to be connected, one can prove that $f \in C^\infty(\mathfrak{g}^*)$ is a Casimir if and only if f is G-invariant, i.e., if and only if $\mathrm{Ad}^\sharp_g f = f$, for all $g \in G$. In this case, f is a Casimir if and only if for all $\alpha \in \mathfrak{g}^*$

$$\mathrm{ad}^\sharp_{df_\alpha}(\alpha) = 0. \tag{40}$$

The vector space of the Casimirs of the Poisson manifold $(P, \{\cdot,\cdot\})$ is denoted by $\mathrm{Cas}(P, \{\cdot,\cdot\})$. It is a commutative Poisson sub-algebra (actually the center) of $(C^\infty(P), \{\cdot,\cdot\})$.

4.1.1 Lax-Type Equations

Let $H \in C^\infty(\mathfrak{g}^*)$. Then $dH_\alpha \in \mathfrak{g}$, for all $\alpha \in \mathfrak{g}^*$, and dH is a (smooth) map between $\mathfrak{g}^*$ and $\mathfrak{g}$. Suppose now that $\mathfrak{g}$ is a quadratic Lie algebra, i.e., a Lie algebra endowed with a non-degenerate, symmetric bilinear form $B : \mathfrak{g} \otimes \mathfrak{g} \to \mathbb{F}$, which is invariant with respect to the adjoint action of $\mathfrak{g}$, i.e., $B(\mathrm{ad}_x\, y, z) + B(y, \mathrm{ad}_{xz}) = 0$, for all x, y, z in $\mathfrak{g}$. Then, if $x_\alpha \in \mathfrak{g}$ is the (unique) vector such that $B(x_\alpha, y) = \langle \alpha, y\rangle$, for all $y \in \mathfrak{g}$, the integral curves of the Hamiltonian vector field X_H correspond to the integral curves of the vector field defined on $\mathfrak{g}$ by the system of ordinary differential equations:

$$\dot{x}_\alpha = [x_\alpha, dH_\alpha], \tag{41}$$

where the bracket on the right-hand side of (41) is the Lie bracket of $\mathfrak{g}$. Evolution equations of type (41) are known as *Lax type equations*, after Peter Lax, or *isospectral flow equations*, since, if $\mathfrak{g}$ is a matrix Lie algebra,[9] then writing $F_k = \frac{\mathrm{tr}\, x_\alpha^k}{k}$, one has

$$\frac{dF_k}{dt} = \mathrm{tr}\left(\frac{dx_\alpha^k}{dt}\right) = k\ \mathrm{tr}\left(\frac{dx_\alpha}{dt}x_\alpha^{k-1}\right) = k\ \mathrm{tr}([x_\alpha, dH_\alpha]x_\alpha^{k-1}) = 0,$$

implying that, generically, the eigenvalues of x_α are conserved quantities along the flow defined by (41).

4.2 *r-matrices, Factorization in Lie Groups and Integrability*

We follow references [20, 33, 34]. Let $\mathfrak{g}$ be a finite dimensional Lie algebra over $\mathbb{F}$, and let $R \in \mathrm{End}_\mathbb{F}(\mathfrak{g})$. Then the bracket

---

is an immersed symplectic manifold, i.e., it is an immersed submanifold of P which carries a symplectic structure. See Example 7, which is defined by the restriction to the leaf of the Poisson structure.

[9]Note that this is not a restriction, since, by the Ado's theorem, every finite dimensional Lie algebra admits a faithful finite dimensional representation.

$$[x, y]_R := \frac{1}{2}([Rx, y] + [x, Ry]), \quad \forall x, y \in \mathfrak{g} \tag{42}$$

is skew-symmetric. Moreover, if:

$$B(x, y) := R([Rx, y] + [x, Ry]) - [Rx, Ry], \tag{43}$$

then $[\cdot, \cdot]_R$ satisfies the Jacobi identity *if and only if*:

$$[B(x, y), z] + [B(z, x), y] + [B(y, z), x] = 0, \quad \forall x, y, z \in \mathfrak{g}. \tag{44}$$

In fact, the Jacobi identity for $[\cdot, \cdot]_R$ is equivalent to:

$$\sum_{\circlearrowleft} \big(\big[R([Rx, y] + [x, Ry]), z\big] + \big[[Rx, y] + [x, Ry], Rz\big]\big) = 0 \tag{45}$$

where $\sum_{\circlearrowleft}$ denote cyclic permutations of (x, y, z). On the left-hand side of the previous equation, the following three-terms sum appears:

$$\big[[Rx, y] + [x, Ry], Rz\big] + \big[[Rz, x] + [z, Rx], Ry\big] + \big[[Ry, z] + [y, Rz], Rx\big]$$

which, using the Jacobi identity for the bracket $[\cdot, \cdot]$, becomes

$$-[[Rx, Ry], z] - [[Rz, Rx], y] - [[Ry, Rz], x].$$

From the previous computation it follows that if, for some $\theta \in \mathbb{F}$, $B(x, y) = \theta[x, y]$, which amount to the following identity

$$[Rx, Ry] = R([Rx, y] + [x, Ry]) - \theta[x, y], \tag{46}$$

for all $x, y \in \mathfrak{g}$, then identity (44) will be fulfilled.

Definition 7 *(Classical r-matrix and modified CYBE)* Equation (46) is called *modified Classical Yang–Baxter Equation* (mCYBE). Its solution is called *classical r-Matrix*. For $\theta = 0$, equation (46) reduces to what is called *classical Yang–Baxter Equation* (CYBE). The Lie algebra with classical r-matrix, $(\mathfrak{g}, R)$, defines a *double Lie algebra*. The Lie algebra with bracket $[\cdot, \cdot]_R$ defined in (42) is denoted $\mathfrak{g}_R$.

Remark 6 Note that on the underlying vector space $\mathfrak{g}$ of a double Lie algebra $(\mathfrak{g}, R)$ are defined two Lie brackets, the original one, $[\cdot, \cdot]$, and the Lie bracket $[\cdot, \cdot]_R$ defined in (42). Correspondingly we have two linear Poisson structures, i.e., the bracket $\{\cdot, \cdot\}$, defined in (37), and the bracket $\{\cdot, \cdot\}_R$, defined by

$$\{f, g\}_R(\alpha) := \langle \alpha, [df_\alpha, dg_\alpha]_R \rangle = \frac{1}{2}\langle \alpha, ([Rdf_\alpha, dg_\alpha] + [df_\alpha, Rdg_\alpha]) \rangle.$$

Furthermore, the two Lie algebra structures yield two co-adjoint actions, $\mathrm{ad}^\sharp$ and $\mathrm{ad}^{\sharp,R}$, defined for all $x, y \in \mathfrak{g}$ and $\alpha \in \mathfrak{g}^*$ by

$$\begin{aligned}\mathrm{ad}_x^\sharp(\alpha)(y) &:= -\langle \alpha, [x, y]\rangle \\ \mathrm{ad}_x^{\sharp,R}(\alpha)(y) &:= -\frac{1}{2}\langle \alpha, [Rx, y] + [x, Ry]\rangle.\end{aligned}$$

The definition of $\mathrm{ad}^{\sharp,R}$, together with a simple calculation shows that

$$\{f, g\}_R(\alpha) = \frac{1}{2}\big(\mathrm{ad}^\sharp_{dg_\alpha}(\alpha)(Rdf_\alpha) - \mathrm{ad}^\sharp_{df_\alpha}(\alpha)(Rdg_\alpha)\big). \tag{47}$$

Let R be a solution of the mCYBE with $\theta = 1$ and let $R_\pm \in \mathrm{End}_{\mathbb{F}}(\mathfrak{g})$ be two maps defined by:

$$R_\pm := \frac{1}{2}(R \pm \mathrm{id}_{\mathfrak{g}}), \tag{48}$$

Proposition 9 *1. The maps $R_\pm : \mathfrak{g} \to \mathfrak{g}$ are homomorphisms of Lie algebras from $\mathfrak{g}_R$ to $\mathfrak{g}$, so that $\mathfrak{g}_\pm = im\ R_\pm$ are Lie sub-algebras of $\mathfrak{g}$.*

2. *Let $\mathfrak{k}_\pm = ker\ R_\mp$. Then $\mathfrak{k}_\pm \subset \mathfrak{g}_\pm$ are ideals, and denoting with $\overline{w}$ the class of the element w, the map $C_R : \mathfrak{g}_+/\mathfrak{k}_+ \to \mathfrak{g}_-/\mathfrak{k}_-$, defined by $C_R\big(\overline{(R + \mathrm{id}_{\mathfrak{g}})x}\big) = \overline{(R - \mathrm{id}_{\mathfrak{g}})x}$, is an isomorphism of Lie algebras. Note that, with a slight abuse of language, we include also the case when $\mathfrak{g}_\pm/\mathfrak{k}_\pm$ are the zero-Lie algebras.*
3. *Let $\Delta : \mathfrak{g} \to \mathfrak{g} \oplus \mathfrak{g}$ be the diagonal morphism, and let $i_R = (R_+, R_-) \circ \Delta : \mathfrak{g} \to \mathfrak{g} \oplus \mathfrak{g}$, defined by $i_R x = (R_+x, R_-x)$, for all $x \in \mathfrak{g}$. Then, if $\mathfrak{g} \oplus \mathfrak{g}$ is endowed with the direct-product Lie algebra structure, i_R is an injective Lie algebra homomorphism of Lie algebras whose image consists of all pairs $(x, y) \in \mathfrak{g} \oplus \mathfrak{g}$ such that $C_R\overline{x} = C_R\overline{y}$.*
4. *Each element $x \in \mathfrak{g}$ has a unique decomposition as $x = x_+ - x_-$, where $(x_+, x_-) = i_R x$.*

Proof We will sketch only the proof of item 1. To this end, observe that, for all $x, y \in \mathfrak{g}$, the following identities hold:

$$[R_\pm x, R_\pm y] = R_\pm\big([R_\pm x, y] + [x, R_\pm y] \mp [x, y]\big). \tag{49}$$

Via a simple computation they yield the equality $R_\pm[x, y]_R = [R_\pm x, R_\pm y]$, for $x, y \in \mathfrak{g}$.

The map C_R is called the *Cayley transform* of R. In the following example we will introduce an important class of solutions of the mCYBE (with $\theta = 1$).

Example 8 Let $\mathfrak{g}_+, \mathfrak{g}_-$ be two Lie subalgebras of $\mathfrak{g}$ such that $\mathfrak{g} = \mathfrak{g}_- \oplus \mathfrak{g}_-$, and let $i_\pm : \mathfrak{g}_\pm \to \mathfrak{g} \oplus \mathfrak{g}$, $i_+ : x \to (x, 0)$ and $i_- : x \to (0, x)$ the two canonical embeddings. In particular $\mathfrak{g}_+$ and $\mathfrak{g}_-$, are Lie sub-algebras of $\mathfrak{g}$, centralizing each other, i.e., $[\mathfrak{g}_+, \mathfrak{g}_-] = 0$. Finally, let $\pi_\pm : \mathfrak{g} \to \mathfrak{g}$ be the corresponding projections, and define

$$R := \pi_+ - \pi_-. \tag{50}$$

First note that

$$R + 2\pi_- = \mathrm{id}_{\mathfrak{g}} = 2\pi_+ - R. \tag{51}$$

Now let us show that R defined in (50) satisfies (46), i.e., the mCYBE with $\theta = 1$. To this end it suffices to observe that $[x, y]_R = [x_+, y_+] - [x_-, y_-]$, which implies

$$R\big([Rx, y] + [x, Ry]\big) - [Rx, Ry] = [x_+, y_+] + [x_-, y_-] + [x_-, y_+] + [x_+, y_-] = [x, y].$$

Since R satisfies the mCYBE, $\pi_\pm := \pm R_\pm$ satisfy the following identity:

$$[\pi_\pm x, \pi_\pm y] = \pi_\pm\big([\pi_\pm x, y] + [x, \pi_\pm y] - [x, y]\big), \quad \forall x, y \in \mathfrak{g},$$

and they are homomorphisms of Lie algebras from $\mathfrak{g}_R$ to $\mathfrak{g}$.

4.3 Factorization in Lie Groups

Let R be a solution of the mCYBE and let $G_\pm \subset G$ be the unique (up to isomorphism), connected and simple Lie groups whose Lie algebras are $\mathfrak{g}_\pm = R_\pm\mathfrak{g}$.

Theorem 5 *Then, every element g' in a suitable neighborhood of the identity element of G admits a factorization as*

$$g' = h_1 h_2^{-1}, \tag{52}$$

where $h_1 \in G_+$ and $h_2 \in G_-$.

Proof Recall that, as vector spaces, $\mathfrak{g}_R = \mathfrak{g}$. Let $\Delta : \mathfrak{g}_R \to \mathfrak{g}_R \oplus \mathfrak{g}_R$ be the diagonal map, i.e., $\Delta(x) = (x, x)$ for all $x \in \mathfrak{g}_R$. Let $i : \mathfrak{g} \oplus \mathfrak{g} \to \mathfrak{g}$ the linear map defined by $i(x, y) = x - y$, for all $x, y \in \mathfrak{g}$.

Consider the linear map defined by:

$$\mathfrak{g}_R \xrightarrow{\ \Delta\ } \mathfrak{g}_R \oplus \mathfrak{g}_R \xrightarrow{(R_+, R_-)} \mathfrak{g} \oplus \mathfrak{g} \xrightarrow{\ i\ } \mathfrak{g}. \tag{53}$$

Then $i \circ (R_+, R_-) \circ \Delta(x) = x$, for all $x \in \mathfrak{g}$. Since $\Delta : \mathfrak{g}_R \to \mathfrak{g}_R \oplus \mathfrak{g}_R$ and $(R_+, R_-) : \mathfrak{g}_R \oplus \mathfrak{g}_R \to \mathfrak{g} \oplus \mathfrak{g}$ are homomorphism of Lie algebras they integrate to homomorphisms of Lie groups, which will be denoted as $\delta : G_R \to G_R \times G_R$ and $(r_+, r_-) : G_R \times G_R \to G \times G$, respectively. In particular, for each $g \in G_R$, $g_\pm = r_\pm g$. Furthermore, note that, even though $i : \mathfrak{g} \oplus \mathfrak{g} \to \mathfrak{g}$ is not a homorphism of Lie algebras, it is the differential (at the identity) of the map $j : G \times G \to G$, defined by $j(g, h) = gh^{-1}$. Then the map defined in (53) is the differential (at the identity $e \in G_R$) of the map $\psi : G_R \to G$ defined by

$$\psi = j \circ (r_+, r_-) \circ \delta. \tag{54}$$

Observe now that since $\psi_{*,e} = \text{id}$, the map ψ is a local diffeomorphism, i.e., there exist neighborhoods V and U of the identities elements of both G_R and G such that $\psi|_V : V \to U$ is a diffeomorphism. Moreover, just applying the definition of the map ψ given in Formula (54), one has that

$$\psi(g) = g_+ g_-^{-1}, \tag{55}$$

for all $g \in G_R$. Taking now any element g' in a suitable neighborhood of the identity of G (eventually contained in U),

$$g' = \psi(\psi^{-1} g') = (\psi^{-1}(g'))_+ (\psi^{-1}(g'))_-^{-1}.$$

The statement now follows taking $h_1 = (\psi^{-1}(g'))_+$ and $h_2 = (\psi^{-1}(g'))_-$.

To simplify statement and notation, let us suppose that the map ψ is a global diffeomorphism. As remarked above, the map ψ is not a homomorphism of Lie groups. On the other hand one can prove that

Corollary 6 *The map $* : G \times G \to G$ defined by*

$$g * h = (\psi^{-1}(g))_+ h (\psi^{-1}(g))_-^{-1} \tag{56}$$

for all $g, h \in G$, defines a new structure of Lie group on the underlying manifold of the Lie group G. Let G_* be the Lie group whose product is $*$ and whose underlying manifold is G. Then $\psi : G_R \to G_*$ is an isomorphism of Lie groups.

Note that the product $*$ defined in the previous corollary can be obtained as the *push-forward* via ψ of the product defined on G_R. More precisely one can prove that

Proposition 10 *For all $g, h \in G$, one has that*

$$g * h = \psi(\psi^{-1}(g)\psi^{-1}(h)).$$

Applications to dynamics and integrability. Classical r-matrices and double Lie algebras play an important role in the theory of classical integrable systems, both finite and infinite dimensional. The relevance of these objects to this theory stems from Theorem 6 further below.

Theorem 6 *Let $\mathfrak{g}$ be a finite dimensional Lie algebra and R a solution of the mCYBE. Let G be the connected and simply-connected Lie group corresponding to $\mathfrak{g}$. Let $\{\cdot, \cdot\}$ and $\{\cdot, \cdot\}_R$ be the linear Poisson brackets defined on $\mathfrak{g}$. Then:*

1. *The elements of $Cas(\mathfrak{g}^*, \{\cdot, \cdot\})$ are in involution with respect to the Poisson bracket $\{\cdot, \cdot\}_R$.*
2. *For every $f \in Cas(\mathfrak{g}^*, \{\cdot, \cdot\})$, the Hamiltonian vector field X_f^R, defined by $\{\cdot, \cdot\}_R$, equals:*

$$X_f^R(\alpha) = -\frac{1}{2} \operatorname{ad}^\sharp_{R d f_\alpha} \alpha. \tag{57}$$

3. *If* $(\mathfrak{g}, (\cdot \mid \cdot))$ *is a quadratic Lie algebra, then to* X_f^R *corresponds a vector field* $\widetilde{X_f^R}$ *on* $\mathfrak{g}$*, defined by the following Lax (type) equation:*

$$\widetilde{X_f^R}(x) = \frac{1}{2}[x, Rdf_x]. \tag{58}$$

The vector field $\widetilde{X_f^R}$ *is obtained using the diffeomorphism between* $\mathfrak{g}$ *and* $\mathfrak{g}^*$ *induced by the bilinear form* $(\cdot \mid \cdot)$.

Proof The proof of the first statement of the theorem follows from (40) and (47). The second statement follows by a direct computation

$$\begin{aligned}\langle X_f^R(\alpha), x\rangle &= -\langle \mathrm{ad}^{\sharp,R}_{df_\alpha}\alpha, x\rangle = \langle \alpha, \mathrm{ad}^R_{df_\alpha} x\rangle \\ &= \frac{1}{2}\langle \alpha, [Rdf_\alpha, x]\rangle + \frac{1}{2}\langle \alpha, [df_\alpha, Rx]\rangle = -\frac{1}{2}\langle \mathrm{ad}^\sharp_{Rdf_\alpha}\alpha, x\rangle,\end{aligned}$$

where we used that $\langle \alpha, [df_\alpha, Rx]\rangle = 0$, see (40). This proves formula (57). Finally, using the non-degenerate, bilinear ad-invariant form $(\cdot \mid \cdot)$ defined on $\mathfrak{g}$ we can write for $\alpha \in \mathfrak{g}^*$ and $y \in \mathfrak{g}$

$$\langle X_f^R(\alpha), y\rangle = \frac{1}{2}\langle \alpha, [Rdf_\alpha, y]\rangle = \frac{1}{2}\big(x_\alpha \mid [Rdf_\alpha, y]\big).$$

In the previous formula, x_α is the element in $\mathfrak{g}$ corresponding to $\alpha \in \mathfrak{g}^*$ via the isomorphism between $\mathfrak{g}^*$ and $\mathfrak{g}$ induced by $(\cdot \mid \cdot)$. Using now the ad-invariance of $(\cdot \mid \cdot)$, the last term of the previous formula can be written as:

$$\frac{1}{2}\big([x_\alpha, Rdf_\alpha] \mid y\big)$$

implying that:

$$\langle X_f^R(\alpha), y\rangle = \frac{1}{2}\big([x_\alpha, Rdf_\alpha] \mid y\big).$$

Using again the isomorphism defined by $(\cdot \mid \cdot)$, we can write:

$$\widetilde{X_f^R}(x_\alpha) = \frac{1}{2}[x_\alpha, Rdf_\alpha],$$

which proves the last part of the theorem.

Remark 7 The Hamiltonian equations corresponding to X_f^R have the following form:

$$\dot{\alpha} = -\frac{1}{2}\mathrm{ad}^\sharp_{R(df_\alpha)}\alpha. \tag{59}$$

Moreover, the Hamiltonian vector field X_f corresponding to a Casimir $f \in C^\infty(\mathfrak{g}^*)$ is identically zero.

Finally, using the notations of Theorem 6, where $G_\pm$ are the connected and simply-connected Lie groups whose Lie algebras are $\mathfrak{g}_\pm = \text{im } R_\pm$, see Theorem 5, one can prove that

Theorem 7 *Let $f \in Cas(\mathfrak{g}, \{\cdot, \cdot\})$ and let $t \to g_\pm(t)$ be two smooth curves in $G_\pm$ solving the factorization problem*

$$\exp(tdf_\alpha) = g_+(t)g_-(t)^{-1}, \quad g_\pm(0) = e.$$

The integral curve $\alpha = \alpha(t)$ of the vector field (57), *solving* (59) *with $\alpha(0) = \alpha$, is*

$$\alpha(t) = \text{Ad}^\sharp_{g_+^{-1}(t)} \alpha = \text{Ad}^\sharp_{g_-^{-1}(t)} \alpha. \tag{60}$$

4.4 Pre-and Post-Lie Algebras from Classical r-matrices

Let $R \in \text{End}_\mathbb{F}(\mathfrak{g})$ be a solution of the CYBE. The next result is well-known.

Proposition 11 *The binary product $\cdot : \mathfrak{g} \otimes \mathfrak{g} \to \mathfrak{g}$ defined by*

$$x \cdot y = [Rx, y], \quad \forall x, y \in \mathfrak{g} \tag{61}$$

defines a left pre-Lie algebra on $\mathfrak{g}$.

Proof Indeed, for all $x, y, z \in \mathfrak{g}$ we have

$$\begin{aligned}(x \cdot y) \cdot z - x \cdot (y \cdot z) &= [R[Rx, y], z] - [Rx, [Ry, z]] \\ &\overset{(a)}{=} [R[Rx, y], z] - [[Rx, Ry], z] - [Ry, [Rx, z]] \\ &\overset{(b)}{=} -[R[x, Ry], z] - [Ry, [Rx, z]] \\ &= [R[Ry, x], z] - [Ry, [Rx, Rz]] \\ &= (y \cdot x) \cdot z - y \cdot (x \cdot z).\end{aligned}$$

In (a) we used the Jacobi identity. In (b) we used (46) with $\theta = 0$.

Note that the Lie bracket (42) defined by a solution R of the CYBE is, up to a numerical factor, subordinate to the pre-Lie product (61). In fact, since $x \cdot y - y \cdot x = [Rx, y] - [Ry, x] = [Rx, y] + [x, Ry]$,

$$[\cdot, \cdot]_R = \frac{1}{2}(x \cdot y - y \cdot x).$$

In particular, the Lie bracket (42) is subordinate to the pre-Lie product $\bullet : \mathfrak{g} \otimes \mathfrak{g} \to \mathfrak{g}$ defined by $x \bullet y = \frac{1}{2} x \cdot y$, for all $x, y \in \mathfrak{g}$. Let R be a solution of the CYBE and let $\bullet : \mathfrak{g} \otimes \mathfrak{g} \to \mathfrak{g}$ be the pre-Lie product defined by $x \bullet y = \frac{1}{2}[Rx, y]$ for all $x, y \in \mathfrak{g}$. Then, if for every $x \in \mathfrak{g}_R = (\mathfrak{g}, [\cdot,\cdot]_R)$ one denotes with X_x the left-invariant vector field on G_R, the unique connected and simply-connected Lie group whose Lie algebra is $\mathfrak{g}_R$, then one can prove the next result.

Proposition 12 *The left-invariant linear connection on G_R defined by*

$$\nabla_{X_x} X_y = \frac{1}{2} X_{[Rx,y]}, \tag{62}$$

is flat and torsion free. In particular, the product

$$X \cdot Y = \nabla_X Y \tag{63}$$

defines a left pre-Lie algebra on $\mathfrak{X}_{G_R}$.

Proof First let us compute the torsion of the connection defined in (62).

$$\begin{aligned} \mathrm{T}(X_x, X_y) &= \nabla_{X_x} X_y - \nabla_{X_y} X_x - [X_x, X_y] \\ &= \frac{1}{2}\big(X_{[Rx,y]} - X_{[Ry,x]}\big) - X_{[x,y]_R} = 0, \end{aligned}$$

for all $x, y \in \mathfrak{g}$. On the other hand, computing the curvature of ∇ one gets:

$$\begin{aligned} \mathrm{R}(X_x, X_y)X_z &= \nabla_{X_x}\nabla_{X_y} X_z - \nabla_{X_y}\nabla_{X_x} X_z - \nabla_{[X_x, X_y]} X_z \\ &= \frac{1}{4}\big(X_{[Rx,[Ry,z]]} - X_{[Ry,[Rx,z]]}\big) - \frac{1}{2}\nabla_{X_{([Rx,y]+[x,Ry])}} X_z \\ &= \frac{1}{4}\big(X_{[Rx,[Ry,z]]-[Ry,[Rx,z]]-[R([Rx,y]+[x,Ry]),z]}\big) \\ &\overset{(a)}{=} \frac{1}{4}\big(X_{[[Rx,Ry]-R([Rx,y]+[x,Ry]),z]}\big) = 0, \end{aligned}$$

for all $x, y, z \in \mathfrak{g}$, proving the first claim. Note that in (a) we used the Jacobi identity, and the last equality follows from R being a solution of CYBE. For the second one, note that it suffices to prove it for the left-invariant vector fields. Then, given $x, y, z \in \mathfrak{g}$, one has that

$$\begin{aligned} a.(X_x, X_y, X_z) &= \nabla_{\nabla_{X_x} X_y} X_z - \nabla_{X_x}\nabla_{X_y} X_z \\ &= \frac{1}{2}(\nabla_{X_{[Rx,y]}} X_z - \nabla_{X_x} X_{[Ry,z]}) \\ &= \frac{1}{4} X_{([R[Rx,y],z]-[Rx[Ry,z]])} \\ &= \frac{1}{4} X_{([R[Rx,y],z]-[[Rx,Ry],z]-[Ry,[Rx,z]])} \end{aligned}$$

$$= \frac{1}{4} X_{[R[Ry,x],z]} - \frac{1}{4} X_{[Ry,[Rx,z]]} = a.(X_y, X_x, X_z),$$

for all $x, y, z \in \mathfrak{g}$.

We will now prove the following result, which completes Example 6. Let $(\mathfrak{g}, B)$ be a quadratic Lie algebra.[10] Then we have the next result.

Proposition 13 (Drinfeld [13]) *The set of* invertible *and* skew-symmetric *solutions of the CYBE on* $(\mathfrak{g}, B)$ *is in one-to-one correspondence with set of the invariant symplectic structures on the corresponding connected and simply-connected Lie group* G_R. *In particular, every invertible solution of the CYBE defines a left pre-Lie algebra structure on* $\mathfrak{X}_{G_R}$.

Proof Let R be an invertible solution of the CYBE on $\mathfrak{g}$ and let $\omega(\cdot, \cdot) = B(R\cdot, \cdot) : \mathfrak{g} \otimes \mathfrak{g} \to \mathfrak{g}$. Then ω is non-degenerate and skew-symmetric. In fact, since R is skew-symmetric one has that

$$\omega(y, x) = B(Ry, x) = -B(y, Rx) = -B(Rx, y) = -\omega(x, y),$$

for all $x, y \in \mathfrak{g}$, and if $x \in \mathfrak{g}$ is such that $\omega(x, y) = 0$ for all $y \in \mathfrak{g}$, then $B(Rx, y) = -B(x, Ry) = 0$ for all $y \in \mathfrak{g}$, which implies that $B(x, z) = 0$ for all $z \in \mathfrak{g}$ since R is invertible. Let us now prove that $\omega \in \mathscr{Z}^2(\mathfrak{g}_R, \mathbb{F})$, i.e., that

$$\omega(x, [y, z]_R) + \omega(z, [x, y]_R) + \omega(y, [z, x]_R) = 0, \quad \forall x, y, z \in \mathfrak{g}.$$

To this end, first compute

$$\begin{aligned} \omega(x, [y, z]_R) &= \frac{1}{2} B(Rx, [Ry, z]) + \frac{1}{2} B(Rx, [y, Rz]) \\ &= \frac{1}{2} B([Rx, Ry], z) - \frac{1}{2} B(R[Rx, y], z), \end{aligned}$$

then compute

$$\begin{aligned} \omega(y, [z, x]_R) &= \frac{1}{2} B(Ry, [Rz, x]) + \frac{1}{2} B(Ry, [z, Rx]) \\ &= \frac{1}{2} B([Rx, Ry], z) - \frac{1}{2} B(R[x, Ry], z). \end{aligned}$$

---

[10]Recall that a quadratic Lie algebra $(\mathfrak{g}, B)$ is a Lie algebra endowed with a non-degenerate, $\mathfrak{g}$-invariant bilinear form $B : \mathfrak{g} \otimes \mathfrak{g} \to \mathfrak{g}$, i.e., B is a bilinear form such that (1) if $x \in \mathfrak{g}$ is such that $B(x, y) = 0$ for all $y \in \mathfrak{g}$, then $x = 0$ and 2) $B([x, y], z) + B(y, [x, z]) = 0$ for all $x, y, z \in \mathfrak{g}$.

On the other hand,

$$\begin{aligned}\omega(z,[x,y]_R) &= B(Rz,[x,y]_R)\\ &= -B(R[x,y]_R,z) = -\frac{1}{2}B\big(R([Rx,y]+[x,Ry]),z\big).\end{aligned}$$

Using the results of these partial computations one has:

$$\begin{aligned}&\omega(x,[y,z]_R)+\omega(z,[x,y]_R)+\omega(y,[z,x]_R)\\ &= \frac{1}{2}B([Rx,Ry],z)-\frac{1}{2}B(R[Rx,y],z)\\ &= \frac{1}{2}B([Rx,Ry],z)-\frac{1}{2}B(R[x,Ry],z)\\ &= -\frac{1}{2}B\big(R([Rx,y]+[x,Ry]),z\big)=0,\end{aligned}$$

since R is a solution of the CYBE. On the other hand, suppose that $\omega \in \mathscr{Z}^2(\mathfrak{g}_R,\mathbb{F})$ is non-degenerate. Then, using B and ω one can define $B^\vee : \mathfrak{g} \to \mathfrak{g}^*$ and, respectively, $\omega^\vee : \mathfrak{g} \to \mathfrak{g}^*$ to be the linear isomorphisms such that $\langle B^\vee(x), y\rangle = B(x,y)$ and $\langle \omega^\vee(x), y\rangle = \omega(x,y)$ for all $x, y \in \mathfrak{g}$. Then if $R := (B^\vee)^{-1} \circ \omega^\vee : \mathfrak{g} \to \mathfrak{g}$, one has that

$$\begin{aligned}B(Ry,x) &= \langle \omega^\vee(y), x\rangle\\ &= \omega(y,x) = -\omega(x,y) = -\langle \omega^\vee(x), y\rangle = -B(Rx,y) = -B(y,Rx),\end{aligned}$$

showing that R is skew-symmetric. On the other hand, if $x \in \mathfrak{g}$ is such that $Rx = 0$, then

$$0 = (Rx,y) = \langle \omega^\vee(x), y\rangle = \omega(x,y),$$

for all $y \in \mathfrak{g}$, which implies that $x = 0$, proving that R is an isomorphism. Furthermore, since $\omega \in \mathscr{Z}^2(\mathfrak{g}_R,\mathbb{F})$,

$$\omega(z,[x,y]_R)+\omega(y,[z,x]_R)+\omega(x,[y,z]_R)=0 \quad \forall x,y,z \in \mathfrak{g},$$

which implies that

$$B(R([Rx,y]+[x,Ry],z) = B([Rx,Ry],z), \quad \forall x,y,z \in \mathfrak{g},$$

proving that R is a solution of the CYBE. Finally, if $\omega \in \mathscr{Z}^2(\mathfrak{g}_R,\mathbb{F})$ is non-degenerate, its extension on G_R by left-translations defines a left-invariant symplectic form on G_R. Then the last part of the statement of the proposition follows now from the discussion in Example 6.

We will now see how given a solution of the mCYBE on $\mathfrak{g}$ one can define a structure of a post-Lie algebra on $\mathfrak{X}_{G_R}$. In spite of the fact that the relation between post-Lie algebra structures on $\mathfrak{X}_{G_R}$ and solutions of the mCYBE is completely analogous to the one just discussed between the solutions of the CYBE and pre-Lie algebra structures on the $\mathfrak{X}_{G_R}$, we will give full details also in this case. Before moving to this more geometrical topic, let us make a few observations of algebraic flavor. Let $R \in \mathrm{End}_{\mathbb{F}}(\mathfrak{g})$ be a solution of the mCYBE, Equation (46), and let $R_\pm$ be defined as in (48). Then:

Theorem 8 ([2]) *The binary product*

$$x \rhd_\pm y := [R_\pm(x), y]. \tag{64}$$

defines a left (right) post-Lie algebra structure on $\mathfrak{g}$.

Proof The axiom (30) holds true since $[R_\pm, \cdot]$ is a derivation with respect to $[\cdot, \cdot]$. The axiom (31) follows from (49) and the Jacobi identity.

Note that

$$x \rhd_- y = [R_- x, y] = [(R_+ - \mathrm{id}_{\mathfrak{g}})x, y] = x \rhd_+ y - [x, y]$$

which is the content of Proposition 4. In particular, a computation shows that:

$$x \rhd_- y - y \rhd_- x + [x, y] = [x, y]_R = x \rhd_+ y - y \rhd_+ x - [x, y],$$

for all $x, y \in \mathfrak{g}$. See Proposition 5, i.e.,

$$[\![\cdot, \cdot]\!] = [\cdot, \cdot]_R. \tag{65}$$

Moreover, one finds the Lie-admissible algebras $(\mathfrak{g}, \succ_\pm)$ with binary compositions

$$x \succ_\pm y := x \rhd_\pm y + \frac{1}{2}[x, y].$$

The Lie bracket (35) is then given by $[\![x, y]\!] = [x, y]_R = x \succ y - y \succ x$, for all $x, y \in \mathfrak{g}$. Writing $\tilde{R} := \frac{1}{2}R$, one can deduce from $[\tilde{R}x, \tilde{R}y] - \tilde{R}\big([\tilde{R}x, y] + [x, \tilde{R}y]\big) = -\frac{1}{4}[x, y]$, that

$$\mathrm{a}_\succ(x, y, z) - \mathrm{a}_\succ(y, x, z) = -\frac{1}{4}[[x, y], z].$$

Let us now move to the geometric side and discuss the post-Lie structure defined on $\mathfrak{X}_{G_R}$ by a any solution of the mCYBE.

Let R be a solution of the mCYBE and let R_+ be as defined in (48). Then denoting with X_x the left-invariant vector field on G_R defined by $x \in \mathfrak{g}$ we have the result.

Theorem 9 *The formula:*

$$\nabla_{X_x} X_y = X_{[R_+ x, y]}, \quad \forall x, y \in \mathfrak{g} \tag{66}$$

defines a flat left-invariant linear connection on G_R with constant torsion. In particular, the product

$$X \triangleright Y = \nabla_X Y \tag{67}$$

defines a left post-Lie algebra on $(\mathfrak{X}_{G_R}, [\cdot,\cdot])$, where $[\cdot,\cdot] : \mathfrak{X}_{G_R} \otimes \mathfrak{X}_{G_R} \to \mathfrak{X}_{G_R}$ is the usual Lie bracket on the set of vector field on the smooth manifold G_R.

Proof The first statement follows from a direct computation. More precisely

$$\begin{aligned}\nabla_{X_x}\nabla_{X_y} X_z - \nabla_{X_y}\nabla_{X_x} X_z &= X_{[R_+ x,[R_+ y,z]]-[R_+ y[R_+ x,z]]}\\ &= X_{[[R_+ x, R_+ y],z]}, \quad \forall x, y, z \in \mathfrak{g}.\end{aligned}$$

On the other hand,

$$\begin{aligned}\nabla_{[X_x,X_y]} X_z = \nabla_{X_{[x,y]_R}} X_z = \frac{1}{2}\nabla_{X_{[Rx,y]+[x,Ry]}} X_z &= X_{[R_+[R_+ x,y]-R_+[x,y]+R_+[x,R_+ y],z]}\\ &= X_{[[R_+ x,R_+ y],z]}\end{aligned}$$

where we used that $R = 2R_+ - \mathrm{id}_{\mathfrak{g}}$ and (49) which, together, prove that $\mathrm{R}(X_x, X_y)X_z = 0$ for all $x, y, z \in \mathfrak{g}$. Let us now compute

$$\begin{aligned}\mathrm{T}(X_x, X_y) &= \nabla_{X_x} X_y - \nabla_{X_y} X_x - [X_x, X_y]\\ &= X_{[R_+ x,y]+[x,R_+ y]-[x,y]_R}\\ &= X_{[x,y]}\end{aligned}$$

for all $x, y, z \in \mathfrak{g}$. Then

$$\begin{aligned}(\nabla_{X_z}\mathrm{T})(X_x, X_y) &= \nabla_{X_z}\mathrm{T}(X_x, X_y) - \mathrm{T}(\nabla_{X_z} X_x, X_y) - \mathrm{T}(X_x, \nabla_{X_z} X_y)\\ &= \nabla_{X_z} X_{[x,y]} - X_{[[R_+ z,x],y]} - X_{[x,[R_+ z,y]]} = 0\end{aligned}$$

Because of its definition, ∇ is left-invariant and since every $X \in \mathfrak{X}_{G_R}$ can be written as $X = \sum_{i=1}^{\dim \mathfrak{g}} f_i X_{x_i}$ where $f_i \in C^\infty(G_R)$ for all $i = 1, \ldots, \dim \mathfrak{g}$ and $x_1, \ldots, x_{\dim \mathfrak{g}}$ is a basis of $\mathfrak{g}$, it follows that ∇ has the properties stated in the theorem. The last part of the statement follows now from Proposition 7 in Sect. 3.

5 Post-Lie Algebras, Factorization Theorems and Isospectral Flows

In this section we will study the properties of the universal enveloping algebra of a post-Lie algebra. For post-Lie algebras coming from classical r-matrices, we will discuss in details the factorization of group-like elements of the relevant I-adic completion. In the last part, we will discuss how this factorization can be applied to find solutions of particular Lax-type equations.

5.1 *The Universal Enveloping Algebra of a Post-Lie Algebra*

Proposition 5 above shows that any post-Lie algebra comes with two Lie brackets, $[\cdot, \cdot]$ and $[[\cdot, \cdot]]$, which are related in terms of the post-Lie product by identity (35). The relation between the corresponding universal enveloping algebras was explored in [17]. In [27] similar results in the context of pre-Lie algebras and the symmetric algebra $S_{\mathfrak{g}}$ appeared.

The next proposition summarizes the results relevant for the present discussion of lifting the post-Lie algebra structure to $\mathscr{U}(\mathfrak{g})$. Denoting the product induced on $\mathscr{U}(\mathfrak{g})$ by the post-Lie product defined on $(\mathfrak{g}, \rhd, [\cdot, \cdot])$ with the same symbol $\rhd$, one can show the next proposition.

Proposition 14 ([17] *Let* $A, B, C \in \mathscr{U}(\mathfrak{g})$ *and* $x, y \in \mathfrak{g} \hookrightarrow \mathscr{U}(\mathfrak{g})$, *then there exists a unique extension of the post-Lie product from* $\mathfrak{g}$ *to* $\mathscr{U}(\mathfrak{g})$, *given by:*

$$1 \rhd A = A \tag{68}$$

$$xA \rhd y = x \rhd (A \rhd y) - (x \rhd A) \rhd y$$

$$A \rhd BC = (A_{(1)} \rhd B)(A_{(2)} \rhd C). \tag{69}$$

Proof The proof of Proposition 14 goes by induction on the length of monomials in $\mathscr{U}(\mathfrak{g})$.

Note that (68) together with (69) imply that the extension of the post-Lie product from $\mathfrak{g}$ to $\mathscr{U}(\mathfrak{g})$ yields a linear map $d : \mathfrak{g} \to \mathrm{Der}\big(\mathscr{U}(\mathfrak{g})\big)$, defined via $d(x)(x_1 \cdots x_n) := \sum_{i=1}^{n} x_1 \cdots (x \rhd x_i) \cdots x_n$, for any word $x_1 \cdots x_n \in \mathscr{U}(\mathfrak{g})$. A simple computation shows that, in general, this map is not a morphism of Lie algebras. Together with Proposition 14 one can prove.

Proposition 15

$$A \rhd 1 = \varepsilon(A), \tag{70}$$

$$\varepsilon(A \rhd B) = \varepsilon(A)\varepsilon(B), \tag{71}$$

$$\Delta(A \rhd B) = (A_{(1)} \rhd B_{(1)}) \otimes (A_{(2)} \rhd B_{(2)}), \tag{72}$$

$$xA \rhd B = x \rhd (A \rhd B) - (x \rhd A) \rhd B, \tag{73}$$

$$A \rhd (B \rhd C) = (A_{(1)}(A_{(2)} \rhd B)) \rhd C. \tag{74}$$

Proof These identities follow by induction on the length of monomials in $\mathscr{U}(\mathfrak{g})$.

It turns out that identity (74) in Proposition 15 can be written $A \rhd (B \rhd C) = (A * B) \rhd C$, where the product $m_* : \mathscr{U}(\mathfrak{g}) \otimes \mathscr{U}(\mathfrak{g}) \to \mathscr{U}(\mathfrak{g})$ is defined by

$$m_*(A \otimes B) = A * B := A_{(1)}(A_{(2)} \rhd B). \tag{75}$$

Theorem 10 [17] *The product defined in* (75) *is non-commutative, associative and unital. Moreover,* $\mathscr{U}_*(\mathfrak{g}) := (\mathscr{U}(\mathfrak{g}), m_*, 1, \Delta, \varepsilon, S_*)$ *is a co-commutative Hopf algebra, whose unit, co-unit and co-product coincide with those defining the usual Hopf algebra structure on* $\mathscr{U}(\mathfrak{g})$*. The antipode* S_* *is given uniquely by the defining equations:*

$$m_* \circ (\mathrm{id} \otimes S_*) \circ \Delta = 1 \circ \varepsilon = m_* \circ (S_* \otimes \mathrm{id}) \circ \Delta.$$

More precisely

$$S_*(x_1 \cdots x_n) = -x_1 \cdots x_n - \sum_{k=1}^{n-1} \sum_{\sigma \in \Sigma_{k,n-k}} x_{\sigma(1)} \cdots x_{\sigma(k)} * S(x_{\sigma(k+1)} \cdots x_{\sigma(n)}), \tag{76}$$

for every $x_1 \cdots x_n \in \mathscr{U}_n(\mathfrak{g})$ *and for all* $n \geq 1$.

Here $\Sigma_{k,n-k} \subset \Sigma_n$ denotes the set of permutations in the symmetric group Σ_n of n elements $[n] := \{1, 2, \ldots, n\}$ such that $\sigma(1) < \cdots < \sigma(k)$ and $\sigma(k+1) < \cdots < \sigma(n)$. Note that since elements $x \in \mathfrak{g}$ are primitive and Δ is a $*$-algebra morphism, one deduces.

Lemma 1

$$\begin{aligned}\Delta(x_1 * \cdots * x_n) &= x_1 * \cdots * x_n \otimes 1 + 1 \otimes x_1 * \cdots * x_n \\ &\quad + \sum_{k=1}^{n-1} \sum_{\sigma \in \Sigma_{k,n-k}} x_{\sigma(1)} * \cdots * x_{\sigma(k)} \otimes x_{\sigma(k+1)} * \cdots * x_{\sigma(n)}.\end{aligned}$$

The relation between the Hopf algebra $\mathscr{U}_*(\mathfrak{g})$ in Theorem 10 and the universal enveloping algebra $\mathscr{U}(\overline{\mathfrak{g}})$ corresponding to the Lie algebra $\overline{\mathfrak{g}}$ is the content of the following theorem.

Theorem 11 [17] $\mathscr{U}_*(\mathfrak{g})$ *is isomorphic, as a Hopf algebra, to* $\mathscr{U}(\overline{\mathfrak{g}})$*. More precisely, the identity map* $\mathrm{id} : \overline{\mathfrak{g}} \to \mathfrak{g}$ *admits a unique extension to an isomorphism of Hopf algebras* $\phi : \mathscr{U}(\overline{\mathfrak{g}}) \to \mathscr{U}_*(\mathfrak{g})$.

Proof First, let us verify the existence of an algebra morphism $\phi : \mathscr{U}(\overline{\mathfrak{g}}) \to \mathscr{U}_*(\mathfrak{g})$. To this end, note that the inclusion map $i : \mathfrak{g} \hookrightarrow \mathscr{U}_*(\mathfrak{g})$, via the universal property of the tensor algebra $T\mathfrak{g}$, guarantees the existence of an algebra morphism $I : T\mathfrak{g} \to \mathcal{U}_*(\mathfrak{g})$ making the following diagram commutative:

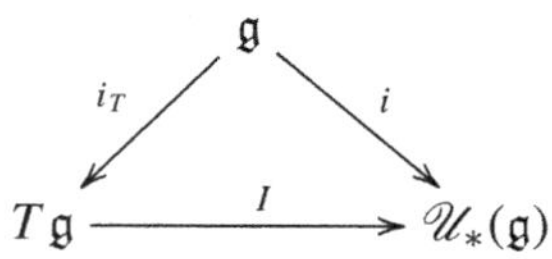

where $i_T : \mathfrak{g} \hookrightarrow T\mathfrak{g}$ is an inclusion map. Note that, since $i(x) = x \in \mathcal{U}_*(\mathfrak{g})$ and $i_T(x) = x \in T\mathfrak{g}$ for all $x \in \mathfrak{g}$, one has $I(x) = x$ for all $x \in \mathfrak{g}$, i.e., the map I restricts to the identity on $\mathfrak{g}$. Then, for all monomials $x_1 \otimes \cdots \otimes x_n \in T\mathfrak{g}$, one has

$$I(x_1 \otimes \cdots \otimes x_n) = x_1 * \cdots * x_n,$$

and, since $x * y - y * x = [\![x, y]\!]$ for $x, y \in \mathfrak{g}$,

$$I(x \otimes y - y \otimes x - [\![x, y]\!]) = 0.$$

It follows then that the map $I : T\mathfrak{g} \to \mathcal{U}_*(\mathfrak{g})$ factors through the (bilateral) ideal $J = \langle x \otimes y - y \otimes x - [\![x, y]\!] \rangle \subset T\mathfrak{g}$, defining a morphism of (filtered) algebras $\phi : \mathcal{U}(\overline{\mathfrak{g}}) \to \mathcal{U}_*(\mathfrak{g})$ which makes the following diagram commutative:

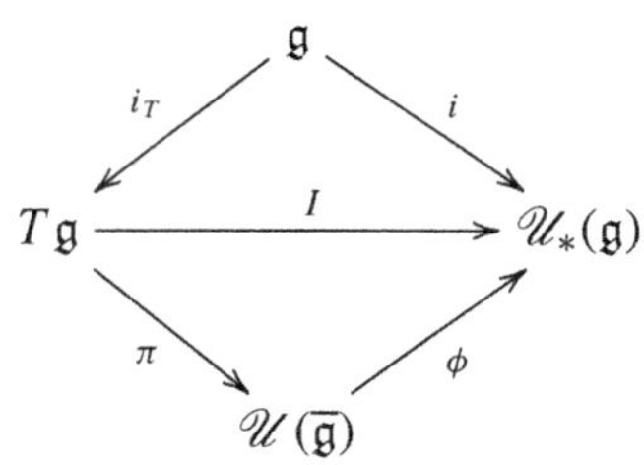

where $\pi : T\mathfrak{g} \to \mathcal{U}(\overline{\mathfrak{g}})$ is the canonical projection, i.e., $\pi(A) = A \bmod J$, for all $A \in T\mathfrak{g}$. Note that since $\pi(x) = x$ for all $x \in \mathfrak{g}$, the map ϕ restricts to the identity on $\mathfrak{g}$. Now, using a simple inductive argument on the length of monomials, one can show that for all $A \in \mathcal{U}_n(\mathfrak{g})$ and $B \in \mathcal{U}_m(\mathfrak{g})$

$$m_*(A \otimes B) = AB \bmod \mathcal{U}_{n+m-1}(\mathfrak{g}),$$

which implies that the graded map $\mathrm{gr}(\phi) : \mathrm{gr}(\mathcal{U}(\overline{\mathfrak{g}})) \to \mathrm{gr}(\mathcal{U}_*(\mathfrak{g}))$, defined, at the level of the homogeneous components, by

$$\mathrm{gr}_n(\phi)\big(x_1 \cdots x_n \bmod \mathcal{U}_{n-1}(\overline{\mathfrak{g}})\big) = \phi(x_1 \cdots x_n) \bmod \mathcal{U}_{*,n-1}(\mathfrak{g})$$

is an isomorphism, proving that $\phi : \mathcal{U}(\overline{\mathfrak{g}}) \to \mathcal{U}_*(\mathfrak{g})$ is an isomorphism of filtered algebras. It is easy now to show that this morphism is compatible with the Hopf algebra structure maps, which implies the statement of the theorem.

In the general case, on the other hand, it is difficult to say more about the isomorphism $\phi : \mathscr{U}(\bar{\mathfrak{g}}) \to \mathscr{U}_*(\mathfrak{g})$. One has the following nice combinatorial description. If $m_. : \mathscr{U}(\overline{\mathfrak{g}}) \otimes \mathscr{U}(\overline{\mathfrak{g}}) \to \mathscr{U}(\overline{\mathfrak{g}})$ denotes the product in $\mathscr{U}(\overline{\mathfrak{g}})$, i.e., $m_.(A \otimes B) = A \cdot B$ for any $A, B \in \mathscr{U}(\overline{\mathfrak{g}})$, then the Hopf algebra isomorphism $\phi : \mathscr{U}(\overline{\mathfrak{g}}) \to \mathscr{U}_*(\mathfrak{g})$ in Theorem 11 can be described as follows. From the proof of Theorem 11 it follows that ϕ restricts to the identity on $\mathfrak{g} \hookrightarrow \mathscr{U}(\mathfrak{g})$. Moreover, for $x_1, x_2, x_3 \in \mathfrak{g}$ we find

$$\phi(x_1 \cdot x_2) = \phi(x_1) * \phi(x_2) = x_1 * x_2 = x_1x_2 + x_1 \rhd x_2,$$

and

$$\begin{aligned}\phi(x_1 \cdot x_2 \cdot x_3) &= x_1 * x_2 * x_3 \\ &= x_1(x_2 * x_3) + x_1 \rhd (x_2 * x_3) \qquad (77)\\ &= x_1x_2x_3 + x_1(x_2 \rhd x_3) + x_2(x_1 \rhd x_3) + (x_1 \rhd x_2)x_3 + x_1 \rhd (x_2 \rhd x_3).\end{aligned}$$

Equality (77) can be generalized to the following simple recursion for words in $\mathscr{U}(\overline{\mathfrak{g}})$ with $n > 0$ letters

$$\phi(x_1 \cdot \cdots \cdot x_n) = x_1\phi(x_2 \cdot \cdots \cdot x_n) + x_1 \rhd \phi(x_2 \cdot \cdots \cdot x_n). \qquad (78)$$

Recall that $x \rhd 1 = 0$ for $x \in \mathfrak{g}$, and $\phi(1) = 1$. From the fact that the post-Lie product on $\mathfrak{g}$ defines a linear map $d : \mathfrak{g} \to \operatorname{Der}\big(\mathscr{U}(\mathfrak{g})\big)$, we deduce that the number of terms on the righthand side of the recursion (78) is given with respect to the length $n = 1, 2, 3, 4, 5, 6$ of the word $x_1 \cdot \cdots \cdot x_n \in \mathscr{U}_*(\mathfrak{g})$ by 1, 2, 5, 15, 52, 203, respectively. These are the Bell numbers B_i, for $i = 1, \ldots, 6$, and for general n, these numbers satisfy the recursion $B_{n+1} = \sum_{i=0}^{n} \binom{n}{i} B_i$. Bell numbers count the different ways the set $[n]$ can be partition into disjoint subsets. From this we deduce the general formula for $x_1 \cdot \cdots \cdot x_n \in \mathscr{U}(\overline{\mathfrak{g}})$

$$\phi(x_1 \cdot \cdots \cdot x_n) = x_1 * \cdots * x_n = \sum_{\pi \in P_n} X_\pi \in \mathscr{U}(\mathfrak{g}), \qquad (79)$$

where P_n is the lattice of set partitions of the set $[n] = \{1, \ldots, n\}$, which has a partial order of refinement ($\pi \leq \kappa$ if π is a finer set partition than κ). Remember that a partition π of the (finite) set $[n]$ is a collection of (non-empty) subsets $\pi = \{\pi_1, \ldots, \pi_b\}$ of $[n]$, called blocks, which are mutually disjoint, i.e., $\pi_i \cap \pi_j = \emptyset$ for all $i \neq j$, and whose union $\cup_{i=1}^{b} \pi_i = [n]$. We denote by $|\pi| := b$ the number of blocks of the partition π, and $|\pi_i|$ is the number of elements in the block π_i. Given $p, q \in [n]$ we will write that $p \sim_\pi q$ if and only if they belong to same block. The partition $\hat{1}_n = \{\pi_1\}$ consists of a single block, i.e., $|\pi_1| = n$. It is the maximum element in P_n. The partition $\hat{0}_n = \{\pi_1, \ldots, \pi_n\}$ has n singleton blocks, and is the minimum partition in P_n.

The element X_π in (79) is defined as follows

$$X_\pi := \prod_{\pi_i \in \pi} x(\pi_i), \tag{80}$$

where $x(\pi_i) := \ell^{\rhd}_{x_{k_1^i}} \circ \ell^{\rhd}_{x_{k_2^i}} \circ \cdots \circ \ell^{\rhd}_{x_{k_{l-1}^i}} (x_{k_l^i})$ for the block $\pi_i = \{k_1^i, k_2^i, \ldots, k_l^i\}$ of the partition $\pi = \{\pi_1, \ldots, \pi_m\}$, and $\ell^{\rhd}_a(b) := a \rhd b$, for a, b elements in the post-Lie algebra $\mathfrak{g} \hookrightarrow \mathscr{U}(\mathfrak{g})$. Recall that $k_l^i \in \pi_i$ is the maximal element in this block.

Remark 8 Defining $m_i := \phi(x^{\cdot i})$ and $d_i := \ell_x^{\rhd i-1}(x) := x \rhd (\ell_x^{\rhd i-2}(x))$, $\ell^{\rhd 0} := \mathrm{id}$, we find that (79) is the i-th-order non-commutative Bell polynomial, $m_i = \mathrm{B}_i^{nc}(d_1, \ldots, d_i)$. See [16, 24] for details.

Next we state a recursion for the compositional inverse $\phi^{-1}(x_1 \cdots x_n)$ of the word $x_1 \cdots x_n \in \mathscr{U}(\mathfrak{g})$. First, it is easy to see that $\phi^{-1}(x_1 x_2) = x_1 \cdot x_2 - x_1 \rhd x_2 \in \mathscr{U}(\overline{\mathfrak{g}})$. Indeed, since ϕ is linear and the identity on $\mathfrak{g} \hookrightarrow \mathscr{U}(\mathfrak{g})$, we have

$$\phi(x_1 \cdot x_2 - x_1 \rhd x_2) = x_1 * x_2 - x_1 \rhd x_2 = x_1 x_2,$$

and

$$\begin{aligned}\phi^{-1}(x_1x_2x_3) &= x_1 \cdot x_2 \cdot x_3 - \phi^{-1}(x_1(x_2 \rhd x_3)) - \phi^{-1}(x_2(x_1 \rhd x_3)) - \phi^{-1}((x_1 \rhd x_2)x_3) \\ &\quad - x_1 \rhd (x_2 \rhd x_3)\end{aligned}$$

which is easy to verify. In general, we find the recursive formula for $\phi^{-1}(x_1 \cdots x_n) \in \mathscr{U}(\overline{\mathfrak{g}})$

$$\phi^{-1}(x_1 \cdots x_n) = x_1 \cdot \cdots \cdot x_n - \sum_{\hat{0}_n < \pi \in P_n} \phi^{-1}(X_\pi). \tag{81}$$

This is well-defined since in the sum on the righthand side all partitions have less than n blocks.

Observe now that since ϕ maps the augmentation ideal of $\mathscr{U}(\bar{\mathfrak{g}})$ to the one of $\mathscr{U}_*(\mathfrak{g})$ it extends to an isomorphism between the completions of the two universal enveloping algebras $\hat{\phi} : \hat{\mathscr{U}}(\bar{\mathfrak{g}}) \to \hat{\mathscr{U}}_*(\mathfrak{g})$, see Subsection 2.2. We are interested in the inverse of the group-like element $\exp(x) \in \mathscr{G}(\hat{\mathscr{U}}(\mathfrak{g}))$, $x \in \mathfrak{g}$, with respect to $\hat{\phi}$. It follows from the inverse of the word $x^n \in \hat{\mathscr{U}}(\mathfrak{g})$, i.e., $\hat{\phi}^{-1}(\exp(x)) = \sum_{n \geq 0} \frac{1}{n!} \hat{\phi}^{-1}(x^n)$.

Theorem 12 *For each $x \in \mathfrak{g}$, there exists an unique element $\chi(x) \in \mathfrak{g}$, such that*

$$\exp(x) = \exp^*(\chi(x)). \tag{82}$$

Proof For $x \in \mathfrak{g}$ the exponential $\exp(x)$ is a group-like element in $\mathscr{G}(\hat{\mathscr{U}}(\mathfrak{g}))$. The proof of Theorem 12 involves calculating the inverse of the group-like element $\exp(x) \in \mathscr{G}(\hat{\mathscr{U}}(\mathfrak{g}))$ with respect to the map $\hat{\phi}$. Indeed, we would like to show that $\hat{\phi}^{-1}(\exp(x)) = \exp^{\cdot}(\chi(x)) \in \mathscr{G}(\hat{\mathscr{U}}(\bar{\mathfrak{g}}))$, from which identity (82) follows

$$\hat{\phi} \circ \hat{\phi}^{-1}(\exp(x)) = \exp(x) = \hat{\phi} \circ \exp^{\cdot}(\chi(x)) = \exp^*(\chi(x)),$$

due to $\hat{\phi}$ being an algebra morphism from $\hat{\mathscr{U}}(\overline{\mathfrak{g}})$ to $\hat{\mathscr{U}}_*(\mathfrak{g})$, which reduces to the identity on $\mathfrak{g}$.

First we show that for $x \in \mathfrak{g}$, the element $\chi(x)$ is defined inductively. For this we consider the expansion $\chi(xt) := xt + \sum_{m>0} \chi_m(x)t^m$ in the parameter t. Comparing $\exp^*(\chi(xt))$ order by order with $\exp(xt)$ yields at second order in t

$$\chi_2(x) := \frac{1}{2}x_1x_2 - \frac{1}{2}x_1 * x_2 = -\frac{1}{2}x \rhd x \in \mathfrak{g}.$$

At third order we deduce from (82) that

$$\begin{aligned}
\chi_3(x) &:= -\frac{1}{3!}\sum_{\hat{0}_3<\pi\in P_3} X_\pi - \frac{1}{2}\chi_2(x) * x - \frac{1}{2}x * \chi_2(x)\\
&= -\frac{1}{3!}\sum_{\hat{0}_3<\pi\in P_3} X_\pi + \frac{1}{4}\big((x \rhd x)x + (x \rhd x) \rhd x\big) + \frac{1}{4}\big(x(x \rhd x) + x \rhd (x \rhd x)\big)\\
&= -\frac{1}{3!}\big(2x(x \rhd x) + (x \rhd x)x + x \rhd (x \rhd x)\big) + \frac{1}{4}\big((x \rhd x)x + (x \rhd x) \rhd x + x(x \rhd x) + x \rhd (x \rhd x)\big)\\
&= \frac{1}{12}[(x \rhd x), x] + \frac{1}{4}(x \rhd x) \rhd x + \frac{1}{12}x \rhd (x \rhd x)\\
&= \frac{1}{6}[\chi_1(x), \chi_2(x)] - \frac{1}{2}\chi_2(x) \rhd x - \frac{1}{6}x \rhd \chi_2(x),
\end{aligned}$$

where we defined $\chi_1(x) := x$. The n-th order term is given by

$$\chi_n(x) := -\frac{1}{n!}\sum_{\hat{0}_n<\pi\in P_n} X_\pi - \sum_{k=2}^{n-1}\frac{1}{k!}\sum_{\substack{p_1+\cdots+p_k=n\\ p_i>0}} \chi_{p_1}(x) * \chi_{p_2}(x) * \cdots * \chi_{p_k}(x) \tag{83}$$

$$= \frac{1}{n!}x^n - \frac{1}{n!}x^{*n} - \sum_{k=2}^{n-1}\frac{1}{k!}\sum_{\substack{p_1+\cdots+p_k=n\\ p_i>0}} \chi_{p_1}(x) * \chi_{p_2}(x) * \cdots * \chi_{p_k}(x). \tag{84}$$

From this we derive an inductive description of the terms $\chi_n(x) \in \hat{\mathscr{U}}_*(\mathfrak{g})$ depending on the $\chi_p(x)$ for $1 \le p \le n-1$

$$\chi_n(x) := \frac{1}{n!}x^n - \sum_{k=2}^{n}\frac{1}{k!}\sum_{\substack{p_1+\cdots+p_k=n\\ p_i>0}} \chi_{p_1}(x) * \chi_{p_2}(x) * \cdots * \chi_{p_k}(x). \tag{85}$$

We have verified directly that the first three terms, $\chi_i(x)$ for $i = 1, 2, 3$, in the expansion $\chi(xt) := xt + \sum_{m>0} \chi_m(x)t^m$ are in $\mathfrak{g}$. However, showing that $\chi_n(x) \in \mathfrak{g}$ for $n > 3$ is more difficult using formula (85). We therefore follow another strategy. At this stage (85) implies that $\chi(x) \in \hat{\mathscr{U}}_*(\mathfrak{g})$ exists. Since $x \in \mathfrak{g}$, we have that $\exp(x)$

is group-like, i.e., $\hat{\Delta}(\exp(x)) = \exp(x)\hat{\otimes}\exp(x)$. Recall that $\hat{\mathscr{U}}_*(\mathfrak{g})$ is a complete Hopf algebra with the same coproduct $\hat{\Delta}$. Hence

$$\hat{\Delta}(\exp^*(\chi(x))) = \hat{\Delta}(\exp(x)) = \exp(x)\hat{\otimes}\exp(x) = \exp^*(\chi(x))\hat{\otimes}\exp^*(\chi(x)).$$

Using $\hat{\phi}$ we can write $\hat{\phi}\,\hat{\otimes}\,\hat{\phi}\circ\hat{\Delta}_{\overline{\mathfrak{g}}}(\exp^{\cdot}(\chi(x))) = \hat{\phi}\hat{\otimes}\hat{\phi}\circ(\exp^{\cdot}(\chi(x))\hat{\otimes}\exp^{\cdot}(\chi(x)))$, which implies that $\exp^{\cdot}(\chi(x))$ is a group-like element in $\hat{\mathscr{U}}(\overline{\mathfrak{g}})$

$$\hat{\Delta}_{\overline{\mathfrak{g}}}(\exp^{\cdot}(\chi(x))) = \exp^{\cdot}(\chi(x))\hat{\otimes}\exp^{\cdot}(\chi(x)).$$

Since $\hat{\mathscr{U}}(\overline{\mathfrak{g}})$ is a complete filtered Hopf algebra, the relation between group-like and primitive elements is one-to-one, see Subsection 2.2. This implies that $\chi(x) \in \overline{\mathfrak{g}} \simeq \mathfrak{g}$, which proves equality (82). Note that $\chi(x)$ actually is an element of the completion of the Lie algebra $\mathfrak{g}$. However, the latter is part of $\hat{\mathscr{U}}(\mathfrak{g})$.

Corollary 7 *Let $x \in \mathfrak{g}$. The following differential equation holds for $\chi(xt) \in \mathfrak{g}[[t]]$*

$$\dot{\chi}(xt) = \mathrm{dexp}^{*-1}_{-\chi(xt)}\Big(\exp^*\big(-\chi(xt)\big) \rhd x\Big). \tag{86}$$

The solution $\chi(xt)$ is called post-Lie Magnus expansion.

Proof Recall the general fact for the dexp-operator [3]

$$\exp^*(-\beta(t)) * \frac{d}{dt}\exp^*(\beta(t)) = \exp^*(-\beta(t)) * \mathrm{dexp}^*_{\beta}(\dot{\beta}) * \exp^*(\beta(t)) = \mathrm{dexp}^*_{-\beta}(\dot{\beta}),$$

where

$$\mathrm{dexp}^*_{\beta}(x) := \sum_{n\geq 0}\frac{1}{(n+1)!}\,\mathrm{ad}^{(*n)}_{\beta}(x) \qquad \text{and} \qquad \mathrm{dexp}^{*-1}_{\beta}(x) := \sum_{n\geq 0}\frac{b_n}{n!}\,\mathrm{ad}^{(*n)}_{\beta}(x).$$

Here b_n are the Bernoulli numbers and $\mathrm{ad}^{(*k)}_a(b) := [a, \mathrm{ad}^{(*k-1)}_a(b)]_*$. This together with the differential equation $\frac{d}{dt}\exp^*(\chi(xt)) = \exp(xt)x$ deduced from (82), implies

$$\begin{aligned}
&\mathrm{dexp}^*_{-\chi(xt)}\big(\dot{\chi}(xt)\big) = \exp^*\big(-\chi(xt)\big) * (\exp(xt)x)\\
&= \exp^*\big(-\chi(xt)\big)\Big(\exp^*\big(-\chi(xt)\big) \rhd (\exp(xt)x)\Big)\\
&= \exp^*\big(-\chi(xt)\big)\Big(\big(\exp^*\big(-\chi(xt)\big) \rhd \exp(xt)\big)\big(\exp^*\big(-\chi(xt)\big) \rhd x\big)\Big)\\
&= \exp^*\big(-\chi(xt)\big)\Big(\big(\exp^*\big(-\chi(xt)\big) \rhd \exp^*\big(\chi(xt)\big)\big)\big(\exp^*\big(-\chi(xt)\big) \rhd x\big)\Big)\\
&= \Big(\exp^*\big(-\chi(xt)\big)\Big(\exp^*\big(-\chi(xt)\big) \rhd \exp^*\big(\chi(ta)\big)\Big)\Big)\big(\exp^*\big(-\chi(xt)\big) \rhd x\big)
\end{aligned}$$

$$= \Big(\exp^*\big(-\chi(xt)\big) * \exp^*\big(\chi(xt)\big)\Big)\big(\exp^*\big(-\chi(xt)\big) \rhd x\big)$$
$$= \exp^*\big(-\chi(xt)\big) \rhd x.$$

The claim in (86) follows after inverting $\mathrm{dexp}^*_{-\chi(xt)}\big(\dot\chi(xt)\big)$. Note that we used successively (75), (69) and (82).

Let us return to point 3. of Remark 4 in Sect. 3, and assume that the post-Lie algebra $(\mathfrak{g}, \rhd, [\cdot,\cdot])$ is equipped with an abelian Lie bracket. This implies that $(\mathfrak{g}, \rhd)$ reduces to a left pre-Lie algebra. The complete universal enveloping algebra $\hat{\mathscr{U}}(\mathfrak{g})$ becomes the complete symmetric algebra $\hat{S}_{\mathfrak{g}}$. This is the setting of [27]. Identity (82) was analyzed in the pre-Lie algebra context in [11].

Corollary 8 *For the pre-Lie algebra $(\mathfrak{g}, \rhd, [\cdot,\cdot]=0)$, identity (82) in $\hat{S}_{\mathfrak{g}}$ is solved by the pre-Lie Magnus expansion*

$$\chi(x) = \frac{\ell^{\rhd}_{-\chi(x)}}{e^{\ell^{\rhd}_{-\chi(x)}}-1}(x) = \sum_{k\geq 0}\frac{(-1)^k b_k}{k!}\ell^{\rhd k}_{\chi(x)}(x),$$

where b_n is the n-th Bernoulli number.

Proof The proof of this result was given in [11] and follows directly from identity (82) in Theorem 12, i.e., by calculating the Lie algebra element $\chi(x)$ as the $\log^*(\exp(x))$ in $\hat{S}_{\mathfrak{g}}$.

The next proposition will be useful in the context of Lie bracket flow equations.

Proposition 16

$$a(t) := \exp^*\big(-\chi(a_0t)\big) \rhd a_0. \tag{87}$$

solves the non-linear post-Lie differential equation with initial value $a(0)=a_0$

$$\dot a(t) = -a(t) \rhd a(t). \tag{88}$$

Proof We calculate

$$\dot a(t) = \Big(-\mathrm{dexp}_{-\chi(a_0t)}\big(\dot\chi(a_0t)\big) * \exp^*\big(-\chi(a_0t)\big)\Big) \rhd a_0$$
$$= -\mathrm{dexp}_{-\chi(a_0t)}\big(\dot\chi(a_0t)\big) \rhd \Big(\exp^*\big(-\chi(a_0t)\big) \rhd a_0\Big)$$
$$= -a(t) \rhd a(t),$$

where we used that $\exp^*\big(-\chi(a_0t)\big) \rhd a_0 = \mathrm{dexp}_{-\chi(a_0t)}\big(\dot\chi(a_0t)\big) = a(t)$.

5.2 Factorization Theorems and r-matrices

In this subsection we will suppose that the post-Lie algebra structure on $\mathfrak{g}$ is defined in terms of a solution of the mCYBE, see Subsection 4.4. Recall that in this case $\bar{\mathfrak{g}} = \mathfrak{g}_R$ implying that $\mathscr{U}(\bar{\mathfrak{g}}) = \mathscr{U}(\mathfrak{g}_R)$ and, correspondingly, that $\hat{\mathscr{U}}(\bar{\mathfrak{g}}) = \hat{\mathscr{U}}(\mathfrak{g}_R)$. In what follows we will prove that for this particular class of post-Lie algebras, the isomorphism ϕ admits an explicit description in terms of the structure of the two Hopf algebras of the universal enveloping algebras $\mathscr{U}(\mathfrak{g}_R)$, $\mathscr{U}_*(\mathfrak{g})$. To this end first we prove the following result.

Proposition 17 *The map $F : \mathscr{U}(\mathfrak{g}_R) \to \mathscr{U}(\mathfrak{g})$ defined by:*

$$F = m_{\mathfrak{g}} \circ (\mathrm{id} \otimes S_{\mathfrak{g}}) \circ (R_+ \otimes R_-) \circ \Delta_{\mathfrak{g}_R}, \tag{89}$$

is a linear isomorphism. Its restriction to $\mathfrak{g}_R \hookrightarrow \mathscr{U}(\mathfrak{g}_R)$ is the identity map.

Proof Note that $m_{\mathfrak{g}}$ and $S_{\mathfrak{g}}$ denote product respectively antipode in $\mathscr{U}(\mathfrak{g})$, whereas $\Delta_{\mathfrak{g}_R}$ denotes the co-product in $\mathscr{U}(\mathfrak{g}_R)$. This slightly more cumbersome notation is applied in order to make the presentation more traceable. Given an element $x \in \mathfrak{g}_R \hookrightarrow \mathscr{U}(\mathfrak{g}_R)$, one has that

$$\begin{aligned}
F(x) &= m_{\mathfrak{g}} \circ (\mathrm{id} \otimes S_{\mathfrak{g}}) \circ (R_+ \otimes R_-) \circ \Delta_{\mathfrak{g}_R}(x) \\
&= m_{\mathfrak{g}} \circ (\mathrm{id} \otimes S_{\mathfrak{g}}) \circ (R_+ \otimes R_-)(x \otimes 1 + 1 \otimes x) \\
&= m_{\mathfrak{g}} \circ (\mathrm{id} \otimes S_{\mathfrak{g}})(R_+(x) \otimes 1 + 1 \otimes R_-(x)) \\
&= m_{\mathfrak{g}}(R_+(x) \otimes 1 - 1 \otimes R_-(x)) \\
&= R_+(x) - R_-(x) = x \in \mathfrak{g} \hookrightarrow \mathscr{U}(\mathfrak{g}),
\end{aligned}$$

showing that F restricts to the identity map between $\mathfrak{g}_R$ and $\mathfrak{g}$. As in Lemma 1 we have

$$\Delta_{\mathfrak{g}_R}(x_1 \cdots x_n) = x_1 \cdots x_n \otimes 1 + 1 \otimes x_1 \cdots x_n + \sum_{k=1}^{n-1} \sum_{\sigma \in \Sigma_{k,n-k}} x_{\sigma(1)} \cdots x_{\sigma(k)} \otimes x_{\sigma(k+1)} \cdots x_{\sigma(n)}.$$

Since $R_\pm$ are homomorphisms of unital associative algebras, one can easily show that for every $x_1 \cdots x_k \in \mathscr{U}_k(\mathfrak{g}_R)$:

$$\begin{aligned}
&F(x_1 \cdots x_k) = R_+(x_1) \cdots R_+(x_k) + (-1)^k R_-(x_k) \cdots R_-(x_1) + \\
&\sum_{l=1}^{k-1} \sum_{\sigma \in \Sigma_{l,k-l}} (-1)^{k-l} R_+(x_{\sigma(1)}) \cdots R_+(x_{\sigma(l)}) R_-(x_{\sigma(k)}) \cdots R_-(x_{\sigma(l+1)}) \in \mathscr{U}_k(\mathfrak{g}),
\end{aligned}$$

which proves that F maps homogeneous elements to homogeneous elements. To verify injectivity of F one can argue as follows. Since $x = R_+(x) - R_-(x)$ for all $x \in \mathfrak{g}$, one can deduce from the previous formula for $x_1 \cdots x_k \in \mathscr{U}_k(\mathfrak{g}_R)$ that

$$F(x_1 \cdots x_k) = x_1 \cdots x_k \bmod \mathscr{U}_{k-1}(\mathfrak{g}),$$

where $x_1 \cdots x_k$ on the righthand side lies in $\mathscr{U}_k(\mathfrak{g})$. For instance

$$F(x_1x_2) = R_+(x_1)R_+(x_2) + R_-(x_2)R_-(x_1) - R_+(x_1)R_-(x_2) - R_+(x_2)R_-(x_1).$$

Using $x + R_-(x) = R_+(x)$ implies in $\mathscr{U}(\mathfrak{g})$ that

$$\begin{aligned} F(x_1x_2) &= (x_1 + R_-(x_1))(x_2 + R_-(x_2)) + R_-(x_2)R_-(x_1) \\ &\quad - (x_1 + R_-(x_1))R_-(x_2) - (x_2 + R_-(x_2))R_-(x_1) \\ &= x_1x_2 + x_1R_-(x_2) + R_-(x_1)x_2 + R_-(x_1)R_-(x_2) \\ &\quad + R_-(x_2)R_-(x_1) - x_1R_-(x_2) - R_-(x_1)R_-(x_2) - x_2R_-(x_1) - R_-(x_2)R_-(x_1) \\ &= x_1x_2 + [R_-(x_1), x_2], \end{aligned}$$

where $x_1x_2 \in \mathscr{U}_2(\mathfrak{g})$ and $[R_-(x_1), x_2] \in \mathscr{U}_1(\mathfrak{g}) \simeq \mathfrak{g}$. Then, if $F(x_1, \cdots x_k) = 0$, one concludes that $x_1 \cdots x_k \in \mathscr{U}_k(\mathfrak{g})$ must be equal to zero, that is, at least one among the elements $x_i \in \mathfrak{g}$ composing the monomial $x_1 \cdots x_k$ is equal to zero. This forces the element $x_1 \cdots x_k \in \mathscr{U}_k(\mathfrak{g}_R)$ to be equal to zero, proving injectivity of F.

To prove that the map F is surjective one can argue by induction on the length of the homogeneous elements of $\mathscr{U}(\mathfrak{g})$. The first step of the induction is provided by the fact that F restricted to $\mathfrak{g}_R$ becomes the identity map, and $\mathfrak{g} \hookrightarrow \mathscr{U}_1(\mathfrak{g})$. Suppose now that every element in $\mathscr{U}_{k-1}(\mathfrak{g})$ is in the image of F and observe that $x_1 \cdots x_k \in \mathscr{U}_k(\mathfrak{g})$ can be written as

$$\begin{aligned} x_1 \cdots x_k &= \prod_{i=1}^{k} \big(R_+(x_i) - R_-(x_i)\big) \\ &= \big(R_+(x_1) - R_-(x_1)\big)\big(R_+(x_2) - R_-(x_2)\big) \prod_{i=3}^{k} \big(R_+(x_i) - R_-(x_i)\big) \\ &= \big(R_+(x_1)R_+(x_2) - R_-(x_1)R_+(x_2) - R_+(x_1)R_-(x_2) \\ &\quad + R_-(x_1)R_-(x_2)\big) \prod_{i=3}^{k} \big(R_+(x_i) - R_-(x_i)\big) \\ &= R_+(x_1) \cdots R_+(x_k) \\ &\quad + \sum_{l=1}^{k-1} \sum_{\sigma \in \Sigma_{l,k-l}} (-1)^{k-l} R_+(x_{\sigma(1)}) \cdots R_+(x_{\sigma(l)}) \cdot R_-(x_{\sigma(k)}) \cdots R_-(x_{\sigma(l+1)}) \\ &\quad + (-1)^k R_-(x_k) \cdots R_-(x_1) \bmod \mathscr{U}_{k-1}(\mathfrak{g}), \end{aligned}$$

which proves the claim, since

$$F(x_1 \cdots x_k) = R_+(x_1) \cdots R_+(x_k) + (-1)^k R_-(x_k) \cdots R_-(x_1)$$
$$+ \sum_{l=1}^{k-1} \sum_{\sigma \in \Sigma_{l,k-l}} (-1)^{k-l} R_+(x_{\sigma(1)}) \cdots R_+(x_{\sigma(l)}) \cdot R_-(x_{\sigma(k)}) \cdots R_-(x_{\sigma(l+1)}).$$

Using the previous computation and the definition of the $*$-product, one can easily see that $F(x_1x_2) = x_1x_2 + [R_-(x_1), x_2] = x_1x_2 + x_1 \rhd_- x_2$, where $\rhd$ is defined in (64) (and lifted to $\mathcal{U}(\mathfrak{g})$), which implies that $F(x_1x_2) = x_1 * x_2 \in \mathcal{U}_*(\mathfrak{g})$. Using a simple induction on the lenght of the monomials, the above calculation extends to all of $\mathcal{U}(\mathfrak{g}_R)$, which is the content of the following.

Corollary 9 *[18] The map F is an isomorphism of unital, filtered algebras from $\mathcal{U}(\mathfrak{g}_R)$ to $\mathcal{U}_*(\mathfrak{g})$. In particular, $F(x_1 \cdots x_n) = x_1 * \cdots * x_n$ for all monomials $x_1 \cdots x_n \in \mathcal{U}(\mathfrak{g}_R)$.*

Comparing this result with the Theorem 11 of the previous section, one has

Proposition 18 *If the post-Lie algebra $(\mathfrak{g}, [\cdot,\cdot], \rhd_-)$ is defined in terms of a r-matrix R via Formula (64), then the isomorphism ϕ of Theorem 11 assumes the explicit form given in Formula (89), i.e., $\phi = F$.*

Proof In fact note that both ϕ and F are isomorphisms of filtered, unital associative algebras taking values in $\mathcal{U}_*(\mathfrak{g})$, restricting to the identity map on $\mathfrak{g}_R$ which is the generating set of $\mathcal{U}(\mathfrak{g}_R)$.

At this point it is worth making the following observation, which will be useful in what follows.

Corollary 10 *Every $A \in \mathcal{U}(\mathfrak{g})$ can be written uniquely as*

$$A = R_+(a_{(1)})S_{\mathfrak{g}}(R_-(a_{(2)})) \tag{90}$$

for a suitable element $a \in \mathcal{U}(\mathfrak{g}_R)$, where we wrote the co-product of this element using the Sweedler's notation, i.e., $\Delta_{\mathfrak{g}_R}(a) = a_{(1)} \otimes a_{(2)}$.

Proof The proof follows from Proposition 17, noticing that for each $a \in \mathcal{U}(\mathfrak{g}_R)$,

$$F(a) = R_+(a_{(1)})S_{\mathfrak{g}}(R_-(a_{(2)})).$$

Finally, in this more specialized context, we can give the following computational proof of the result contained in Theorem 11.

Theorem 13 *The map $F : \mathcal{U}(\mathfrak{g}_R) \to \mathcal{U}_*(\mathfrak{g})$ is an isomorphism of Hopf algebras.*

Proof F is a linear isomorphism which sends a monomial of length k to (a linear combination of) monomials of the same length. For this reason the compatibility of F with the co-units is verified. Since $F : \mathscr{U}(\mathfrak{g}_R) \to \mathscr{U}_*(\mathfrak{g})$ is an isomorphism of filtered, unital, associative algebras, the product $*$ defined in Formula (75) can be defined as the push-forward to $\mathscr{U}(\mathfrak{g})$, via F, of the associative product of $\mathscr{U}(\mathfrak{g}_R)$, i.e

$$A * B = F(m_{\mathfrak{g}_R}(F^{-1}(A) \otimes F^{-1}(B))), \tag{91}$$

for all monomials $A, B \in \mathscr{U}(\mathfrak{g})$, which immediately implies the compatibility of F with the algebra units. Let us show that F is a morphism of co-algebras, i.e., that

$$\Delta_{\mathfrak{g}} \circ F = (F \otimes F) \circ \Delta_{\mathfrak{g}_R}. \tag{92}$$

Since $F(x_1 \cdots x_n) = x_1 * \cdots * x_n$, see Corollary 9, using the formula in Lemma 1, one gets

$$\begin{aligned}\Delta_{\mathfrak{g}}\big(F(x_1 \cdots x_n)\big) &= x_1 * \cdots * x_n \otimes 1 + 1 \otimes x_1 * \cdots * x_n \\ &+ \sum_{k=1}^{n-1} \sum_{\sigma \in \Sigma_{k,n-k}} x_{\sigma(1)} * \cdots * x_{\sigma(k)} \otimes x_{\sigma(k+1)} * \cdots * x_{\sigma(n)},\end{aligned}$$

which turns out to be equal to $(F \otimes F) \circ \Delta_{\mathfrak{g}_R}(x_1 \cdots x_n)$. The only thing that is left to be checked is that F is compatible with the antipodes of the two Hopf algebras, i.e., that $F \circ S_{\mathfrak{g}_R} = S_* \circ F$, where for $x_1 \cdots x_n \in \mathscr{U}(\mathfrak{g}_R)$, $S_{\mathfrak{g}_R}(x_1 \cdots x_n) = (-1)^n x_n \cdots x_1$. To this end first recall that the antipode is an algebra anti-homomorphism, i.e., $S_*(A * B) = S_*(B) * S_*(A)$, for all $A, B \in \mathscr{U}(\mathfrak{g})$. From this and from the property that $S_{\mathfrak{g}_R}(x) = -x$ for all $x \in \mathfrak{g}_R$, using a simple induction on the length of the monomials, one obtains

$$S_*(x_1 * \cdots * x_n) = (-1)^n x_n * \cdots * x_1.$$

From this observation follows now easily that $F \circ S_{\mathfrak{g}_R} = S_* \circ F$.

We conclude this section with the following observation, see Remark 9.

Proposition 19 *For all $A, B \in \mathscr{U}(\mathfrak{g})$, one has that:*

$$A * B = R_+(a_{(1)}) B S_{\mathfrak{g}}(R_-(a_{(2)})), \tag{93}$$

where $a \in \mathscr{U}(\mathfrak{g}_R)$ is the unique element, such that $A = F(a)$, see Corollary 10.

Proof Let $a, b \in \mathscr{U}(\mathfrak{g}_R)$ such that $F(a) = A$ and $F(b) = B$. We will use Sweedler's notation for the co-product $\Delta_{\mathfrak{g}_R}(a) = a_{(1)} \otimes a_{(2)}$, and write $m_{\mathfrak{g}_R}(a \otimes b) := a \cdot b$ for the product in $\mathscr{U}(\mathfrak{g}_R)$.

$$\begin{aligned}A * B = F(a \cdot b) &= m_{\mathfrak{g}} \circ (\mathrm{id} \otimes S_{\mathfrak{g}}) \circ (R_+ \otimes R_-) \circ \Delta_{\mathfrak{g}_R}(a \cdot b) \\ &= m_{\mathfrak{g}} \circ (\mathrm{id} \otimes S_{\mathfrak{g}}) \circ (R_+ \otimes R_-)(a_{(1)} \otimes a_{(2)}) \cdot (b_{(1)} \otimes b_{(2)})\end{aligned}$$

$$
\begin{aligned}
&= m_{\mathfrak{g}} \circ (\mathrm{id} \otimes S_{\mathfrak{g}}) \circ (R_+ \otimes R_-)(a_{(1)} \cdot b_{(1)}) \otimes (a_{(2)} \cdot b_{(2)}) \\
&= m_{\mathfrak{g}} \circ (\mathrm{id} \otimes S_{\mathfrak{g}})\big(R_+(a_{(1)})R_+(b_{(1)}) \otimes R_-(a_{(2)})R_-(b_{(2)})\big) \\
&\overset{(*)}{=} m_{\mathfrak{g}}\big(R_+(a_{(1)})R_+(b_{(1)}) \otimes S_{\mathfrak{g}}(R_-(b_{(2)}))S_{\mathfrak{g}}(R_-(a_{(2)}))\big) \\
&= R_+(a_{(1)})R_+(b_{(1)})S_{\mathfrak{g}}(R_-(b_{(2)}))S_{\mathfrak{g}}(R_-(a_{(2)})) \\
&= R_+(a_{(1)})F(b)S_{\mathfrak{g}}(R_-(a_{(2)}) \\
&= R_+(a_{(1)})BS_{\mathfrak{g}}(R_-(a_{(2)}),
\end{aligned}
$$

which proves the statement. In equality $(*)$ we applied that $S_{\mathfrak{g}}(\xi\eta) = S_{\mathfrak{g}}(\eta)S_{\mathfrak{g}}(\xi)$.

Remark 9 (Link to the work of Semenov-Tian-Shansky and Reshetikhin) In this remark we aim to link the previous results to the ones described in the references [35] and [31]. The map (89) was first defined in [35] (see also [31]), where it was used to push-forward to $\mathscr{U}(\mathfrak{g})$ the associative product of $\mathscr{U}(\mathfrak{g}_R)$ using the formula (91). From the equality between the maps ϕ and F, see Proposition 18, it follows at once that the associative product defined in $\mathscr{U}(\mathfrak{g})$ by the authors of [31, 35], *is* the product defined in Formula (75). Moreover, at the best knowledge of the authors of the present note, in the references [31, 35], the Hopf algebra structure induced on (the underlying vector space of) $\mathscr{U}(\mathfrak{g})$, by the push-forward of the associative product of $\mathscr{U}(\mathfrak{g}_R)$ was not disclosed. In this way, via the theory of the post-Lie algebras, on one hand we could extend (part of) the results of [31, 35] to an Hopf algebraic theoretical framework, while one the one other one, we could get a more computable formula for the product defined in [31, 35]. In particular note that, in spite of the result in Proposition 19 was stated in [31, 35], the product in Formula (93) is not easily computable since it supposes the knowledge of the inverse of the map F. On the other hand, Formula (75) provides an explicit way to compute the $*$-product between any two monomials of $\mathscr{U}(\mathfrak{g})$.

In this final part we discuss an application of the result presented above to the problem of the factorization of the group like-elements of the completed universal enveloping algebra of $\mathfrak{g}_R$. This result should be compared with the one in Theorem 5 in Sect. 4.2. We start observerving that, since $R_\pm : \mathscr{U}(\mathfrak{g}_R) \to \mathscr{U}(\mathfrak{g})$ are algebra morphisms, they map the augmentation ideal of $\mathscr{U}(\mathfrak{g}_R)$ to the augmentation ideal of $\mathscr{U}(\mathfrak{g})$ and, for this reason, both these morphisms extend to morphisms $R_\pm : \hat{\mathscr{U}}(\mathfrak{g}_R) \to \hat{\mathscr{U}}(\mathfrak{g})$. In particular, the map F extends to an isomorphism of (complete) Hopf algebras $\hat{F} : \hat{\mathscr{U}}(\mathfrak{g}_R) \to \hat{\mathscr{U}}_*(\mathfrak{g})$, defined by

$$\hat{F} = \hat{m}_{\mathfrak{g}} \circ (\mathrm{id}\,\hat{\otimes}\hat{S}_{\mathfrak{g}}) \circ (R_+\hat{\otimes}R_-) \circ \hat{\Delta}_{\mathfrak{g}_R},$$

where, $\hat{\Delta}_{\mathfrak{g}_R}$ denotes the coproduct of $\hat{\mathscr{U}}(\mathfrak{g}_R)$, and with $\hat{m}_{\mathfrak{g}}$, $\hat{S}_{\mathfrak{g}}$ the product respectively the antipode of $\hat{\mathscr{U}}(\mathfrak{g})$ are denoted. Let $\exp^{\cdot}(x) \in \mathscr{G}(\hat{\mathscr{U}}(\mathfrak{g}_R))$, $\exp^*(x) \in \mathscr{G}(\hat{\mathscr{U}}_*(\mathfrak{g}))$ and $\exp(x) \in \mathscr{G}(\hat{\mathscr{U}}(\mathfrak{g}))$, the respective exponentials.

Following [31] we now compare identity (94) with (52). At the level of the universal enveloping algebra, the main result of Theorem 5 can be rephrased as follows.

Theorem 14 *Every element* $\exp^*(x) \in \mathcal{G}(\hat{\mathcal{U}}_*(\mathfrak{g}))$ *admits the following factorization:*

$$\exp^*(x) = \exp(x_+)\exp(-x_-). \tag{94}$$

Proof To simplify notation, write $m_{\mathfrak{g}_R}(x \otimes y) = x \cdot y$, for all $x, y \in \mathfrak{g}_R$, so that for each $x \in \mathfrak{g}_R$, $x^{\cdot n} := x \cdots x$. Then observe that, for each $n \geq 0$, one has

$$\hat{F}(x^{\cdot n}) = R_+(x)^n + \sum_{l=1}^{n-1}(-1)^{n-l}\binom{n}{l}R_+(x)^l R_-(x)^{n-l} + (-1)^n R_-(x)^n.$$

Then, after reordering the terms, one gets $\hat{F}(\exp^{\cdot}(x)) = e^{x_+}e^{-x_-}$. On the other hand, since $\hat{F} : \hat{\mathcal{U}}(\mathfrak{g}_R) \to \hat{\mathcal{U}}_*(\mathfrak{g})$ is an algebra morphism, one obtains for each $n \geq 0$

$$\hat{F}(x^{\cdot n}) = \hat{F}(x) * \cdots * \hat{F}(x) = x^{*n},$$

from which it follows that

$$\hat{F}(\exp^{\cdot}(x)) = \hat{F}(1) + \hat{F}(x) + \frac{\hat{F}(x^{\cdot 2})}{2!} + \cdots + \frac{\hat{F}(x^{\cdot n})}{n!} + \cdots = \exp^*(x),$$

giving the result

The observation in Theorem 12 implies for group-like elements in $\mathcal{G}(\hat{\mathcal{U}}(\mathfrak{g}))$ and $\mathcal{G}(\hat{\mathcal{U}}_*(\mathfrak{g}))$ that $\exp(x) = \exp^*(\chi(x))$, from which we deduce.

Corollary 11 *Group-like elements* $\exp(x) \in \mathcal{G}(\hat{\mathcal{U}}(\mathfrak{g}))$ *factorize*

$$\exp(x) = \exp(\chi_+(x))\exp(-\chi_-(x)). \tag{95}$$

Proof The proof follows from Theorem 12 and Theorem 14.

Remark 10 Looking at $\chi(x)$ in the context of $\hat{\mathcal{U}}(\mathfrak{g})$, i.e., with the post-Lie product on $\mathfrak{g}$ defined in terms of the r-matrix, $x \rhd_- y = [R_-(x), y]$, we find that $\chi_2(x) = -\frac{1}{2}[R_-(x), x]$ and

$$\chi_3(x) = \frac{1}{4}[R_-([R_-(x), x]), x] + \frac{1}{12}([[R_-(x), x], x] + [R_-(x), [R_-(x), x]]).$$

This should be compared with Eq. (7) in [15], as well as with the results in [18].

5.3 Applications to Isospectral Flow Equations

Recall Proposition 16. In the context of the post-Lie product $x \rhd_- y := [R_-(x), y]$ induced on $\mathfrak{g}$ by an r-matrix R, this proposition says that the Lie bracket flow

$$\dot{x}(t) = [x, R_-(x)], \quad x(0) = x_0$$

has solution

$$\begin{aligned} x(t) &= \exp^*(-\chi(x_0 t)) \rhd_- x_0 \\ &= \exp\big(-R_-(\chi(x_0 t))\big)x_0 \exp\big(R_-(\chi(x_0 t))\big). \end{aligned}$$

The last equality follows from general results of post-Lie algebra. Since, $-\chi(x_0 t) \in \mathfrak{g}$ we have

$$\begin{aligned} \exp^*(-\chi(x_0 t)) \rhd_- x_0 &= x_0 - \chi(x_0 t) \rhd_- x_0 + \frac{1}{2!}(\chi(x_0 t) * \chi(x_0 t)) \rhd_- x_0 + \cdots \\ &= X_0 - \chi(X_0 t) \rhd_- X_0 + \frac{1}{2!}\chi(X_0 t) \rhd_- (\chi(x_0 t) \rhd_- x_0) + \cdots \\ &= \sum_{n \geq 0} \frac{(-1)^n}{n!} \operatorname{ad}^{(n)}_{R_-(\chi(x_0 t))} x_0. \end{aligned}$$

Acknowledgements The first author acknowledges support from the Spanish government under the project MTM2013-46553-C3-2-P.

References

1. Babelon, O., Bernard, D., Talon, M.: Introduction to classical integrable systems. Cambridge Monographs on Mathematical Physics. Cambridge University Press (2007)
2. Bai, C., Guo, L., Ni, X.: Nonabelian generalized Lax pairs, the classical Yang-Baxter equation and PostLie algebras. Commun. Math. Phys. **297**(2), 553 (2010)
3. Blanes, S., Casas, F., Oteo, J.A., Ros, J.: Magnus expansion: mathematical study and physical applications. Phys. Rep. **470**, 151 (2009)
4. Burde, D.: Left-symmetric algebras, or pre-Lie algebras in geometry and physics. Central Eur. J. Math. **4**(3), 323 (2006)
5. Cartan, E., Schouten, J.A.: On the Geometry of the Group-manifold of simple and semi-simple groups. In: Proceedings of the Royal Academy of Amsterdam, vol. XXIX (1926)
6. Cartier, P.: A primer of Hopf algebras. In: Frontiers in Number Theory, Physics, and Geometry II, pp. 537–615. Springer, Berlin, Heidelberg (2007)
7. Cartier, P.: Vinberg algebras. Lie groups and combinatorics. Clay Math. Proc. **11**, 107 (2011)
8. Casas, F.: Numerical integration methods for the double-bracket flow. J. Comput. Appl. Math. **166**(2), 477 (2004)
9. Casas, F., Iserles, A.: Explicit Magnus expansions for nonlinear equations. J. Phys. A: Math. Gen. **39**, 5445 (2006)
10. Chapoton, F., Livernet, M.: Pre-Lie algebras and the rooted trees operad. Int. Math. Res. Not. **2001**, 395 (2001)
11. Chapoton, F., Patras, F.: Enveloping algebras of preLie algebras, Solomon idempotents and the Magnus formula. Int. J. Algebra Comput. **23**(4), 853 (2013)
12. Chu, M.T., Norris, L.K.: Isospectral flows and abstract matrix factorizations. SIAM J. Numer. Anal. **25**, 1383 (1988)
13. Drinfeld, V.G.: Hamiltonian structures on Lie groups, Lie bialgebras and the geometric meaning of the classical Yang–Baxter equation. Sov. Math. Dokl. **27**(1) (1983)

14. Duistermaat, J.J., Kolk, J.A.: Lie Groups, Universitext. Springer, Berlin (2000)
15. Ebrahimi-Fard, K., Guo, L., Manchon, D.: Birkhoff type decompositions and the Baker-Campbell-Hausdorff recursion. Commun. Math. Phys. **267**, 821 (2006)
16. Ebrahimi-Fard, K., Lundervold, A., Manchon, D.: Noncommutative Bell polynomials, quasideterminants and incidence Hopf algebras. Int. J. Algebra Comput. **24**(5), 671 (2014)
17. Ebrahimi-Fard, K., Lundervold, A., Munthe-Kaas, H.Z.: On the Lie enveloping algebra of a post-Lie algebra. J. Lie Theory **25**(4), 1139 (2015)
18. Ebrahimi-Fard, K., Lundervold, A., Mencattini, I., Munthe-Kaas, H.Z.: Post-Lie algebras and isospectral flows. SIGMA **25**(11), 093 (2015)
19. Ebrahimi-Fard, K., Mencattini, I., Munthe-Kaas, H.Z.: Post-Lie algebras and factorization theorems. arXiv:1701.07786
20. Faybusovich, L.E.: QR-type factorizations, the Yang-Baxter equation and an eigenvalue problem of control theory. Linear Algebra Appl. **122–124**, 943 (1989)
21. Fuks, D.B.: Cohomology of Infinite Dimensional Lie Algebras Consultant Bureau, New York (1986)
22. Lax, P.D.: Outline of a theory of KdV equation in Recent mathematical methods in nonlinear wave propagation (Montecatini Terme 1994). Lecture Notes in Mathematics, vol. 1640, pp. 70–102. Springer, Berlin (1996)
23. Lundervold, A., Munthe-Kaas, H.Z.: On post-Lie algebras, Lie–Butcher series and moving frames. Found. Comput. Math. **13**(4), 583 (2013)
24. Lundervold, A., Munthe-Kaas, H.Z.: On algebraic structures of numerical integration on vector spaces and manifolds. In: Ebrahimi-Fard, K., Fauvet, F. (eds.) Faà di Bruno Hopf Algebras, Dyson–Schwinger Equations, and Lie–Butcher Series. IRMA Lectures in Mathematics and Theoretical Physics, vol. 21 (2015). Eur. Math. Soc. Strasbourg, France
25. Laurent-Gengoux, C., Pichereau, A., Vanhaecke, P.: Poisson Structures, Grundlehren der Mathematischen Wissenschaften [Fundamental Principles of Mathematical Sciences]. Springer, Heidelberg (2013)
26. Manchon, D.: A short survey on pre-Lie algebras, in Noncommutative Geometry and Physics: renormalisation, motives, index theory. In: Carey A. (ed) E. Schrödinger Institut Lectures in Mathematics and Physics. European Mathematical Society (2011)
27. Oudom, J.-M., Guin, D.: On the Lie enveloping algebra of a pre-Lie algebra. J. K-theory: K-theory Appl. Algebra Geom. Topol. **2**(1), 147 (2008)
28. Postnikov, M.M.: Lie Groups and Lie Algebras. Lectures in Geometry. Semester V. Mir, Moscow (1986)
29. Postnikov, M.M.: Geometry VI. Riemannian geometry. Encyclopaedia of Mathematical Sciences, vol. 9. Springer, Berlin (2001)
30. Quillen, D.: Rational homotopy theory. Ann. Math. **90**(2), 205 (1969)
31. Reshetikhin, Y.N., Semenov-Tian-Shansly, M.A.: Quantum R-matrices and factorization problems. J. Geom. Phys. **5**(4), 533 (1988)
32. Reutenauer, C.: Free Lie Algebras, London Mathematical Society Monographs. New Series, 7, Oxford Science Publications, The Clarendon Press, Oxford University Press, New York (1993)
33. Semenov-Tian-Shansky, M.A.: What is a classical r-matrix? Funct. Ana. Appl. **17**, 254 (1983)
34. Semenov Tian-Shansky, M.A.: Lectures on R-matrices, Poisson-Lie groups and integrable systems. In: Lectures on Integrable Systems, Sophia-Antipolis, vol. 1994, pp. 269–317. World Scientific Publishing, River Edge, NJ (1991)
35. Semenov-Tian-Shansky, M.A.: Classical r-matrix and quantization. Journal of Soviet Mathematics **31**(6), 3411 (1985)
36. Semenov-Tian-Shansky, M.A.: Quantum and classical integrable systems. In: Integrability of Nonlinear Systems (Pondicherry, 1996). Lecture Notes in Physics, vol. 495, pp. 314–377. Springer, Berlin (1997)
37. Suris, Y.B.: The Problem of Integrable Discretization: Hamiltonian Approach, Progress in Mathematics. Birkhäuser, Basel (2003)
38. Sweedler, M.E.: Hopf algebras, Mathematical Lectures Notes. W.A. Benjamin, Inc., New York (1969)

39. Vallette, B.: Homology of generalized partition posets. J. Pure Appl. Algebra **208**(2), 699 (2007)
40. Warner, F.W.: Foundations of differentiable manifolds and Lie groups, Graduate Texts in Mathematics. Springer, New York, Berlin (1983)
41. Watkins, D.S.: Isospectral flows. SIAM Rev. **26**, 379 (1984)

Overview of (pro-)Lie Group Structures on Hopf Algebra Character Groups

Geir Bogfjellmo, Rafael Dahmen and Alexander Schmeding

Abstract Character groups of Hopf algebras appear in a variety of mathematical and physical contexts. To name just a few, they arise in non-commutative geometry, renormalisation of quantum field theory, and numerical analysis. In the present article we review recent results on the structure of character groups of Hopf algebras as infinite-dimensional (pro-)Lie groups. It turns out that under mild assumptions on the Hopf algebra or the target algebra the character groups possess strong structural properties. Moreover, these properties are of interest in applications of these groups outside of Lie theory. We emphasise this point in the context of two main examples:

- the Butcher group from numerical analysis and
- character groups which arise from the Connes–Kreimer theory of renormalisation of quantum field theories.

Keywords Infinite-dimensional Lie group · Hopf algebra
Locally convex algebra · Butcher group · Weakly complete space · pro-Lie group
Regular Lie group

MSC 2010 22E65 (primary) · 16T05 · 43A40 · 58B25 · 46H30 · 22A05 (Secondary)

G. Bogfjellmo
Chalmers Technical University & Gothenburg University, Chalmers Tvä rgata 3, Gothenburg, Sweden
e-mail: geir.bogfjellmo@gu.se

R. Dahmen
TU Darmstadt, Schloßgartenstr. 7, Darmstadt, Germany
e-mail: dahmen@mathematik.tu-darmstadt.de

A. Schmeding (✉)
Department of Mathematical Sciences, Norwegian University of Science and Technology—NTNU, Trondheim, Norway
e-mail: schmeding@tu-berlin.de

K. Ebrahimi-Fard and M. Barbero Liñán (eds.), *Discrete Mechanics, Geometric Integration and Lie–Butcher Series*, Springer Proceedings in Mathematics & Statistics 267, https://doi.org/10.1007/978-3-030-01397-4_8

1 Introduction

Character groups of Hopf algebras appear in a variety of mathematical and physical contexts. To name just a few, they arise in non-commutative geometry, renormalisation of quantum field theory (see [1]) and numerical analysis (cf. [2]). In these contexts the arising groups are often studied via an associated Lie algebra or by assuming an auxiliary topological or differentiable structure on these groups. For example, in the context of the Connes–Kreimer theory of renormalisation of quantum field theories, the group of characters of a Hopf algebra of Feynman graphs is studied via its Lie algebra (cf. [1]). Moreover, it turns out that this group is a projective limit of finite-dimensional Lie groups. The (infinite-dimensional) Lie algebra and the projective limit structure are important tools to analyse these character groups.

Another example for a character group is the so called Butcher group from numerical analysis which can be realised as a character group of a Hopf algebra of trees. In [3] we have shown that the Butcher group can be turned into an infinite-dimensional Lie group. All of these results can be interpreted in the general framework of (infinite-dimensional) Lie group structures on character groups of Hopf algebras developed in [4].

The present article provides an introduction to the theory developed in [4] with a view towards applications and further research. To be as accessible as possible, we consider the results from the perspective of examples arising in numerical analysis and mathematical physics. Note that the Lie group structures discussed here will in general be infinite-dimensional and their modelling spaces will be more general than Banach spaces. Thus we base our investigation on a concept of C^r-maps between locally convex spaces known as Bastiani calculus or Keller's C^r_c-theory. However, we recall and explain the necessary notions as we do not presuppose familiarity with the concepts of infinite-dimensional analysis and Lie theory.

We will always suppose that the target algebra supports a suitable topological structure, i.e. it is a locally convex algebra (e.g. a Banach algebra). The infinite-dimensional structure of a character group is then determined by the algebraic structure of the source Hopf algebra and the topological structure of the target algebra. If the Hopf algebra is graded and connected, it turns out that the character group (with values in a locally convex algebra) can be made into an infinite-dimensional Lie group. Hopf algebras with these properties and their character groups appear often in combinatorial contexts[1] and are connected to certain operads and applications in numerical analysis. For example in [5] Hopf algebras related to numerical integration methods and their connection to pre-Lie and post-Lie algebras are studied.

If the Hopf algebra is not graded and connected, then the character group can in general not be turned into an infinite-dimensional Lie group. However, it turns out that the character group of an arbitrary Hopf algebra (with values in a finite-dimensional algebra) is always a topological group with strong structural properties, i.e. it is always the projective limit of finite-dimensional Lie groups. Groups with

[1] In these contexts the term "combinatorial Hopf algebra" is often used though there seems to be no broad consensus on the meaning of this term.

these properties—so called pro-Lie groups—are accessible to Lie theoretic methods (cf. [6]) albeit they may not admit a differential structure.

Finally, let us summarise the broad perspective taken in the present article by the slogan: Many constructions on character groups can be interpreted in the framework of infinite-dimensional (pro-)Lie group structures on these groups.

2 Hopf Algebras and Characters

In this section, we recall the language of Hopf algebras and their character groups. The material here is covered by almost every introduction to Hopf algebras (see [7–12]). Thus we restrict the exposition here to fix only some notation.

Notation 1 Denote by $\mathbb{N}$ the set of natural numbers (without 0) and $\mathbb{N}_0 := \mathbb{N} \cup \{0\}$. By $\mathbb{K}$ we denote either the field of real numbers $\mathbb{R}$ or of complex numbers $\mathbb{C}$.

Definition 1 (*Hopf algebra*) A Hopf algebra $\mathcal{H} = (\mathcal{H}, m_{\mathcal{H}}, u_{\mathcal{H}}, \Delta_{\mathcal{H}}, \varepsilon_{\mathcal{H}}, S_{\mathcal{H}})$ (over $\mathbb{K}$) is a bialgebra with compatible antipode $S\colon \mathcal{H} \rightarrow \mathcal{H}$, i.e. $(\mathcal{H}, m_{\mathcal{H}}, u_{\mathcal{H}})$ is a unital $\mathbb{K}$-algebra, $(\mathcal{H}, \Delta_{\mathcal{H}}, \varepsilon_{\mathcal{H}})$ is a unital $\mathbb{K}$-coalgebra and the following holds:

1. the maps $\Delta_{\mathcal{H}} : \mathcal{H} \rightarrow \mathcal{H} \otimes \mathcal{H}$ and $\varepsilon_{\mathcal{H}} : \mathcal{H} \rightarrow \mathbb{K}$ are algebra morphisms
2. S is $\mathbb{K}$-linear with $m_{\mathcal{H}} \circ (\mathrm{id} \otimes S) \circ \Delta_{\mathcal{H}} = u_{\mathcal{H}} \circ \varepsilon_{\mathcal{H}} = m_{\mathcal{H}} \circ (S \otimes \mathrm{id}) \circ \Delta_{\mathcal{H}}$.

A Hopf algebra $\mathcal{H}$ is ($\mathbb{N}_0$-)*graded*, if it admits a decomposition $\mathcal{H} = \oplus_{n\in\mathbb{N}_0} \mathcal{H}_n$ such that $m_{\mathcal{H}}(\mathcal{H}_n \otimes \mathcal{H}_m) \subseteq \mathcal{H}_{n+m}$ and $\Delta_{\mathcal{H}}(\mathcal{H}_n) \subseteq \bigoplus_{k+l=n} \mathcal{H}_l \otimes \mathcal{H}_k$ hold for all $n, m \in \mathbb{N}_0$. If in addition $\mathcal{H}_0 \cong \mathbb{K}$ is satisfied $\mathcal{H}$ is called *graded and connected* Hopf algebra.

In the Connes–Kreimer theory of renormalisation of quantum field theory one considers the Hopf algebra $\mathcal{H}_{FG}$ of Feynman graphs.[2] As it is quite involved to define this Hopf algebra we refer to [1, 1.6] for details. Below we describe in Example 1 a related but simpler Hopf algebra. This Hopf algebra of rooted trees arises naturally in numerical analysis, renormalisation of quantum field theories and non-commutative geometry (see [2] for a survey). We recall the definition of the Hopf algebra in broad strokes as it is somewhat prototypical for Hopf algebras from numerical analysis. To construct the Hopf algebra, recall some notation first.

Notation 2 1. A *rooted tree* is a connected *finite* graph without cycles with a distinguished node called the *root*. We identify rooted trees if they are graph isomorphic via a root preserving isomorphism.
Let $\mathcal{T}$ be *the set of all rooted trees* and write $\mathcal{T}_0 := \mathcal{T} \cup \{\emptyset\}$ where $\emptyset$ denotes the empty tree. The *order* $|\tau|$ of a tree $\tau \in \mathcal{T}_0$ is its number of vertices.
2. An *ordered subtree*[3] of $\tau \in \mathcal{T}_0$ is a subset s of all vertices of τ which satisfies

---

[2]This Hopf algebra depends on the quantum field theory under consideration, but we will suppress this dependence in our notation.

[3]The term "ordered" refers to that the subtree remembers from which part of the tree it was cut.

(i) s is connected by edges of the tree τ,
(ii) if s is non-empty, then it contains the root of τ.

The set of all ordered subtrees of τ is denoted by $\mathrm{OST}(\tau)$. Further, s_τ denotes the tree given by vertices of s with root and edges induced by τ.

3. A *partition* p of a tree $\tau \in \mathscr{T}_0$ is a subset of edges of the tree. We denote by $\mathscr{P}(\tau)$ the set of all partitions of τ (including the empty partition).

Associated to $s \in \mathrm{OST}(\tau)$ is a forest $\tau\backslash s$ (collection of rooted trees) obtained from τ by removing the subtree s and its adjacent edges. Similarly, to a partition $p \in \mathscr{P}(\tau)$ a forest $\tau\backslash p$ is associated as the forest that remains when the edges of p are removed from the tree τ. In either case, we let $\#\tau\backslash p$ be the number of trees in the forest.

Example 1 (*The Connes–Kreimer Hopf algebra of rooted trees* [13]) Consider the algebra $\mathscr{H}_{CK}^{\mathbb{K}} := \mathbb{K}[\mathscr{T}]$ of polynomials which is generated by the trees in $\mathscr{T}$. We denote the structure maps of this algebra by m (multiplication) and u (unit). Indeed $\mathscr{H}_{CK}^{\mathbb{K}}$ becomes a bialgebra whose coproduct we define on the trees (which generate the algebra) via

$$\Delta\colon \mathscr{H}_{CK}^{\mathbb{K}} \to \mathscr{H}_{CK}^{\mathbb{K}} \otimes \mathscr{H}_{CK}^{\mathbb{K}}, \quad \tau \mapsto \sum_{s \in \mathrm{OST}(\tau)} (\tau\backslash s) \otimes s_\tau .$$

Then the counit $\varepsilon\colon \mathscr{H}_{CK}^{\mathbb{K}} \to \mathbb{K}$ is given by $\varepsilon(1_{\mathscr{H}_{CK}^{\mathbb{K}}}) = 1$ and $\varepsilon(\tau) = 0$ for all $\tau \in \mathscr{T}$. Finally, the antipode is defined on the trees (which generate $\mathscr{H}_{CK}^{\mathbb{K}}$) via

$$S\colon \mathscr{H}_{CK}^{\mathbb{K}} \to \mathscr{H}_{CK}^{\mathbb{K}}, \quad \tau \mapsto \sum_{p \in \mathscr{P}(\tau)} (-1)^{\#\tau\backslash p} (\tau\backslash p)$$

and one can show that $\mathscr{H}_{CK}^{\mathbb{K}} = (\mathscr{H}_{CK}^{\mathbb{K}}, m, u, \Delta, \varepsilon, S)$ is a $\mathbb{K}$-Hopf algebra (see [14, 5.1] for more details and references).

Furthermore, $\mathscr{H}_{CK}^{\mathbb{K}}$ is a graded and connected Hopf algebra with respect to the *number of nodes grading*: For each $n \in \mathbb{N}_0$ we define the nth degree via

$$\text{for } \tau_i \in \mathscr{T}, 1 \leq i \leq k, k \in \mathbb{N}_0 \qquad \tau_1 \cdot \tau_2 \cdot \ldots \cdot \tau_k \in (\mathscr{H}_{CK}^{\mathbb{K}})_n \text{ if and only if } \sum_{r=1}^{k} |\tau_k| = n$$

Remark 1 Recently modifications of the Hopf algebra of rooted trees have been studied in the context of numerical analysis. In particular, non-commutative analogues to the Hopf algebra $\mathscr{H}_{CK}^{\mathbb{K}}$ (where the algebra is constructed as the non-commutative polynomial algebra of planar trees) have been studied in the context of Lie–Butcher theory. Their groups of characters, which we define in a moment, are of particular interest to develop techniques for numerical integration on manifolds (see [15]).

We briefly mention another example of a "combinatorial" Hopf algebra:

Example 2 (*The shuffle Hopf algebra* [16]) Fix a non-empty set A called the *alphabet*. A *word* in the alphabet A is a finite (possibly empty) sequence of elements in A. We denote by A^* the set of all words in A and consider the vector space $\mathbb{C}\langle A\rangle$ freely generated by the elements in A^*. Let $w = a_1 \dots a_n$ and $w' = a_{n+1} \dots a_{n+m}$ be words of length n and m, respectively. Define the *shuffle product* by

$$w \sqcup\!\sqcup w' := \sum_{\{i_1<i_2<\cdots<i_n\},\{j_1<j_2<\cdots<j_m\}} a_{i_1} \cdots a_{i_n} a_{j_1} \cdots a_{j_m},$$

where the summation runs through all disjoint sets which satisfy

$$\{i_1 < i_2 < \cdots < i_n\} \sqcup \{j_1 < j_2 < \cdots < j_m\} = \{1, \dots, m+n\}.$$

The (bilinear extensions of the) shuffle product and the deconcatenation of words turns $\mathbb{C}\langle A\rangle$ into a complex bialgebra. Together with the antipode $S(w) = (-1)^n w$ (for a word $w = a_n \dots a_1$ of length n) and the grading by word length, we obtain a graded and connected Hopf algebra $\mathrm{Sh}(A)$. We call this Hopf algebra the *shuffle Hopf algebra*.

Recall that the shuffle Hopf algebra appears in diverse applications connected to numerical analysis, see e.g. [15, 17, 18].

We will now consider groups of characters of Hopf algebras:

Definition 2 Let $\mathscr{H}$ be a Hopf algebra and B a commutative algebra. A linear map $\phi\colon \mathscr{H} \to B$ is called

1. (B-valued) *character* if it is a homomorphism of unital algebras, i.e.

$$\phi(ab) = \phi(a)\phi(b) \text{ for all } a, b \in \mathscr{H} \text{ and } \phi(1_{\mathscr{H}}) = 1_B. \tag{1}$$

   The set of characters is denoted by $\mathscr{G}(\mathscr{H}, B)$.
2. *infinitesimal character* if

$$\phi(ab) = \phi(a)\varepsilon_{\mathscr{H}}(b) + \varepsilon_{\mathscr{H}}(a)\phi(b) \text{ for all } a, b \in \mathscr{H} \tag{2}$$

   We denote by $\mathfrak{g}(\mathscr{H}, B)$ the set of all infinitesimal characters.

Lemma 1 ([10, 4.3 Propositions 21 and 22]) *Let $\mathscr{H}$ be a Hopf algebra and B a commutative algebra with multiplication map $\mu_B\colon B \otimes B \to B$. Then we obtain the algebra of $\mathbb{K}$-linear maps $\mathrm{Hom}_{\mathbb{K}}(\mathscr{H}, B)$ with the* convolution product

$$\phi \star \psi := \mu_B \circ (\phi \otimes \psi) \circ \Delta_{\mathscr{H}}$$

and the following holds:

1. *$\mathscr{G}(\mathscr{H}, B)$ is a subgroup of the group of units $(\mathrm{Hom}_{\mathbb{K}}(\mathscr{H}, B)^\times, \star)$. Inversion in $\mathscr{G}(\mathscr{H}, B)$ is given by $\phi \mapsto \phi \circ S_{\mathscr{H}}$ and $1_A := u_B \circ \varepsilon_{\mathscr{H}} : \mathscr{H} \to B, x \mapsto \varepsilon_{\mathscr{H}}(x)1_B$ is the unit element.*

2. $\mathfrak{g}(\mathcal{H}, B)$ *is a Lie subalgebra of* $(\mathrm{Hom}_{\mathbb{K}}(\mathcal{H}, B), [\cdot,\cdot])$, *where* $[\cdot,\cdot]$ *is the commutator bracket of* $(A, \star)$.

Example 3 (*Character groups of Hopf algebras*)

1. The universal enveloping algebra $\mathcal{U}(\mathfrak{g})$ of a Lie algebra $\mathfrak{g}$ over $\mathbb{K}$ is a Hopf algebra (see [7, 3.6]).[4] Every character $\phi \in \mathcal{G}(\mathcal{U}(\mathfrak{g}), \mathbb{K})$ corresponds to a Lie algebra homomorphism $\phi|_{\mathfrak{g}} : \mathfrak{g} \to \mathbb{K}$ which in turn factors naturally through a linear map $\phi^{\sim} : \mathfrak{g}/([\mathfrak{g}, \mathfrak{g}]) \to \mathbb{K}$, yielding a group isomorphism
$$\Phi : \mathcal{G}(\mathcal{U}(\mathfrak{g}), \mathbb{K}) \to \big((\mathfrak{g}/[\mathfrak{g}, \mathfrak{g}])^*, +\big), \quad \phi \mapsto \big(\phi^{\sim} : v + [\mathfrak{g}, \mathfrak{g}] \mapsto \phi(v)\big).$$
Thus we can identify the character group with the additive group of the dual vector space of the abelianisation of $\mathfrak{g}$. (The group $\mathcal{G}(\mathcal{U}(\mathfrak{g}), \mathbb{K}) \cong \mathfrak{g}/[\mathfrak{g}, \mathfrak{g}]$ is also naturally isomorphic to the first cohomology group $H^1(\mathfrak{g}, \mathbb{K})$ of the Lie algebra $\mathfrak{g}$ with values in the trivial module $\mathbb{K}$, cf. [19, p. 168].)
2. Character groups of the Hopf algebra of Feynman graphs $\mathcal{H}_{FG}$ play an important role in renormalisation of quantum field theories. Namely in the mathematical formulation of the renormalisation procedure one considers the groups $\mathcal{G}(\mathcal{H}_{FG}, \mathbb{C})$ and $\mathcal{G}(\mathcal{H}_{FG}, \mathcal{K})$, where $\mathcal{K}$ is the algebra of germs of meromorphic functions. We will return to these groups in Examples 6 and 13.
3. The character group $\mathcal{G}(\mathcal{H}^{\mathbb{R}}_{CK}, \mathbb{R})$ of the real Hopf algebra of rooted trees $\mathcal{H}^{\mathbb{R}}_{CK}$ from Example 1 turns out to be the Butcher group G_{TM} from numerical analysis. This group is connected to numerical integration theory (see [2]).
4. The character group of the shuffle algebra $\mathrm{Sh}(A)$ (Example 2) appears for example in [18]. There, the character group has been studied in the context of dynamical systems and their discretisation. We will provide more details below in Example 7. Moreover, in [17] these character groups are considered in the context of Lie-Butcher theory.

Note that the concepts recalled in this section were purely algebraic and combinatorial in nature. In particular, we have neither referred to a topology nor to a differentiable structure on these groups. We will introduce the necessary tools (differential calculus on locally convex spaces) to endow the character groups of Hopf algebras with an infinite-dimensional Lie group structure in the next section.

3 A Primer on Infinite-Dimensional Differential Calculus and Lie Groups

In this section basic facts on the differential calculus in infinite-dimensional spaces are recalled. The general setting for our calculus are locally convex spaces (see [20, 21]).

---

[4]Note that $\mathfrak{g}$ can be recovered from the Hopf algebra $\mathcal{U}(\mathfrak{g})$ (see [7, Theorem 3.6.1]) and the Hopf algebras which arise in this way are characterised by the Milnor–Moore theorem (cf. Remark 7).

Definition 3 Let E be a vector space over $\mathbb{K} \in \{\mathbb{R}, \mathbb{C}\}$ together with a topology T.

1. (E, T) is called *topological vector space*, if the vector space operations are continuous with respect to T and the metric topology on $\mathbb{K}$.
2. A *seminorm* on E is a map $p\colon E \to [0, \infty[$ which satisfies for $x, y \in E, \lambda \in \mathbb{K}$ the identities $p(x + y) \leq p(x) + p(y)$ and $p(\lambda x) = |\lambda| p(x)$, but may have $p^{-1}(0) \neq \{0\}$. We denote by $\mathrm{pr}_p\colon E \to E_p := E/p^{-1}(0)$ the canonical projection to the normed space associated to p.
3. A topological vector space (E, T) is called *locally convex space* if there is a family $\{p_i \mid i \in I\}$ of continuous seminorms for some index set I, such that

   i. the topology T is the initial topology with respect to $\{\mathrm{pr}_{p_i} : E \to E_{p_i} \mid i \in I\}$, i.e. the E-valued map f is continuous if and only if $\mathrm{pr}_i \circ f$ is continuous for each $i \in I$,
   ii. if $x \in E$ with $p_i(x) = 0$ for all $i \in I$, then $x = 0$ (i.e. T is Hausdorff).

   In this case, the topology T is *generated by the family of seminorms* $\{p_i\}_{i \in I}$. Usually we suppress T and write $(E, \{p_i\}_{i \in I})$ or simply E instead of (E, T).

Many familiar results from finite-dimensional calculus carry over to infinite dimensions if we assume that all spaces are locally convex. Hence we will only consider topological vector spaces which are locally convex. Note that the term "locally convex" comes from the fact that the semi-norm balls form convex neighbourhoods of the points.

Example 4 1. Normed spaces are locally convex spaces (see [21, Chap. I 6.2]).
2. Let $(E, \{p_i\}_{i \in I})$ be a locally convex vector space and X a set. Consider the space E^X of all mappings from X to E. For $x \in X$ we define the point-evaluation $\mathrm{ev}_x\colon E^X \to E, f \mapsto f(x)$. The *topology of pointwise convergence* on E^X is the locally convex topology generated by the seminorms

$$q_{i,x} := p_i \circ \mathrm{ev}_x \text{ where } (i, x) \text{ runs through } I \times X.$$

With the pointwise vector space operations and the topology of pointwise convergence E^X becomes a locally convex vector space. By definition, this topology is completely determined by the topology of the target vector space.
If X is a linear space, we consider the subspace $\mathrm{Hom}_{\mathbb{K}}(X, E)$ of all linear maps in E^X with the induced topology (also called topology of pointwise convergence).

As we are working beyond the realm of Banach spaces, the usual notion of Fréchet-differentiability cannot be used.[5] Moreover, there are several inequivalent notions of differentiability on locally convex spaces (see [23]). For more information on our setting of differential calculus we refer the reader to [23, 24]. The notion of differentiability we adopt is natural and quite simple, as the derivative is defined via directional derivatives.

[5]The problem here is that the bounded linear operators do not admit a good topological structure if the spaces are not normable. In particular, the chain rule will not hold for Fréchet-differentiability in general for these spaces (cf. [22, p. 73] or [23]).

Definition 4 Let $\mathbb{K} \in \{\mathbb{R}, \mathbb{C}\}$, $r \in \mathbb{N} \cup \{\infty\}$ and E, F locally convex $\mathbb{K}$-vector spaces and $U \subseteq E$ open. Moreover we let $f : U \to F$ be a map. If it exists, we define for $(x, h) \in U \times E$ the directional derivative

$$df(x,h) := D_h f(x) := \lim_{\mathbb{K}^\times \ni t \to 0} t^{-1}(f(x+th) - f(x)).$$

We say that f is $C^r_{\mathbb{K}}$ if the iterated directional derivatives

$$d^{(k)} f(x, y_1, \ldots, y_k) := (D_{y_k} D_{y_{k-1}} \cdots D_{y_1} f)(x)$$

exist for all $k \in \mathbb{N}_0$ such that $k \leq r$, $x \in U$ and $y_1, \ldots, y_k \in E$ and define continuous maps $d^{(k)} f : U \times E^k \to F$. If it is clear which $\mathbb{K}$ is meant, we simply write C^r for $C^r_{\mathbb{K}}$. If f is $C^\infty_{\mathbb{C}}$, we say that f is *holomorphic* and if f is $C^\infty_{\mathbb{R}}$ we say that f is *smooth*.

On Fréchet spaces (i.e. complete metrisable locally convex spaces) our notion of differentiability coincides with that from the "convenient setting" of global analysis outlined in [25]. Note that differentiable maps in our setting are continuous by default (which is in general not true in the convenient setting). Later on we will need analytic mappings between infinite-dimensional spaces. Note first:

Remark 2 A map $f : U \to F$ is of class $C^\infty_{\mathbb{C}}$ if and only if it is *complex analytic* i.e., if f is continuous and locally given by a series of continuous homogeneous polynomials (cf. [26, Propositions 7.4 and 7.7]). We then also say that f is of class $C^\omega_{\mathbb{C}}$.

To introduce real analyticity, we have to generalise a suitable characterisation from the finite-dimensional case: A map $\mathbb{R} \to \mathbb{R}$ is real analytic if it extends to a complex analytic map $\mathbb{C} \supseteq U \to \mathbb{C}$ on an open $\mathbb{R}$-neighbourhood U in $\mathbb{C}$. We proceed analogously for locally convex spaces by replacing $\mathbb{C}$ with a suitable complexification.

Definition 5 (*Complexification of a locally convex space*) Let E be a real locally convex topological vector space. Endow $E_{\mathbb{C}} := E \times E$ with the following operation

$$(x+iy) \cdot (u,v) := (xu - yv, xv + yu) \quad \text{for } x, y \in \mathbb{R}, u, v \in E.$$

The complex vector space $E_{\mathbb{C}}$ with the product topology is called the *complexification* of E. We identify E with the closed real subspace $E \times \{0\}$ of $E_{\mathbb{C}}$.

Definition 6 Let E, F be real locally convex spaces and $f : U \to F$ defined on an open subset U. Following [27] and [24], we call f *real analytic* (or $C^\omega_{\mathbb{R}}$) if f extends to a $C^\infty_{\mathbb{C}}$-map $\tilde{f} : \tilde{U} \to F_{\mathbb{C}}$ on an open neighbourhood $\tilde{U}$ of U in the complexification $E_{\mathbb{C}}$.[6]

[6] If E and F are Fréchet spaces, real analytic maps in the sense just defined coincide with maps which are continuous and can be locally expanded into a power series. See [28, Proposition 4.1].

Note that many of the usual results of differential calculus carry over to our setting. In particular, maps on connected domains whose derivative vanishes are constant as a version of the fundamental theorem of calculus holds. Moreover, the chain rule holds in the following form:

Lemma 2 (Chain Rule [24, Propositions 1.12, 1.15, 2.7 and 2.9]) *Fix $k \in \mathbb{N}_0 \cup \{\infty, \omega\}$ and $\mathbb{K} \in \{\mathbb{R}, \mathbb{C}\}$ together with $C^k_{\mathbb{K}}$-maps $f\colon E \supseteq U \to F$ and $g\colon H \supseteq V \to E$ defined on open subsets of locally convex spaces. Assume that $g(V) \subseteq U$. Then $f \circ g$ is of class $C^k_{\mathbb{K}}$ and the first derivative of $f \circ g$ is given by*

$$d(f \circ g)(x; v) = df(g(x); dg(x, v)) \quad \textit{for all } x \in U,\ v \in H.$$

The differential calculus developed so far extends easily to maps which are defined on non-open sets. This situation occurs frequently in the context of differential equations on closed intervals (see [29] for an overview).

Having the chain rule at our disposal we can define manifolds and related constructions which are modelled on locally convex spaces.

Definition 7 Fix a Hausdorff topological space M and a locally convex space E over $\mathbb{K} \in \{\mathbb{R}, \mathbb{C}\}$. An $(E\text{-})$manifold chart (U_κ, κ) on M is an open set $U_\kappa \subseteq M$ together with a homeomorphism $\kappa\colon U_\kappa \to V_\kappa \subseteq E$ onto an open subset of E. Two such charts are called C^r-compatible for $r \in \mathbb{N}_0 \cup \{\infty, \omega\}$ if the change of charts map $\nu^{-1} \circ \kappa\colon \kappa(U_\kappa \cap U_\nu) \to \nu(U_\kappa \cap U_\nu)$ is a C^r-diffeomorphism. A C^r-atlas of M is a set of pairwise C^r-compatible manifold charts, whose domains cover M. Two such C^r-atlases are equivalent if their union is again a C^r-atlas.

A *locally convex C^r-manifold* M modelled on E is a Hausdorff space M with an equivalence class of C^r-atlases of $(E\text{-})$manifold charts.

Direct products of locally convex manifolds, tangent spaces and tangent bundles as well as C^r-maps of manifolds may be defined as in the finite dimensional setting (see [30, I.3]). The advantage of this construction is that we can now give a very simple answer to the question, what an infinite-dimensional Lie group is:

Definition 8 A (locally convex) *Lie group* is a group G equipped with a $C^\infty_{\mathbb{K}}$-manifold structure modelled on a locally convex space, such that the group operations are smooth. If the manifold structure and the group operations are in addition ($\mathbb{K}$-) analytic, then G is called a ($\mathbb{K}$-) *analytic* Lie group.

Later on, we will encounter special classes of infinite-dimensional Lie groups. We recommend [30] for a survey on the theory of locally convex Lie groups.

4 The Lie Group Structure of Character Groups of Hopf Algebras

In this section we recall and explain how to construct a Lie group structure on character groups of Hopf algebras (cf. [4]). Let us first construct a suitable topology on the character groups which will in the end turn the character group into a Lie group.

As our notion of differentiability is built on top of continuity (i.e. every differentiable map is continuous), the topology needs to turn the character group into a topological group.

Remark 3 Consider the character group $\mathcal{G}(\mathcal{H}, B)$ of a Hopf algebra $\mathcal{H}$ with values in the commutative algebra B. If B is a locally convex vector space, we endow the space $\mathrm{Hom}_{\mathbb{K}}(\mathcal{H}, B)$ with the topology of pointwise convergence (see Example 4). This topology turns $\mathfrak{g}(\mathcal{H}, B)$ and $\mathcal{G}(\mathcal{H}, B)$ into closed subsets of $\mathrm{Hom}_{\mathbb{K}}(\mathcal{H}, B)$ which will always be endowed with the topology of pointwise convergence.

To see that this topology turns the character groups into topological groups we need the group product, the convolution $\star$, to be continuous. The product $\star$ is defined via the multiplication of the algebra B. Thus we will see that $\star$ is continuous if the locally convex structure on B is compatible with the algebra structure, i.e. the algebra B is a locally convex algebra.

Definition 9 Let B be an algebra over $\mathbb{K} \in \{\mathbb{R}, \mathbb{C}\}$. We call B *locally convex algebra* if B is a locally convex vector space such that the algebra product $\mu_B \colon B \times B \to B$ is a continuous bilinear map.

Remark 4 We have not asked the algebra product to be a continuous linear map on $B \otimes B$ since this would require us to choose a topology on the tensor product. As there are different valid choices for topological tensor products between locally convex spaces we refrain from doing this. In fact, the whole point of the definition is to avoid topological tensor products.[7] Though we frequently write $B \otimes B$ also for a locally convex algebra B this will always denote the algebraic tensor product.

Example 5 Every Banach algebra and thus in particular every finite-dimensional algebra (over $\mathbb{K} \in \{\mathbb{R}, \mathbb{C}\}$) is a locally convex algebra. Further examples are constructed in Example 6 below.

Lemma 3 *Let $\mathcal{H}$ be a Hopf algebra and B be a locally convex algebra. Then the topology of pointwise convergence turns* $(\mathrm{Hom}_{\mathbb{K}}(\mathcal{H}, B), \star)$ *into a locally convex algebra,* $(\mathcal{G}(\mathcal{H}, B))$ *into a topological group and* $(\mathfrak{g}(\mathcal{H}, B), [\cdot, \cdot])$ *into a topological Lie algebra.*

Proof The multiplication $\star$ of the character group and the Lie bracket of $\mathfrak{g}(\mathcal{H}, B)$ will be continuous if the algebra product $\star$ on $\mathrm{Hom}_{\mathbb{K}}(\mathcal{H}, B)$ is continuous. Since this algebra is endowed with the topology of pointwise convergence, it suffices to test continuity of $(f, g) \mapsto f \star g$ by evaluating in an element $c \in \mathcal{H}$. Using Sweedler notation (see [11]) for the coproduct we write $\Delta(c) = \sum_{(c)} c_1 \otimes c_2$, whence $f \star g(c) = \sum_{(c)} f(c_1) \cdot g(c_2)$ where the dot is the multiplication in B. Since

[7] Note that it is well known that the algebra multiplication as a bilinear map will be continuous if we require it to be continuous with respect to the projective topological tensor product $B \otimes_\pi B$ (see e.g. [31]). However, one does not gain more information in doing so, whence we avoid discussing this tensor product (or any topology on the tensor product). For this reason one of the authors has been accused of "cheating".

point evaluations are continuous on $\mathrm{Hom}_{\mathbb{K}}(\mathcal{H}, B)$ and the multiplication on B is continuous, we deduce that $\star$ is continuous.

Observe that inversion in the character group $(\mathcal{G}(\mathcal{H}, B), \star)$ is given by precomposition with the antipode. Clearly this map is continuous with respect to the topology of pointwise convergence, whence $\mathcal{G}(\mathcal{H}, B)$ is a topological group. □

Note that as a by-product of Lemma 3 we also obtain that $(\mathrm{Hom}_{\mathbb{K}}(\mathcal{H}, B), \star)$ is a locally convex algebra if B is a locally convex algebra. The last proof is a nice example of how continuity can be determined by "testing" it at a given point. In particular, the topology on the group relegates continuity questions to the topology of the target algebra B.

Our aim is now to turn the topological group $\mathcal{G}(\mathcal{H}, B)$ into a Lie group modelled on the infinitesimal characters. To do this we will need some additional assumptions on the source Hopf algebra:

Definition 10 Let $\mathcal{H} = \bigoplus_{n\in\mathbb{N}_0} \mathcal{H}_n$ be a graded and connected Hopf algebra and B be a locally convex algebra. Then we define $\mathcal{I} := \mathrm{Hom}_{\mathbb{K}}(\bigoplus_{n\in\mathbb{N}} \mathcal{H}_n, B)$ and

$$\exp\colon \mathcal{I} \to \mathrm{Hom}_{\mathbb{K}}(\mathcal{H}, B),\ \psi \mapsto \sum_{k=0}^{\infty} \frac{\psi^{\star k}}{k!} \quad \text{where } \psi^{\star k} := \underbrace{\psi \star \psi \star \cdots \star \psi}$$

Lemma 4 *For each* $\psi \in \mathcal{I}$, $\exp(\psi)$ *is contained in* $\mathrm{Hom}_{\mathbb{K}}(\mathcal{H}, B)$. *Hence it makes sense to define* $\exp\colon \mathrm{Hom}_{\mathbb{K}}(\bigoplus_{n\in\mathbb{N}} \mathcal{H}_n, B) \to \mathrm{Hom}_{\mathbb{K}}(\mathcal{H}, B)$ *and this mapping is* $\mathbb{K}$*-analytic. Furthermore, it restricts to a bijection* $\exp\colon \mathfrak{g}(\mathcal{H}, B) \to \mathcal{G}(\mathcal{H}, B)$.

Proof We have the following isomorphisms of locally convex spaces

$$\mathrm{Hom}_{\mathbb{K}}(\mathcal{H}, B) = \mathrm{Hom}_{\mathbb{K}}\Big(\bigoplus_{n\in\mathbb{N}_0} \mathcal{H}_n, B\Big) \cong \prod_{n\in\mathbb{N}_0} \mathrm{Hom}_{\mathbb{K}}(\mathcal{H}_n, B).$$

Note that $\mathrm{Hom}_{\mathbb{K}}(\mathcal{H}, B)$ is **not** a graded algebra (as a grading requires it to decompose into a direct sum). However, the convolution satisfies

$$\mathrm{Hom}_{\mathbb{K}}(\mathcal{H}_n, B) \star \mathrm{Hom}_{\mathbb{K}}(\mathcal{H}_m, B) \subseteq \mathrm{Hom}_{\mathbb{K}}(\mathcal{H}_{n+m}, B) \quad \text{for all } m, n \in \mathbb{N}_0.$$

We call an algebra with this property a *densely graded algebra* (to express that it decomposes as a product and not as a direct sum).

Now for each $n \in \mathbb{N}$ the formula for exp yields only finitely many summands in $\mathrm{Hom}_{\mathbb{K}}(\mathcal{H}_n, B)$ (as elements in its domain contain no contribution of $\mathrm{Hom}_{\mathbb{K}}(\mathcal{H}_0, B)$). We deduce that exp makes sense as a mapping to $\prod_{n\in\mathbb{N}_0} \mathrm{Hom}_{\mathbb{K}}(\mathcal{H}_n, B)$ whence it makes sense as a mapping into $\mathrm{Hom}_{\mathbb{K}}(\mathcal{H}, B)$. To see that exp is $\mathbb{K}$-analytic one uses functional calculus for densely graded algebras (see [4, Lemma B.6] for details).

That exp restricts to a bijection $\mathfrak{g}(\mathcal{H}, B) \to \mathcal{G}(\mathcal{H}, B)$ is well known in the literature. We refer to [4, Lemma B.11] for a proof. □

Remark 5 Densely graded algebras as discussed in the proof of Lemma 4 are an important tool in the investigation of character groups of Hopf algebras. Indeed they can be used to link the investigation of character groups to the Lie theory for unit groups of continuous inverse algebras (see [32, 33]). For more details we refer to [4].

Using exp as a (global) parametrisation for $\mathcal{G}(\mathcal{H}, B)$ we obtain a manifold structure on the character group and it turns out that this renders the group a Lie group.

Theorem 1 ([4, Theorem 2.7]) *Let $\mathcal{H}$ be a graded and connected Hopf algebra and B be a locally convex algebra. Then the manifold structure induced by the global parametrisation* $\exp\colon \mathfrak{g}(\mathcal{H}, B) \to \mathcal{G}(\mathcal{H}, B)$ *turns* $(\mathcal{G}(\mathcal{H}, B), \star)$ *into a* $\mathbb{K}$*-analytic Lie group.*

The Lie algebra associated to this Lie group is $(\mathfrak{g}(\mathcal{H}, B), [\cdot, \cdot])$*. Moreover, the Lie group exponential map is given by the real analytic diffeomorphism* exp *from Lemma 4.*

Remark 6 The assumption that the Hopf algebra $\mathcal{H}$ is graded and connected is crucial for the proof of Theorem 1. If we drop this assumption then character groups need not be infinite-dimensional Lie groups (see [4, Example 4.11] for an explicit counterexample). However, the construction does not depend on the choice of the grading, i.e. two connected gradings on a Hopf algebra will yield the same Lie group structure on the character group. See for example [1, Proposition 1.30] where different (connected) gradings on the Hopf algebra of Feynman graphs are discussed.

In the literature, the infinitesimal characters are not the only Lie algebra associated with the group $\mathcal{G}(\mathcal{H}, B)$. Recall that for certain Hopf algebras there is a Lie algebra which is associated to the Hopf algebra by the Milnor–Moore theorem.

Remark 7 (*The Lie algebra* $\mathfrak{g}(\mathcal{H}, \mathbb{K})$ *of* $\mathcal{G}(\mathcal{H}, \mathbb{K})$ *and* the Milnor–Moore theorem) Consider a graded, connected and commutative Hopf algebra $\mathcal{H} = \bigoplus_{n\in\mathbb{N}_0} \mathcal{H}_n$ such that each $\mathcal{H}_n$ is a finite-dimensional vector space. Then the Milnor–Moore theorem [34] asserts that there is a Lie algebra $\mathrm{Lie}(\mathcal{H})$ such that $\mathcal{H}^\vee \cong \mathcal{U}(\mathrm{Lie}(\mathcal{H}))$, where $\mathcal{H}^\vee \cong \bigoplus_{n\in\mathbb{N}_0} \mathrm{Hom}_{\mathbb{K}}(\mathcal{H}_n, \mathbb{K})$ is the restricted dual. This Lie algebra is often associated with the Lie algebra of the character group $\mathcal{G}(\mathcal{H}, \mathbb{K})$ and indeed one can identify

$$\mathrm{Lie}(\mathcal{H}) \cong \mathfrak{g}(\mathcal{H}, \mathbb{K}) \cap \bigoplus_{n\in\mathbb{N}_0} \mathrm{Hom}_{\mathbb{K}}(\mathcal{H}_n, \mathbb{K}) = \mathfrak{g}(\mathcal{H}, \mathbb{K}) \cap \mathcal{H}^\vee.$$

Note that the Lie algebra $\mathfrak{g}(\mathcal{H}, \mathbb{K})$ will in general be strictly larger than $\mathrm{Lie}(\mathcal{H})$ but with respect to the topology of pointwise convergence $\mathrm{Lie}(\mathcal{H})$ is a dense subalgebra of $\mathfrak{g}(\mathcal{H}, \mathbb{K})$. This fact is frequently exploited (cf. [1, 13, 35]) as it connects the Hopf algebra $\mathcal{H}$ with the Lie algebra $\mathfrak{g}(\mathcal{H}, \mathbb{K})$.

Before we continue with the general theory let us develop some examples which arise from applications in numerical analysis and mathematical physics. In Example 3, we have already seen character groups of (graded and connected) Hopf algebras which arise naturally in applications. Most of these groups are constructed with the ground field $\mathbb{K}$ as target algebra. However, in the theory of renormalisation of quantum field theories one also considers characters with values in an infinite-dimensional locally convex algebra B. We construct this algebra in the next example.

Example 6 Consider the *set of germs of meromorphic functions* in $0 \in \mathbb{C}$

$$\mathscr{K} = \{\mathrm{germ}_0 f \mid f : U \to \mathbb{C} \cup \{\infty\} \text{ meromorphic and } 0 \in U \subseteq \mathbb{C} \text{ open}\},$$

where as usual $\mathrm{germ}_0 f = \mathrm{germ}_0 g$ if f and g coincide on some 0-neighbourhood. Then $\mathscr{K}$ can be made a locally convex space as an inductive limit of Banach spaces such that

1. $\mathscr{K}$ is a complete locally convex algebra with respect to multiplication of germs,
2. as a closed subspace of $\mathscr{K}$, the set $\mathscr{O}$ of germs of holomorphic functions is the inductive limit of the Banach spaces $\{(\mathrm{Hol}_b(U_n, \mathbb{C}), \|\cdot\|_\infty)\}_{n\in\mathbb{N}}$ of bounded holomorphic function on open, relatively compact sets U_n which form a base of zero neighbourhoods. In particular, this structure turns $\mathscr{O}$ into a locally convex subalgebra.

Character groups of graded connected Hopf algebras with values in $\mathscr{K}$ and $\mathscr{O}$ are studied in the context of the Connes–Kreimer theory of renormalisation (cf. [1, p. 83]). Note that we can apply Theorem 1 to turn these character groups (cf. Example 3) into Lie groups.

Proof (Construction of the topology on $\mathscr{K}$)

Step 1: *$\mathscr{O}$ as a locally convex algebra.*
Fix $U_n = B^{\mathbb{C}}_{\frac{1}{n}}(0)$ for $n \in \mathbb{N}$ (open ball in $\mathbb{C}$ around 0 of radius $\frac{1}{n}$). For later use observe that U_n is relatively compact, $\overline{U_{n+1}} \subseteq U_n$ for all n and $\{U_n\}_{n\in\mathbb{N}}$ is a base of zero-neighbourhoods. Recall that the space $\mathrm{Hol}_b(U_n, \mathbb{C})$ of bounded holomorphic functions is a Banach algebra with respect to the supremum norm. The space $\mathscr{O}$ of germs of holomorphic functions around 0 can be realised as the inductive locally convex limit of the spaces $\mathrm{Hol}_b(U_n, \mathbb{C})$. The bonding maps of the inductive system are given by

$$\iota_{n,m} : \mathrm{Hol}_b(U_n, \mathbb{C}) \to \mathrm{Hol}_b(U_m, \mathbb{C}), \quad f \mapsto f|_{U_m}$$

and these maps are compact operators for $m > n$ (see [25, Sect. 8] or [36, Appendix A]). Hence $\mathscr{O}$ is a so called *Silva space*. These spaces have many nice properties (cf. [25, 52.37] and [37]), in particular, they are Hausdorff and complete. Bilinear mappings on Silva spaces are continuous if they are continuous on each step of the limit (see e.g. [38, Proposition 4.5]). Thus multiplication of germs is continuous and $\mathscr{O}$ is a locally convex algebra.

Step 2: *Polynomials $\mathscr{P}^\infty(X)$ without constant term as a Silva space.*
Consider the space of polynomials $\mathscr{P}^\infty(X) := \mathbb{C}[X] = \mathrm{span}\{X^0, X^1, X^2, \ldots\}$ in the formal variable X and let us denote by $\mathscr{P}^\infty_* := \mathrm{span}\{X^1, X^2, \ldots\}$ the subspace of polynomials without constant term. This last space is a Silva space as

the direct union $\mathscr{P}_*^\infty(X) = \bigcup_{n\in\mathbb{N}} \mathscr{P}_*^n(X)$ of finite dimensional spaces $\mathscr{P}_*^n(X) := \operatorname{span}(X^1, \dots, X^n)$. The bonding maps $\mathscr{P}_*^n(X) \to \mathscr{P}_*^m(X)$ are compact operators for $m \geq n$ as the corresponding spaces are finite dimensional.

Step 3: *The topology on* $\mathscr{K}$.
Consider a meromorphic function $g\colon \Omega \to \mathbb{C} \cup \{\infty\}$ defined on a 0-neighbourhood Ω. Since the set of poles is discrete, we find an $n \in \mathbb{N}$ such that g has no poles in $\overline{U_n} \subseteq \Omega$, except possibly at 0. Furthermore, write $g = f + p(1/z)$ where $f \in \operatorname{Hol}_b(U_n, \mathbb{C})$ and $p \in \mathscr{P}_*^\infty(1/z)$ is a polynomial in the variable $1/z$ (without constant term).

As g was arbitrary, the vector space $\mathscr{K}$ is a direct sum of the two vector subspaces $\mathscr{O}$ and $\mathscr{P}_*^\infty(1/z)$. Both of the summands have a natural Silva space structure, which can be used to topologise $\mathscr{K} = \mathscr{O} \oplus \mathscr{P}_*^\infty(1/z)$. The locally convex space

$$\mathscr{K} = \varinjlim \left(\operatorname{Hol}_b(U_n, \mathbb{C}) \oplus \mathscr{P}_*^n(1/z)\right)$$

thus becomes a Silva space as a sum of two Silva spaces (see [39, Corollary 8.6.9]).

This construction can also be found in [40] where also spaces of meromorphic functions (rather than just germs of such functions) are given a topology using a similar construction. For meromorphic functions the multiplication turns out to be only separately continuous (see [40, Theorem 5]) and so these spaces are no locally convex algebras in the sense of Definition 9.

Step 4: $\mathscr{K}$ *as a locally convex algebra.*
We now exploit that the space of germs $\mathscr{K}$ is a Silva space by Step 3 to prove that multiplication is (jointly) continuous. Computing on the steps of the limit, fix $n \in \mathbb{N}$ and set $E_n := \operatorname{Hol}_b(U_n, \mathbb{C}) \oplus \mathscr{P}_*^n(1/z)$. Let us now prove that the multiplication map (co-)restricts to a continuous map $\mu_n\colon E_n \times E_n \to E_{2n}$. As a tool we use the operator

$$\Phi_n\colon \operatorname{Hol}_b(U_n, \mathbb{C}) \oplus \mathscr{P}_*^n(1/z) \to \operatorname{Hol}_b(U_n, \mathbb{C}), \quad f + p(1/z) \mapsto z^n f + z^n p(1/z)$$

which is easily seen to be linear continuous and bijective. Hence the Open Mapping Theorem for Banach spaces implies that Φ_n is a topological isomorphism. Using Φ_n, we write μ_n as composition of continuous maps:

$$\mu_n = \Phi_{2n}^{-1} \circ \iota_{n,2n} \circ \mu_{\operatorname{Hol}_b(U_n,\mathbb{C})} \circ (\Phi_n \times \Phi_n)$$

where $\mu_{\operatorname{Hol}_b(U_n,\mathbb{C})}$ is the continuous multiplication in $\operatorname{Hol}_b(U_n, \mathbb{C})$. This shows that $\mathscr{K}$ is a locally convex algebra. □

Remark 8 It should be noted that although $\mathscr{K}$ is algebraically a field and a locally convex algebra, inversion is *not* continuous with respect to the topology just described, hence $\mathscr{K}$ is not a topological field. This however is no defect of this particular construction but follows from the (locally convex) Gelfand Mazur Theorem: There is no complex locally convex division algebra—except $\mathbb{C}$ (see e.g. [32, Remark 4.15] or [41, Theorem 1]).

We have already encountered the character group of the shuffle Hopf algebra (see Examples 2 and 3). Recently, these groups were used as building blocks for a group of extended word series which is of interest in the discretisation of dynamical systems and in the computation of normal forms for these systems (see [18, 42]). In the following example we will revisit this construction.

Example 7 Let $\mathscr{G} := \mathscr{G}(\mathrm{Sh}(A), \mathbb{C})$ be the complex valued character group of the shuffle Hopf algebra $\mathrm{Sh}(A)$. Since $\mathrm{Sh}(A)$ is a graded and connected Hopf algebra by Example 2, Theorem 1 implies that $\mathscr{G}$ is a Lie group with the topology of pointwise convergence. The goal pursued in [42] is to study a certain class of ordinary differential equation. This leads one to consider so called "extended word series". These series are elements in a semidirect product $\mathscr{G} \rtimes_{\Xi} \mathbb{C}^d$ of the groups $\mathscr{G}$ and $\mathbb{C}^d$ for some $d \in \mathbb{N}$. Here the group morphism $\Xi \colon \mathbb{C}^d \to \mathrm{Aut}(\mathscr{G}), z \mapsto \Xi_z$ is given for $\delta \in \mathscr{G}, z := (z_1, \ldots, z_d) \in \mathbb{C}^d$ and $w = a_n \ldots a_1 \in A^*$ by

$$\Xi_z(\delta)(w) = \exp\left(\sum_{k=1}^{d} z_k(\nu_{k,a_1} + \ldots + \nu_{k,a_n})\right) \cdot \delta(w)$$

where the numbers $\nu_{k,a} \in \mathbb{C}$ are fixed for all $1 \leq k \leq d$ and $a \in A$.[8] Note that the map Ξ takes its image indeed in $\mathrm{Aut}(\mathscr{G})$ by virtue of [42, 4.2]. Moreover, it is easy to see that the mapping $\Xi^\wedge \colon \mathbb{C}^d \times \mathscr{G} \to \mathscr{G}, (z, \delta) \mapsto \Xi_z(\delta)$ is smooth, i.e. $\Xi^\wedge$ is a Lie group action. Hence we obtain a semidirect product of Lie groups $\mathscr{G} \rtimes_{\Xi} \mathbb{C}^d$. In [42] the authors then proceed to study the Lie algebra $\mathbf{L}(\mathscr{G} \rtimes_{\Xi} \mathbb{C}^d) = \mathbf{L}(\mathscr{G}) \rtimes_{\mathbf{L}(\Xi)} \mathbb{C}^d$, differential equations on $\mathscr{G} \rtimes_{\Xi} \mathbb{C}^d$ and properties of the Lie group exponential.

Finally, one can show that certain closed subgroups of $\mathscr{G}(\mathscr{H}, B)$ which are associated to Hopf ideals are Lie subgroups.[9]

Definition 11 (*Hopf ideal and annihilator*) Let $\mathscr{H}$ be a Hopf algebra. We say $\mathscr{J} \subseteq \mathscr{H}$ is a *Hopf ideal* if the subset $\mathscr{J}$ is

1. a two-sided (algebra) ideal,
2. a coideal, i.e. $\varepsilon(\mathscr{J}) = 0$ and $\Delta(\mathscr{J}) \subseteq \mathscr{J} \otimes \mathscr{H} + \mathscr{H} \otimes \mathscr{J}$ and
3. stable under the antipode, i.e. $S(\mathscr{J}) \subseteq \mathscr{J}$.

Let B be a locally convex algebra, then we define the *annihilator of* $\mathscr{J}$

$$\mathrm{Ann}(\mathscr{J}, B) = \{\phi \in \mathrm{Hom}_{\mathbb{K}}(\mathscr{H}, B) \mid \phi(\mathscr{J}) = 0_B\},$$

which is a closed unital subalgebra of $\mathrm{Hom}_{\mathbb{K}}(\mathscr{H}, B)$.

Proposition 1 ([4, Theorem 3.4]) *Let $\mathscr{H}$ be a graded connected Hopf algebra, $\mathscr{J} \subseteq \mathscr{H}$ be a Hopf ideal and B be a commutative locally convex algebra.*

---

[8] These numbers depend on the structure of a certain ordinary differential equation. We refer to [42] for more details.

[9] Contrarily to the situation for finite-dimensional Lie groups, not every closed subgroup of an infinite-dimensional Lie group is again a Lie subgroup. See [30, Remark IV.3.17] for an example.

Then $\operatorname{Ann}(\mathcal{J}, B) \cap \mathcal{G}(\mathcal{H}, B) \subseteq \mathcal{G}(\mathcal{H}, B)$ *is a closed Lie subgroup whose Lie algebra is* $\operatorname{Ann}(\mathcal{J}, B) \cap \mathfrak{g}(\mathcal{H}, B) \subseteq \mathfrak{g}(\mathcal{H}, B)$. *Moreover,* $\operatorname{Ann}(\mathcal{J}, B) \cap \mathcal{G}(\mathcal{H}, B)$ *is even an exponential BCH–group.*

Example 8 (*Annihilator subgroups*)

1. In [43] Hopf ideals of the Hopf algebra of Feynman graphs $\mathcal{H}_{FG}$ (cf. Example 3 (b)) are studied. In the physical theory these ideals implement the so called "Ward–Takahashi" and "Slavnov–Taylor" identities. Then in [44] character groups related to the annihilator subgroups of these Hopf ideals are studied.
2. It is well known that the tree maps associated to numerical integration schemes which preserve certain first order integrals form a subgroup of the Butcher group, called the *group of symplectic tree maps* $S^{\mathbb{K}}_{\mathrm{TM}}$. This subgroup is the annihilator subgroup of a certain Hopf ideal in the Hopf algebra of rooted trees $\mathcal{H}^{\mathbb{K}}_{CK}$ (see Example 1). We refer to [4, Example 4/9] for more details.
3. Recently in [45] an even smaller subgroup of the group of symplectic tree maps (as discussed in 2.) has been considered. This group $\widehat{\mathcal{G}}$ consists of all elements in the Butcher group such that the operations forming B-series and changing variables in the vector field commute (see [45, Proposition 3.1] for the detailed statement). Moreover, following [45, Eq. (56)] one can prove that elements in $\widehat{\mathcal{G}}$ admit a more "compact" expansion as B-series.
   Due to loc. cit. and [46, Remark 24] this group is the annihilator of a Hopf ideal $\mathcal{I}$ in $\mathcal{H}^{\mathbb{K}}_{CK}$, whence a Lie group by Proposition 1. Further, we remark that the ideal $\mathcal{I}$ is of interest in its own right: As pointed out in [46, Sect. 7] the ideal $\mathcal{I}$ appears as the kernel of a Hopf algebra morphism used to interpret the theory developed in loc. cit. in the context of Hopf algebras.

5 Notes on the Topology of the Character Groups

By definition of the topology of pointwise convergence, the topology and the differentiable structure of the character group $\mathcal{G}(\mathcal{H}, B)$ is completely determined by the target algebra B. Of course the algebraic structure of $\mathcal{H}$ determines the set of characters, e.g. it controls whether the character group is an abelian group. However, we do not need a topology on the Hopf algebra to turn its character groups into Lie groups. In a nutshell, the main idea behind the construction can be described as: The Lie group structure is controlled by the combinatorial data of the Hopf algebra and the topological data of the target algebra.

Remark 9 Note that the topology of pointwise convergence is a very natural choice for a topology on the character groups of Hopf algebras. In this respect certain character groups with this topology have already been studied as topological groups in the literature.

1. For example in the Connes–Kreimer theory of renormalisation (cf. [1, Proposition 1.47]) the structural properties of certain character groups as topological groups with this topology are exploited. We refer to Sect. 7 for further information on this structure.
2. As a further example we mention [5, 3.2] where the projective limit topology on the character groups coincides with the topology of pointwise convergence with respect to the *discrete* topology on the ground field $\mathbb{K}$. Hence it does not coincide with our topology which induces on every one-dimensional subspace the (natural) metric topology of $\mathbb{R}$ (or $\mathbb{C}$). Since [5] works with an arbitrary field $\mathbb{K}$ of characteristic 0 the discrete topology on $\mathbb{K}$ is the right choice in that setting.

The topology of pointwise convergence is rather coarse, i.e. compared to other natural function space topologies it has few open sets. In particular, open sets only control the behaviour of characters in a finite number of points at once. Especially in applications in numerical analysis one would like to have finer topologies which enable a better control. Let us illustrate this in the example of the Butcher group:

Example 9 The Butcher group G_{TM} coincides with the group $\mathscr{G}(\mathscr{H}_{CK}^{\mathbb{K}}, \mathbb{K})$ of $\mathbb{K}$-valued characters of the Connes–Kreimer Hopf algebra of rooted trees (see Example 3 (c)). This Hopf algebra is graded and connected, whence Theorem 1 allows G_{TM} to be turned into a Lie group with the pointwise topology. Recall that as an algebra $\mathscr{H}_{CK}^{\mathbb{K}}$ is the polynomial algebra $\mathbb{K}[\mathscr{T}]$ (where $\mathscr{T}$ is the set of rooted trees). The elements in the Butcher group correspond to numerical integration schemes as they are linked to a certain type of (formal) series called B-series. In applications one now wants to restrict the growth of the series coefficients to achieve convergence of the series at least on a small disk. To this end, one commonly imposes an exponential growth bound to the elements in the Butcher group, i.e. the growth of a character is restricted by an exponential bound in every tree. The topology of pointwise convergence does not contain open sets which allow one to control infinitely many coefficients at once, whence it is too coarse for some applications. However, no suitable replacement for the topology on G_{TM} to circumvent these problems is presently known (see the discussion of topologies on the Butcher group in [3, Remark 2.5]).

Though there seems to be no candidate for a finer topology on character groups which turns these into topological groups, the situation is better if one considers only certain subgroups. Again we specialise to the case of the Butcher group.

Example 10 (The tame Butcher group) Let $\mathscr{B}$ be the subgroup of $\mathscr{G}(\mathscr{H}_{CK}^{\mathbb{K}}, \mathbb{K})$ of all elements φ which satisfy

$$\text{there exist } C,\ K > 0 \text{ such that } \|\varphi(\tau)\| \leq CK^{|\tau|} \text{ for all } \tau \in \mathscr{T}_0.$$

Adapting a result of Hairer and Lubich [47, Lemma 9], one can show that the B-series associated to elements in $\mathscr{B}$ with respect to an analytic map f converge at least locally. We call $\mathscr{B}$ the *tame Butcher group.*

One can show that the tame Butcher group is a Lie group modelled on an inductive limit of Banach spaces. Note that the resulting topology is strictly finer than the topology of pointwise convergence.

Albeit the differential structure is more complicated than the one of the Butcher group, the tame Butcher group is closely related to the Butcher group. The key property of the tame Butcher group is that it provides a better control for the purposes of numerical analysis. Furthermore, consider on a Banach space E the differential equation

$$\begin{cases} \frac{\mathrm{d}}{\mathrm{d}t}x(t) &= f(x(t)) \\ x(0) &= y_0 \end{cases} \quad \text{with } f\colon E \to E \text{ analytic.}$$

Then the map sending an element of $\mathscr{B}$ to the B-series with respect to f induces a Lie group (anti)morphism

$$B_f\colon \mathscr{B} \to \mathrm{DiffGerm}_{(y_0,0)}(E \times \mathbb{C}) := \left\{ \mathrm{germ}_{(y_0,0)}\phi \,\middle|\, {}^{\phi\colon E\times\mathbb{K}\supseteq U\to V \text{ is a diffeomorphism}}_{\text{with } \phi(y_0,0)=(y_0,0)} \right\},$$

where the group operation of $\mathrm{DiffGerm}_{(y_0,0)}(E \times \mathbb{C})$ is composition (cf. [48, Sect. 3]). Then the map B_f can be seen as the Lie theoretic realisation of the mechanism which passes from a numerical integrator to the associated numerical solution of the differential equation. We refer to [49] for further details.

Conceivably one can adapt the idea of the tame Butcher group to the general setting of character groups of Hopf algebras.

Problem 1 It would be interesting to see whether the construction of the tame Butcher group can be generalised to obtain a Lie theory for "groups of exponentially bounded characters" of graded connected Hopf algebras. The groups we envisage here should arise as subgroups of character groups (albeit with a finer topology/different differential structure).

We expect these groups to be useful in several applications. In particular, one would hope that such a theory is applicable in the following situations:

1. In the context of Lie–Butcher theory of numerical analysis, i.e. for numerical integrators on manifolds and Lie groups (see e.g. [50] and the references therein).
2. Often in the theory of numerical analysis bounds appear naturally. For an example of such a situation see e.g. [51] where bounds on a function and its Fourier coefficients are used to derive error estimates.
3. In control theory so called output feedback equations are studied. Recently connections of problems related to these equations with character groups of Hopf algebras have been discovered (see [52] and [53]). In particular, one is interested in the convergence of certain formal series, which is assured by similar growth conditions as imposed in the tame Butcher group (cf. [52, Sect. 2.2]).

Note added in print (2018): Problem 1 has been partially solved as the construction of the tame Butcher group was generalised in [54].

Another candidate for a topology on the Butcher group G_{TM} appears implicitly in Butcher's 1972 paper.

Theorem 2 ([55, Theorem 6.9]) *If $a \in G_{\mathrm{TM}}$ and $\mathcal{T}_f$ is any finite subset of $\mathcal{T}$ then there is a $b \in G_0$ such that $a|_{\mathcal{T}_f} = b|_{\mathcal{T}_f}$.*

The above theorem is sometimes paraphrased as saying that the set of Runge–Kutta methods is dense in G_{TM}.[10]

Indeed, [55, Theorem 6.9] implies that the set of Runge–Kutta methods is dense in the Butcher group equipped with the topology of pointwise convergence. However, it also implies that the set of Runge–Kutta methods stays dense if one uses a much finer topology on $G_{\mathrm{TM}} \subset \mathrm{Hom}_{\mathbb{K}}(\mathcal{H}_{CK}^{\mathbb{K}}, \mathbb{K})$, the ultrametric topology.

Definition 12 (*Ultrametric topology*) When $\mathcal{H}$ is graded, the ultrametric topology on $\mathrm{Hom}_{\mathbb{K}}(\mathcal{H}, B)$ is generated by the *ultrametric*[11]

$$d(\phi, \psi) = 2^{-\mathrm{ord}(\phi-\psi)},$$

where $\mathrm{ord}(\phi)$ is the largest $N \in \mathbb{N}_0 \cup \{\infty\}$ such that

$$\phi(x) = 0 \quad \text{for all } x \in \bigoplus_{n=0}^{N} \mathcal{H}_n.$$

For $G_{\mathrm{TM}} \subset \mathrm{Hom}_{\mathbb{K}}(\mathcal{H}_{CK}^{\mathbb{K}}, \mathbb{K})$, define the ultrametric topology as the subspace topology. In the ultrametric topology on G_{TM}, a sequence $(a_n)_{n\in\mathbb{N}}$ will converge to a only if, for every tree $\tau \in \mathcal{T}$, there is an N such that $N \leq n$ implies $a_n(\tau) = a(\tau)$. Note that the ultrametric topology and the topology of pointwise convergence behave quite differently: Let $\mathcal{H}$ be a $\mathbb{K}$-Hopf algebra with $\mathbb{K} \in \{\mathbb{R}, \mathbb{C}\}$. If we embed $\mathbb{K}$ as a linear subspace into $\mathrm{Hom}_{\mathbb{K}}(\mathcal{H}, B)$, the ultrametric topology induces the discrete topology on $\mathbb{K}$ whereas the topology of pointwise convergence induces the usual (metric) topology on $\mathbb{K}$. In particular, this shows that the ultrametric topology does not turn $\mathrm{Hom}_{\mathbb{K}}(\mathcal{H}, B)$ (and $\mathfrak{g}(\mathcal{H}, B)$) into locally convex spaces over $\mathbb{K}$.

The ultrametric topology can be useful in numerics; for instance, the order of a B-series method given by $a \in G_{\mathrm{TM}}$ can be read off directly from the ultrametric via $d(a, e) = 2^{-p}$, where e is the "exact" method and p is the order of a. We record a further difference between both topologies in the following example:

Example 11 Let a_h denote the B-series $a_h(\emptyset) = 1$, $a_h(\bullet) = h$, $a_h(\tau) = 0$ for all trees with $|\tau| \geq 2$, and consider the sequence of B-series

$$\{b_n\}_{n\in\mathbb{N}}, \quad b_n = a_{1/n}^{\star n}.$$

b_n corresponds to the numerical method obtained by taking n steps with the forward Euler method with stepsize $\frac{1}{n}$. It is possible to show that

[10] G_0 in [55, Theorem 6.9] is larger than the set of Runge–Kutta methods, however, the statement still holds if G_0 is the group of Runge–Kutta methods. See also [56, Theorem 317A].

[11] A metric d is an ultrametric if $d(\phi, \psi) \leq \max\{d(\phi, \chi), d(\chi, \psi)\}$.

$$\lim_{n\to\infty} b_n(\tau) = e(\tau), \quad \text{for all}\, \tau \in \mathscr{T}.$$

Therefore, in the topology of pointwise convergence, $\lim_{n\to\infty} b_n = e$, reflecting that the Euler method is consistent.

It is also possible to show that $b_n([\bullet]) = \frac{n-1}{2n}$, whereas $e([\bullet]) = \frac{1}{2}$. Therefore $d(b_n, e) = 2^{-1}$, reflecting that the numerical methods corresponding to b_n are all first order methods. However $d(b_n, e) = 2^{-1}$ also shows that, in the ultrametric topology, $\{b_n\}_{n\in\mathbb{N}}$ does not converge to e (or at all).

6 The Exponential Map and Regularity of Character Groups

In this section we discuss Lie theoretic properties of the Lie group of characters of a graded and connected Hopf algebra. Namely we consider the Lie group exponential map and a property called regularity in the sense of Milnor. The common theme of both properties is that they are related to the solution of certain differential equations on the Lie group.

We begin with a discussion of the properties of the Lie group exponential map. Recall that the interplay between the Lie algebra and the Lie group for infinite-dimensional Lie groups is more delicate than in the finite-dimensional case. For example there are infinite-dimensional Lie groups whose exponential map does not define a local diffeomorphism in a neighbourhood of the unit. See the survey in [30] for more information. However, it turns out that the situation for character groups of Hopf algebras is much better as they belong to certain well behaved classes of Lie groups which we define now.

Definition 13 Let G be a Lie group with smooth exponential map $\exp_G : \mathbf{L}(G) \to G$. The Lie group G is called a

1. *(locally) exponential* Lie group if $\exp_G$ induces a (local) diffeomorphism,
2. *Baker–Campbell–Hausdorff*-Lie group (or BCH-Lie group for short), if G is a $\mathbb{K}$-analytic (locally) exponential Lie group and $\exp_G$ induces a local $\mathbb{K}$-analytic diffeomorphism at 0.

In a BCH-Lie group the Baker–Campbell–Hausdorff-series converges on an open zero-neighbourhood and defines an analytic multiplication on this neighbourhood. The BCH-formula is often used in applications of Lie algebras to numerical analysis (see e.g. [57, III.4 and III.5]). However, there are even applications of the BCH-formula associated to character groups of certain Hopf algebras.

Remark 10 In [46] computations of the BCH-formula in an arbitrary Hall basis using labelled rooted trees are presented. It turns out that the BCH-formula discussed there is related to the BCH-formula on the Lie algebra of infinitesimal characters on certain Hopf algebras of labelled trees (cf. [46, Sect. 7] and in particular loc. cit. Example 10). Also see [58] for further information and the references contained therein.

From a Lie theoretic point of view BCH-Lie groups have very strong structural properties. For example the automatic smoothness theorem [59, Proposition 2.4] implies that every continuous group homomorphism between BCH-Lie groups is automatically $\mathbb{K}$-analytic. In particular this entails that the structure of a BCH-Lie group as a topological group uniquely determines the Lie group structure.

Proposition 2 ([4]) *Let $\mathscr{H}$ be a graded and connected Hopf algebra and B be a commutative locally convex algebra.*

1. *Then the Lie group $\mathscr{G}(\mathscr{H}, B)$ is a BCH-Lie group,*
2. *the exponential map $\exp_{\mathscr{G}(\mathscr{H},B)}$ is a $\mathbb{K}$-analytic diffeomorphism.*

Remark 11 (*Applications of the character group exponential map*)

1. The Lie group exponential of the Butcher group (cf. Example 3 (c)) has been used in computations in numerical analysis. Namely, in backward error analysis one exploits that this map is a diffeomorphism (see e.g. [46, Proposition 8]). The inverse of the exponential map is closely related to so called Lie derivatives of B-series (see [57, IX.1]). In particular, the formula in [57, Lemma 9.1] for the Lie derivative can be identified with the recursion formula for the inverse of the exponential map (cf. [3, Sect. 6]). However, albeit the term "Lie derivative" is used in the literature, only algebraic properties of these maps are exploited.
2. In [5] related pairs of exponential maps of character groups have been studied in the context of the universal enveloping algebra of a post-Lie algebra. These mappings play a role in the numerical integration on post-Lie algebras. However, we should mention at this point that the topology on the character groups used in loc.cit. differs from the one we used in the construction of the Lie group structure.

Problem 2 We have already mentioned several times that the Hopf algebras considered in numerical analysis are connected to so called pre- and post-Lie algebras. Recently these structures have gathered a lot of interest, see e.g. the work by Munthe-Kaas and collaborators [5, 50]. It would be interesting to see whether these additional structures induce more structure which is visible in the Lie group structure of the character groups.

We now turn to regularity properties of character groups. Roughly speaking, regularity (in the sense of Milnor) means that a certain class of (ordinary) differential equations can be solved on the Lie group. Note that it is highly non-trivial to solve ordinary differential equations on locally convex spaces beyond the realm of Banach spaces. For example there are linear differential equations without solution or which admit infinitely many different solutions (see [60] or [61, Sect. 5.5]).

Definition 14 (*Regularity (in the sense of Milnor)*) Let G be a Lie group modelled on a locally convex space, with identity element $\mathbf{1}$, and $r \in \mathbb{N}_0 \cup \{\infty\}$. We use the tangent map of the left translation $\lambda_g \colon G \to G, x \mapsto gx$ by $g \in G$ to define $g \,.\, v := T_1\lambda_g(v) \in T_gG$ for $v \in T_1(G) =: \mathbf{L}(G)$. Following [62], G is called *C^r-semiregular* if for each C^r-curve $\gamma \colon [0, 1] \to \mathbf{L}(G)$ the initial value problem

$$\begin{cases} \eta'(t) &= \eta(t) \,.\, \gamma(t) \\ \eta(0) &= \mathbf{1} \end{cases}$$

has a (necessarily unique) C^{r+1}-solution $\mathrm{Evol}(\gamma) := \eta \colon [0, 1] \to G$. If furthermore the map

$$\mathrm{evol} \colon C^r([0, 1], \mathbf{L}(G)) \to G, \quad \gamma \mapsto \mathrm{Evol}(\gamma)(1)$$

is smooth, G is called *C^r-regular.*[12] If G is C^r-regular and $r \leq s$, then G is also C^s-regular. A C^∞-regular Lie group G is called *regular (in the sense of Milnor)*—a property first defined in [27]. Every finite-dimensional Lie group is C^0-regular (cf. [30]).

Several important results in Lie theory are only available for regular Lie groups (see [27, 30, 62], cf. also [25] and the references therein). Up to this point all known Lie groups modelled on sufficiently complete spaces (i.e. on Mackey complete spaces, see [25, Chap. I.2]) are regular.

Example 12 Consider again the Butcher group G_{TM}, discussed in Example 3 (c). The differential equation for regularity of this Lie group takes the form of a countable system of differential equations: For a continuous curve $\mathbf{a} \colon [0, 1] \to \mathbf{L}(G_{\mathrm{TM}})$ we seek a differentiable curve $\gamma \colon [0, 1] \to G_{\mathrm{TM}}$ such that

$$\begin{cases} \gamma'(t)(\tau) &= \mathbf{a}(t)(\tau) + \sum\limits_{\substack{\theta \in \mathcal{T}_0 \\ |\theta| < |\tau|}} A_{\theta,\tau}(t, \mathbf{a}) \gamma(t)(\theta), \\ \gamma(0)(\tau) &= 0 \end{cases} \qquad \forall \tau \in \mathcal{T}$$

where $A_{\theta,\tau}(t, \mathbf{a})$ is a polynomial in $\mathbf{a}(t)(\sigma)$, $\sigma \in \mathcal{T}_0$ with $|\sigma| < |\tau|$. These differential equations form an infinite lower diagonal system of differential equations. As shown in [3, Sect. 5] this system can be solved inductively, i.e. via a projective limit argument.[13] Hence G_{TM} is a C^0-regular Lie group.

Proposition 3 ([4]) *Let $\mathcal{H}$ be a graded and connected Hopf algebra and B be a commutative and complete locally convex algebra. Then $\mathcal{G}(\mathcal{H}, B)$ is a C^0-regular Lie group.*

Note that we had to require B to be a complete algebra in Proposition 3. One can weaken this requirement, as it turns out that $\mathcal{G}(\mathcal{H}, B)$ is still a regular Lie group for so called "Mackey complete" locally convex algebras. We refer to [4] for further information.

---

[12] Here we consider $C^r([0, 1], \mathbf{L}(G))$ as a locally convex vector space with the pointwise operations and the topology of uniform convergence of the function and its derivatives on compact sets.

[13] This is possible since the differentiable structure of the Butcher group turns it into a projective limit of finite-dimensional Lie groups. We will return to this phenomenon in Sect. 7.

Remark 12 The Lie theory of character groups is closely connected to the Lie theory for unit groups of continuous inverse algebras (see [32, 33]). In fact, each character group of a graded and connected Hopf algebra can be identified with a closed Lie subgroup of the unit group of a suitable continuous inverse algebra. The details of this construction are recorded in [4, Proof of Theorem 2.10], where we have exploited this link to derive the regularity of the character group from the regularity of the ambient unit group.

The differential equations of regularity also occur in natural questions connected to character groups of Hopf algebras. Let us illustrate this with two examples.

Our first example is from applications in numerical analysis. In [45, p. 8] these differential equations appear naturally in the investigation of higher order averaging.

The second example appears in the theory of renormalisation of quantum field theories:

Example 13 (*Birkhoff decomposition and time ordered exponentials*) Consider the Hopf algebra of Feynman graphs $\mathcal{H}_{FG}$ with its group $\mathcal{G}(\mathcal{H}_{FG}, \mathbb{C})$ of $\mathbb{C}$ valued characters. A crucial step in the Connes–Kreimer theory of renormalisation—the so called BPHZ procedure—can be formulated as a Birkhoff factorisation in the Lie group $\mathcal{G}(\mathcal{H}_{FG}, \mathbb{C})$. To this end, one wants to decompose a smooth loop, i.e. a smooth map $\gamma \colon C \to \mathcal{G}(\mathcal{H}_{FG}, \mathbb{C})$ defined on a circle $C \subseteq \mathbb{C}$ as $\gamma(z) = \gamma_-^{-1}(z)\gamma_+(z)$. Here γ_-, γ_+ are boundary values of certain holomorphic functions (see [1, Definition 1.37]). Then the negative part γ_- of the Birkhoff decomposition yields the counterterms one seeks to compute in the renormalisation procedure (as explained in [1, 1.6.4], cf. explicitly [1, Theorem 1.40]).

To prove some desirable properties of the Birkhoff decomposition one defines a *time-ordered exponential*, i.e. for a smooth curve $\alpha \colon [a, b] \to \mathfrak{g}(\mathcal{H}_{FG}, \mathbb{C})$ define

$$\mathrm{T}e^{\int_a^b \alpha(t)\mathrm{d}t} := \mathbf{1}_{\mathcal{G}(\mathcal{H}_{FG},\mathbb{C})} + \sum_{n=1}^{\infty} \int_{a \leq s_1 \leq \cdots \leq s_n \leq b} \alpha(s_1) \cdots \alpha(s_n) \mathrm{d}s_1 \cdots \mathrm{d}s_n .$$

Then it turns out that the time ordered exponentials solve the differential equation associated to regularity in $\mathcal{G}(\mathcal{H}_{FG}, \mathbb{C})$ for the curve α (see [1, Proposition 1.51 (3)]). Time-ordered exponentials are important since they determine the Birkhoff decomposition in $\mathcal{G}(\mathcal{H}_{FG}, \mathbb{C})$. Explicitly, for a loop γ_μ (on an infinitesimal punctured disk in $\mathbb{C}$) one has as negative part of the Birkhoff decomposition

$$\gamma_-(z) = \mathrm{T}e^{-\frac{1}{z}\int_0^\infty \theta_{-t}(\beta)\mathrm{d}t}$$

where β is the so called β-function of the theory (cf. [1, Theorem 1.58]) and θ a certain one-parameter family generated by the grading operator. In particular, the time ordered exponentials determine the counterterms of perturbative renormalisation and one concludes that these depend only on the β-function of the theory. We are deliberately hiding the technical details here and refer instead to [1].

7 Character Groups as pro-Lie Groups

A crucial requirement to turn character groups of Hopf algebras into Lie groups has been that the Hopf algebra is needed to be graded and connected. As we have already mentioned, character groups of Hopf algebras which do not satisfy these conditions will in general not be infinite-dimensional Lie groups (with the topology of pointwise convergence). Note however that we have already seen in Lemma 3 that character groups of arbitrary Hopf algebras (with values in a locally convex algebra) are topological groups. Similarly the Lie algebra of infinitesimal characters is a topological Lie algebra regardless of a grading on $\mathcal{H}$.

In the present section we investigate the structure of the topological group $\mathcal{G}(\mathcal{H}, B)$ and its relation to the topological Lie algebra $\mathfrak{g}(\mathcal{H}, B)$ for Hopf algebras which are not necessarily graded. It turns out that for certain target algebras these topological groups admit a strong structure theory which is reminiscent of finite-dimensional Lie theory. To phrase our results let us recall the notion of a pro-Lie group (see the extensive monograph [6] or the recent survey [63]).

Definition 15 (*pro-Lie group*) A topological group G is called a *pro-Lie group* if one of the following equivalent conditions holds:

1. G is isomorphic (as a topological group) to a closed subgroup of a product of finite-dimensional (real) Lie groups.
2. G is the projective limit of a projective system of finite-dimensional (real) Lie groups (taken in the category of topological groups).

The class of pro-Lie groups contains all compact groups (see e.g. [64, Corollary 2.29]) and all connected locally compact groups (Yamabe's Theorem, see [65]). However, this does not imply that all pro-Lie groups are locally compact and the pro-Lie groups in the present paper will almost never be locally compact.

The structure theory of pro-Lie groups mirrors to a surprising degree the structure theory of finite-dimensional Lie groups (details are recorded in [6]). Most importantly for us, every pro-Lie group is connected to a Lie algebra. We recall its construction now. Note that in absence of a differential structure a Lie algebra to the group can not be constructed as a tangent space.

Definition 16 (*The pro-Lie algebra of a pro-Lie group*) Let G be a pro-Lie group and $\mathcal{L}(G)$ the space of all continuous G-valued one-parameter subgroups, endowed with the compact-open topology. As a projective limit of finite-dimensional Lie algebras it is naturally a locally convex topological Lie algebra over $\mathbb{R}$ (see [6, Definition 2.11]).

The character group of an arbitrary Hopf algebra (with values in a finite-dimensional algebra) turns out to be a pro-Lie group. In these cases the pro-Lie algebra can also be identified.

Theorem 3 (Character groups as pro-Lie groups [4, Theorem 5.6]) *Let $\mathcal{H}$ be a Hopf algebra and B be a commutative finite-dimensional $\mathbb{K}$-algebra (e.g. $B := \mathbb{K}$). Then the group of B-valued characters $\mathcal{G}(\mathcal{H}, B)$ endowed with the topology of pointwise convergence is pro-Lie group.*

Its pro-Lie algebra is isomorphic to the locally convex Lie algebra $\mathfrak{g}(\mathcal{H}, B)$ *of infinitesimal characters via the canonical isomorphism*

$$\mathfrak{g}(\mathcal{H}, B) \to \mathcal{L}(\mathcal{G}(\mathcal{H}, B)), \quad \phi \mapsto (t \mapsto \exp(t\phi)).$$

Proof (*Sketch of ideas*) Due to the fundamental theorem of coalgebras (see [66, Theorem 4.12]), one can write $\mathcal{H}$ as a directed union of finite-dimensional coalgebras $\{C_i\}_{i\in I}$. On each of the spaces $\mathrm{Hom}_{\mathbb{K}}(C_i, B)$ the convolution induces a locally convex algebra structure such that the unit groups satisfy

$$\mathrm{Hom}_{\mathbb{K}}(\mathcal{H}, B)^\times = \mathrm{Hom}_{\mathbb{K}}(\lim_{\rightarrow} C_i, B)^\times \cong \lim_{\leftarrow} \mathrm{Hom}_{\mathbb{K}}(C_i, B)^\times$$

where the limit on the right hand side is taken in the category of topological groups. In particular, the groups $\mathrm{Hom}_{\mathbb{K}}(C_i, B)^\times$ are finite-dimensional Lie groups, whence $\mathrm{Hom}_{\mathbb{K}}(\mathcal{H}, B)^\times$ is a pro-Lie group and $\mathcal{G}(\mathcal{H}, B)$ inherits this structure. The remaining assertions follow from direct computations involving the exponential map. □

Remark 13 1. If we consider character groups of graded connected Hopf algebras with values in finite-dimensional algebras, we obtain two structures on the character group: The locally convex Lie group structure and the pro-Lie group structure. Fortunately the results in [67] affirms that these two structures are compatible with each other. This means that the only additional information provided by the pro-Lie structure in this case is that the infinite-dimensional Lie group already constructed is a projective limit of finite-dimensional Lie groups.
2. One can generalise the preceding result beyond the realm of finite-dimensional target algebras to so called "weakly complete algebras". See [4, Sect. 6] for the definition and detailed statements.

In applications of character groups of Hopf algebras the pro-Lie property has frequently been of crucial importance. Most importantly, it allows one to work with the projective limit structures of the Lie algebra of infinitesimal characters and the character group.

Example 14 1. For the Hopf algebra of Feynman diagrams $\mathcal{H}_{FG}$ one considers flat $\mathfrak{g}(\mathcal{H}_{FG}, \mathbb{C})$-valued connections. To prove some desirable properties for maps depending on these connections one has to employ projective limit arguments, e.g. see [1, Proposition 1.52]. Furthermore, one can use the projective limit property to construct geometric data for elements in certain interesting algebras of infinitesimal characters. Here we mean the monodromy representation and their limit constructed in [1, Lemma 1.54] for elements in $\mathfrak{g}(\mathcal{H}_{FG}, \mathbb{C})$.
Note that in [1] the projective limit structure was deduced from the fact that every affine group scheme is a projective limit of linear algebraic groups. This requires the source Hopf algebra to be commutative, whereas the pro-Lie structure established in Theorem 3 does not depend on the commutativity of $\mathcal{H}_{FG}$.
2. In numerical analysis, properties of series are often studied by truncating to a finite number of terms, see e.g. the treatment of modified equations in [47] or

[18, Sects. 5.2, 5.4]. This corresponds in the pro-Lie picture to passing from the projective limit to one of the (finite-dimensional) steps.

The above list shows that it is quite useful to have the structure of a pro-Lie group to analyse character groups of Hopf algebras. Conversely, one can ask which pro-Lie groups can be obtained as character groups.

Problem 3 Characterize all pro-Lie groups that can be obtained as $\mathbb{R}$-valued character groups of (in general non-graded) Hopf algebras.

At least all compact abelian groups appear as character groups of certain Hopf algebras (this follows from Pontryagin duality). On the other hand, one can show that an uncountable discrete group is never a character group of a Hopf algebra.

Acknowledgements The research on this paper was partially supported by the project *Topology in Norway* (NRC project 213458) and *Structure Preserving Integrators, Discrete Integrable Systems and Algebraic Combinatorics* (NRC project 231632). We thank J. M. Sanz-Serna for pointing out references to results from numerical analysis which the authors were unaware of. Finally, we thank the anonymous referees for many useful comments which helped to improve the manuscript.

References

1. Connes, A., Marcolli, M.: Noncommutative geometry, quantum fields and motives. In: American Mathematical Society Colloquium Publications, vol. 55. American Mathematical Society, Providence, RI; Hindustan Book Agency, New Delhi (2008)
2. Brouder, C.: Trees, renormalization and differential equations. BIT Num. Anal. **44**, 425–438 (2004)
3. Bogfjellmo, G., Schmeding, A.: The Lie group structure of the Butcher group. Found. Comput. Math. **17**(1), 127–159 (2017)
4. Bogfjellmo, G., Dahmen, R., Schmeding, A.: Character groups of Hopf algebras as infinite-dimensional Lie groups. Ann. Inst. Fourier (Grenoble) **66**(5), 2101–2155 (2016)
5. Ebrahimi-Fard, K., Lundervold, A., Munthe-Kaas, H.Z.: On the Lie enveloping algebra of a post-Lie algebra. J. Lie Theory **25**(4), 1139–1165 (2015)
6. Hofmann, K.H., Morris, S.A.: The Lie theory of connected pro-Lie groups. In: EMS Tracts in Mathematics, vol. 2. EMS, Zürich (2007)
7. Cartier, P.: A primer of Hopf algebras. Frontiers in number theory, physics, and geometry. II, pp. 537–615. Springer, Berlin (2007)
8. Kassel, C.: Quantum groups. In: Graduate Texts in Mathematics, vol. 155. Springer-Verlag, New York (1995)
9. Majid, S.: Foundations of quantum group theory. Cambridge University Press, Cambridge (1995)
10. Manchon, D.: Hopf algebras in renormalisation. In: Handbook of Algebra, vol. 5, pp. 365–427. Elsevier/North-Holland, Amsterdam (2008)
11. Sweedler, M.E.: Hopf algebras. In: Mathematics Lecture Note Series. W. A. Benjamin Inc, New York (1969)
12. Waterhouse, W.C.: Introduction to affine group schemes. In: Graduate Texts in Mathematics, vol. 66. Springer-Verlag, New York-Berlin (1979)
13. Connes, A., Kreimer, D.: Hopf algebras, renormalization and noncommutative geometry. Commun. Math. Phys. **199**, 203–242 (1998)
14. Chartier, P., Hairer, E., Vilmart, G.: Algebraic structures of B-series. Found. Comput. Math. **10**(4), 407–427 (2010)

15. Munthe-Kaas, H.Z., Wright, W.M.: On the Hopf algebraic structure of Lie group integrators. Found. Comput. Math. **8**(2), 227–257 (2008)
16. Reutenauer, C.: Free Lie algebras. In: London Mathematical Society Monographs. New Series, vol. 7. The Clarendon Press, Oxford University Press, New York. Oxford Science Publications (1993)
17. Lundervold, A., Munthe-Kaas, H.: Hopf algebras of formal diffeomorphisms and numerical integration on manifolds. In: Combinatorics and Physics. Contemporary Mathematics, vol. 539, pp. 295–324. American Mathematical Society, Providence, RI (2011)
18. Murua, A., Sanz-Serna, J.M.: Word series for dynamical systems and their numerical integrators. Found. Comput. Math. (2015). https://doi.org/10.1007/s10208-015-9295-3
19. Knapp, A.W.: Lie groups, Lie algebras, and cohomology. In: Mathematical Notes, vol. 34. Princeton University Press, Princeton, NJ (1988)
20. Jarchow, H.: Locally convex spaces. In: Lecture Notes in Mathematics, vol. 417. Teubner, Stuttgart (1981)
21. Schaefer, H.H.: Topological vector spaces. Springer-Verlag, New York-Berlin. Third printing corrected, Graduate Texts in Mathematics, vol. 3 (1971)
22. Michor, P.W.: Manifolds of differentiable mappings. In: Shiva Mathematics Series, vol. 3. Shiva Publishing Ltd., Nantwich (1980)
23. Keller, H.: Differential calculus in locally convex spaces. In: Lecture Notes in Mathematics, vol. 417. Springer Verlag, Berlin (1974)
24. Glöckner, H.: Infinite-dimensional Lie groups without completeness restrictions. In: Geometry and analysis on finite- and infinite-dimensional Lie groups (Będlewo, 2000), Banach Center Publications, vol. 55, pp. 43–59. Polish Academy of Sciences, Warsaw (2002)
25. Kriegl, A., Michor, P.W.: The convenient setting of global analysis. In: Mathematical Surveys and Monographs, vol. 53. American Mathematical Society, Providence, RI (1997)
26. Bertram, W., Glöckner, H., Neeb, K.-H.: Differential calculus over general base fields and rings. Expo. Math. **22**(3), 213–282 (2004)
27. Milnor, J.: Remarks on infinite-dimensional Lie groups. In: Relativity groups and topology, II (Les Houches, 1983), pp. 1007–1057. North-Holland, Amsterdam (1984)
28. Glöckner, H.: Instructive examples of smooth, complex differentiable and complex analytic mappings into locally convex spaces. J. Math. Kyoto Univ. **47**(3), 631–642 (2007)
29. Alzaareer, H., Schmeding, A.: Differentiable mappings on products with different degrees of differentiability in the two factors. Expo. Math. **33**(2), 184–222 (2015)
30. Neeb, K.-H.: Towards a Lie theory of locally convex groups. Jpn. J. Math. **1**(2), 291–468 (2006)
31. Trèves, F.: Topological vector spaces, distributions and kernels. Dover Publications, Inc., Mineola, NY. Unabridged republication of the 1967 original (2006)
32. Glöckner, H.: Algebras whose groups of units are Lie groups. Studia Math. **153**(2), 147–177 (2002)
33. Glöckner, H., Neeb, K.-H.: When unit groups of continuous inverse algebras are regular Lie groups. Studia Math. **211**(2), 95–109 (2012)
34. Milnor, J.W., Moore, J.C.: On the structure of Hopf algebras. Ann. Math. **2**(81), 211–264 (1965)
35. Ebrahimi-Fard, K., Gracia-Bondía, J.M., Patras, F.: A Lie theoretic approach to renormalization. Comm. Math. Phys. **276**(2), 519–549 (2007)
36. Dahmen, R., Schmeding, A.: The Lie group of real analytic diffeomorphisms is not real analytic. Studia Math. **229**(2), 141–172 (2015)
37. Floret, K.: Lokalkonvexe Sequenzen mit kompakten Abbildungen. J. Reine Angew. Math. **247**, 155–195 (1971)
38. Glöckner, H.: Direct limits of infinite-dimensional Lie groups. In: Developments and trends in infinite-dimensional Lie theory, Progress in Mathematics, vol. 288, pp. 243–280. Birkhäuser Boston, Inc., Boston, MA (2011)
39. Bonet, J., Pérez Carreras, P.: Barrelled locally convex spaces. In: North-Holland Mathematics Studies, vol. 131. North-Holland Publishing Co., Amsterdam. Notas de Matemática [Mathematical Notes], 113 (1987)

40. Grosse-Erdmann, K.-G.: The locally convex topology on the space of meromorphic functions. J. Aust. Math. Soc. Ser. A **59**(3), 287–303 (1995)
41. Arens, R.: Linear topological division algebras. Bull. Am. Math. Soc. **53**, 623–630 (1947)
42. Murua, A., Sanz-Serna, J.M.: Computing normal forms and formal invariants of dynamical systems by means of word series. Nonlinear Anal. **138**, 326–345 (2016)
43. van Suijlekom, W.D.: Renormalization of gauge fields: a Hopf algebra approach. Comm. Math. Phys. **276**(3), 773–798 (2007)
44. van Suijlekom, W.D.: The structure of renormalization Hopf algebras for gauge theories. I. representing Feynman graphs on BV-algebras. Comm. Math. Phys. **290**(1), 291–319 (2009)
45. Chartier, P., Murua, A., Sanz-Serna, J.M.: Higher-order averaging, formal series and numerical integration II: the quasi-periodic case. Found. Comput. Math. **12**(4), 471–508 (2012)
46. Murua, A.: The Hopf algebra of rooted trees, free Lie algebras, and Lie series. Found. Comput. Math. **6**(4), 387–426 (2006)
47. Hairer, E., Lubich, C.: The life-span of backward error analysis for numerical integrators. Numerische Math. **76**(4), 441–462 (1997)
48. Dahmen, R.: Analytic mappings between LB-spaces and applications in infinite-dimensional Lie theory. Math. Z. **266**(1), 115–140 (2010)
49. Bogfjellmo, G., Schmeding, A.: The tame Butcher group. J. Lie Theor. **26**(4), 1107–1144 (2016). cf. arXiv:1509.03452v3
50. Munthe-Kaas, H.Z., Lundervold, A.: On post-Lie algebras, Lie-Butcher series and moving frames. Found. Comput. Math. **13**(4), 583–613 (2013)
51. Chartier, P., Murua, A., Sanz-Serna, J.M.: Higher-order averaging, formal series and numerical integration III: error bounds. Found. Comput. Math. **15**(2), 591–612 (2015)
52. Gray, W.S., Duffaut, L.A.: Espinosa, and K. Ebrahimi-Fard. Faà di Bruno Hopf algebra of the output feedback group for multivariable Fliess operators. Syst. Control Lett. **74**, 64–73 (2014)
53. Ebrahimi-Fard, K., Gray, W.S.: Center problem, Abel equation and the Faà di Bruno Hopf algebra for output feedback. Int. Math. Res. Not. IMRN **17**, 5415–5450 (2017)
54. Dahmen, R., Schmeding, A.: Lie groups of controlled characters of combinatorial Hopf algebras (2018). arXiv:1609.02044v4
55. Butcher, J.C.: An algebraic theory of integration methods. Math. Comp. **26**, 79–106 (1972)
56. Butcher, J.C.: Numerical methods for ordinary differential equations, 2nd edn. Wiley, Chichester (2008)
57. Hairer, E., Lubich, C., Wanner, G.: Geometric numerical integration. Structure-preserving algorithms for ordinary differential equations In: Springer Series in Computational Mathematics vol. 31, 2nd edn. Springer-Verlag, Berlin (2006)
58. Casas, F., Murua, A.: An efficient algorithm for computing the Baker-Campbell-Hausdorff series and some of its applications. J. Math. Phys. **50**(3), 033513, 23 (2009)
59. Glöckner, H.: Lie group structures on quotient groups and universal complexifications for infinite-dimensional Lie groups. J. Funct. Anal. **194**(2), 347–409 (2002)
60. Milnor, J.: On infinite-dimensional Lie groups, 1982. unpublished preprint
61. Hamilton, R.S.: The inverse function theorem of Nash and Moser. Bull. Amer. Math. Soc. (N.S.), **7**(1), 65–222 (1982)
62. Glöckner, H.: Regularity properties of infinite-dimensional Lie groups, and semiregularity (2015). arXiv:1208.0715v3
63. Hofmann, K.H., Morris, S.A.: Pro-Lie groups: a survey with open problems. Axioms **4**(3), 294–312 (2015)
64. Hofmann, K.H., Morris, S.A.: The structure of compact groups. In: De Gruyter Studies in Mathematics, vol. 25, 3rd edn. De Gruyter, Berlin (2013)
65. Yamabe, H.: On the conjecture of Iwasawa and Gleason. Ann. Math. **2**(58), 48–54 (1953)
66. Michaelis, W.: Coassociative coalgebras. In: Handbook of Algebra, vol. 3, pp. 587–788. North-Holland, Amsterdam (2003)
67. Hofmann, K.H., Neeb, K.-H.: Pro-Lie groups which are infinite-dimensional Lie groups. Math. Proc. Cambridge Philos. Soc. **146**(2), 351–378 (2009)

Bäcklund Transformations in Discrete Variational Principles for Lie-Poisson Equations

María Barbero Liñán and David Martín de Diego

Abstract We consider a dynamical system on the dual of a Lie algebra. On the dual of that algebra there is another linear Poisson structure. This system is integrable for one of the Poisson structures because it admits a suitable Lax representation. The discrete variational principle is applied to the problem given by the *non-usual* linear Poisson structure to obtain Lie-Poisson integrators which preserve all the Casimir functions of the system. In the 19th century Bäcklund transformations were introduced in the area of partial differential equations as transformations that map solutions to solutions. It is known that Bäcklund transformations satisfy some specific properties such as commutativity. We geometrically define Bäcklund transformations associated with the obtained Lie-Poisson integrators under some invariance assumptions. The existence of an invariant scalar product that identifies the Lie algebra and the dual of the Lie algebra allows to establish the connection with the results proved in Suris' book on the Lie algebra. We will make clear the constructions by looking at the Toda Lattice example.

Keywords Variational integrators · Lie-Poisson equations
Bäcklund transformations

MSC 2010 37K10 · 65P10 · 70G45 · 70H06

M. Barbero Liñán (✉)
Departamento de Matemática Aplicada, Universidad Politécnica de Madrid (UPM),
Avd. Juan Herrera 4, 28040 Madrid, Spain
e-mail: m.barbero@upm.es

M. Barbero Liñán
Instituto de Ciencias Matemáticas (CSIC-UAM-UC3M-UCM), Madrid, Spain

D. Martín de Diego
Instituto de Ciencias Matemáticas (CSIC-UAM-UC3M-UCM),
C/ Nicolás Cabrera 13-15, 28049 Madrid, Spain
e-mail: david.martin@icmat.es

K. Ebrahimi-Fard and M. Barbero Liñán (eds.), *Discrete Mechanics, Geometric Integration and Lie–Butcher Series*, Springer Proceedings in Mathematics & Statistics 267, https://doi.org/10.1007/978-3-030-01397-4_9

1 Introduction

Bäcklund transformations (BTs) appear at the end of the XIXth century. They are introduced by the Swedish mathematician and physicist called Albert Victor Bäcklund as transformations of partial differential equations bringing solutions into solutions in a differential geometric context [3]. Independently, Darboux [7] discovered similar transformations in the context of second order ordinary differential equations.

Since then, literature has provided BTs with specific properties such as preservation of the commuting Hamiltonians, canonicity, spectrality when there exists a dependence on a spectral parameter [23, 31]. In the nineties it was discovered how BTs appear in discretization of finite-dimensional integrable systems such as Euler top, Lagrange top, etc., [26, 30, 31].

Ge-Marsden [9] and Channel-Scovel [6] developed the notion of integrators for Lie-Poisson dynamical systems. These systems include the Euler equations for the rigid body, the Vlasov-Poisson equations, the Vlasov- Maxwell equations, and others (see references in [6]).

Ge-Marsden [9] developed numerical integrators that exactly preserve momentum maps and linear Poisson brackets. As a consequence, the integrators preserve the Lie-Poisson structure for finite dimensional systems. The results are obtained by using the theory of generating functions in Hamilton-Jacobi theory. Channel-Scovel [6] developed Lie-Poisson integrators using the exponential map that relates the elements in the Lie group with the elements in the Lie algebra.

Here we focus on constructing Lie-Poisson integrators for completely integrable systems by using discrete variational principles [15] on the dual of the Lie algebra. As the elements are defined on the dual of the Lie algebra, no reduction is needed prior to discretize the Lagrangian. Our contribution is to consider two different structures of Lie algebra to obtain Lie-Poisson integrators for completely integrable systems. There is a debate on what a discrete integrable system is [19]. Here we search for geometric integrators that exactly preserve the integrability of the system. Other approaches consider the preservation of integrability up to certain order [2].

The existence of a Lax pair does not guarantee the complete integrability of the system [12, 21] in the sense of Liouville. However, the systems under study are completely integrable and they admit a Lax pair representation with enough conserved quantities being involutive, that is, mutually commuting. To guarantee the involutivity of the conserved quantities obtained from the Lax pair some additional conditions are needed.

The paper is organized as follows. Section 2 covers all the preliminary contents about Poisson structures and completely integrable systems that admit a Lax pair. Section 3 introduces the object under study: completely integrable Lie-Poisson equations admiting an R-matrix approach. Our objective is to obtain numerical methods by applying a discrete variational principle to the Lie-Poisson equations as described in Sect. 4 using the two Poisson structures defined on the same set. In Sect. 5 we establish a connection between the obtained numerical methods and the use of Bäcklund transformation to discretize completely integrable systems. Finally we apply the methods in Sects. 4 and 5 to the Toda lattice example and the QR algorithm in Sect. 6 to connect with the constructions in [26] for matrix Lie groups.

2 Preliminaries

In this section we will quickly introduce the notions of Poisson structures and integrable systems necessary for developing the main contributions of this paper.

2.1 On Poisson Structures

Many dynamical systems are described by means of Poisson structures, such as Euler equations for the rigid body, Maxwell equations, etc. More details on Poisson structures can be found, for instance, in [10, 28, 29].

A *Poisson structure* on a m-dimensional manifold M is defined as a Lie algebra structure $\{\cdot,\cdot\}$ on the set $\mathcal{C}^\infty(M)$ of smooth functions on M satisfying the Leibniz identity $\{FG, H\} = F\{G, H\} + \{F, H\}G$ for every F, G, H in $\mathcal{C}^\infty(M)$. Thus the bracket operation is a derivation in each entry. In particular for each function H there is a vector field ξ_H such that $\xi_H(F) = \{F, H\}$ for all F in $\mathcal{C}^\infty(M)$. This vector field ξ_H is called the *Hamiltonian vector field* generated by H.

Note that at every point x in M the value of the bracket $\{\cdot, H\}$ depends only on the derivatives of H, not on the function itself. So there is a bundle map $\sharp\colon T^*M \to TM$, from the cotangent bundle T^*M to the tangent bundle TM, such that there exists a Hamiltonian vector field defined as follows $\xi_H = \sharp \circ \mathrm{d}H$ for all H. The map $\sharp$ can also be interpreted as a contravariant skew symmetric 2-tensor Λ in M for which $\{F, G\} = \Lambda(\mathrm{d}F, \mathrm{d}G)$. The bivector field Λ does not always define a Poisson structure because the bracket defined by Λ satisfies the Jacobi identity if and only if the so-called Schouten bracket $[\Lambda, \Lambda]$ vanishes [24]. In local coordinates $(x_1, \dots, x_m)$, the condition $[\Lambda, \Lambda] = 0$ is equivalent to $\Lambda_{ij} = -\Lambda_{ji}$ and

$$\sum_{l=1}^{m} \left(\Lambda_{lj} \frac{\partial \Lambda_{ik}}{\partial x_l} + \Lambda_{li} \frac{\partial \Lambda_{kj}}{\partial x_l} + \Lambda_{lk} \frac{\partial \Lambda_{jl}}{\partial x_l} \right) = 0 .$$

Whenever $[\Lambda, \Lambda] = 0$, the Poisson structure is determined by providing the bracket relations satisfied by the coordinate functions $\{x_i, x_j\} = \Lambda_{ij}(x)$. Then,

$$\{F, G\} = \sum_{i,j=1}^{m} \Lambda_{ij} \frac{\partial F}{\partial x_i} \frac{\partial G}{\partial x_j} .$$

Any symplectic manifold (M, ω) is a Poisson manifold, but not all the Poisson manifolds are symplectic. In fact, any Poisson manifold is a union of symplectic manifolds, generally of varying dimensions, called the leaves of the symplectic foliation of the Poisson manifold.

A Poisson structure on a vector space V is called *linear* if the Poisson bracket of any two linear functions on V is again linear. This makes the dual space V^* into a Lie algebra which we denote by $\mathfrak{g}$. The bracket on $V \equiv \mathfrak{g}^*$ is given by

$$\{F, G\}(\mu) = \left\langle \mu, \left[\frac{\delta F}{\delta \mu}, \frac{\delta G}{\delta \mu}\right]\right\rangle, \tag{1}$$

where $[\cdot, \cdot]$ is the Lie algebra operation in $\mathfrak{g}$, $\langle \cdot, \cdot \rangle$ is the pairing of $\mathfrak{g}^*$ with $\mathfrak{g}$, and $\delta F/\delta \mu$ is the differential of F as an element of $\mathfrak{g}$ instead of $\mathfrak{g}^{**}$.

The linear Poisson structure on the dual space $\mathfrak{g}^*$ defined by (1) is called *Lie-Poisson structure* because it was discovered by Lie [11] who wrote down the structure in local coordinates. If $X_1, \ldots, X_m$ is a basis of $\mathfrak{g}$ satisfying $[X_i, X_j] = \sum_{k=1}^m c_{ijk} X_k$, then the components of the Poisson structure are $\Lambda_{ij}(x) = \sum_{k=1}^m c_{ijk} x_k$. These are linear functions of x. That is why it is called a *linear Poisson structure*.

Let us recall some operations on Lie groups and Lie algebras that will be needed in the sequel. A real Lie group is a real, analytic manifold G equipped with a product which defines a group structure on G. For any $g \in G$, the left and right translation of G by g are diffeomorphisms defined as follows:

$$l_g(x) = gx, \quad r_g(x) = xg .$$

These maps satisfy the following properties:

$$(l_g)^{-1} = l_{g^{-1}}, \quad (r_{g^{-1}})^{-1} = r_g, \quad l_{g_1} \circ l_{g_2} = l_{g_1 g_2}, \quad r_{g_1} \circ l_{g_2} = l_{g_2} \circ r_{g_1} .$$

For each $x,\ g \in G$, $\xi \in \mathfrak{g}^*$, $X,\ Y \in \mathfrak{g} \equiv T_e G$, we define:

- the automorphism of the group G by conjugation or inner automorphism:

$$\mathrm{I}_g \colon G \to G, \quad \mathrm{I}_g(x) = (l_g \circ r_{g^{-1}})(x) = g x g^{-1} ;$$

- the differential of the conjugation or the group adjoint representation of G on $\mathfrak{g}$:

$$\mathrm{Ad}_g \colon \mathfrak{g} \to \mathfrak{g}, \quad \mathrm{Ad}_g(X) = T_e \mathrm{I}_g(X) = \left(T_{g^{-1}} l_g \circ T_e r_{g^{-1}}\right)(X) ;$$

- the adjoint representation of the Lie algebra:

$$\mathrm{ad}_X Y = [X, Y] ;$$

- the dual of Ad_g:

$$\mathrm{Ad}_g^*\colon \mathfrak{g}^* \to \mathfrak{g}^*, \quad \langle \mathrm{Ad}_g^*\xi, X\rangle = \langle \xi, \mathrm{Ad}_g X\rangle$$

- the coadjoint representation of the Lie algebra $\mathfrak{g}^*$:

$$\mathrm{ad}_X^*\colon \mathfrak{g}^* \to \mathfrak{g}^*, \quad \langle \mathrm{ad}_X^*\xi, Y\rangle = -\langle \xi\, , \mathrm{ad}_X Y\rangle,$$

where $\langle \cdot, \cdot \rangle$ denotes the natural pairing between elements in the Lie algebra $\mathfrak{g}$ and in the dual of the Lie algebra, $\mathfrak{g}^*$. More details on these operations can be found, for instance, in [14].

2.2 Lax Pair and Integrability

When first integrals of a Hamiltonian system are known, one may use the method of reduction to facilitate the determination of the integral curves of the system. A Hamiltonian system is said to be *completely integrable* in the Liouville sense if it has m differentiable first integrals $f_1, \ldots, f_m$, defined on the whole manifold M, which are pairwise in involution and whose differentials are linearly independent on a dense open subset of M. However, this characterization does not make easy to identify the completely integrable systems because all the first integrals must be found. Almost all known integrable systems possess Lax representations, that is, the equations of motion can be rewritten as

$$\dot{L} = [L, B], \tag{2}$$

where $L, B\colon M \to \mathfrak{g}$ are maps from the phase space M into some Lie algebra $\mathfrak{g}$. The Lax pair given by (L, B) allows to obtain integrals of motion, but not necessarily enough of them to guarantee the complete integrability. However, R-matrix approach [16, 20] incorporates the involutivity property to the Lax pair. In these cases the matrix B can be written as $B = R(f(L))$, where $R\colon \mathfrak{g} \to \mathfrak{g}$ is a linear operator and $f\colon \mathfrak{g} \to \mathfrak{g}$ is an Ad-covariant function, that is,

$$\mathrm{Ad}_g f(L) = f(\mathrm{Ad}_g L) \quad \forall\, g \in G\,.$$

On the sequel we focus on linear R-matrix brackets that define Lie-Poisson bracket on the dual of the Lie algebra, $\mathfrak{g}^*$, identified with the Lie algebra $\mathfrak{g}$ by means of an invariant scalar product. In this framework it is possible to define a trace form on the Lie algebra. In the finite dimensional case the traces of powers of L are conserved quantities in involution. The R-matrix approach connects with the Yang-Baxter equations coming from physics under some conditions, see [20, 25] for details. The trace operator can also be described for infinite-dimensional Lie algebras, see for instance the Adler's notion [1]. See [22] for a review on the Lie-Poisson equations in infinite dimensional spaces.

3 Integrable Lie-Poisson Equations

The bi-hamiltonian structure introduced by Magri [12] has been a powerful tool to construct integrable systems on finite and infinite dimensional Lie algebras. Here we consider the situation where the Lie-Poisson equations are completely integrable. That is why we only consider Lie-Poisson structures associated with a linear R-matrix structure so that the complete integrability of the equations of motion rewritten in a Lax pair form is guaranteed [16, 20]. Let $(\mathfrak{g}, [\cdot, \cdot])$ be a Lie algebra. Consider a linear map $R\colon \mathfrak{g} \to \mathfrak{g}$ that defines a different Lie bracket on $\mathfrak{g}$ as follows:

$$[\xi, \eta]_R = [R\xi, \eta] + [\xi, R\eta], \quad \xi,\ \eta \in \mathfrak{g}.$$

The linear map R must satisfy the modified Yang-Baxter equation,

$$[R\xi, R\eta] - R([\xi, \eta]_R) = -[\xi, \eta],$$

so that the R-bracket verifies the Jacobi identity [25]. Under that circumstance, $(\mathfrak{g}, [\cdot, \cdot]_R)$ is also a Lie algebra. Thus the set G has two different structures of Lie groups: $(G, \cdot)$ is the Lie group with the Lie algebra $(\mathfrak{g}, [\cdot, \cdot])$ and $(G, \star)$ is the Lie group with the Lie algebra $(\mathfrak{g}, [\cdot, \cdot]_R)$.

There are also two linear Poisson structures on the dual of the algebra, $\mathfrak{g}^*$, associated with the two Lie algebras: $(\mathfrak{g}^*, \{\cdot, \cdot\})$ and $(\mathfrak{g}^*, \{\cdot, \cdot\}_R)$. The brackets are defined as in (1).

Let $H\colon \mathfrak{g}^* \to \mathbb{R}$ be Ad*-invariant, that is, $H(Ad_g^*\alpha) = H(\alpha)$ for all α in $\mathfrak{g}^*$ and g in G. The Hamiltonian function is Casimir only for the bracket $\{\cdot, \cdot\}$.

Our objective is to integrate the differential equation on $(\mathfrak{g}^*, \{\cdot, \cdot\}_R)$ given by

$$\frac{\mathrm{d}\alpha}{\mathrm{d}t} = \left(ad_R^*\right)_{\mathrm{d}H(\alpha)} \alpha \,. \tag{3}$$

These are the corresponding *Lie-Poisson equations* for the Hamiltonian system $(H, \{\cdot, \cdot\}_R)$. Note that the analogous differential equation $\dfrac{\mathrm{d}\alpha}{\mathrm{d}t} = ad^*_{\mathrm{d}H(\alpha)}\alpha \equiv 0$ on $(\mathfrak{g}^*, \{\cdot, \cdot\})$ vanishes identically because the Hamiltonian function $H : \mathfrak{g}^* \to \mathbb{R}$ is a Casimir function for the bracket $\{\cdot, \cdot\}$.

Take $\xi \in \mathfrak{g}$. The Eq. (3) can be written in terms of the original bracket as follows

$$\begin{aligned}\langle \left(ad_R^*\right)_{\mathrm{d}H(\alpha)} \alpha, \xi\rangle &= \langle \alpha, [\mathrm{d}H(\alpha), \xi]_R\rangle = \langle \alpha, [R\mathrm{d}H(\alpha), \xi] + [\mathrm{d}H(\alpha), R\xi]\rangle \\ &= \langle \alpha, [R\mathrm{d}H(\alpha), \xi]\rangle = \langle \left(ad^*\right)_{R\mathrm{d}H(\alpha)} \alpha, \xi\rangle \,. \end{aligned} \tag{4}$$

Note that $\langle \alpha, [\mathrm{d}H(\alpha), R\xi]\rangle = 0$ because H is Casimir for $\{\cdot, \cdot\}$.

As an example of a linear R-matrix structure, let us assume now we have a splitting of the Lie algebra $\mathfrak{g} = \mathfrak{g}_+ \oplus \mathfrak{g}_-$, with projections $\pi_+\colon \mathfrak{g} \to \mathfrak{g}_+$, $\pi_-\colon \mathfrak{g} \to \mathfrak{g}_-$ onto the corresponding factors. Take the linear map $R = \frac{1}{2}(\pi_+ - \pi_-)$, which can be equivalently written as $R = \pi_+ - \frac{\mathrm{Id}}{2} = \frac{\mathrm{Id}}{2} - \pi_-$.

For such a linear map the Eq. (3) becomes

$$\begin{aligned}\frac{\mathrm{d}\alpha}{\mathrm{d}t} &= \left(ad_R^*\right)_{\mathrm{d}H(\alpha)}\alpha = \left(ad^*\right)_{RH(\alpha)}\alpha = \left(ad^*\right)_{\pi_+\mathrm{d}H(\alpha)-\frac{\mathrm{Id}}{2}\mathrm{d}H(\alpha)}\alpha \\ &= \left(ad^*\right)_{\pi_+\mathrm{d}H(\alpha)}\alpha = -\left(ad^*\right)_{\pi_-\mathrm{d}H(\alpha)}\alpha\,, \end{aligned} \tag{5}$$

because of (4), plus the fact that H is Casimir for $\{\cdot,\cdot\}$.

Remember that the Eq. (3) is integrable because the linear map R admits a Lax representation generating enough conserved quantities in involution. Our next step is to find a discretization of the Eq. (3) by discretizing Eq. (5).

4 Discrete Variational Principle for Integrable Lie-Poisson Equations

Let us apply a discrete variational principle [5, 13, 15] on the Lie group $(G,\star)$ for a Lagrangian $L\colon \mathfrak{g}\to\mathbb{R}$. The starting point of this discretization is not a Lagrangian on TG whose discretization is a function on $G\times G$ being G-invariant under an action to reduce it to G as usual. We start from a Lagrangian already on $\mathfrak{g}$ whose discretization is a function on G. Thus, no G-invariance of the Lagrangian is needed.

As mentioned in the example in Sect. 3, the splitting $\mathfrak{g} = \mathfrak{g}_+ \oplus \mathfrak{g}_-$ at the level of the Lie algebra induces a factorization of the Lie group, $G = G_+\cdot G_-$, in two Lie subgroups $(G_+,\cdot)$ and $(G_-,\cdot)$. On the Lie group $(G,\star)$, the $\star$ product and the inverse element, $(g)^{-1_\star}$, are defined as follows using the usual product of the Lie group $(G,\cdot)$:

$$g_1\star g_2 = g_1^+ g_2 g_1^-\,,\quad (g)^{-1_\star} = (g^+)^{-1}(g^-)^{-1}\,,$$

where g_1^+ is the factor of g_1 in G_+ and g_1^- is the factor of g_1 in G_-.

The variational principle consists of minimizing a functional over all the admissible curves on G. Instead of minimizing $\int_0^T L(\xi(t))\,\mathrm{d}t$, we consider a discrete Lagrangian $L_d\colon G\to\mathbb{R}$ and we seek to minimize $\sum_{k=1}^N L_d(g_k)$ over all sequences of points $\{g_1,\ldots,g_N\}$ on G such that $g_1\star g_2\star\ldots\star g_N = g$, where g is a fixed point in G. The discrete variational principle on Lie groups considered here is not on the usual Lie group as in [5, 13], but on the Lie group $(G,\star)$. The admissible curves considered in the classical variational principles are replaced by sequence of points in the discrete variational principle. To make things easier let us consider sequences with only two points, g_1 and g_2, but everything can be extended to longer sequences. In order to apply a discrete variational principle it is necessary to define the admissible perturbations, that is, those perturbations fulfilling the end-point conditions. All this must be done in terms of the $\star$-product. We consider curves $h\colon\mathbb{R}\to G$ satisfying:

1. $h(0) = \mathrm{e}$, the identity element in G;
2. $\dot{h}(0) = \delta h \in \mathfrak{g}$;
3. $g_1 \star h(t) \star h(t)^{-1_\star} \star g_2 = g$,

where δh is the initial velocity for the curve in G, $g^{-1_\star}$ denotes the inverse with respect to the $\star$ product. We want to extremize the functional $L_d(g_1) + L_d(g_2)$ over all admissible perturbations. A necessary condition for minimization comes from taking the first derivative of $L_d(g_1 \star h(t)) + L_d(h(t)^{-1_\star} \star g_2)$ with respect to time because the discrete Lagrangian is evaluated in a sequence of two points that define an admissible perturbation according to the property 3 above.

$$
\begin{aligned}
0 = \left.\frac{\mathrm{d}}{\mathrm{d}t}\right|_{t=0} \left(L_d(g_1 \star h(t)) + L_d(h(t)^{-1_\star} \star g_2)\right) &= \left.\frac{\mathrm{d}}{\mathrm{d}t}\right|_{t=0} \left(L_d(g_1^+ h(t) g_1^-)\right) \\
&+ \left.\frac{\mathrm{d}}{\mathrm{d}t}\right|_{t=0} \left(L_d((h(t)^{-1_\star})^+ g_2 (h(t)^{-1_\star})^-)\right) \\
&= \langle (l_{g_1^+} \circ r_{g_1^-})^* \mathrm{d}L_d(g_1), \dot{h}(0)\rangle - \langle (r_{g_2})^* \mathrm{d}L_d(g_2), \dot{h}^+(0)\rangle \\
&- \langle (l_{g_2})^* \mathrm{d}L_d(g_2), \dot{h}^-(0)\rangle\,,
\end{aligned}
$$

where $\dot{h}^-(0) = \pi_-(\dot{h}(0))$, $\dot{h}^+(0) = \pi_+(\dot{h}(0))$. Thus

$$
(l_{g_1^+} \circ r_{g_1^-})^* \mathrm{d}L_d(g_1) - \pi_+^*(r_{g_2})^* \mathrm{d}L_d(g_2) - \pi_-^*(l_{g_2})^* \mathrm{d}L_d(g_2) = 0\,. \tag{6}
$$

If the Lagrangian is invariant under conjugation, we will be able to rewrite the above equation in a different way. The invariance property means that

$$
L_d(\mathrm{I}_{h(t)}(g)) = L_d(h(t) g h(t)^{-1}) = L_d(g)
$$

for all $h\colon \mathbb{R} \to G$ such that $h(0) = \mathrm{e}$. Taking derivative with respect to time in this equality we obtain

$$
(l_g)^* \mathrm{d}L_d(g) = (r_g)^* \mathrm{d}L_d(g), \quad \text{equivalently} \quad \mathrm{Ad}_g^* \mathrm{d}L_d(g) = \mathrm{d}L_d(g).
$$

Using these equalities plus the fact that $\pi_- = \mathrm{Id} - \pi_+$ where $\pi_\pm\colon \mathfrak{g} \to \mathfrak{g}_\pm$ are the projections, Eq. (6) becomes

$$
(l_{g_1^+} \circ r_{g_1^-})^* \mathrm{d}L_d(g_1) - r_{g_2}^* \mathrm{d}L_d(g_2) = 0. \tag{7}
$$

Alternatively, using the fact that $\pi_+ = \mathrm{Id} - \pi_-$, Eq. (6) becomes

$$
(l_{g_1^+} \circ r_{g_1^-})^* \mathrm{d}L_d(g_1) - l_{g_2}^* \mathrm{d}L_d(g_2) = 0. \tag{8}
$$

Take $\mu_1 = (r_{g_1})^* \mathrm{d}L_d(g_1)$ and $\mu_2 = (r_{g_2})^* \mathrm{d}L_d(g_2)$ so that Eq. (7) can be rewritten as follows

$$\begin{aligned}0 =&(l_{g_1^+} \circ r_{g_1^-})^* \mathrm{d}L_d(g_1) - \mu_2 = (l_{g_1^+} \circ r_{g_1^-})^* (r_{g_1^{-1}})^* (r_{g_1})^* \mathrm{d}L_d(g_1) - \mu_2 \\ =&(r_{g_1^{-1}} \circ l_{g_1^+} \circ r_{g_1^-})^* \mu_1 - \mu_2 = \mathrm{Ad}^*_{g_1^+} \mu_1 - \mu_2 \,,\end{aligned}$$

because for ξ in $\mathfrak{g}$ we have $\mathrm{T}_e(r_{g_1^{-1}} \circ l_{g_1^+} \circ r_{g_1^-})\xi = (\mathrm{T}_{g_1^+ g_1^-} r_{g_1^{-1}} \circ \mathrm{T}_{g_1^-} l_{g_1^+} \circ \mathrm{T}_e r_{g_1^-})\xi = g_1^+ \xi g_1^- (g_1^+ g_1^-)^{-1} = g_1^+ \xi g_1^- (g_1^-)^{-1} (g_1^+)^{-1} = g_1^+ \xi (g_1^+)^{-1} = \mathrm{Ad}_{g_1^+} \xi$. As a result, we have obtained the following scheme to solve the integrable Lie-Poisson Eq. (3):

$$\mu_1 = (r_{g_1})^* \mathrm{d}L_d(g_1), \tag{9}$$

$$\mu_2 = \mathrm{Ad}^*_{g_1^+} \mu_1 \,. \tag{10}$$

The method acts as follows: μ_1 is the known information. Once the initial condition $\mu_1 \in \mathfrak{g}^*$ is given, the implicit Eq. (9) must be solved to obtain g_1 in G. Then, we obtain g_1^+ from g_1 by factorization and (10) gives us μ_2 in $\mathfrak{g}^*$. We start the process again. This is a Lie-Poisson method for $[\cdot, \cdot]_R$ and it is the main contribution of this paper.

Similarly, using (8) instead of (7) we obtain another characterization of the Lie-Poisson method for $[\cdot, \cdot]_R$:

$$\mu_1 = (l_{g_1})^* \mathrm{d}L_d(g_1), \tag{11}$$

$$\mu_2 = \mathrm{Ad}^*_{(g_1^-)^{-1}} \mu_1 \,. \tag{12}$$

5 Bäcklund Transformations and Lie-Poisson Methods

Let us establish the connection between the numerical methods (9)–(10) on $\mathfrak{g}^*$, and (11)–(12), and the Bäcklund transformations in Suris' book [26] associated with numerical methods on $\mathfrak{g}$. First, we prove some general results that can be applied to the Lie-Poisson methods in Sect. 4.

Proposition 1 *Let $\tilde{F}: G \to \mathfrak{g}^*$ and $F: \mathfrak{g}^* \to G$ be local diffeomorphisms such that $\tilde{F} \circ F = \mathrm{Id}_{\mathfrak{g}^*} = F \circ \tilde{F}$. The equivariance of $\tilde{F}$, that is, $\tilde{F}(\mathrm{I}_{h^{-1}}(g)) = \mathrm{Ad}^*_h \tilde{F}(g)$ for all g, h in G is equivalent to the equivariance of F, that is,*

$$F(\mathrm{Ad}^*_{h^{-1}} \mu) = \mathrm{I}_h F(\mu) \,. \tag{13}$$

Proof For every μ in $\mathfrak{g}^*$ it is satisfied that $(\tilde{F} \circ F)(\mu) = \mu$. Denote $F(\mu)$ by g in G. Thus $\tilde{F}(g) = \mu$ and

$$F(\mathrm{Ad}^*_{h^{-1}} \mu) = F(\mathrm{Ad}^*_{h^{-1}} \tilde{F}(g)) = F(\tilde{F}(\mathrm{I}_h(g))) = \mathrm{I}_h(g) = \mathrm{I}_h F(\mu),$$

because of the equivariance of $\tilde{F}$. Analogous reasoning applies to the inverse result. □

The equivariance property is enough to prove the following commutativity property:

Proposition 2 *Let* $\tilde{F}_1, \tilde{F}_2 \colon G \to \mathfrak{g}^*$ *be equivariant with equivariant inverse functions* $F_1, F_2 \colon \mathfrak{g}^* \to G$. *It holds*

$$(F_1 F_2)(\mu) = (F_2 F_1)(\mu) \quad \forall\, \mu \in \mathfrak{g}^* \,.$$

Proof The property of commutativity for F_1 and F_2 is equivalent to

$$F_1(\mu) = F_2(\mu) F_1(\mu)(F_2(\mu))^{-1} = \mathrm{I}_{F_2(\mu)} F_1(\mu) = F_1(\mathrm{Ad}^*_{(F_2(\mu))^{-1}}\mu),$$

because of (13). Now, we apply $\tilde{F}_1$ in the equality $F_1(\mu) = F_1(\mathrm{Ad}^*_{(F_2(\mu))^{-1}}\mu)$ to obtain

$$\mu = \mathrm{Ad}^*_{(F_2(\mu))^{-1}}\mu \,. \tag{14}$$

If we take $g = h$ in the property of adjoint invariance of $\tilde{F}_1$ and $\tilde{F}_2$,

$$\tilde{F}_a(\mathrm{I}_h(g)) = \tilde{F}_a(hgh^{-1}) = \mathrm{Ad}^*_{h^{-1}} \tilde{F}_a(g) \quad \forall\, h, g \in G, \quad \text{for } a = 1, 2,$$

then

$$\tilde{F}_a(h) = \mathrm{Ad}^*_{h^{-1}} \tilde{F}_a(h) \quad \forall\, h \in G, \quad \text{for } a = 1, 2. \tag{15}$$

In the equation for $a = 2$, we take $\mu = \tilde{F}_2(h)$, so $F_2(\mu) = h$. This gives us the Eq. (14) and the commutativity of F_1 and F_2 is proved.

Analogous proof can be written if $F_2(\mu)$ is isolated instead of $F_1(\mu)$ in the first step and we use

$$\mathrm{Ad}^*_h \tilde{F}_a(h) = \tilde{F}_a(h) \quad \forall\, h \in G, \quad \text{for } a = 1, 2,$$

instead of (15). □

The theoretical results we have just proved can be applied to the numerical methods (9)–(10) and (11)–(12). It is only necessary to define an equivariant function $\tilde{F}$ associated with (9) and find *an* inverse function. Note that we have used *an* instead of *the* because the inverse function is not necessarily unique. This is connected with the fact that even if the factorization of the Lie group exists, that is not necessarily unique [19]. We define $\tilde{F}(g) = r_g^* \mathrm{d}L_d(g)$ so that the method can be rewritten as

$$\mu_2 = \mathrm{Ad}^*_{g_1^+} \tilde{F}(g_1) \,.$$

Let us prove the equivariance of the chosen function in the following result.

Proposition 3 *Assume that the discrete Lagrangian is adjoint invariant, that is,* $L_d(\mathrm{I}_h(g)) = L_d(g)$ *for all* h, g *in* G. *Let* $\tilde{F} \colon G \to \mathfrak{g}^*$ *be defined by* $\tilde{F}(g)$

$= r_g^* \mathrm{d}L_d(g)$. *Then the function* $\tilde{F}$ *is equivariant, that is,* $\tilde{F}(\mathrm{I}_h(g)) = \tilde{F}(hgh^{-1}) = \mathrm{Ad}^*_{h^{-1}} \tilde{F}(g)$ *for all* g, h *in* G.

Proof Take ξ in $\mathfrak{g}$. Using the properties of the different elements that appear,

$$
\begin{aligned}
\langle \mathrm{Ad}^*_{h^{-1}} \tilde{F}(g), \xi \rangle &= \langle \tilde{F}(g), \mathrm{Ad}_{h^{-1}} \xi \rangle = \langle \tilde{F}(g), h^{-1}\xi h \rangle \\
&= \langle r_g^* \mathrm{d}L_d(g), h^{-1}\xi h \rangle = \langle \mathrm{d}L_d(g), h^{-1}\xi h g \rangle = \langle l^*_{h^{-1}} \mathrm{d}L_d(g), \xi h g \rangle \\
&= \langle l^*_{h^{-1}} l^*_h r^*_{h^{-1}} \mathrm{d}L_d(hgh^{-1}), \xi h g \rangle = \langle r^*_{h^{-1}} \mathrm{d}L_d(hgh^{-1}), \xi h g \rangle \\
&= \langle \mathrm{d}L_d(hgh^{-1}), \xi hgh^{-1} \rangle = \langle r^*_{hgh^{-1}} \mathrm{d}L_d(hgh^{-1}), \xi \rangle \\
&= \langle \tilde{F}(hgh^{-1}), \xi \rangle ,
\end{aligned}
$$

the result follows. □

To define a Bäcklund transformation on the dual of the Lie algebra $\mathfrak{g}^*$ we need an inverse function of the particular $\tilde{F}$ associated with (9), which is not necessarily unique as mentioned earlier. Analogously, an equivariant function $\tilde{F}$ can be associated with (11) to obtain a corresponding Bäcklund transformation on $\mathfrak{g}^*$.

For each inverse function F we define the corresponding *Bäcklund transformation* as follows:

$$
\begin{aligned}
BT_F \colon \mathfrak{g}^* &\to \mathfrak{g}^* \\
\mu &\mapsto \mathrm{Ad}^*_{\Pi_+(F(\mu))} \mu
\end{aligned}
\tag{16}
$$

where $\Pi_\pm \colon G \to G_\pm$ are the factorizations at the level of the Lie group. This definition agrees with the idea that such transformations map solutions to solutions because Eq. (16) defines a sequence solution to the discrete variational principle applied to the integrable Lie-Poisson Eq. (3) in Sect. 4. Propositions 2 and 3 guarantee that the Bäcklund transformations defined as in (16) commute.

Corollary 1 *Let* $F_1, F_2 \colon \mathfrak{g}^* \to G$ *be equivariant function, that is, satisfying* (13), *we have*

$$BT_{F_2} \circ BT_{F_1} = BT_{F_2 F_1} = BT_{F_1 F_2} .$$

Proof Have in mind that $F_1 F_2$ in $BT_{F_1 F_2}$ stands for the product in the group. To shorten the expressions we use the notation $\mu_1 = \mathrm{Ad}^*_{\Pi_+(F_1(\mu))}\mu$. By Proposition 1,

$$F_2(\mu_1) = F_2\left(\mathrm{Ad}^*_{\Pi_+(F_1(\mu))}\mu\right) = \mathrm{Ad}_{(\Pi_+(F_1(\mu)))^{-1}} F_2(\mu) .$$

It can be proved using the operations of $(G, \star)$ that $\Pi_+(g_1 \star g_2) = \Pi_+(g_1)\Pi_+(g_2)$ because $(G_+, \cdot)$ is also a Lie group.

Take $\xi \in \mathfrak{g}$. We first use that $\Pi_+(g_1)\Pi_+(g_2) = \Pi_+(g_1 \star g_2)$, then the definition of the $\star$-product.

$$
\begin{aligned}
\langle BT_{F_2} \circ BT_{F_1}(\mu), \xi \rangle &= \left\langle BT_{F_2}\left(\mathrm{Ad}^*_{\Pi_+(F_1(\mu))}\mu\right), \xi \right\rangle = \langle BT_{F_2}(\mu_1), \xi \rangle \\
&= \langle \mathrm{Ad}^*_{\Pi_+(F_2(\mu_1))}\mu_1, \xi \rangle = \langle \mu_1, \mathrm{Ad}_{\Pi_+(F_2(\mu_1))}\xi \rangle \\
&= \langle \mathrm{Ad}^*_{\Pi_+(F_1(\mu))}\mu, \mathrm{Ad}_{\Pi_+(F_2(\mu_1))}\xi \rangle = \langle \mu, \mathrm{Ad}_{\Pi_+(F_1(\mu))}\mathrm{Ad}_{\Pi_+(F_2(\mu_1))}\xi \rangle \\
&= \langle \mu, \mathrm{Ad}_{\Pi_+(F_1(\mu))\Pi_+(F_2(\mu_1))}\xi \rangle = \langle \mu, \mathrm{Ad}_{\Pi_+(F_1(\mu))\Pi_+(\mathrm{I}_{(\Pi_+(F_1(\mu)))^{-1}}(F_2(\mu)))}\xi \rangle \\
&= \langle \mu, \mathrm{Ad}_{\Pi_+(F_1(\mu))\Pi_+((\Pi_+(F_1(\mu)))^{-1}F_2(\mu)\Pi_+(F_1(\mu)))}\xi \rangle \\
&= \langle \mu, \mathrm{Ad}_{\Pi_+(F_1(\mu)\star((\Pi_+(F_1(\mu)))^{-1}F_2(\mu)\Pi_+(F_1(\mu))}\xi \rangle \\
&= \langle \mu, \mathrm{Ad}_{\Pi_+(\Pi_+(F_1(\mu))(\Pi_+(F_1(\mu)))^{-1}F_2(\mu)\Pi_+(F_1(\mu))\Pi_-(F_1(\mu)))}\xi \rangle \\
&= \langle \mu, \mathrm{Ad}_{\Pi_+(F_2(\mu)F_1(\mu))}\xi \rangle = \langle \mathrm{Ad}^*_{\Pi_+(F_2(\mu)F_1(\mu))}\mu, \xi \rangle \\
&= \langle BT_{F_2F_1}(\mu), \xi \rangle .
\end{aligned}
$$

□

In order to build the bridge between the Bäcklund transformations on $\mathfrak{g}^*$ defined here and the Bäcklund transformations on $\mathfrak{g}$ in Suris' book [26], we need to introduce a bi-invariant scalar product $(\cdot, \cdot)$ on $\mathfrak{g}$ so that there is an identification between $\mathfrak{g}$ and $\mathfrak{g}^*$. The invariance of the scalar product implies that

$$(\xi, \mathrm{Ad}_{g^{-1}}\eta) = (\mathrm{Ad}_g\xi, \eta) \quad \forall\, \xi, \eta \in \mathfrak{g}, \quad \forall\, g \in G, \tag{17}$$

$$(\xi, \mathrm{ad}_\zeta \eta) + (\mathrm{ad}_\zeta \xi, \eta) = 0 \quad \forall\, \xi, \eta, \zeta \in \mathfrak{g}. \tag{18}$$

The identification between $\mathfrak{g}$ and $\mathfrak{g}^*$ given by the scalar product is defined as follows for every ξ in $\mathfrak{g}$,

$$\langle \mu_\xi, \eta \rangle = (\xi, \eta) \quad \forall\, \eta \in \mathfrak{g}. \tag{19}$$

Remember that

$$\langle \mathrm{ad}^*_\eta \mu, \xi \rangle = \langle \mu, \mathrm{ad}_\eta \xi \rangle \quad \forall\, \eta, \xi \in \mathfrak{g}, \ \mu \in \mathfrak{g}^*. \tag{20}$$

Proposition 4 *Let $(\cdot, \cdot)$ be an invariant scalar product on $\mathfrak{g}$, the following diagram is commutative:*

$$
\begin{array}{ccc}
\mathfrak{g}^* & \xrightarrow{\mathrm{Ad}^*_{g^{-1}}} & \mathfrak{g}^* \\
\uparrow & & \uparrow \\
\mathfrak{g} & \xrightarrow{\mathrm{Ad}_g} & \mathfrak{g}
\end{array}
$$

Equivalently, $\mathrm{Ad}^*_{g^{-1}}(\mu_\xi) = \mu_{\mathrm{Ad}_g\xi}$ *for every* ξ *in* $\mathfrak{g}$.

Proof Take η in $\mathfrak{g}$.

$$
\begin{aligned}
\langle \mathrm{Ad}^*_{g^{-1}}(\mu_\xi), \eta \rangle &= \langle \mu_\xi, \mathrm{Ad}_{g^{-1}}(\eta) \rangle = (\xi, \mathrm{Ad}_{g^{-1}}(\eta)) \\
&= (\mathrm{Ad}_g(\xi), \eta) = \langle \mu_{\mathrm{Ad}_g(\xi)}, \eta \rangle .
\end{aligned}
$$

This proves the commutativity of the diagram. □

Proposition 4 establishes the relation between the Bäcklund transformations on $\mathfrak{g}^*$ described here and the ones on $\mathfrak{g}$ in Suris's book [26].

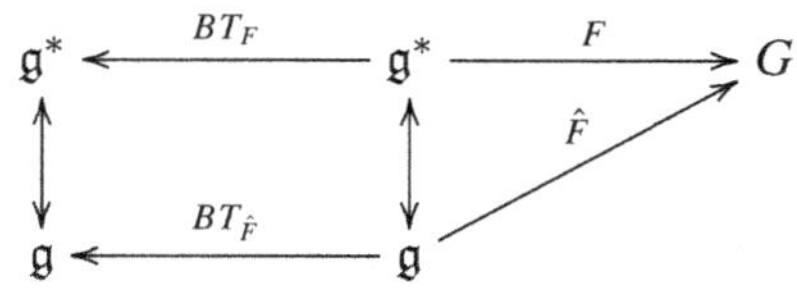

Given the Bäcklund transformation BT_F on $\mathfrak{g}^*$, we define the Bäcklund transformation $BT_{\hat{F}}$ on $\mathfrak{g}$ for a function $\hat{F}\colon \mathfrak{g} \to G$ as follows:

$$BT_{\hat{F}}(\xi) = \eta \text{ such that} \mu_\eta = BT_F(\mu_\xi) .$$

Take ζ in $\mathfrak{g}$ to obtain the precise expression of $BT_{\hat{F}}$ when $BT_F(\mu_\xi) = \mathrm{Ad}^*_{\Pi_+(F(\mu_\xi))}\mu_\xi$:

$$\begin{aligned}(\eta, \zeta) &= \langle \mu_\eta, \zeta \rangle = \langle BT_F(\mu_\xi), \zeta \rangle = \langle \mathrm{Ad}^*_{\Pi_+(F(\mu_\xi))}\mu_\xi, \zeta \rangle \\ &= \langle \mu_\xi, \mathrm{Ad}_{\Pi_+(F(\mu_\xi))}\zeta \rangle = (\xi, \mathrm{Ad}_{\Pi_+(F(\mu_\xi))}\zeta) = (\mathrm{Ad}_{(\Pi_+(F(\mu_\xi)))^{-1}}\xi, \zeta) .\end{aligned}$$

Thus $\eta = \mathrm{Ad}_{(\Pi_+(F(\mu_\xi)))^{-1}}\xi$ and the Bäcklund transformation on $\mathfrak{g}$ is given by

$$BT_{\hat{F}}(\xi) = \mathrm{Ad}_{(\Pi_+(F(\mu_\xi)))^{-1}}\xi \tag{21}$$

because of Proposition 4. Note that $F(\mu_\xi) = \hat{F}(\xi)$. As expected, all properties proved for the Bäcklund transformations BT_F on $\mathfrak{g}^*$ in Proposition 2 and Corollary 1 can also be proved by inheritance from the corresponding Bäcklund transformations $BT_{\hat{F}}$ on $\mathfrak{g}$ because of the identification between $\mathfrak{g}^*$ and $\mathfrak{g}$ given by the bi-invariant scalar product.

6 Example

We apply the formalism developed in the above sections to discretize the completely integrable system called Toda lattice, in particular, the finite nonperiodic one as described in [27]. That system is Hamiltonian and represents the dynamics of n particles of unit mass, constrained to move on the line with positions $Q_i(t)$ and momenta $P_i(t)$ under the influence of exponential repulsive forces. Moser showed in [18] that the Toda lattice is a completely integrable system, that is, there exist smooth functions $H_1, \ldots, H_n$ defined on the phase space, which are functionally independent and in involution, that is, $\{H_i, H_j\} = 0$ for all $i, j = 1, \ldots, n$. In particular, $H_1, \ldots, H_n$ are constants of motion for the Toda lattice [17]. In [4] it is studied the integrability of lattice systems that admit a R-matrix approach.

The Lie group G for the Toda lattice consists of the $n \times n$ invertible matrices denoted by $GL(n)$. The corresponding Lie algebra $\mathfrak{gl}(n)$ admits different splittings, in particular, we consider $\mathfrak{gl}(n) = \mathfrak{g}_+ \oplus \mathfrak{g}_-$ where $\mathfrak{g}_+$ is the set of lower matrices and $\mathfrak{g}_-$ is the set of skew symmetric matrices.

As described at the end of Sect. 3, we take $R = \frac{1}{2}(\pi_+ - \pi_-)$, where $\pi_+ : \mathfrak{g} \to \mathfrak{g}_+$ and $\pi_- : \mathfrak{g} \to \mathfrak{g}_-$ are the projections onto the corresponding factors given by $\pi_+(L) = (L_u)^T + L_d + L_l$ and $\pi_-(L) = L_u - (L_u)^T$. The notation L_u, L_d and L_l stands for the strictly upper triangular part, the diagonal and the strictly lower triangular part of L, respectively. The transpose of a matrix L is denoted by L^T. It is clear that $\pi_+(L)$ is a lower triangular matrix, $\pi_-(L)$ is a skew symmetric matrix and $\pi_+(L) + \pi_-(L) = L$. Equivalently, the linear map $R = \dfrac{1}{2}(\pi_+ - \pi_-)$ can be written as $R = \pi_+ - \frac{\mathrm{Id}}{2} = \frac{\mathrm{Id}}{2} - \pi_-$.

The identification between ξ in $\mathfrak{g}$ and α_ξ in $\mathfrak{g}^*$ given by the scalar product is defined as follows

$$\langle \alpha_\xi, \eta \rangle = (\xi, \eta) \quad \forall\, \eta \in \mathfrak{g}. \tag{22}$$

Now, we associate the differential equation $\dfrac{\mathrm{d}\alpha_\xi}{\mathrm{d}t} = (ad^*)_{\pi_+ \mathrm{d}H(\alpha)}\, \alpha_\xi = -\,(ad^*)_{\pi_- \mathrm{d}H(\alpha)}$ α_ξ on $\mathfrak{g}^*$ with one on $\mathfrak{g}$. On the one hand, take η in $\mathfrak{g}$

$$\begin{aligned}\left\langle \frac{\mathrm{d}\alpha_\xi}{\mathrm{d}t}, \eta \right\rangle &= \langle (ad^*)_{\pi_+ \mathrm{d}H(\alpha)}\, \alpha_\xi, \eta \rangle = \langle \alpha_\xi, (ad)_{\pi_+ \mathrm{d}H(\alpha)}\, \eta \rangle \\ &= \left(\xi, (ad)_{\pi_+ \mathrm{d}H(\alpha)}\, \eta\right) = -\left((ad)_{\pi_+ \mathrm{d}H(\alpha)}\, \xi, \eta\right),\end{aligned}$$

using the invariance properties (18) and (20) of the scalar product. On the other hand,

$$\frac{\mathrm{d}}{\mathrm{d}t}\langle \alpha_\xi, \eta \rangle = \frac{\mathrm{d}}{\mathrm{d}t}(\xi, \eta) = \left(\frac{\mathrm{d}}{\mathrm{d}t}\xi, \eta\right).$$

Thus,

$$\frac{\mathrm{d}}{\mathrm{d}t}\xi = -[\pi_+ \mathrm{d}H(\alpha), \xi] = [\pi_- \mathrm{d}H(\alpha), \xi].$$

The identification between $\mathfrak{g}$ and $\mathfrak{g}^*$ in (22) in the example under consideration is given by the trace of the product of matrices, that is,

$$\langle \mu_\xi, \eta \rangle = \mathrm{tr}(\xi\, \eta)\,.$$

The Hamiltonian for the Toda lattice is $H(\alpha_\xi) = \dfrac{1}{2}\mathrm{tr}(\xi\, \xi^T)$ so that $\langle \mathrm{d}H(\alpha_\xi), \eta \rangle =$ $\mathrm{tr}(\xi\, \eta)$ and $\mathrm{d}H(\alpha_\xi)$ is identified with ξ. The Toda lattice equations can be rewritten as the following matrix equation:

$$\frac{\mathrm{d}\xi}{\mathrm{d}t} = [\pi_-(\xi), \xi] = [(\xi)_u - (\xi_u)^T, \xi] = [\xi, (\xi_u)^T - \xi_u] \tag{23}$$

as appears in [8, 27]. We have just shown that the system admits a Lax pair representation coming from a R-matrix approach satisfying the modified Yang-Baxter equation and we can conclude it is integrable. It can also be proved that the flow of such a system admits a unique QR factorization [8]: $e^{tL_0} = Q(t)R(t)$, where $Q(t)$ is an orthogonal matrix and $R(t)$ is an upper triangular matrix. The discrete flow becomes $e^{L(1)} = R(1)Q(1)$.

To apply the discrete variational principle described in Sect. 4 we consider the following discrete Lagrangian for the Toda lattice defined on the Lie group of the set of invertible square matrices:

$$\begin{aligned} L_d\colon GL(n) &\longrightarrow \mathbb{R} \\ g &\longrightarrow \frac{1}{h}\,\mathrm{tr}(g)\,, \end{aligned}$$

where $\mathrm{tr}(g)$ denotes the trace of g. For η in $\mathfrak{g}$

$$\langle \mathrm{d}L_d(g), \eta\rangle = \frac{1}{h}\mathrm{tr}(\eta).$$

Thus the element $\mathrm{d}L_d(g)$ in the dual Lie algebra $\mathfrak{gl}(n)^*$ is identified with $\dfrac{1}{h}$ Id in $\mathfrak{gl}(n)$, where Id is the identity matrix.

To write the Lie-Poisson method described in (9)–(10) we must first compute

$$\langle (r_{g_1})^*\mathrm{d}L_d(g_1), \eta\rangle = \langle \mathrm{d}L_d(g_1), \mathrm{T}_e r_{g_1}\eta\rangle = \frac{1}{h}\,\mathrm{tr}(\eta g_1) = \mathrm{tr}\left(\frac{1}{h}\, g_1\eta\right).$$

As $\langle \mu_\xi, \eta\rangle = \mathrm{tr}(\xi\,\eta)$, we have $\xi = \dfrac{1}{h}g_1$ and $\mu_1 = \mu_{g_1/h}$. Now, using the cyclic properties of the trace, we have

$$\begin{aligned} \langle \mu_2, \eta\rangle &= \langle \mathrm{Ad}^*_{g_1^+}\mu_1, \eta\rangle = \langle \mu_1, \mathrm{Ad}_{g_1^+}\eta\rangle = \langle \mu_{g_1/h}, \mathrm{Ad}_{g_1^+}\eta\rangle \\ &= \mathrm{tr}\left(\frac{1}{h}\, g_1 g_1^+\eta (g_1^+)^{-1}\right) = \frac{1}{h}\,\mathrm{tr}\left(g_1^+ g_1^- g_1^+\eta(g_1^+)^{-1}\right) \\ &= = \frac{1}{h}\,\mathrm{tr}\left(g_1^- g_1^+\eta(g_1^+)^{-1} g_1^+\right) = \frac{1}{h}\,\mathrm{tr}\left(g_1^- g_1^+\eta\right) = \left(\frac{1}{h}\, g_1^- g_1^+, \eta\right). \end{aligned}$$

Hence, the Lie-Poisson method is given by

$$\begin{aligned} \mu_1 &= \mu_{g_1/h} \\ \mu_2 &= \mu_{g_1^- g_1^+/h}\,. \end{aligned}$$

This method agrees with the ones known in the literature, see for instance [27], because at the level of the Lie group starting from g_1 the next step is $g_1^- g_1^+$.

To recover the Bäcklund transformations in [26] in the case of matrix Lie groups we must consider the following discrete Lagrangian:

$$\begin{aligned} L_d\colon GL(n) &\longrightarrow \mathbb{R} \\ g &\longrightarrow \frac{1}{h}\,\mathrm{tr}(g - \log(g^T))\,, \end{aligned}$$

where $\log\colon G \to \mathfrak{g}$ is the inverse map of the exponential map $\exp\colon \mathfrak{g} \to G$. Using the properties of matrices and Taylor series evaluated at matrices we have

$$\langle \mathrm{d}L_d(g), \eta\rangle = \mathrm{tr}\left(\frac{1}{h}\left(\mathrm{Id} - g^{-1}\right)\eta\right).$$

Thus $\mathrm{d}L_d(g)$ in $\mathfrak{gl}(n)^*$ can be identified with $\frac{1}{h}\left(\mathrm{Id} - g^{-1}\right)$ in $\mathfrak{gl}(n)$. The starting point in $\mathfrak{g}^*$ is given by

$$\begin{aligned} \langle \mu_1, \eta\rangle &= \langle (r_{g_1})^*\mathrm{d}L_d(g_1), \eta\rangle = \langle \mathrm{d}L_d(g_1), (\mathrm{T}r_{g_1})\eta\rangle \\ &= \mathrm{tr}\left(\frac{1}{h}\left(\mathrm{Id} - g_1^{-1}\right)\eta g_1\right) = \mathrm{tr}\left(\frac{1}{h}\left(g_1 - \mathrm{Id}\right)\eta\right). \end{aligned}$$

On the other hand, by the binvariant inner product there exists ξ_1 in $\mathfrak{g}$ such that $\langle \mu_{\xi_1}, \eta\rangle = \mathrm{tr}(\xi_1\eta) = \langle \mu_1, \eta\rangle$ for all η in $\mathfrak{g}$. Hence,

$$\xi_1 = \frac{1}{h}\left(g_1 - \mathrm{Id}\right), \text{ equivalently, } g_1 = \mathrm{Id} + h\xi_1\,,$$

as appears in Suris' book [26] as a choice for the arbitrary conjugation covariant function $\widehat{F}\colon \mathfrak{g} \to G$, $\widehat{F}(\xi) = \mathrm{Id} + h\xi$.

Similarly, the next step in the Lie-Poisson method is given by

$$\begin{aligned} \langle \mu_2, \eta\rangle &= \langle \mathrm{Ad}^*_{g_1^+}\mu_1, \eta\rangle = \langle \mu_1, \mathrm{Ad}_{g_1^+}\eta\rangle = \langle \mu_{(g_1-\mathrm{Id})/h}, \mathrm{Ad}_{g_1^+}\eta\rangle \\ &= \mathrm{tr}\left(\frac{1}{h}\left(g_1 - \mathrm{Id}\right)g_1^+\eta(g_1^+)^{-1}\right) = \frac{1}{h}\,\mathrm{tr}\left((g_1^-g_1^+ - \mathrm{Id})\eta\right) = \left(\frac{1}{h}\left(g_1^-g_1^+ - \mathrm{Id}\right), \eta\right). \end{aligned}$$

Thus,

$$\begin{aligned} \mu_1 &= \mu_{(g_1-\mathrm{Id})/h} \\ \mu_2 &= \mu_{(g_1^-g_1^+-\mathrm{Id})/h}\,. \end{aligned}$$

We show now that the Bäcklund transformation $BT_{\widehat{F}}\colon \mathfrak{g} \to \mathfrak{g}$,

$$BT_{\widehat{F}}(\xi) = \Pi_+^{-1}(\widehat{F}(\xi))\xi\Pi_+(\widehat{F}(\xi)) = \Pi_-(\widehat{F}(\xi))\xi\Pi_-^{-1}(\widehat{F}(\xi))$$

for an arbitrary conjugation covariant function $\widehat{F}: \mathfrak{g} \to G$ given in [26] agrees with the Lie algebra element identified with μ_2. The choice of $\widehat{F}(\xi)$ is *a transcedent problem*, according to Suris, which happens to be solved by taking $\widehat{F}(\xi) = \mathrm{Id} + hf(\xi)$ in most of the known integrable cases, as long as the expression makes sense and lies in the Lie group G.

Here we take

$$g_1 = \widehat{F}(\xi_1) = h\xi_1 + \mathrm{Id}, \tag{24}$$

so that

$$\begin{aligned} BT_{\widehat{F}}(\xi_1) &= (h\xi_1 + \mathrm{Id})_+^{-1}\xi_1(h\xi_1 + \mathrm{Id})_+ = (g_1^+)^{-1}\xi_1 g_1^+ = \frac{1}{h}(g_1^+)^{-1}(g_1 - \mathrm{Id})g_1^+ \\ &= \frac{1}{h}((g_1^+)^{-1}g_1 g_1^+ - \mathrm{Id}) = \frac{1}{h}(g_1^- g_1^+ - \mathrm{Id}) \end{aligned}$$

Effectively, $BT_{\widehat{F}}(\xi_1) = \xi_2 = \frac{1}{h}(g_1^- g_1^+ - \mathrm{Id})$ that nicely closes the example.

It is important to highlight that our discrete variational approach replaces the "simplest possible choice" of (24) in [26] by the choice of a discrete Lagrangian as shown in the example.

Acknowledgements This work has been supported by a grant (010-ABEL-CM-2014A) from Iceland, Liechtenstein and Norway through the EEA Financial Mechanism, operated by Universidad Complutense de Madrid, by the ICMAT Severo Ochoa project SEV-2011-0087, MTM2013-42870-P and MTM2016-76072-P. MBL developed this research mainly at Universidad Carlos III de Madrid as Assistant Professor.

References

1. Adler, M.: On a trace functional for formal pseudodifferential operators and symplectic structure of the korteweg-de vries type equations. Invent. Math. **50** (1979)
2. Austin, M., Krishnaprasad, P.S., Wang, L.-S.: Almost poisson integration of rigid body systems. J. Comput. Phys. **107**, 105–117 (1993)
3. Bäcklund, A.V.: Einiges über curven und flächen transformations. Lunds Univ. Arsskr. **10**, 1–12 (1874)
4. Blaszak, M., Marciniak, K.: R-matrix approach to lattice integrable systems. J. Math. Phys. **35**(9) (1994)
5. Bobenko, A.I., Suris, Y.B.: Discrete time lagrangian mechanics on lie groups, with an application to the lagrange top. Comm. Math. Phys. **204**, 147–188 (1999)
6. Channell, P.J., Scovel, J.C.: Integrators for lie-poisson dynamical systems. Phys. D **50**, 80–85 (1991)
7. Darboux, G.: Sur une proposition relative aux équations linéaires. Compt. Rend. Acad. Sci., Ser. I Math., **94**, 1456–1459 (1882)
8. Deift, P.: Integrable hamiltonian systems, in dynamical systems and probabilistic methods in partial differential equations. Amer. Math. Soc. Lectures in Applied Mathematics, **31**, 103–138 (1996)
9. Ge, Z., Marsden, J.E.: Lie-poisson hamilton-jacobi theory and lie-poisson integrators. Phys. Lett. A **133**(3), 134–139 (1988)

10. Libermann, P., Marle, C.-M.: Symplectic Geometry and Analytical Mechanics. Mathematics and its Applications. Springer, Netherlands (1987)
11. Lie, S.: Theorie der Transformationsgruppen. (Zweiter Abschnitt, unter mitwirkung von Friedrich Engel), Teubner, Leipzig (1890)
12. Magri, F.: A simple model of the integrable Hamiltonian equation. J. Math. Phys. **19**, 1156 (1978)
13. Marsden, J.E., Pekarsky, S., Shkoller, S.: Discrete euler-poincaré and lie-poisson equations. Nonlinearity **12**, 1647–1662 (1999)
14. Marsden, J.E., Ratiu, T.S.: Introduction to mechanics and symmetry. In: Texts in Applied Mathematics, vol. 17. Springer-Verlag (1994)
15. Marsden, J.E., West, M.: Discrete mechanics and variational integrators. Acta Numer. **10**, 357–514 (2001)
16. Morosi, C.: The R-matrix theory and the reduction of poisson manifolds. J. Math. Phys. **33**, 941 (1992)
17. Morosi, C., Pizzocchero, L.: R-matrix theory formal Casimirs and the periodic Toda lattice. J. Math. Phys. **37**, 4484 (1996)
18. Moser, J.: Finitely many mass points on the line under the influence of an exponential potential-an integrable system. Dynamical Systems, Theory and Applications (Rencontres, Battelle Res. Inst., Seatlle, Wash., 1974). Lecture Notes in Physis, vol. 38, pp. 467–497. Springer, Berlin (1975)
19. Moser, J., Veselov, A.P.: Discrete versions of some classical integrable systems and factorization of matrix polynomials. Commun. Math. Phys. **139**, 243–271 (1991)
20. Oevel, W.: R structures Yang-Baxter equations, and related involution theorems. J. Math. Phys. **30**, 1140 (1989)
21. Okubo, S., Das, A.: The integrability condition for dynamical systems. Phys. Lett. B **209**, 311 (1988)
22. Ratiu, T.S.: Coadjoint orbits and the beginnings of a geometric representation theory, in developments and trends in infinite-dimensional lie theory. Progr. Math., 288, 417–457, Birkhäuser Boston, Inc., Boston, MA (2011)
23. Sasaki, R.: Canonical structure of bäcklund transformations. Phys. Lett. A **78**, 7–10 (1980)
24. Schouten, J.A.: Über differentialkomitanten zweier kontravarianter Grössen. Nederl. Akad. Wetensch. Proc. Ser. A. **43**, 449–452 (1940)
25. Semenov-Tyan-Shanskii, M.A.: What is a classical R-matrix? Funct. Anal. Appl. **17**(4), 259–272 (1983)
26. Suris, Y.B.: The problem of integrable discretization: Hamiltonian approach. In: Progress in mathematics, vol. 219, Birkhäuser, Basel (2003)
27. Symes, W.W.: The QR algorithm and scattering for the finite nonperiodic Toda lattice. Phys. D. **4**(2), 275–280 (1981/82)
28. Vaisman, I.: Lectures on the Geometry of Poisson Manifolds, Birkhäuser (2012)
29. Weinstein, A.: The local structure of poisson manifolds. J. Differ. Geom. **18**, 523–557 (1983)
30. Zullo, F.: Bäcklund transformations and exact time-discretizations for Gaudin and related models. PhD Thesis, Universitá degli Studi Roma Tre (2010)
31. Zullo, F.: Bäcklund transformations and hamiltonian flows. J. Phys. A Math. Theor. **46**(14) (2013)

Numerical Precession in Variational Discretizations of the Kepler Problem

Mats Vermeeren

Abstract Kepler's first law states that the orbit of a point mass with negative energy in a classical gravitational potential is an ellipse with one of its foci at the gravitational center. In numerical simulations of this system one often observes a slight precession of the ellipse around the gravitational center. Using the Lagrangian structure of modified equations and a perturbative version of Noether's theorem, we provide leading order estimates of this precession for the implicit MidPoint rule (MP) and the Störmer-Verlet method (SV). Based on those estimates we construct some new numerical integrators that perform significantly better than MP and SV on the Kepler problem.

Keywords Variational integrators · Modified equations · Kepler problem Orbital precession

MSC 2010: 65L12 · 65P10 · 70F05 · 70H33

1 Introduction

The Kepler problem models a point mass moving in a classical gravitational potential. Its Lagrangian is

$$\mathscr{L}(x, \dot{x}) = \frac{1}{2}|\dot{x}|^2 + \frac{1}{|x|},$$

where $|x|$ denotes the Euclidean norm on $\mathbb{R}^N$. The equations of motion are

M. Vermeeren (✉)
Institut für Mathematik, Technische Universität Berlin, MA 7-1 Str. des 17. Juni 136, 10623 Berlin, Germany
e-mail: vermeeren@math.tu-berlin.de

K. Ebrahimi-Fard and M. Barbero Liñán (eds.), *Discrete Mechanics, Geometric Integration and Lie–Butcher Series*, Springer Proceedings in Mathematics & Statistics 267, https://doi.org/10.1007/978-3-030-01397-4_10

$$\ddot{x} = -\frac{x}{|x|^3}. \tag{1}$$

It is well known that the orbits of the Kepler problem with negative energy are ellipses with one of their foci at the origin. Since every orbit lies in a plane, it is sufficient to study this problem in $\mathbb{R}^2$.

In this work we are interested in numerical integration of the Kepler problem. Very good integrators for this problem are already available, see for example [3] and the references therein. Our main objective here is to illustrate methods to analyze and improve numerical integrators. For the sake of clarity we start from simple methods. Accordingly, the improved methods we construct will not be competitive compared with specialized methods available in the literature.

Central in our treatment will be the *precession* or *perihelion advance* of the numerical orbits, i.e. the slow rotation of the ellipse that the solution traces. For the exact solution there is no precession, but no common numerical method integrates the Kepler problem without precession. Using the theory of modified equations, we will provide leading order estimates of the precession for the Störmer-Verlet method and the implicit midpoint rule. We will use those estimates to construct some new methods which are superior for the Kepler problem. This procedure is similar in spirit to the concept of *modifying integrators* [1].

Throughout this paper we use the Lagrangian formulation of classical mechanics. We will describe the modified equations using modified Lagrangians and use a version of Noether's theorem to analyze the perturbation. We start by mentioning a few well-known properties of the Kepler problem that will be useful later on.

Proposition 1 *The angular momentum $\mathbb{L} = x_1\dot{x}_2 - \dot{x}_1 x_2$ and the total energy $\mathbb{E} = \frac{1}{2}|\dot{x}|^2 - \frac{1}{|x|}$ are constants of motion of the Kepler problem in $\mathbb{R}^2$. Furthermore, the angular momentum satisfies*

$$\mathbb{L}^2 = |x||\dot{x}|^2 - \langle x, \dot{x}\rangle^2,$$

where the brackets $\langle\cdot,\cdot\rangle$ denote the standard scalar product on $\mathbb{R}^N$.

Proposition 2 *Let a and b denote the semimajor and semiminor axes of an orbit respectively. Then*

- *the square of the angular momentum equals $\mathbb{L}^2 = \frac{b^2}{a}$,*
- *the energy equals $\mathbb{E} = \frac{-1}{2a}$,*
- *the period equals $T = 2\pi a^{3/2}$,*
- *the eccentricity equals $e = \sqrt{1 - \frac{b^2}{a^2}}$.*

A thorough analytical study of the Kepler problem, including proofs of these properties, can be found for example in [7, Chap. 3].

2 Modified Lagrangians

To study the behavior of a numerical method it is often useful to consider the modified equation, a perturbation of the original differential equation whose solutions interpolate the discrete solutions. Generally, modified equations are found as formal power series in the step size of the method. Here we will truncate these power series after the first nontrivial term. For an introduction to this subject, see [8, Chap. IX] and the references therein.

It is well-known that the modified equation of a symplectic integrator applied to a Hamiltonian system is again Hamiltonian. This means that the modified equation of a variational integrator applied to a Lagrangian system is Lagrangian as well. We will use a Lagrangian for the modified equation as the basis of our analysis. For its construction we refer to [15].

The modified equation of a numerical integrator for the Kepler problem describes a perturbed Kepler problem. Perturbed Kepler problems are very relevant in celestial mechanics. In particular, one of the classical tests of general relativity is that its perturbation in the Kepler potential accounts for the precession of the orbit of the planet Mercury [16] (along with perturbations caused by the gravitational pull of the other planets). A Hamiltonian treatment of perturbed Kepler problems can be found for example in [7] or [3].

2.1 *Störmer-Verlet Method*

The Störmer-Verlet (SV) discretization with step size h of a second order differential equation $\ddot{x} = f(x)$ is

$$x_{k+1} - 2x_k + x_{k-1} = h^2 f(x_k).$$

If $f(x) = -\frac{\mathrm{d}}{\mathrm{d}x} U(x)$, this is the discrete Euler-Lagrange equation for

$$L_{SV}(x_k, x_{k+1}) = \frac{1}{2}\left|\frac{x_{k+1} - x_k}{h}\right|^2 - \frac{1}{2}U(x_k) - \frac{1}{2}U(x_{k+1}).$$

As shown in [15], the modified Lagrangian of second order accuracy is

$$\mathscr{L}_{\mathrm{mod},2}(x, \dot{x}) = \frac{1}{2}|\dot{x}|^2 - U(x) + \frac{h^2}{24}\Big(\big\langle U'(x), U'(x)\big\rangle - 2\big\langle \dot{x}, U''(x)\dot{x}\big\rangle\Big).$$

By definition its Euler-Lagrange equation agrees with the modified equation with a defect of order $\mathscr{O}(h^4)$. In the particular case of the Kepler problem this becomes

$$\mathscr{L}_{\mathrm{mod},2}(x, \dot{x}) = \frac{1}{2}|\dot{x}|^2 + \frac{1}{|x|} + \frac{h^2}{24}\left(\frac{1}{|x|^4} - 2\frac{|\dot{x}|^2}{|x|^3} + 6\frac{\langle x, \dot{x}\rangle^2}{|x|^5}\right). \tag{2}$$

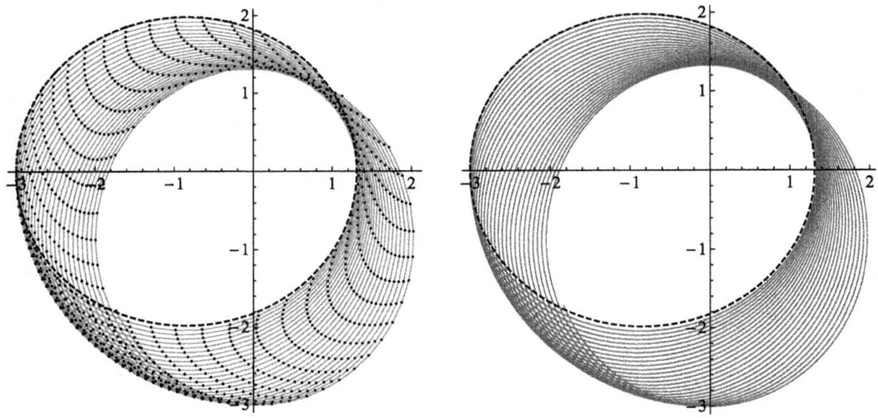

Fig. 1 Störmer-Verlet method with 1000 steps of size $h = 0.5$. Left: numerical solution. Right: exact solution of the modified equation of second order accuracy. In both images the dashed ellipse is the exact solution. The initial values are chosen as described in Sect. 6.1

A comparison of the numerical solution and the solution of the modified equation of second order accuracy is shown in Fig. 1.

2.2 Implicit Midpoint Rule

The second order formulation of the implicit midpoint rule (MP) applied to the differential equation $\ddot{x} = f(x)$ is

$$x_{k+1} - 2x_k + x_{k-1} = \frac{h^2}{2} f\left(\frac{x_k + x_{k+1}}{2}\right) + \frac{h^2}{2} f\left(\frac{x_{k-1} + x_k}{2}\right).$$

If $f(x) = -\frac{\mathrm{d}}{\mathrm{d}x} U(x)$, this is the discrete Euler-Lagrange equation for

$$L_{MP}(x_k, x_{k+1}) = \frac{1}{2}\left|\frac{x_{k+1} - x_k}{h}\right|^2 - U\left(\frac{x_k + x_{k+1}}{2}\right).$$

The modified Lagrangian of second order accuracy is

$$\mathscr{L}_{\mathrm{mod},2}(x, \dot{x}) = \frac{1}{2}|\dot{x}|^2 + \frac{h^2}{24}\Big(\big\langle U'(x), U'(x)\big\rangle + \big\langle \dot{x}, U''(x)\dot{x}\big\rangle\Big).$$

For the Kepler problem we have

$$\mathscr{L}_{\mathrm{mod},2}(x, \dot{x}) = \frac{1}{2}|\dot{x}|^2 + \frac{1}{|x|} + \frac{h^2}{24}\left(\frac{1}{|x|^4} + \frac{|\dot{x}|^2}{|x|^3} - 3\frac{\langle x, \dot{x}\rangle^2}{|x|^5}\right).$$

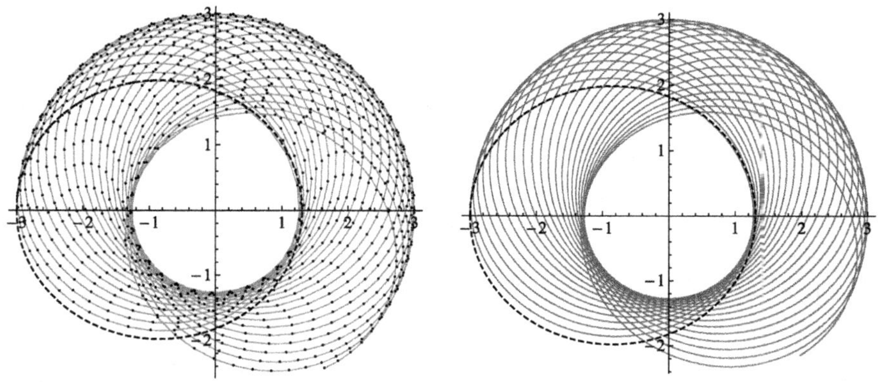

Fig. 2 Implicit midpoint rule with 1000 steps of size $h = 0.5$. Left: numerical solution. Right: exact solution of the modified equation of second order accuracy. In both images the dashed ellipse is the exact solution. The initial values are chosen as described in Sect. 6.1

A comparison of the numerical solution and the solution of the modified equation of second order accuracy is shown in Fig. 2.

3 Noether's Theorem with Perturbations

The key observation in our study of the perturbed Kepler problem is that Noether's theorem [12, 13] can be extended to describe how perturbations affect conserved quantities.

Theorem 1 *Consider a Lagrange function $\mathscr{L} : T\mathbb{R}^2 \to \mathbb{R}$ and a horizontal vector field ξ on $T\mathbb{R}^2$, i.e. $\xi = \xi_1 \frac{\partial}{\partial x_1} + \xi_2 \frac{\partial}{\partial x_2}$ with coefficients ξ_i that are functions $T\mathbb{R}^2 \to \mathbb{R}$. Let*

$$\xi^{(1)} = \sum_{i=1}^{2} \left(\xi_i \frac{\partial}{\partial x_i} + \dot{\xi}_i \frac{\partial}{\partial \dot{x}_i} \right)$$

be the first prolongation of ξ, evaluated on solutions of the Euler-Lagrange equations, i.e. with

$$\dot{\xi}_i = \left\langle \frac{\partial \xi_i}{\partial x}, \dot{x} \right\rangle + \left\langle \frac{\partial \xi_i}{\partial \dot{x}}, \left(\frac{\partial^2 \mathscr{L}}{\partial \dot{x}^2} \right)^{-1} \left(\frac{\partial \mathscr{L}}{\partial x} - \frac{\partial^2 \mathscr{L}}{\partial x \partial \dot{x}} \dot{x} \right) \right\rangle.$$

If

$$\xi^{(1)} \mathscr{L} = \left\langle \frac{\partial G}{\partial x}, \dot{x} \right\rangle + \varepsilon F$$

for some functions $F : T\mathbb{R}^2 \to \mathbb{R}$ and $G : \mathbb{R}^2 \to \mathbb{R}$, and a (small) parameter $\varepsilon \in \mathbb{R}$, then on solutions of the Euler-Lagrange equations we have

$$\frac{\mathrm{d}}{\mathrm{d}t}\left(\left\langle\frac{\partial\mathscr{L}}{\partial\dot{x}},\xi\right\rangle - G\right) = \varepsilon F,$$

where by abuse of notation $\xi = (\xi_1, \xi_2)$*. In particular, if* $\varepsilon F = 0$*, we have a conserved quantity* $A := \frac{\partial\mathscr{L}}{\partial\dot{x}_1}\xi_1 + \frac{\partial\mathscr{L}}{\partial\dot{x}_2}\xi_2 - G$*.*

Proof We have

$$\begin{aligned}\frac{\mathrm{d}}{\mathrm{d}t}\left(\left\langle\frac{\partial\mathscr{L}}{\partial\dot{x}},\xi\right\rangle - G\right) &= \left\langle\frac{\mathrm{d}}{\mathrm{d}t}\frac{\partial\mathscr{L}}{\partial\dot{x}},\xi\right\rangle + \left(\xi^{(1)}\mathscr{L} - \left\langle\frac{\partial\mathscr{L}}{\partial x},\xi\right\rangle\right) - \left\langle\frac{\partial G}{\partial x},\dot{x}\right\rangle \\ &= \xi^{(1)}\mathscr{L} - \left\langle\frac{\partial G}{\partial x},\dot{x}\right\rangle - \left\langle\frac{\partial\mathscr{L}}{\partial x} - \frac{\mathrm{d}}{\mathrm{d}t}\frac{\partial\mathscr{L}}{\partial\dot{x}},\xi\right\rangle = \varepsilon F.\end{aligned}$$

□

3.1 The Laplace-Runge-Lenz Vector

Following [9] we consider the Kepler problem and the vector field ξ defined by

$$\xi_1 = -\frac{1}{2}x_2\dot{x}_2 \qquad \text{and} \qquad \xi_2 = x_1\dot{x}_2 - \frac{1}{2}\dot{x}_1 x_2. \tag{3}$$

On solutions we have

$$\dot{\xi}_1 = -\frac{1}{2}\dot{x}_2^2 + \frac{1}{2}\frac{x_2^2}{|x|^3} \qquad \text{and} \qquad \dot{\xi}_2 = \frac{1}{2}\dot{x}_1\dot{x}_2 - \frac{1}{2}\frac{x_1 x_2}{|x|^3}.$$

A straightforward calculation then shows that

$$\xi^{(1)}\mathscr{L} = \left\langle\frac{\partial\mathscr{L}}{\partial x},\xi\right\rangle + \left\langle\frac{\partial\mathscr{L}}{\partial\dot{x}},\dot{\xi}\right\rangle = \frac{\dot{x}_1}{|x|} - \frac{\langle x,\dot{x}\rangle\, x_1}{|x|^3} = \frac{\mathrm{d}}{\mathrm{d}t}\left(\frac{x_1}{|x|}\right).$$

Hence we can apply the unperturbed Noether theorem (i.e. $\varepsilon F = 0$) with $G(x) = \frac{x_1}{|x|}$ and find that

$$A(x,\dot{x}) = -\dot{x}_1 x_2\dot{x}_2 + x_1\dot{x}_2^2 - \frac{x_1}{|x|} = |\dot{x}|^2 x_1 - \langle x,\dot{x}\rangle\,\dot{x}_1 - \frac{x_1}{|x|}$$

is a conserved quantity.

The conserved quantity A is the first component of the *Laplace-Runge-Lenz (LRL) vector*, which points from the gravitational center to the perihelion and has a magnitude equal to the eccentricity e of the orbit. The second component of the LRL vector is

$$B(x,\dot{x}) = |\dot{x}|^2 x_2 - \langle x,\dot{x}\rangle\,\dot{x}_2 - \frac{x_2}{|x|}$$

and can be obtained by setting $\xi_1 = x_2\dot{x}_1 - \frac{1}{2}x_1\dot{x}_2$ and $\xi_2 = -\frac{1}{2}x_1\dot{x}_1$. We denote by $\omega = \arctan\left(\frac{B}{A}\right)$ the angle of the LRL vector with the first coordinate axis.

Remark 1 The existence of this conserved quantity is related to the fact that the three-dimensional Kepler problem possesses an $SO(4)$-symmetry, rather than just the obvious $SO(3)$-symmetry. In suitable coordinates a solution can be "rotated" into other solutions with the same energy but different angular momentum [11, 14].

3.2 Precession in the Perturbed Kepler Problem

Now consider the perturbed Kepler problem, $\mathscr{L} = \frac{1}{2}|\dot{x}|^2 + \frac{1}{|x|} + \varepsilon\overline{\mathscr{L}}(x, \dot{x})$. Note that this also induces a perturbation in the prolonged vector field, which now reads $\xi^{(1)} + \varepsilon\overline{\xi^{(1)}}$, because the quantities $\dot{\xi}_1$ and $\dot{\xi}_2$ contain second derivatives which are evaluated using the perturbed equations of motion. We call the change in angle of the LRL vector over one period of the unperturbed system the *precession rate*.

Proposition 3 *If the major axis of an orbit is $\mathscr{O}(\varepsilon)$-close to the x_2-axis, then the precession rate is*

$$\Delta\omega = -\frac{2\varepsilon T}{e}\left[\left\langle \mathrm{EL}(\overline{\mathscr{L}}), \xi\right\rangle\right] + \mathscr{O}(\varepsilon^2), \tag{4}$$

where T is the period of the unperturbed orbit, $\mathrm{EL}(\overline{\mathscr{L}}) = \frac{\partial\overline{\mathscr{L}}}{\partial x} - \frac{\mathrm{d}}{\mathrm{d}t}\frac{\partial\overline{\mathscr{L}}}{\partial\dot{x}}$ *is the Euler-Lagrange expression for $\overline{\mathscr{L}}$, $\xi = (\xi_1, \xi_2)$ is defined by Eq. 3, and $[\,\cdot\,]$ denotes the average over one period.*

Proof Set $G = \frac{x_1}{|x|}$ and $F = \overline{\xi^{(1)}}\mathscr{L} + \xi^{(1)}\overline{\mathscr{L}}$, then

$$\left(\xi^{(1)} + \varepsilon\overline{\xi^{(1)}}\right)\left(\mathscr{L} + \varepsilon\overline{\mathscr{L}}\right) = \left\langle \frac{\partial G}{\partial \dot{x}}, \dot{x}\right\rangle + \varepsilon F + \mathscr{O}(\varepsilon^2),$$

where $\xi^{(1)} + \varepsilon\overline{\xi^{(1)}}$ is the first prolongation of ξ on solutions of the Euler Lagrange equations of the perturbed Lagrangian $\mathscr{L} + \varepsilon\overline{\mathscr{L}}$. Hence by Theorem 1 it follows that

$$\frac{\mathrm{d}}{\mathrm{d}t}\left(\left\langle \frac{\partial(\mathscr{L} + \varepsilon\overline{\mathscr{L}})}{\partial\dot{x}}, \xi\right\rangle - G\right) = \varepsilon F + \mathscr{O}(\varepsilon^2),$$

from which we conclude that

$$\begin{aligned} \frac{\mathrm{d}A}{\mathrm{d}t} &= \varepsilon\left(F - \frac{\mathrm{d}}{\mathrm{d}t}\left\langle \frac{\partial\overline{\mathscr{L}}}{\partial\dot{x}}, \xi\right\rangle\right) + \mathscr{O}(\varepsilon^2) \\ &= \varepsilon\left(\overline{\xi^{(1)}}\mathscr{L} + \xi^{(1)}\overline{\mathscr{L}} - \frac{\mathrm{d}}{\mathrm{d}t}\left\langle \frac{\partial\overline{\mathscr{L}}}{\partial\dot{x}}, \xi\right\rangle\right) + \mathscr{O}(\varepsilon^2). \end{aligned} \tag{5}$$

Now observe that

$$\begin{aligned}\xi^{(1)}\overline{\mathscr{L}} - \frac{\mathrm{d}}{\mathrm{d}t}\left\langle \frac{\partial \overline{\mathscr{L}}}{\partial \dot{x}}, \xi \right\rangle &= \left\langle \frac{\partial \overline{\mathscr{L}}}{\partial x}, \xi \right\rangle + \left\langle \frac{\partial \overline{\mathscr{L}}}{\partial \dot{x}}, \dot{\xi} \right\rangle - \frac{\mathrm{d}}{\mathrm{d}t}\left\langle \frac{\partial \overline{\mathscr{L}}}{\partial \dot{x}}, \xi \right\rangle \\ &= \left\langle \mathrm{EL}(\overline{\mathscr{L}}), \xi \right\rangle + \mathscr{O}(\varepsilon),\end{aligned}$$

where the error term comes from the fact that $\dot{\xi}$ is evaluated on the unperturbed system. We also have that

$$\overline{\xi^{(1)}\mathscr{L}} = \left\langle \frac{\partial \xi_1}{\partial \dot{x}}, \mathrm{EL}(\overline{\mathscr{L}}) \right\rangle \dot{x}_1 + \left\langle \frac{\partial \xi_2}{\partial \dot{x}}, \mathrm{EL}(\overline{\mathscr{L}}) \right\rangle \dot{x}_2 = \left\langle \frac{\partial \xi_1}{\partial \dot{x}}\dot{x}_1 + \frac{\partial \xi_2}{\partial \dot{x}}\dot{x}_2, \mathrm{EL}(\overline{\mathscr{L}}) \right\rangle.$$

For our choice of ξ, defined in Eq. 3, we have $\frac{\partial \xi_1}{\partial \dot{x}}\dot{x}_1 + \frac{\partial \xi_2}{\partial \dot{x}}\dot{x}_2 = (\xi_1, \xi_2) = \xi$, hence Eq. 5 simplifies to

$$\frac{\mathrm{d}A}{\mathrm{d}t} = 2\varepsilon \left\langle \mathrm{EL}(\overline{\mathscr{L}}), \xi \right\rangle + \mathscr{O}(\varepsilon^2).$$

The change in angle of the Laplace-Runge-Lenz vector is given by

$$\dot{\omega} = \frac{\mathrm{d}}{\mathrm{d}t}\left(\arctan \frac{B}{A} \right) = \frac{1}{A^2 + B^2}\left(A\frac{\mathrm{d}B}{\mathrm{d}t} - B\frac{\mathrm{d}A}{\mathrm{d}t} \right).$$

Choose a coordinate system such that $A = \mathscr{O}(\varepsilon)$ and $B \geq 0$. Then B approximately equals the eccentricity e and the derivative of the angle of the LRL vector is

$$\dot{\omega} = -\frac{1}{B}\frac{\mathrm{d}A}{\mathrm{d}t} + \mathscr{O}(\varepsilon^2) = -\frac{2\varepsilon}{e}\left\langle \mathrm{EL}(\overline{\mathscr{L}}), \xi \right\rangle + \mathscr{O}(\varepsilon^2).$$

□

4 Numerical Precession

We now apply Proposition 3 to the modified Lagrangians from Sect. 2. This gives us a leading order estimate of the precession rates of the integrators.

4.1 Störmer-Verlet Scheme

The perturbation term of the truncated modified Lagrangian (Eq. 2) is

$$\varepsilon\overline{\mathscr{L}} = \frac{h^2}{24}\left(\frac{1}{|x|^4} - 2\frac{|\dot{x}|^2}{|x|^3} + 6\frac{\langle x, \dot{x}\rangle^2}{|x|^5} \right).$$

In the following we identify $\varepsilon = \frac{h^2}{24}$. We want to evaluate Eq. 4. Using the leading order equations of motion (Eq. 1), which are valid up to an error of order $\mathscr{O}(h^2)$, we find

$$\mathrm{EL}(\overline{\mathscr{L}}) = 4\frac{x}{|x|^6} - 6\frac{|\dot{x}|^2 x}{|x|^5} + 30\frac{\langle x, \dot{x}\rangle^2 x}{|x|^7} - 12\frac{\langle x, \dot{x}\rangle \dot{x}}{|x|^5} + \mathscr{O}(h^2).$$

Using the fact that $\langle x, \xi\rangle = \frac{1}{2}(x_1\dot{x}_2 - \dot{x}_1 x_2)x_2 = \frac{1}{2}\mathbb{L}x_2$ and $\langle \dot{x}, \xi\rangle = \mathbb{L}\dot{x}_2$, the leading order equations of motion, and Proposition 1 we obtain

$$\begin{aligned}\left[\left\langle \mathrm{EL}(\overline{\mathscr{L}}), \xi\right\rangle\right] &= \left[2\frac{x_2}{|x|^6} - 3\frac{|\dot{x}|^2 x_2}{|x|^5} + 15\frac{\langle x, \dot{x}\rangle^2 x_2}{|x|^7} - 12\frac{\langle x, \dot{x}\rangle \dot{x}_2}{|x|^5}\right]\mathbb{L} + \mathscr{O}(h^2)\\ &= \left[30\frac{x_2}{|x|^6} + 24\mathbb{E}\frac{x_2}{|x|^5} - 15\mathbb{L}^2\frac{x_2}{|x|^7} + 4\frac{\mathrm{d}}{\mathrm{d}t}\frac{\dot{x}_2}{|x|^3}\right]\mathbb{L} + \mathscr{O}(h^2). \quad (6)\end{aligned}$$

The average $[\cdot]$ is taken along the unperturbed orbit, which is periodic, so $\left[\frac{\mathrm{d}}{\mathrm{d}t}\frac{\dot{x}_2}{|x|^3}\right] = 0$. For the other terms we have the following Lemma, which corresponds to the computation of the $C_n(e)$ of [3].

Lemma 1 *On solutions of the unperturbed Kepler problem for which the major axis is the x_2-axis there holds*

(a) $\left[\frac{x_2}{|x|^5}\right] = \frac{a}{b^5}e,$

(b) $\left[\frac{x_2}{|x|^6}\right] = \frac{a^2}{b^7}\left(\frac{3}{2}e + \frac{3}{8}e^3\right),$

(c) $\left[\frac{x_2}{|x|^7}\right] = \frac{a^3}{b^9}\left(2e + \frac{3}{2}e^3\right),$

where a and b are the semimajor and semiminor axes of the orbit respectively, and e is the eccentricity.

Proof Introduce polar coordinates $x_1 = -r\sin\theta$, $x_2 = r\cos\theta$, where $\theta = 0$ corresponds to the positive x_2-axis. We have

$$\left[\frac{x_2}{|x|^k}\right] = \left[\frac{cos\theta}{|x|^{k-1}}\right] = \frac{1}{T}\int_0^T \frac{cos\theta}{|x|^{k-1}}\,\mathrm{d}t.$$

Using Proposition 2 and Kepler's laws as in [5], we can rewrite this as

$$\begin{aligned}\left[\frac{x_2}{|x|^k}\right] &= \frac{b^{5-2k}}{\pi a^{4-k}}\int_0^\pi (1 + e\cos\theta)^{k-3}\cos\theta\,\mathrm{d}\theta\\ &= \frac{b^{5-2k}}{\pi a^{4-k}}\int_0^\pi \sum_j \binom{k-3}{j} e^j \cos^{j+1}\theta\,\mathrm{d}\theta.\end{aligned}$$

Whenever, j is even, we have $\int_0^\pi \cos^{j+1}\theta \, \mathrm{d}\theta = 0$. For $j = 1$ and $j = 3$ we find $\int_0^\pi \cos^2\theta \, \mathrm{d}\theta = \frac{\pi}{2}$ and $\int_0^\pi \cos^4\theta \, \mathrm{d}\theta = \frac{3\pi}{8}$. Hence

$$\left[\frac{x_2}{|x|^k}\right] = \frac{b^{5-2k}}{\pi a^{4-k}} \left(\frac{\pi}{2} \binom{k-3}{1} e + \frac{3\pi}{8} \binom{k-3}{3} e^3 + \ldots \right).$$

The claims now follow by evaluating this expression for $k = 5, 6, 7$. □

Combining Proposition 3, Eq. 6, and Lemma 1 we find that the precession per revolution is given by

$$\begin{aligned}
&-4\pi a^{3/2} \frac{h^2}{24} \left(30 \frac{a^2}{b^7} \left(\frac{3}{2} + \frac{3}{8} e^2 \right) + 24 \frac{-1}{2a} \frac{a}{b^5} - 15 \frac{b^2}{a} \frac{a^3}{b^9} \left(2 + \frac{3}{2} e^2 \right) \right) \frac{b}{\sqrt{a}} \operatorname{sgn}(\mathbb{L}) \\
&\quad + \mathcal{O}(h^4) \\
&= -4\pi a b \frac{h^2}{24} \left(30 \frac{a^2}{b^7} \left(\frac{15}{8} - \frac{3}{8} \frac{b^2}{a^2} \right) + 24 \frac{-1}{2a} \frac{a}{b^5} - 15 \frac{b^2}{a} \frac{a^3}{b^9} \left(\frac{7}{2} - \frac{3}{2} \frac{b^2}{a^2} \right) \right) \operatorname{sgn}(\mathbb{L}) \\
&\quad + \mathcal{O}(h^4) \\
&= -\frac{\pi h^2}{24} \left(15 \frac{a^3}{b^6} - 3 \frac{a}{b^4} \right) \operatorname{sgn}(\mathbb{L}) + \mathcal{O}(h^4),
\end{aligned}$$

assuming the major axis of the orbit is $\mathcal{O}(h^2)$-close to the x_2-axis. However, since both this expression and the perturbed Kepler problem are rotationally symmetric, we can conclude that statement holds regardless of the orientation of the major axis.

In summary we have the following:

Theorem 2 *The numerical precession rate of the Störmer-Verlet method with step size h is*

$$-\operatorname{sgn}(\mathbb{L}) \frac{\pi}{24} \left(15 \frac{a^3}{b^6} - 3 \frac{a}{b^4} \right) h^2 + \mathcal{O}(h^4),$$

where a and b denote the semimajor and semiminor axes of the orbit of the exact solution and sgn *is the sign function. In particular, the precession and the motion are in opposite directions.*

For the example shown in Fig. 1, the precession rate predicted by Theorem 2 is 0.067 radians per revolution and the observed numerical precession rate is 0.064 radians per revolution.

4.2 Implicit Midpoint Rule

In exactly the same way as for the Störmer-Verlet method, we obtain the following result:

Theorem 3 *The numerical precession rate of the midpoint rule with step size h is*

$$\operatorname{sgn}(\mathbb{L})\frac{\pi}{12}\left(15\frac{a^3}{b^6}-3\frac{a}{b^4}\right)h^2+\mathcal{O}(h^4).$$

In particular, the precession is in the same direction as the motion.

Note that in the leading order this expression differs by exactly a factor -2 from the expression for the Störmer-Verlet method. We will exploit this in the next section to construct new integrators.

For the example shown in Fig. 2, the precession rate predicted by Theorem 3 is -0.13 radians per revolution and the observed numerical precession rate is -0.16 radians per revolution.

5 New Integrators

Based on Theorems 2 and 3 we propose three new integrators. They all have a precession rate of order $\mathcal{O}(h^4)$ instead of $\mathcal{O}(h^2)$.

5.1 Linear Combination of the Lagrangians

Consider the discrete Lagrangian

$$\begin{aligned}L(x_j,x_{j+1})&=\frac{2}{3}L_{SV}(x_j,x_{j+1})+\frac{1}{3}L_{MP}(x_j,x_{j+1})\\&=\frac{1}{2}\left|\frac{x_{j+1}-x_j}{h}\right|^2-\frac{1}{3}U(x_j)-\frac{1}{3}U(x_{j+1})-\frac{1}{3}U\left(\frac{x_j+x_{j+1}}{2}\right).\end{aligned}$$

Its Euler-Lagrange equations define an implicit method,

$$x_{j+1}-2x_j+x_{j-1}=-\frac{2h^2}{3}U'(x_j)-\frac{h^2}{6}U'\left(\frac{x_{j-1}+x_j}{2}\right)-\frac{h^2}{6}U'\left(\frac{x_j+x_{j+1}}{2}\right).$$

We refer to this integrator as the *mixed Lagrangian* (ML) method. By construction, this is a variational integrator.

5.2 Lagrangian Composition

Consider the discrete Lagrangians

$$L_j(x_k, x_{k+1}) = \begin{cases} L_{MP}(x_k, x_{k+1}) = \frac{1}{2}\left|\frac{x_{k+1}-x_k}{h}\right|^2 - U\left(\frac{x_k+x_{k+1}}{2}\right) & \text{if } 3|j, \\ L_{SV}(x_k, x_{k+1}) = \frac{1}{2}\left|\frac{x_{k+1}-x_k}{h}\right|^2 - \frac{1}{2}U(x_k) - \frac{1}{2}U(x_{k+1}) & \text{otherwise.} \end{cases}$$

We look for a discrete curve $(x_j)_j$ that extremizes the action

$$\sum_{j=1}^{N} L_j(x_{j-1}, x_j) = L_{SV}(x_0, x_1) + L_{SV}(x_1, x_2) + L_{MP}(x_2, x_3) + \cdots .$$

This gives us three different Euler-Lagrange equations which are applied for different values of j mod 3. Indeed $\mathrm{D}_2 L_j(x_{j-1}, x_j) + \mathrm{D}_1 L_{j+1}(x_j, x_{j+1})$ simplifies to

$$\begin{cases} x_{j+1} - 2x_j + x_{j-1} = -\frac{h^2}{2}U'\left(\frac{x_{j-1}+x_j}{2}\right) - \frac{h^2}{2}U'(x_j) & \text{if } j \equiv 0 \mod 3, \\ x_{j+1} - 2x_j + x_{j-1} = -h^2 U'(x_j) & \text{if } j \equiv 1 \mod 3, \\ x_{j+1} - 2x_j + x_{j-1} = -\frac{h^2}{2}U'\left(\frac{x_j+x_{j+1}}{2}\right) - \frac{h^2}{2}U'(x_j) & \text{if } j \equiv 2 \mod 3. \end{cases}$$

Hence to determine the evolution we alternate between the Störmer-Verlet method (for $j \equiv 1$ mod 3) and two new difference equations. We refer to this integrator as the *Lagrangian composition* (LC) method. Strictly speaking the LC method should be considered as an integrator with step size $3h$, but for fair comparison with the other methods we will still refer to the internal step h as the step size.

This method of composing variational integrators is equivalent to composing the corresponding symplectic maps [10, Sect. 2.5].

5.3 Composition of the Difference Equations

Alternatively we can compose the difference equations obtained by the implicit midpoint rule and the Störmer-Verlet method respectively,

$$\begin{cases} x_{j+1} - 2x_j + x_{j-1} = -\frac{h^2}{2}U'\left(\frac{x_{j-1}+x_j}{2}\right) - \frac{h^2}{2}U'\left(\frac{x_j+x_{j+1}}{2}\right) & \text{if } j \equiv 2 \mod 3, \\ x_{j+1} - 2x_j + x_{j-1} = -h^2 U'(x_j) & \text{otherwise.} \end{cases}$$

We refer to this integrator as the *difference equation composition* (DEC) method. Just like for the LC method, we will abuse terminology and call the internal step h the step size.

It is not clear if this construction yields a variational method, but numerical experiments show long-term near-conservation of energy and angular momentum. This seems to be a general phenomenon: also for other potentials U and other variational integrators, the corresponding DEC method shows the long-term behavior one expects from a variational integrator.

6 Numerical Results

In this section we compare the new methods of Sect. 5 numerically with the Störmer-Verlet scheme, the implicit midpoint rule, and two fourth order symplectic methods: the well-known integrator of Forest and Ruth [6] and Chin's "C" algorithm which is especially well-suited for the Kepler problem [2, 4].

6.1 Choice of Initial Values

In all our examples we use the initial values

$$x(0) = (-3, 0) \qquad \text{and} \qquad \dot{x}(0) = (0, 0.45).$$

For the discretizations we need specify $x_0 = x(0)$ and $x_1 \approx x(h)$. Our convention is to choose x_1 such that the discrete momentum $p_0 = -\mathrm{D}_1 L(x_0, x_1)$ equals the initial velocity $\dot{x}(0)$.

For the composition of difference equations no discrete Lagrangian and hence no discrete momentum is known. To determine the second initial point x_1 in this case we use the momentum p_0 corresponding to the Störmer-Verlet method, because this is the method we would have used to calculate x_1 if x_0 was not the first point.

The choice of the initial value x_1 does not affect the precession behavior. However, it can have a significant effect on the error over time. If the initial condition has a slightly wrong energy, then the period of the numerical solution will have a slight error as well. This will cause a linearly growing phase shift.

6.2 Precession

Figure 3 shows the precession rates on a logarithmic scale for all five methods and a few choices of step size. It shows that the precession rates of the new methods behave like h^4, compared to h^2 for the methods from Sect. 2.

As for the three new methods, the mixed Lagrangian method beats the Lagrangian composition method, but the surprising winner is the composition of difference equations.

All our new methods have smaller precession rates than the fourth order symplectic integrator of Forest and Ruth [6]. On the other hand, Chin's fourth order symplectic "C" algorithm [2, 4] outperforms our methods.

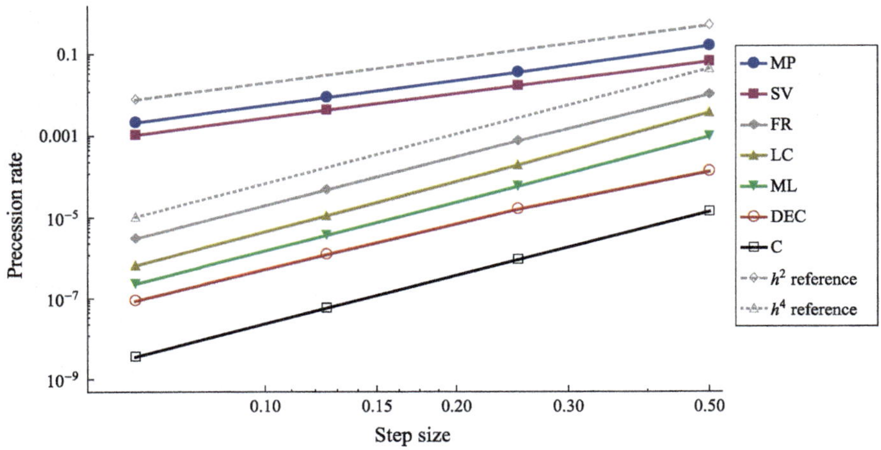

Fig. 3 Precession rate in radians per revolution for the different methods with step sizes $h = 0.0625$, $h = 0.125$, $h = 0.25$ and $h = 0.5$

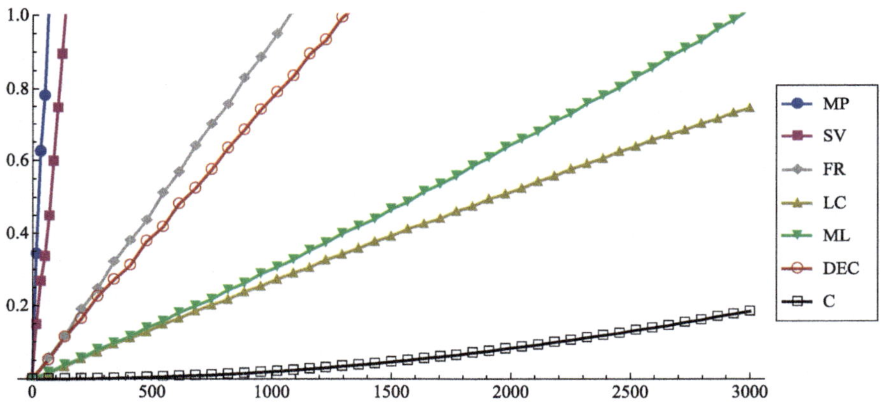

Fig. 4 Smoothed graph of the error in position over a time interval of length 3000 with step size $h = 0.45$. The markers are only for the purpose of identifying the methods, they do not correspond to individual time steps

6.3 Total Error

The precession rate is not as closely related to the total error as one might expect. In many cases the numerical solution has a phase shift which contributes significantly to the total error. For the composition methods LC and DEC this phase shift is highly dependent on the step size and the initial conditions. Hence the total error growth for these methods is also sensitive to the choice of step size and initial condition. This can be seen by comparing Figs. 4 and 5. In these figures we show a long time calculation with a large step size, leading to large errors. This means that the result

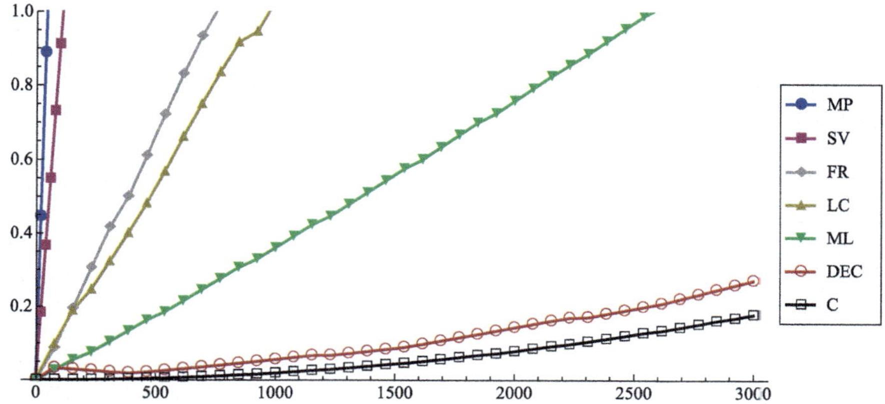

Fig. 5 The evolution of the error with step size $h = 0.5$

is useless for practical purposes, but it allows us to visualize the rate of error growth of the different methods relative to each other.

6.4 Speed

To give a rough comparison of the computational effort required for the different methods, we list the relative running times of a long time calculation (20 000 steps):

Störmer-Verlet	(SV)	0.67s	Mixed Lagrangian	(ML)	23s
MidPoint rule	(MP)	22s	Difference Equation composition	(DEC)	7.9s
Forest-Ruth	(FR)	2.0s	Lagrangian Composition	(LC)	8.2s
Chin C	(C)	2.2s			

We made a limited effort towards optimizing our implementation, so the given running times should only be taken as a rough indication. As expected the explicit methods SV, FR, and C are the fastest. Between those, SV is about three times faster than the other two. For the composition methods DEC and LC only one out of every three steps is implicit, hence they are roughly three times faster than MP and ML.

7 Conclusion

Using a modified equation approach, we have studied the precession rates of the implicit midpoint rule and the Störmer-Verlet method applied to the Kepler problem. We used the Lagrangian point of view, which lends itself to the use of a perturbed version of Noether's Theorem. The leading order estimates of the precession rate motivated the construction of three new integrators. They are significantly better than the methods we started from, but they are clearly outperformed by known specialized methods.

Our main goal was to elucidate methodology, rather than to obtain competitive methods. The techniques we used to analyze the integrators can be applied to any variational integrator and generalized to any order. However, it is not clear in general if we can use a similar procedure to write the resulting expressions in terms of the semi-axes of the orbits. Hence further research is needed in order to convert these ideas into a scheme to improve more advanced methods.

Acknowledgements This research is supported by the DFG Collaborative Research Center TRR 109 "Discretization in Geometry and Dynamics". The author would like to thank the organizers and participants of the *Brainstorming Workshop on New Developments in Discrete Mechanics, Geometric Integration and Lie-Butcher Series* for making it an inspiring event.

References

1. Chartier, P., Hairer, E., Vilmart, G.: Numerical integrators based on modified differential equations. Math. Comput. **76**(260), 1941–1953 (2007)
2. Chin, S.A.: Symplectic integrators from composite operator factorizations. Phys. Lett. A **226**(6), 344–348 (1997)
3. Chin, S.A.: Physics of symplectic integrators: Perihelion advances and symplectic corrector algorithms. Phys. Rev. E **75**(3), 036–701 (2007)
4. Chin, S.A., Kidwell, D.W.: Higher-order force gradient symplectic algorithms. Phys. Rev. E **62**(6), 8746 (2000)
5. Curtis, L.J., Haar, R.R., Kummer, M.: An expectation value formulation of the perturbed Kepler problem. Am. J. Phys. **55**(7), 627–631 (1987)
6. Forest, E., Ruth, R.D.: Fourth order symplectic integration. Physica **43**(LBL-27662), 105–117 (1989)
7. Goldstein, H.: Classical Mechanics. Addison-Wesley Pub, Co (1980)
8. Hairer, E., Lubich, C., Wanner, G.: Geometric Numerical Integration: Structure-Preserving Algorithms for Ordinary Differential Equations, vol. 31. Springer (2006)
9. Lévy-Leblond, J.M.: Conservation laws for gauge-variant Lagrangians in classical mechanics. Am. J. Phys. **39**(5), 502–506 (1971)
10. Marsden, J.E., West, M.: Discrete mechanics and variational integrators. Acta Num. **2001**(10), 357–514 (2001)
11. Morehead, J.: Visualizing the extra symmetry of the Kepler problem. Am. J. Phys. **73**(3), 234–239 (2005)
12. Noether, E.: Nachrichten von der Gesellschaft der Wissenschaften zu Göttingen, mathematisch-physikalische Klasse, Invariante Variationsprobleme **1918**, 235–257 (1918)
13. Olver, P.J.: Applications of Lie Groups to Differential Equations, vol. 107. Springer Science & Business Media (2000)
14. Rogers, H.H.: Symmetry transformations of the classical Kepler problem. J. Math. Phys. **14**(8), 1125–1129 (1973)
15. Vermeeren, M.: Modified Equations for Variational Integrators Numer. Math. **137**(4), 1007–1037 (2017)
16. Will, C.M.: Theory and experiment in gravitational physics, vol. 1. Cambridge University Press (1981)

Full Affine Equivariance and Weak Natural Transformations in Numerical Analysis—The Case of B-Series

Olivier Verdier

Abstract Many algorithms in numerical analysis are affine equivariant: they are immune to changes of affine coordinates. This is because those algorithms are defined using affine invariant constructions. There is, however, a crucial ingredient missing: most algorithms are in fact defined regardless of the underlying dimension. As a result, they are also invariant with respect to non-invertible affine transformation from spaces of different dimensions. We formulate this property precisely: these algorithms fall short of being natural transformations between affine functors. We give a precise definition of what we call a *weak natural transformation* between functors, and illustrate the point using examples coming from numerical analysis, in particular B-Series.

Keywords Affine · Allegory · Equivariance · Natural transformation

MSC codes 18B10 · 58J70 · 65L06

1 Affine Equivariance

We define an *algorithm* as a function F from a *data space* $\mathcal{D}$ to a *computation space* $\mathcal{C}$:

$$F : \mathcal{D} \to \mathcal{C}.$$

In most of the examples, $\mathcal{D}$ and $\mathcal{C}$ are manifolds.

O. Verdier (✉)
Department of Computing, Mathematics and Physics, Western Norway University of Applied Sciences, Bergen, Norway
e-mail: olivier.verdier@hvl.no

O. Verdier
Department of Mathematics, KTH - Royal Institute of Technology, Stockholm, Sweden

K. Ebrahimi-Fard and M. Barbero Liñán (eds.), *Discrete Mechanics, Geometric Integration and Lie–Butcher Series*, Springer Proceedings in Mathematics & Statistics 267, https://doi.org/10.1007/978-3-030-01397-4_11

Recall that for a given dimension d, the affine group $\mathrm{Aff}(d)$ is defined as the semi-direct product $\mathrm{Aff}(d) := \mathrm{GL}(d) \ltimes \mathbf{R}^d$. The action of $\mathrm{Aff}(d)$ on $\mathbf{R}^d$ is defined as follows. An element

$$g = \begin{bmatrix} A & b \\ 0 & 1 \end{bmatrix}, \qquad A \in \mathrm{GL}(d), \quad b \in \mathbf{R}^d,$$

acts on an element

$$\begin{bmatrix} x \\ 1 \end{bmatrix}$$

by matrix multiplication. The action is thus

$$g \cdot x = A\,x + b.$$

Definition 1.1 Suppose that the group $\mathrm{Aff}(d)$ acts on the spaces $\mathcal{D}$ and $\mathcal{C}$. An algorithm $F : \mathcal{D} \to \mathcal{C}$ is *affine equivariant* if the following diagram commutes, for any $\mathsf{a} \in \mathrm{Aff}(d)$.

$$\begin{array}{ccc} D_1 & \xrightarrow{\ \mathsf{a}\ } & D_2 \\ \Big\downarrow F & & \Big\downarrow F \\ F(D_1) & \xrightarrow{\ \mathsf{a}\ } & F(D_2) \end{array}$$

In practice, affine equivariance means invariance with respect to a change of affine coordinates. It means in particular invariance with respect to

- translations (change of origin)
- anisotropic scalings (change of units)
- rotations
- shearing

We can rephrase Definition 1.1 in order to prepare for Sect. 2. We regard the group $\mathrm{Aff}(d)$ as a *category* with one object $\star$ [3, Sect. 4.3]. We regard $\mathcal{C}$ and $\mathcal{D}$ as objects in the category of smooth manifolds and smooth maps. The actions of $\mathrm{Aff}(d)$ on $\mathcal{D}$ and $\mathcal{C}$ now define *functors*. With a slight abuse of notation we note these functors $\mathcal{D}$ and $\mathcal{C}$, so the actual data and computational spaces are $\mathcal{D}_\star := \mathcal{D}(\star)$ and $\mathcal{C}_\star := \mathcal{C}(\star)$. Affine equivariance (Definition 1.1) now expresses that the algorithm F is a *natural transformation* from the functor $\mathcal{D}$, to the functor $\mathcal{C}$. Indeed, Definition 1.1 can be rewritten as

$$F \circ \mathcal{D}(\mathsf{a}) = \mathcal{C}(\mathsf{a}) \circ F \tag{1.1}$$

or, using a commuting diagram,

$$\begin{array}{ccc} \mathcal{D}_\star & \xrightarrow{\mathcal{D}(\mathsf{a})} & \mathcal{D}_\star \\ \downarrow F & & \downarrow F \\ \mathcal{C}_\star & \xrightarrow{\mathcal{C}(\mathsf{a})} & \mathcal{C}_\star \end{array} .$$

We give many examples of such functors $\mathcal{D}$ and $\mathcal{C}$ in this note.

For convenience, for an affine map $\mathsf{a}(x) = Ax + b$, we introduce the corresponding *tangent map*

$$\mathsf{Ta} := A.$$

Note that the notion of a tangent map is defined for any nonlinear map: indeed, most of what we present in this section can be generalised to any group action (see Remark 1.7).

Example 1.2 (Quadrature) The data domain consists of all intervals and continuous functions on those intervals. This is the union of the spaces $C^0([\alpha, \beta])$, where we keep track of the interval $[\alpha, \beta]$, so a piece of data is $D = (\alpha, \beta, f)$, where $f \in C^0([\alpha, \beta])$. Formally, the data domain $\mathcal{D}$ is thus a fibre bundle.

The action of Aff(1) on a pair (α, β) is the diagonal action

$$\mathsf{a} \cdot (\alpha, \beta) = (\mathsf{a}(\alpha), \mathsf{a}(\beta))$$

and the action of Aff(1) on $C^0([\alpha, \beta])$ is defined by

$$\mathsf{a} \cdot f := f \circ \mathsf{a}^{-1},$$

so the total action is a fibre bundle mapping (it preserves the fibres).

The action of Aff(1) on the computational domain $\mathcal{C} = \mathbf{R}$ is the linear action

$$\mathsf{a} \cdot x = \mathsf{Ta}\, x.$$

Affine equivariance is now the requirement that the quadrature formula should fulfil

$$F\big(\mathsf{a} \cdot (I, f)\big) = \mathsf{a} \cdot F(I, f). \tag{1.2}$$

Explicitly, this corresponds to the requirement that F should behave as the exact integral under affine transformations. Indeed, the exact integral fulfils

$$\int_{a\alpha+b}^{a\beta+b} f\big((y-\beta)/\alpha\big)\mathrm{d}y = a \int_{\alpha}^{\beta} f(x)\mathrm{d}x$$

for any real numbers $\alpha \neq 0$ and β, which is equivalent to (1.2).

If we interpret the group actions as functors, then the data functor $\mathcal{D}$ maps the group object $\star$ to $\mathcal{D}_\star$, the fibre bundle defined above, and an invertible one-dimensional affine map a is mapped to $\mathcal{D}(\mathsf{a})$ defined by

$$\mathcal{D}(\mathsf{a})(\alpha, \beta, C^0([\alpha, \beta], \mathbf{R})) = (\mathsf{a}\,\alpha, \mathsf{a}\,\beta, f \circ \mathsf{a}^{-1}).$$

Similarly, the computational functor $\mathcal{C}$ maps the group object $\star$ to $\mathcal{C}_\star = \mathbf{R}$, and an invertible one-dimensional affine map a is mapped to $\mathcal{C}(\mathsf{a})$ defined by

$$\mathcal{C}(\mathsf{a}) := \mathsf{Ta}.$$

Example 1.3 (Numerical integrators) We consider numerical integration of ordinary differential equations (ODEs). The data domain $\mathcal{D}$ is the set of compactly supported vector fields on an affine space $\mathbf{A}_d$ of fixed dimension d:

$$\mathcal{D} = \mathfrak{X}_0(\mathbf{A}_d).$$

The affine action of the group $\mathrm{Aff}(d)$ on a vector field is defined by

$$\mathsf{a} \cdot f := \mathsf{Ta} \circ f \circ \mathsf{a}^{-1}.$$

The computational domain $\mathcal{C}$ is the set of diffeomorphisms:

$$\mathcal{C} = \mathrm{Diff}(\mathbf{A}_d).$$

The action on a diffeomorphism is the adjoint action

$$\mathsf{a} \cdot \Phi := \mathsf{a} \circ \Phi \circ \mathsf{a}^{-1}.$$

The equivariance assumption is thus

$$F(\mathsf{a} \cdot f) = \mathsf{a} \cdot F(f).$$

Again, that requirement makes sense as the exact solution fulfils that property for any invertible mappings (not only the affine ones) [12, Sect. 2.4]. Enforcing equivariance with respect to all invertible transformation would leave us with the exact solution alone.

For an initial condition x_0, the condition means that

$$F(\mathsf{Ta} f \circ \mathsf{a}^{-1})(\mathsf{a}x_0) = \mathsf{a}F(f)(x_0). \tag{1.3}$$

The layman description of the invariance of an integrator is that if one moves both the initial condition and the vector field with an affine transformation, then the computed point is also moved by the same affine transformation.

Let us see what that definition becomes for a concrete example of an integrator, the forward Euler method. In that case,

$$F(f) = \big[x \mapsto x + f(x)\big].$$

Writing $\mathsf{a}\,x = Ax + b$, we can check that

$$\begin{aligned} F\Big(Af\big(A^{-1}(y-\beta)\big)\Big)(Ax_0+\beta) &= (Ax_0+\beta) + Af(x_0) \\ &= A(x_0 + f(x_0)) + \beta \\ &= \mathsf{a}(F(f)(x_0)), \end{aligned}$$

which was condition (1.3).

Using the functor description, the data domain is now $\mathcal{D}_\star = \mathcal{X}_0(\mathbf{A}_d)$, and an invertible affine transformation a is mapped to $\mathcal{D}(\mathsf{a}) \in \mathrm{Hom}(\mathcal{D}_\star, \mathcal{D}_\star)$ defined by

$$\mathcal{D}(\mathsf{a})(f) = \mathsf{Ta} \circ f \circ \mathsf{a}^{-1}.$$

The computational functor $\mathcal{C}$ maps the group object to $\mathcal{C}_\star = \mathrm{Diff}(\mathbf{A}_d)$, and an invertible map a is mapped to $\mathcal{C}(\mathsf{a}) \in \mathrm{Hom}(\mathcal{C}_\star, \mathcal{C}_\star)$ as

$$\mathcal{C}(\mathsf{a})(\Phi) = \mathsf{a} \circ \Phi \circ \mathsf{a}^{-1}.$$

Remark 1.4 It turns out that all Runge–Kutta methods are affine equivariant. It has therefore been conjectured that Runge–Kutta methods, or more precisely, B-Series methods, were the only integrators enjoying that property. A recent result shows that this is not the case [12]. An example of an integrator which is affine equivariant but not a B-Series method is

$$F(f) := \big[x \mapsto f(x)\big(1 + \mathrm{div}(f)(x)\big)\big].$$

See Example 2.6 for a complete characterisation of B-Series.

Example 1.5 (Polynomial interpolation and splines) Here the domain is $\mathbf{A}_d^n$, the data of n points P_i in an affine space $\mathbf{A}_d$ of dimension d. The computation is a curve $C^\infty(\mathbf{R}, \mathbf{A}_d)$, which interpolates the points P_i in a variety of generalised meanings: exact interpolation, splines of various smoothness, etc.

The actions of the affine group $\mathrm{Aff}(d)$ are particularly simple in this case. As we shall see in Example 2.4, this simplicity reflects the fact that the algorithm is in this case a natural transformation between affine functors.

On the domain, $\mathcal{D} = \mathbf{A}_d^n$, the action is the standard diagonal action

$$\mathsf{a} \cdot (P_1, \ldots, P_n) := (\mathsf{a}(P_1), \ldots, \mathsf{a}(P_n)).$$

The action is essentially the same on $\mathcal{C}$:

$$\mathsf{a} \cdot \gamma := \mathsf{a} \circ \gamma.$$

The equivariance condition is thus simply that

$$F\big(\mathsf{a}\, P_1, \ldots, \mathsf{a}\, P_n\big) = \mathsf{a}\, F(P_1, \ldots, P_n).$$

The interpretation is particularly intuitive: first moving the control points and then computing the interpolation curves gives the same result as first computing the interpolation curve and then moving it with the same displacement.

Examples of interpolation algorithms which are affine equivariant are

- exact interpolation
- Bézier splines [14, Sect. 2.2]
- B-Splines [14, Sect. 5.7]

Again, we give the functorial point of view for completeness. The data functor $\mathcal{D}$ maps the group object to $\mathcal{D}_\star = \mathbf{A}_d^n$, and an invertible affine map a is mapped to $\mathcal{D}(\mathsf{a}) \in \mathrm{Hom}(\mathcal{D}_\star, \mathcal{D}_\star)$ defined by $\mathcal{D}(\mathsf{a})(P_i) = (\mathsf{a}\, P_i)$. The computational domain is $\mathcal{C}_\star = C^\infty(\mathbf{R}, \mathbf{A}_d)$, and the functor $\mathcal{C}$ maps an invertible affine map a to $\mathcal{C}(\mathsf{a}) \in \mathrm{Hom}(\mathcal{C}_\star, \mathcal{C}_\star)$ defined by $\mathcal{C}(\mathsf{a})(\gamma) = \mathsf{a} \circ \gamma$.

Example 1.6 (Downhill simplex minimization algorithm (Nelder–Mead)) The space is $\mathbf{A}_d$, and the data is a function $\varphi \in C^0(\mathbf{A}_d)$ to minimise, as well as a set of n starting points. The algorithm F then produces a new set of n points.

Usually there are $n = d + 1$ points, which span a simplex (hence the name of the algorithm), but this is too restrictive, as we shall see in Example 2.5.

There are several variants to that algorithm, but the crucial aspect here is that they are all affine invariant [6].

The action on functions is given by $\mathsf{a} \cdot \varphi := \varphi \circ \mathsf{a}^{-1}$, and the action on points is again the diagonal one: $\mathsf{a} \cdot x_i = \mathsf{a}(x_i)$. The requirement of equivariance is thus $F\big(\mathsf{a} \cdot (\varphi, x_i)\big) = \mathsf{a} \cdot F(\varphi, x_i)$. Of course, the actual Nelder–Mead algorithm consists of N iterations of the function F until convergence, and the iterated function F^N inherits the equivariance property of F.

What is the meaning of affine equivariance in that case? It is the idea that if one transforms the function to minimise with an affine transformation, and if one transforms the initial simplex by the same transformation, the final result will be the same as if one had run the algorithm directly, only transforming the last simplex.

The functorial point of view is $\mathcal{D}_\star = \mathbf{A}_d^n \times C^0(\mathbf{R}^d)$, with corresponding action $\mathcal{D}(\mathsf{a})(\varphi, X) = (\varphi \circ \mathsf{a}^{-1}, \mathsf{a}\, X)$. The computational domain is $\mathcal{C}_\star = \mathbf{A}_d^n$ and $\mathcal{C}(\mathsf{a})(X) = \mathsf{a}\, X$. In both cases, $\mathsf{a}\, X$ denotes the diagonal action on an element $X \in \mathbf{A}_d^n$.

Remark 1.7 Even though the main focus of this section is the affine group, it is legitimate to ask which algorithms are equivariant with respect to another group. Note that Definition 1.1 is unchanged: we only replace the affine group with another Lie group, with suitable actions on the data domain $\mathcal{D}$ and computational domain $\mathcal{C}$.

There are already some answers if we restrict the discussion to *numerical integrators on homogeneous spaces*, where equivariance is described along the lines of

Example 1.3. If the underlying homogeneous space is *reductive*, then all the standard extensions of Runge–Kutta methods on homogeneous spaces (Crouch–Grossman, RKMK, commutator-free) are equivariant with respect to the group at hand [13].

When the homogeneous space is symplectic, there are no general way to construct equivariant, symplectic integrators. In some cases, when the symplectic homogeneous space is the coadjoint orbit of a Lie group, the construction is still possible using appropriate *symplectic realisations*, i.e., Poisson maps from a symplectic vector space into the Lie–Poisson space at hand. The integrators thus obtained are automatically equivariant with respect to the Lie group at hand [9, 10].

Remark 1.8 Affine transformations play also a fundamental role in finite element methods. The families of polynomial differential forms discovered by Raviart, Thomas, Nédélec, later put in a common framework by Hiptmair, all have a common point: they are all affine invariant spaces: they are mapped to themselves by invertible affine maps (see [1, Sect. 1.3] and references therein). Remarkably, one can describe *all* such spaces [2, Th. 3.6]. The techniques used are very similar to those used to describe all the affine equivariant integrators in [12].

Remark 1.9 We conclude this section by mentioning that there are many algorithms which are not affine equivariant.

First of all, some algorithm are not equivariant because there are no obvious groups acting on the data or computation spaces. An example would be the ODE integrators known as splitting methods [11] because the algorithm is dependent of a particular structure of the problem.

Second of all, many algorithms are not scaling invariant. Examples are optimisation methods such as gradient descent, or even conjugate gradient. These methods are, however, invariant with respect to rigid motions.

When the translation group acts in a meaningful way on the data space, to the knowledge of the author, all algorithms used in practice are translation equivariant. However, it would be easy to artificially construct an algorithm which is not.

Finally, some algorithms used in numerical linear algebra are invariant with some group, again, if the action on the data space is meaningful. For instance, Arnoldi iterations and GMRES are equivariant with respect to the action of various groups [15, Th. 34.2, Th. 35.1]. Other algorithms, such as the QR algorithm, can be regarded as a discretization of a corresponding continuous dynamical system [4]. It is therefore possible that, in this case, equivariance properties could be derived, although it is clear that, the QR decomposition is not equivariant in any obvious way. Unfortunately, to the knowledge of the author, a systematic study of the equivariance properties of algorithms in numerical linear algebra is currently not available.

2 Full Affine Equivariance

In almost all the examples of Sect. 1, the algorithms are in fact defined *in any dimension*. So we have instead a sequence of algorithms F_d for every dimension d, mapping a data domain $\mathcal{D}_d$ into a computational domain $\mathcal{C}_d$.

For instance, an interpolation algorithm is defined for any dimension d, takes n points in $\mathbf{A}_d$ as input, and returns a curve in $\mathbf{A}_d$. As a result, the data domain is $\mathcal{D}_d := \mathbf{A}_d^n$ and the computational domain is $\mathcal{C}_d := C^\infty(\mathbf{R}, \mathbf{A}_d)$.

The crucial observation is that *these functions F_d must be related.* What we proceed to do now is to express this precisely. This will sometimes lead to surprising results (see the characterisation of Example 2.6).

We first motivate on an examples why full affine equivariance is needed.

Example 2.1 (Downhill simplex minimization algorithm revisited) We revisit Example 1.6. For each natural number d, the space is $\mathbf{A}_d$. As we saw in Example 1.6, the data is a function $\varphi\colon \mathbf{R}^d \to \mathbf{R}$ to minimise, as well as a set of n starting point, and the algorithm F then produces a new set of n points.

We say that two data points D_1 and D_2 in $C^\infty(\mathbf{A}_d) \times \mathbf{A}_d^n$, respectively equal to φ_1, X_1 and φ_2, X_2 are *related*, which we denote by

$$D_1 \overset{\mathsf{a}}{\rightsquigarrow} D_2$$

if

$$\varphi_1 = \varphi_2 \circ \mathsf{a} \qquad \text{and} \qquad X_2 = \mathsf{a} X_1,$$

where the operation $\mathsf{a} X_1$ is the diagonal action.

The difference with Example 1.6 is that the relation between φ_1 and φ_2 can no longer be expressed as $\varphi_2 = \varphi_1 \circ \mathsf{a}^{-1}$, as a is no longer required to be invertible.

The requirement that the algorithm F is affine equivariant in a stronger sense, is now that

$$D_1 \overset{\mathsf{a}}{\rightsquigarrow} D_2 \implies F(D_1) \overset{\mathsf{a}}{\rightsquigarrow} F(D_2).$$

We will examine the consequences of such a stronger requirement in Example 2.5, but we first put it in a formal setting.

Remark 2.2 We use the word "equivariance", which is not completely correct. Indeed, equivariance is usually applied to natural transformations between functors from a group object. It may then perhaps be applied to any natural transformation. However, as we shall see in Definition 2.3, the algorithms which are fully affine equivariant fall short of being natural transformations, they are instead *weak* natural transformations (Definition 2.3).

We first describe the *affine category*. It consists of finite dimensional affine spaces as objects, and affine maps between affine spaces as morphisms. To simplify the notations, we will identify all the affine spaces of the same dimensions, so the objects of the affine category are identified with the natural numbers:

$$d \equiv \mathbf{A}_d. \tag{2.1}$$

As a result, $\mathrm{Hom}(m, n)$ denotes the affine maps between the affine spaces $\mathbf{A}_m$ and $\mathbf{A}_n$.

The data and computational domain are now replaced by the relevant functors. These functors map an object in the affine category (hence, a natural number) to an object in a category of smooth manifolds.

It turns out that we need the category of *relations* associated to that of manifolds and smooth maps. An object in that category is still a manifold, but a morphism between manifolds $\mathcal{M}$ and $\mathcal{N}$ is now a submanifold of $\mathcal{M} \times \mathcal{N}$. If that submanifold is a graph, then this corresponds to a smooth map between $\mathcal{M}$ and $\mathcal{N}$, but this is otherwise a *relation*. We refer to [5] for a complete treatment of relation categories, also called *allegories*. For the general definition of full equivariance (Definition 2.3), we only need to assume that $\mathcal{D}$ and $\mathcal{C}$ are functors from the affine category to an allegory.

The data functor $\mathcal{D}$ maps an affine space d (identified with its dimension according to (2.1)) to some manifold $\mathcal{D}_d$. A morphism $\mathsf{a} \in \mathrm{Hom}(m, n)$ is mapped to a morphism $\mathcal{D}(\mathsf{a}) \in \mathrm{Hom}\left(\mathcal{D}_m, \mathcal{D}_n\right)$ in the above allegory.

The data and computation objects are now indexed by an integer (the dimension). We denote by $x_1 \overset{\mathsf{a}}{\rightsquigarrow} x_2$ the fact that x_1 is related to x_2 by the affine map a, as in Example 2.1. This is also denoted by $(D_1, D_2) \in \mathcal{D}(\mathsf{a})$.

The full equivariance condition is expressed as

$$D_1 \overset{\mathsf{a}}{\rightsquigarrow} D_2 \implies F_i(D_1) \overset{\mathsf{a}}{\rightsquigarrow} F_j(D_2), \tag{2.2}$$

which can also be written as

$$(D_1, D_2) \in \mathcal{D}(\mathsf{a}) \implies \big(F(D_1), F(D_2)\big) \in \mathcal{C}(\mathsf{a}).$$

What is the meaning of the full equivariance in the context of allegories? The answer is that such a method is *almost* a natural transformation between the functors $\mathcal{D}$ and $\mathcal{C}$. To understand this, we can reformulate condition (2.2) using composition in the allegory.

But we must first address a small problem: the algorithm is a *function*, and is thus not a morphism in the allegory (a relation). But a function F naturally gives rise to a relation given by its graph, and we denote the corresponding relation by $\overline{F}$:

$$\overline{F} := \big\{(D, C) \,\big|\, C = F(D)\big\}.$$

We compute the composition

$$\overline{F}_n \circ \mathcal{D}(\mathsf{a}) = \big\{(D_1, F(D_2)) \,\big|\, (D_1, D_2) \in \mathcal{D}(\mathsf{a})\big\},$$

and the composition

$$\mathcal{C}(\mathsf{a}) \circ \overline{F}_m = \big\{(D_1, C_2) \,\big|\, (F(D_1), C_2) \in \mathcal{C}(\mathsf{a})\big\}.$$

For an affine map $\mathsf{a} \in \mathrm{Hom}(m, n)$, condition (2.2) is thus

$$\overline{F}_n \circ \mathcal{D}(\mathsf{a}) \subset \mathcal{C}(\mathsf{a}) \circ \overline{F}_m.$$

Note that if the relation above was *equality* instead of subset, it would be exactly the requirement that F be a *natural transformation* between the functors $\mathcal{D}$ and $\mathcal{C}$. This leads us to the definition of a weaker notion of a natural transformations in allegories.

Definition 2.3 Given two functors $\mathcal{D}$ and $\mathcal{C}$ from a category A to an allegory B, a *weak natural transformation* is the data, for any object M in the category A, of a morphism $F_M \in \mathrm{Hom}(\mathcal{D}_M, \mathcal{C}_M)$, and such that for any object M and N in the category A, and any morphism $a \in \mathrm{Hom}(M, N)$ we have

$$F_N \circ \mathcal{D}(a) \subset \mathcal{C}(a) \circ F_M. \tag{2.3}$$

The reader should compare (2.3) with (1.1).

One can examine the meaning of full affine equivarience by breaking it into particular cases. Indeed, the affine category has two important subcategories: the category of *injective* affine maps, and the category of *surjective* affine maps. We will call *injective equivariance* and *surjective equivariance* the property of being a weak natural transformation with respect to the corresponding subcategories. For each dimension d, there is also a subcategory containing only the object d, and the invertible affine maps on that object: this is the category that we studied in Sect. 1. For each dimension, we will denote the corresponding equivariance by *bijective equivariance*.

1. *Injective* equivariance generally means that if the data of the algorithm happens to lie in an affine subspace, then the result of the computation not only will lie on the subspace, but will also work exactly as if the lower dimensional version of the algorithms was used with that data.
2. Projective equivariance generally indicates how the algorithm behaves with certain degenerate data. It highly depends on the algorithm. In the case of ODE integrators, it has a very understandable meaning (see Example 2.6).
3. Bijective equivariance, is what we covered in Sect. 1.

Example 2.4 (Polynomial interpolation and splines) We revisit Example 1.5. Now the dimension d is arbitrary, and the algorithm works in any dimension.

The domain is $\mathbf{A}_d^n$, the data of n points in an affine space of dimension d. The data functor $\mathcal{D}$ maps the object d of the affine category to $\mathcal{D}_d = \mathbf{A}_d^n$. An affine map $\mathsf{a} \in \mathrm{Hom}(m, n)$ is mapped to the relation $\mathcal{D}(\mathsf{a})$ which we identify to the *map* $\mathcal{D}(\mathsf{a})(X) := \mathsf{a}\, X$, where we used the diagonal action. The computational functor $\mathcal{C}$

maps the object d to $C^\infty(\mathbf{R}, \mathbf{A}_d)$, and the relation $\mathcal{C}(\mathsf{a})$ is identified to the function $\mathcal{C}(\mathsf{a})(f) := \mathsf{a} \circ f$.

Note that in this case, both compositions occurring in (2.3) are *graphs*, so the inclusion is in fact an equality, and F is in this case a *natural transformation* between the functors $\mathcal{D}$ and $\mathcal{C}$.

Injective equivariance here is related to a well known property of splines and interpolation: if the control points actually lie in a subspace, then the whole interpolating curve or spline, also lies in that subspace. What is more, that curve is exactly the same as if the calculation had been done in a lower dimensional space instead.

Surjective equivariance is perhaps less intuitive: it means that interpolation commutes with affine projections. Computing the interpolation of projected points on a smaller subspace gives the same result as projecting the interpolated curve instead.

Example 2.5 (Downhill simplex) We now revisit Example 1.6. The data functor is $\mathcal{D}_d = C^\infty(\mathbf{A}_d) \times \mathbf{A}_d^n$ and a map $\mathsf{a} \in \mathrm{Hom}(m, n)$ is mapped to $\mathcal{D}(\mathsf{a}) \in \mathrm{Hom}(\mathcal{D}_m, \mathcal{D}_n)$ defined as the relation

$$\mathcal{D}(\mathsf{a}) = \big\{((\varphi_1, X_1), (\varphi_2, X_2)) \,\big|\, \varphi_1 = \varphi_2 \circ \mathsf{a} \qquad X_2 = \mathsf{a}\, X_1\big\}.$$

The computational functor is simply

$$\mathcal{C}(\mathsf{a}) = \big\{(X_1, X_2) \,\big|\, X_2 = \mathsf{a}\, X_1\big\}.$$

Note that $\mathcal{C}(\mathsf{a})$ is in fact a graph, so it is associated to a function.

What does injective equivariance mean in that case? We consider an *injective* affine map $\mathsf{a} \in \mathrm{Hom}(m, n)$. The relation $\varphi_1 = \varphi_2 \circ \mathsf{a}$ means that the function $\varphi_1 \in C^\infty(\mathbf{A}_m)$ to minimise is the *restriction* of the function $\varphi_2(\mathbf{A}_n)$, along the subspace given by the image of $\mathsf{a} \in \mathrm{Hom}(m, n)$. Equivariance means in this case is that: *if one starts with a degenerate simplex*, i.e., if all the points lie in the subspace above, then the simplex algorithm will find the minimum *in that subspace*, i.e., the minimum of the function φ_1. Picture here?

Let us examine surjective equivariance. We consider a *surjective* affine map $\mathsf{a} \in \mathrm{Hom}(m, n)$. The relation $\varphi_1 = \varphi_2 \circ \mathsf{a}$ now means that φ_1 is equal to φ_2 and is constant on the fibres (i.e., the level sets) of a. This is an example of degenerate data. What equivariance means in this case is that the simplex algorithms works *fibre-wise*, i.e., the result will not depend on where the initial points X were chosen inside the fibres.

Example 2.6 We now look at the example that we understand perhaps best of all: numerical integration of ODEs. The data functor $\mathcal{D}$ maps an affine space of dimension d to $\mathcal{D}_d = \mathcal{X}_0(\mathbf{A}_d)$, the space of compactly supported vector fields on $\mathbf{A}_d$. An affine map $\mathsf{a} \in \mathrm{Hom}(m, n)$ is mapped to the relation $\mathcal{D}(\mathsf{a}) \in \mathrm{Hom}(\mathcal{D}_m, \mathcal{D}_n)$ defined by

$$\mathcal{D}(\mathsf{a}) = \big\{(f_1, f_2) \,\big|\, f_2 \circ \mathsf{a} = \mathsf{Ta} \circ f_1\big\}.$$

The computational functor $\mathcal{C}$ maps an object d to $\mathcal{C}_d := \mathrm{Diff}(\mathbf{A}_d)$. The relation $\mathcal{C}(\mathsf{a}) \in \mathrm{Hom}(\mathcal{C}_m, \mathcal{C}_n)$ is defined by

$$\mathcal{C}(\mathsf{a}) = \big\{(\Phi_1, \Phi_2) \,\big|\, \Phi_2 \circ \mathsf{a} = \mathsf{a} \circ \Phi_1\big\}.$$

The meaning of injective equivariance is known in numerical analysis as *preservation of weak (affine) invariants* [7, Sect. IV.4]. An affine weak invariant is an affine subspace which is preserved by the flow of the vector field. If that subspace is the image by an affine map a of an affine space of smaller dimension (one can choose a to be injective), the requirement of weak invariance is exactly that of being in relation with another vector field. Injective equivariance thus means that: if a vector field has a weak invariant subspace, not only is it preserved by the numerical flow, but that numerical flow is the same as if computed in the lower dimensional subspace instead.

The meaning of surjective equivariance is particularly interesting. Suppose that $\mathsf{a} \in \mathrm{Hom}(m, n)$ is a surjective affine map. The requirement that $(f_1, f_2) \in \mathcal{D}(\mathsf{a})$ is that the flow of f_1 descends to the flow of f_2. After change of variable, this can be rewritten as the differential equation

$$\begin{aligned} x' &= g_1(x) \\ y' &= g_2(x, y). \end{aligned}$$

The property of surjective equivariance is that the numerical integrator behaves like the exact solution: the numerical flow descends to the numerical flow of f_2.

We fully understand that case, as we can give a complete characterisation of the fully affine equivariant integrators in the sense above: these are exactly the integrators which have a *B-Series* [8].

3 Conclusion

One of the biggest open questions is how weak natural transformations extend to other group actions. Indeed, there are many examples of integrators on homogeneous spaces, which generalise the equivariance with respect to a *group*. However, as we saw, the equivariance with respect to the affine *category* seems to be of the utmost importance. We do not know of any other category for which a range of numerical algorithms are equivariant.

A particularly acute question is the characterisation of numerical integrators on homogeneous spaces: as [12] shows, group equivariance is not sufficient. So what is the nonlinear equivalent of the affine category? This is an area of ongoing research, but we speculate that they may be related to free Lie algebras, or possibly the related structures of post-Lie algebras

References

1. Arnold, D., Falk, R., Winther, R.: Finite element exterior calculus: from Hodge theory to numerical stability. Bull. Am. Math. Soc. **47**(2), 281–354 (New Series) (2010). https://doi.org/10.1090/S0273-0979-10-01278-4
2. Arnold, D., Falk, R., Winther, R.: Finite element exterior calculus, homo- logical techniques, and applications. Acta Numer. **15**, 1–155 (2006). https://doi.org/10.1017/S0962492906210018
3. Awodey, S.: Category Theory. Oxford Logic Guides. OUP, Oxford (2010). ISBN: 9780191612558
4. Chu, M.T.: Linear algebra algorithms as dynamical systems. Acta Numer. **17**, 1–86 (2008). ISSN: 0962-4929; 1474-0508/e. https://doi.org/10.1017/S0962492906340019
5. Freyd, P.J., Scedrov, A.: Categories, Allegories. North-Holland Mathematical Library. Elsevier Science (1990). ISBN: 9780080887012
6. Goodman, J., Weare, J.: Ensemble samplers with affine invariance. Commun. Appl. Math. Comput. Sci. **5**(1), 65–80 (2010). https://doi.org/10.2140/camcos.2010.5.65
7. Hairer, E., Lubich, C., Wanner, G.: Geometric Numerical Integration: Structure-preserving Algorithms for Ordinary Differential Equations. Springer series in computational mathematics. Springer (2006). ISBN: 9783540306634
8. McLachlan, R.I., Modin, K., Munthe-Kaas, H., Verdier, O.: B-series methods are exactly the affine equivariant methods. Numerische Matematik (2015). https://doi.org/10.1007/s00211-015-0753-2. arXiv:1409.1019
9. McLachlan, R.I., Modin, K., Verdier, O.: Collective Lie–Poisson integrators on R3. IMA J. Numer. Anal. **35**(2), 546–560 (2015). https://doi.org/10.1093/imanum/dru013. arXiv:1307.2387
10. McLachlan, R.I., Modin, K., Verdier, O.: Collective symplectic integrators. Nonlinearity **27**(6), 1525 (2014). https://doi.org/10.1088/0951-7715/27/6/1525. arXiv:1308.6620
11. McLachlan, R.I., Quispel, G.R.W.: Splitting methods. Acta Numerica **11**, 341–434 (2002). ISSN: 0962-4929; 1474-0508/e. https://doi.org/10.1017/S0962492902000053
12. Munthe-Kaas, H., Verdier, O.: Aromatic Butcher series. Found. Comput. Math. **16**(1), 183–215 (2016). https://doi.org/10.1007/s10208-015-9245-0. arXiv:1308.5824
13. Munthe-Kaas, H., Verdier, O.: Integrators on homogeneous spaces: isotropy choice and connections. Found. Comput. Math. (2015). https://doi.org/10.1007/s10208-015-9267-7. arXiv:1402.6981
14. Prautzsch, H., Boehm, W., Paluszny, M.: Bézier and B-spline techniques. Mathematics and Visualization. Springer, Berlin, Heidelberg (2013). ISBN: 9783662049198
15. Trefethen, L.N., Bau, D.: Numerical Linear Algebra. SIAM (1997). ISBN: 9780898719574

Printed by Printforce, the Netherlands